Evolution Equations

PURE AND APPLIED MATHEMATICS

A Program of Monographs, Textbooks, and Lecture Notes

LECTURE NOTES IN PURE AND APPLIED MATHEMATICS

1. *N. Jacobson,* Exceptional Lie Algebras
2. *L.-Å. Lindahl and F. Poulsen,* Thin Sets in Harmonic Analysis
3. *I. Satake,* Classification Theory of Semi-Simple Algebraic Groups
4. *F. Hirzebruch et al., Differentiable Manifolds and Quadratic Forms*
5. *I. Chavel,* Riemannian Symmetric Spaces of Rank One
6. *R. B. Burckel,* Characterization of C(X) Among Its Subalgebras
7. *B. R. McDonald et al.,* Ring Theory
8. *Y.-T. Siu,* Techniques of Extension on Analytic Objects
9. *S. R. Caradus et al.,* Calkin Algebras and Algebras of Operators on Banach Spaces
10. *E. O. Roxin et al.,* Differential Games and Control Theory
11. *M. Orzech and C. Small,* The Brauer Group of Commutative Rings
12. *S. Thomier,* Topology and Its Applications
13. *J. M. Lopez and K. A. Ross,* Sidon Sets
14. *W. W. Comfort and S. Negrepontis,* Continuous Pseudometrics
15. *K. McKennon and J. M. Robertson,* Locally Convex Spaces
16. *M. Carmeli and S. Malin,* Representations of the Rotation and Lorentz Groups
17. *G. B. Seligman,* Rational Methods in Lie Algebras
18. *D. G. de Figueiredo,* Functional Analysis
19. *L. Cesari et al.,* Nonlinear Functional Analysis and Differential Equations
20. *J. J. Schäffer,* Geometry of Spheres in Normed Spaces
21. *K. Yano and M. Kon,* Anti-Invariant Submanifolds
22. *W. V. Vasconcelos,* The Rings of Dimension Two
23. *R. E. Chandler,* Hausdorff Compactifications
24. *S. P. Franklin and B. V. S. Thomas,* Topology
25. *S. K. Jain,* Ring Theory
26. *B. R. McDonald and R. A. Morris,* Ring Theory II
27. *R. B. Mura and A. Rhemtulla,* Orderable Groups
28. *J. R. Graef,* Stability of Dynamical Systems
29. *H.-C. Wang,* Homogeneous Branch Algebras
30. *E. O. Roxin et al.,* Differential Games and Control Theory II
31. *R. D. Porter,* Introduction to Fibre Bundles
32. *M. Altman,* Contractors and Contractor Directions Theory and Applications
33. *J. S. Golan,* Decomposition and Dimension in Module Categories
34. *G. Fairweather,* Finite Element Galerkin Methods for Differential Equations
35. *J. D. Sally,* Numbers of Generators of Ideals in Local Rings
36. *S. S. Miller,* Complex Analysis
37. *R. Gordon,* Representation Theory of Algebras
38. *M. Goto and F. D. Grosshans,* Semisimple Lie Algebras
39. *A. I. Arruda et al.,* Mathematical Logic
40. *F. Van Oystaeyen,* Ring Theory
41. *F. Van Oystaeyen and A. Verschoren,* Reflectors and Localization
42. *M. Satyanarayana,* Positively Ordered Semigroups
43. *D. L Russell,* Mathematics of Finite-Dimensional Control Systems
44. *P.-T. Liu and E. Roxin,* Differential Games and Control Theory III
45. *A. Geramita and J. Seberry,* Orthogonal Designs
46. *J. Cigler, V. Losert, and P. Michor,* Banach Modules and Functors on Categories of Banach Spaces
47. *P.-T. Liu and J. G. Sutinen,* Control Theory in Mathematical Economics
48. *C. Byrnes,* Partial Differential Equations and Geometry
49. *G. Klambauer,* Problems and Propositions in Analysis
50. *J. Knopfmacher,* Analytic Arithmetic of Algebraic Function Fields
51. *F. Van Oystaeyen,* Ring Theory
52. *B. Kadem,* Binary Time Series
53. *J. Barros-Neto and R. A. Artino,* Hypoelliptic Boundary-Value Problems
54. *R. L. Sternberg et al.,* Nonlinear Partial Differential Equations in Engineering and Applied Science
55. *B. R. McDonald,* Ring Theory and Algebra III
56. *J. S. Golan,* Structure Sheaves Over a Noncommutative Ring
57. *T. V. Narayana et al.,* Combinatorics, Representation Theory and Statistical Methods in Groups
58. *T. A. Burton,* Modeling and Differential Equations in Biology
59. *K. H. Kim and F. W. Roush,* Introduction to Mathematical Consensus Theory

60. *J. Banas and K. Goebel,* Measures of Noncompactness in Banach Spaces
61. *O. A. Nielson,* Direct Integral Theory
62. *J. E. Smith et al.,* Ordered Groups
63. *J. Cronin,* Mathematics of Cell Electrophysiology
64. *J. W. Brewer,* Power Series Over Commutative Rings
65. *P. K. Kamthan and M. Gupta,* Sequence Spaces and Series
66. *T. G. McLaughlin,* Regressive Sets and the Theory of Isols
67. *T. L. Herdman et al.,* Integral and Functional Differential Equations
68. *R. Draper,* Commutative Algebra
69. *W. G. McKay and J. Patera,* Tables of Dimensions, Indices, and Branching Rules for Representations of Simple Lie Algebras
70. *R. L. Devaney and Z. H. Nitecki,* Classical Mechanics and Dynamical Systems
71. *J. Van Geel,* Places and Valuations in Noncommutative Ring Theory
72. *C. Faith,* Injective Modules and Injective Quotient Rings
73. *A. Fiacco,* Mathematical Programming with Data Perturbations I
74. *P. Schultz et al.,* Algebraic Structures and Applications
75. *L Bican et al.,* Rings, Modules, and Preradicals
76. *D. C. Kay and M. Breen,* Convexity and Related Combinatorial Geometry
77. *P. Fletcher and W. F. Lindgren,* Quasi-Uniform Spaces
78. *C.-C. Yang,* Factorization Theory of Meromorphic Functions
79. *O. Taussky,* Ternary Quadratic Forms and Norms
80. *S. P. Singh and J. H. Burry,* Nonlinear Analysis and Applications
81. *K. B. Hannsgen et al.,* Volterra and Functional Differential Equations
82. *N. L. Johnson et al.,* Finite Geometries
83. *G. I. Zapata,* Functional Analysis, Holomorphy, and Approximation Theory
84. *S. Greco and G. Valla,* Commutative Algebra
85. *A. V. Fiacco,* Mathematical Programming with Data Perturbations II
86. *J.-B. Hiriart-Urruty et al.,* Optimization
87. *A. Figa Talamanca and M. A. Picardello,* Harmonic Analysis on Free Groups
88. *M. Harada,* Factor Categories with Applications to Direct Decomposition of Modules
89. *V. I. Istrătescu,* Strict Convexity and Complex Strict Convexity
90. *V. Lakshmikantham,* Trends in Theory and Practice of Nonlinear Differential Equations
91. *H. L. Manocha and J. B. Srivastava,* Algebra and Its Applications
92. *D. V. Chudnovsky and G. V. Chudnovsky,* Classical and Quantum Models and Arithmetic Problems
93. *J. W. Longley,* Least Squares Computations Using Orthogonalization Methods
94. *L. P. de Alcantara,* Mathematical Logic and Formal Systems
95. *C. E. Aull,* Rings of Continuous Functions
96. *R. Chuaqui,* Analysis, Geometry, and Probability
97. *L. Fuchs and L. Salce,* Modules Over Valuation Domains
98. *P. Fischer and W. R. Smith,* Chaos, Fractals, and Dynamics
99. *W. B. Powell and C. Tsinakis,* Ordered Algebraic Structures
100. *G. M. Rassias and T. M. Rassias,* Differential Geometry, Calculus of Variations, and Their Applications
101. *R.-E. Hoffmann and K. H. Hofmann,* Continuous Lattices and Their Applications
102. *J. H. Lightbourne III and S. M. Rankin III,* Physical Mathematics and Nonlinear Partial Differential Equations
103. *C. A. Baker and L. M. Batten,* Finite Geometrics
104. *J. W. Brewer et al.,* Linear Systems Over Commutative Rings
105. *C. McCrory and T. Shifrin,* Geometry and Topology
106. *D. W. Kueke et al.,* Mathematical Logic and Theoretical Computer Science
107. *B.-L. Lin and S. Simons,* Nonlinear and Convex Analysis
108. *S. J. Lee,* Operator Methods for Optimal Control Problems
109. *V. Lakshmikantham,* Nonlinear Analysis and Applications
110. *S. F. McCormick,* Multigrid Methods
111. *M. C. Tangora,* Computers in Algebra
112. *D. V. Chudnovsky and G. V. Chudnovsky,* Search Theory
113. *D. V. Chudnovsky and R. D. Jenks,* Computer Algebra
114. *M. C. Tangora,* Computers in Geometry and Topology
115. *P. Nelson et al.,* Transport Theory, Invariant Imbedding, and Integral Equations
116. *P. Clément et al.,* Semigroup Theory and Applications
117. *J. Vinuesa,* Orthogonal Polynomials and Their Applications
118. *C. M. Dafermos et al.,* Differential Equations
119. *E. O. Roxin,* Modern Optimal Control
120. *J. C. Díaz,* Mathematics for Large Scale Computing

121. *P. S. Milojević* Nonlinear Functional Analysis
122. *C. Sadosky,* Analysis and Partial Differential Equations
123. *R. M. Shortt,* General Topology and Applications
124. *R. Wong,* Asymptotic and Computational Analysis
125. *D. V. Chudnovsky and R. D. Jenks,* Computers in Mathematics
126. *W. D. Wallis et al.,* Combinatorial Designs and Applications
127. *S. Elaydi,* Differential Equations
128. *G. Chen et al.,* Distributed Parameter Control Systems
129. *W. N. Everitt,* Inequalities
130. *H. G. Kaper and M. Garbey,* Asymptotic Analysis and the Numerical Solution of Partial Differential Equations
131. *O. Arino et al.,* Mathematical Population Dynamics
132. *S. Coen,* Geometry and Complex Variables
133. *J. A. Goldstein et al.,* Differential Equations with Applications in Biology, Physics, and Engineering
134. *S. J. Andima et al.,* General Topology and Applications
135. *P Clément et al.,* Semigroup Theory and Evolution Equations
136. *K. Jarosz,* Function Spaces
137. *J. M. Bayod et al.,* *p*-adic Functional Analysis
138. *G. A. Anastassiou,* Approximation Theory
139. *R. S. Rees,* Graphs, Matrices, and Designs
140. *G. Abrams et al.,* Methods in Module Theory
141. *G. L. Mullen and P. J.-S. Shiue,* Finite Fields, Coding Theory, and Advances in Communications and Computing
142. *M. C. Joshi and A. V. Balakrishnan,* Mathematical Theory of Control
143. *G. Komatsu and Y. Sakane,* Complex Geometry
144. *I. J. Bakelman,* Geometric Analysis and Nonlinear Partial Differential Equations
145. *T. Mabuchi and S. Mukai,* Einstein Metrics and Yang–Mills Connections
146. *L. Fuchs and R. Göbel,* Abelian Groups
147. *A. D. Pollington and W. Moran,* Number Theory with an Emphasis on the Markoff Spectrum
148. *G. Dore et al.,* Differential Equations in Banach Spaces
149. *T. West,* Continuum Theory and Dynamical Systems
150. *K. D. Bierstedt et al.,* Functional Analysis
151. *K. G. Fischer et al.,* Computational Algebra
152. *K. D. Elworthy et al.,* Differential Equations, Dynamical Systems, and Control Science
153. *P.-J. Cahen, et al.,* Commutative Ring Theory
154. *S. C. Cooper and W. J. Thron,* Continued Fractions and Orthogonal Functions
155. *P. Clément and G. Lumer,* Evolution Equations, Control Theory, and Biomathematics
156. *M. Gyllenberg and L. Persson,* Analysis, Algebra, and Computers in Mathematical Research
157. *W. O. Bray et al.,* Fourier Analysis
158. *J. Bergen and S. Montgomery,* Advances in Hopf Algebras
159. *A. R. Magid,* Rings, Extensions, and Cohomology
160. *N. H. Pavel,* Optimal Control of Differential Equations
161. *M. Ikawa,* Spectral and Scattering Theory
162. *X. Liu and D. Siegel,* Comparison Methods and Stability Theory
163. *J.-P. Zolésio,* Boundary Control and Variation
164. *M. Křížek et al.,* Finite Element Methods
165. *G. Da Prato and L. Tubaro,* Control of Partial Differential Equations
166. *E. Ballico,* Projective Geometry with Applications
167. *M. Costabel et al.,* Boundary Value Problems and Integral Equations in Nonsmooth Domains
168. *G. Ferreyra, G. R. Goldstein, and F. Neubrander,* Evolution Equations
169. *S. Huggett,* Twistor Theory
170. *H. Cook et al.,* Continua
171. *D. F. Anderson and D. E. Dobbs,* Zero-Dimensional Commutative Rings
172. *K. Jarosz,* Function Spaces
173. *V. Ancona et al.,* Complex Analysis and Geometry
174. *E. Casas,* Control of Partial Differential Equations and Applications
175. *N. Kalton et al.,* Interaction Between Functional Analysis, Harmonic Analysis, and Probability
176. *Z. Deng et al.,* Differential Equations and Control Theory
177. *P. Marcellini et al.* Partial Differential Equations and Applications
178. *A. Kartsatos,* Theory and Applications of Nonlinear Operators of Accretive and Monotone Type
179. *M. Maruyama,* Moduli of Vector Bundles
180. *A. Ursini and P. Aglianò,* Logic and Algebra
181. *X. H. Cao et al.,* Rings, Groups, and Algebras
182. *D. Arnold and R. M. Rangaswamy,* Abelian Groups and Modules
183. *S. R. Chakravarthy and A. S. Alfa,* Matrix-Analytic Methods in Stochastic Models

184. *J. E. Andersen et al.*, Geometry and Physics
185. *P.-J. Cahen et al.*, Commutative Ring Theory
186. *J. A. Goldstein et al.*, Stochastic Processes and Functional Analysis
187. *A. Sorbi*, Complexity, Logic, and Recursion Theory
188. *G. Da Prato and J.-P. Zolésio*, Partial Differential Equation Methods in Control and Shape Analysis
189. *D. D. Anderson*, Factorization in Integral Domains
190. *N. L. Johnson*, Mostly Finite Geometries
191. *D. Hinton and P. W. Schaefer*, Spectral Theory and Computational Methods of Sturm–Liouville Problems
192. *W. H. Schikhof et al.*, *p*-adic Functional Analysis
193. *S. Sertöz*, Algebraic Geometry
194. *G. Caristi and E. Mitidieri*, Reaction Diffusion Systems
195. *A. V. Fiacco*, Mathematical Programming with Data Perturbations
196. *M. Křížek et al.*, Finite Element Methods: Superconvergence, Post-Processing, and A Posteriori Estimates
197. *S. Caenepeel and A. Verschoren*, Rings, Hopf Algebras, and Brauer Groups
198. *V. Drensky et al.*, Methods in Ring Theory
199. *W. B. Jones and A. Sri Ranga*, Orthogonal Functions, Moment Theory, and Continued Fractions
200. *P. E. Newstead*, Algebraic Geometry
201. *D. Dikranjan and L. Salce*, Abelian Groups, Module Theory, and Topology
202. *Z. Chen et al.*, Advances in Computational Mathematics
203. *X. Caicedo and C. H. Montenegro*, Models, Algebras, and Proofs
204. *C. Y. Yıldırım and S. A. Stepanov*, Number Theory and Its Applications
205. *D. E. Dobbs et al.*, Advances in Commutative Ring Theory
206. *F. Van Oystaeyen*, Commutative Algebra and Algebraic Geometry
207. *J. Kakol et al.*, *p*-adic Functional Analysis
208. *M. Boulagouaz and J.-P. Tignol*, Algebra and Number Theory
209. *S. Caenepeel and F. Van Oystaeyen*, Hopf Algebras and Quantum Groups
210. *F. Van Oystaeyen and M. Saorin*, Interactions Between Ring Theory and Representations of Algebras
211. *R. Costa et al.*, Nonassociative Algebra and Its Applications
212. *T.-X. He*, Wavelet Analysis and Multiresolution Methods
213. *H. Hudzik and L. Skrzypczak*, Function Spaces: The Fifth Conference
214. *J. Kajiwara et al.*, Finite or Infinite Dimensional Complex Analysis
215. *G. Lumer and L. Weis*, Evolution Equations and Their Applications in Physical and Life Sciences
216. *J. Cagnol et al.*, Shape Optimization and Optimal Design
217. *J. Herzog and G. Restuccia*, Geometric and Combinatorial Aspects of Commutative Algebra
218. *G. Chen et al.*, Control of Nonlinear Distributed Parameter Systems
219. *F. Ali Mehmeti et al.*, Partial Differential Equations on Multistructures
220. *D. D. Anderson and I. J. Papick*, Ideal Theoretic Methods in Commutative Algebra
221. *Á. Granja et al.*, Ring Theory and Algebraic Geometry
222. *A. K. Katsaras et al.*, *p*-adic Functional Analysis
223. *R. Salvi*, The Navier-Stokes Equations
224. *F. U. Coelho and H. A. Merklen*, Representations of Algebras
225. *S. Aizicovici and N. H. Pavel*, Differential Equations and Control Theory
226. *G. Lyubeznik*, Local Cohomology and Its Applications
227. *G. Da Prato and L. Tubaro*, Stochastic Partial Differential Equations and Applications
228. *W. A. Carnielli et al.*, Paraconsistency
229. *A. Benkirane and A. Touzani*, Partial Differential Equations
230. *A. Illanes et al.*, Continuum Theory
231. *M. Fontana et al.*, Commutative Ring Theory and Applications
232. *D. Mond and M. J. Saia*, Real and Complex Singularities
233. *V. Ancona and J. Vaillant*, Hyperbolic Differential Operators
234. *A. Giambruno et al.*, Polynomial Identities and Combinatorial Methods
235. *G. R. Goldstein et al.*, Evolution Equations

Additional Volumes in Preparation

Evolution Equations

proceedings in honor of J. A. Goldstein's 60th birthday

edited by

Gisèle Ruiz Goldstein
University of Memphis
Memphis, Tennessee, U.S.A.

Rainer Nagel
University of Tübingen
Tübingen, Germany

Silvia Romanelli
University of Bari
Bari, Italy

MARCEL DEKKER, INC. NEW YORK • BASEL

Although great care has been taken to provide accurate and current information, neither the author(s) nor the publisher, nor anyone else associated with this publication, shall be liable for any loss, damage, or liability directly or indirectly caused or alleged to be caused by this book. The material contained herein is not intended to provide specific advice or recommendations for any specific situation.

Trademark notice: Product or corporate names may be trademarks or registered trademarks and are used only for identification and explanation without intent to infringe.

Library of Congress Cataloging-in-Publication Data
A catalog record for this book is available from the Library of Congress.

ISBN: 0-8247-0975-6

This book is printed on acid-free paper.

Headquarters
Marcel Dekker, Inc.
270 Madison Avenue, New York, NY 10016, U.S.A.
tel: 212-696-9000; fax: 212-685-4540

Distribution and Customer Service
Marcel Dekker, Inc.
Cimarron Road, Monticello, New York 12701, U.S.A.
tel: 800-228-1160; fax: 845-796-1772

Eastern Hemisphere Distribution
Marcel Dekker AG
Hutgasse 4, Postfach 812, CH-4001 Basel, Switzerland
tel: 41-61-260-6300; fax: 41-61-260-6333

World Wide Web
http://www.dekker.com

The publisher offers discounts on this book when ordered in bulk quantities. For more information, write to Special Sales/Professional Marketing at the headquarters address above.

Current printing (last digit):

10 9 8 7 6 5 4 3 2 1

PRINTED IN THE UNITED STATES OF AMERICA

Preface

During the week of June 11-17, 2001, more than one hundred mathematicians gathered at the Heinrich-Fabri-Institut in Blaubeuren, Germany, to participate in a conference in honor of the 60th birthdays of Philippe Bénilan, Jerome A. Goldstein and Rainer Nagel.

It is a remarkable coincidence that three of the fundamental contributors to the theory of *Semigroups of Operators and Evolution Equations* would all celebrate their 60th birthdays within one calendar year. These three mathematicians have all played an essential role in semigroup theory not only by their own contributions to the theory and its applications, but also by creating a new generation of researchers in this field through their numerous research articles, monographs and books, and the many graduate students whom they have inspired and whose theses they have directed.

Sadly, Philippe Bénilan passed away on 17th February 2001, and the mathematical world mourned a great loss. In New Orleans there is the tradition of a "jazz funeral". In this tradition when a jazz musician dies, the funeral procession makes its way from the church to the cemetery accompanied by somber music. After the burial, the somber music ends and the band begins to play upbeat jazz to celebrate the deceased's time on earth in the same way that he loved to live his life. And so the purpose of the conference became twofold – to celebrate the 60th birthdays of Jerry Goldstein and Rainer Nagel and to serve as a mathematical version of a jazz funeral for Philippe Bénilan.

The conference was organized by Wolfgang Arendt, Gregor Nickel, Susanna Piazzera, Roland Schnaubelt and Lutz Weis. It began with a tribute to Philippe Bénilan and his work in the form of plenary lectures by Michel Pierre, Jan Prüss, and Juan-Luis Vázquez. In the following days many interesting lectures were given on a wide variety of topics in linear and nonlinear semigroup theory, linear and nonlinear partial differential equations, functional analysis and operator theory, as well as applications of the theory to problems from physics, biology and finance. The speakers in semigroup theory were Franco Altomare, Yuri Latushkin, Günter Lumer, Jaroslav Milota, Enrico Priola, Silvia Romanelli, Sen-Yen Shaw, Sachi Srivastava, Jan Van Neerven, and Hendrik Vogt. Bob Dorroh, Eduard Feireisl, Gisèle Goldstein, Davide Guidetti, José Mazón, John Neuberger, Frédérique Simondon,

Gieri Simonett, Kazuaki Taira, Peter Takáč, and Qi Zhang spoke on partial differential equations. Lectures on functional analysis and operator theory were given by Giovanni Dore, John Erdos, Eva Fašangová, Nigel Kalton and Peter Volkmann. The final day of the conference began with an extraordinary lecture by Hans-Otto Peitgen on *Harnessing Chaos*. This was an utterly memorable end to the scientific part of the conference. A series of social events followed, each one a jewel by itself.

From the Fabri-Institut the participants walked to the Kloster, a fabulous 16th century monastery, where they enjoyed a concert by the *Ensemble ISEM* composed entirely of mathematicians. The concert included a string quartet and quintet comprised of Anne Bückling (viola), Michael Gutfleisch (viola, violin), Gregor Nickel (violin), Jan Polland (cello), Antje Reichart (viola) and László Székelyhidi (violin). The next performance was an incredible piano concerto by Ingrid Carbone bringing the audience to their feet. The celebration then moved to the Klostersaal for a *cena sociale*. As dinner music, *The Hille-Yosida Cantata*, written by Friedrich Wille (music) and Gregor Nickel (lyrics) was performed by participants of the conference. The words and music are included on the following pages. A spectacular dinner buffet organized by Susanne Nagel was enjoyed by all. Participants, family and friends danced until the early morning hours with music provided by the band *First Class*. Many reported sightings of mathematicians dancing who had never before seemed so inclined.

The conference was a truly unique experience in the mathematical community and one that will not be soon forgotten by the participants. It was indeed a fitting celebration for three people who have given so much to the mathematical community.

A special issue of the *Journal of Evolution Equations* which will appear in 2003 is dedicated to Philippe Bénilan. The students of Rainer Nagel prepared a personal Festschrift in his honor. This volume is dedicated to Jerry Goldstein in celebration of his 60th birthday.

Gisèle Goldstein

Silvia Romanelli

Contents

Contributors

Eugene G. Belokolos Institute of Magnetism, National Academy of Sciences, Ukraine, Kiev, Ukraine

Jerry L. Bona University of Illinois at Chicago, Chicago, Illinois, U.S.A.

Radu C. Cascaval University of Missouri, Columbia, Missouri, U.S.A.

Alfonso Castro The University of Texas at San Antonio, San Antonio, Texas, U.S.A.

Hongqiu Chen The University of Memphis, Memphis, Tennessee, U.S.A.

Fabrizio Colombo Politecnico di Milano, Milan, Italy

David Cramer University of Missouri, Columbia, Missouri, U.S.A.

M. G. Crandall University of California, Santa Barbara, California, U.S.A.

G. Da Prato Scuola Normale Superiore di Pisa, Pisa, Italy

Gabriella Di Blasio Università di Roma "La Sapienza", Rome, Italy

Klaus-Jochen Engel Università di L'Aquila, Roio Poggio (AQ), Italy

Joachim Escher Institute for Applied Mathematics, University of Hannover, Hannover, Germany

Angelo Favini Università degli Studi di Bologna, Bologna, Italy

Fritz Gesztesy University of Missouri, Columbia, Missouri, U.S.A.

Maria Girardi University of South Carolina, Columbia, South Carolina, U.S.A.

Ronald Grimmer Southern Illinois University at Carbondale, Carbondale, Illinois, U.S.A.

Davide Guidetti Università degli Studi di Bologna, Bologna, Italy

Matthias Hieber TU Darmstadt, Darmstadt, Germany

Robert M. Kaufmann University of Alabama at Birmingham, Birmingham, Alabama, U.S.A.

M. K. Kwong Lucent Technologies, Lisle, Illinois, U.S.A.

Wilson Lamb University of Strathclyde, Glasgow, UK

Yuri Latushkin University of Missouri, Columbia, Missouri, U.S.A.

Alfredo Lorenzi Università degli Studi di Milano, Milan, Italy

Konstantin A. Makarov University of Missouri, Columbia, Missouri, U.S.A.

Lahcen Maniar Faculté des Sciences Semlalia, Université Cadi Ayyad, Marrakech, Morocco

Adam C. McBride University of Strathclyde, Glasgow, UK

J. W. Neuberger University of North Texas, Denton, Texas, U.S.A.

Michel Pierre Antenne de Bretagne de l'ENS Cachan et Institut de Recherche Mathématique de Rennes, Bruz, France

Jan Prüss Martin-Luther-Universität Halle-Wittenberg, Halle, Germany

M. M. Rao University of California at Riverside, Riverside, California, U.S.A.

Mounir Rihani Faculté des Sciences Ben M'Sik, Université Hassan II – Mohammédia, Casablanca, Morocco

Steve Rosencrans Tulane University, New Orleans, Louisiana, U.S.A.

Lev A. Sakhnovich Brooklyn, New York, U.S.A.

Roland Schnaubelt Martin-Luther-Universität Halle-Wittenberg, Halle, Germany

Junping Shi College of William and Mary, Williamsburg, Virginia, U.S.A., and Harbin Normal University, Harbin, Heilongjiang, P.R. China

Ratnasingham Shivaji Mississippi State University, Mississippi State, Mississippi, U.S.A.

Gieri Simonett Vanderbilt University, Nashville, Tennessee, U.S.A.

Eugenio Sinestrari Università di Roma "La Sapienza", Rome, Italy

Yudi Soeharyadi Institut Teknologi Bandung, Bandung, Indonesia

A. Święch School of Mathematics, Georgia Institute of Technology, Atlanta, Georgia, U.S.A.

Vincenzo Vespri Università degli Studi di Firenze, Florence, Italy

Jürgen Voigt Technische Universität Dresden, Dresden, Germany

Xuefeng Wang Tulane University, New Orleans, Louisiana, U.S.A.

Lutz Weis Mathematisches Institut I, Universität Karlsruhe, Karlsruhe, Germany

Ian Wood TU Darmstadt, Darmstadt, Germany

Anton Zettl Northern Illinois University, DeKalb, Illinois, U.S.A.

Hongkun Zhang University of Alabama at Birmingham, Birmingham, Alabama, U.S.A.

Qi S. Zhang University of California at Riverside, Riverside, California, U.S.A.

J. A. Goldstein

Biography

Jerry Goldstein got all his degrees at Carnegie Tech (now called Carnegie Mellon University). After finishing his PhD in 1967, he accepted a postdoctoral fellowship at the Institute for Advanced Study. Jerry went to Tulane University and remained there for 24 years. He then spent five years at Louisiana State University before moving to the University of Memphis in 1996. Jerry has visited and lectured at numerous universities in many countries. He was a visiting professor at the Mathematical Sciences Research Institute, Carnegie Mellon University, Stanford University, Strathclyde University, the Universities of Bari, Bologna, Brasilia, Graz, Padova, Rio de Janeiro, Tuebingen, and Westfield College. In 1985 he was the first recipient of the Faculty Excellence in Research Award from Tulane University. In 2002 Jerry was awarded the Distinguished Research and Creative Achievement Award from the University of Memphis.

His thesis problem involved second order Itô processes. Markov processes led him to operator semigroups, and that path naturally led him to partial differential equations. In 1969 Jerry attended a four-week summer seminar on Scattering Theory in Flagstaff, Arizona. He was attracted by the courses given by Kato, Lax and Phillips; this meeting had a tremendous impact on his future work. His interest in abstract scattering theory led to a study of quantum mechanics, and later to other parts of physics. Much of Jerry's earlier work involved linear evolution equations with time dependent generators, and as a young mathematician he accepted the prevailing wisdom that mathematics was an "individual sport". That viewpoint changed later, and Jerry has had extensive long term collaborations, including those with Jim Sandefur (on equipartition of energy), with Mel Levy (on quantum chemistry), with Chin-Yuan Lin (on nonlinear parabolic problems via nonlinear semigroups), with Eugene Eckstein (on suspension flows), with Angelo Favini, Gisèle Goldstein, Enrico Obrecht, and Silvia Romanelli (on many issues involving analyticity, asymptotics, and general boundary conditions), and with Gisèle Goldstein (on Thomas-Fermi theory, regularity for nonlinear problems). He also collaborated with Wofgang Arendt, Philippe Bénilan, Earl Berkson, Haim Brézis, Paul Chernoff, Jim Donaldson, John Erdos, Rainer Nagel, Steve Rosencrans, and many others.

Jerry has written or edited ten books, is on the editorial board of thirteen journals, and has written more than one hundred and fifty research papers.

Jerry has advised the PhD theses of Charles Monlezun, James Sandefur, Joseph Henrickson, Bruno Wichnoski, Clay Burch, Janet Hughes, Kuong Lin Ou Young, Michael Ballotti, Giséle Ruiz Rieder, Lige Li, Chin-Yuan Lin, Douglas Pickett, Mi Ai Park, Gabriella Segalla, Denise Kirshner, Chien-An lung, Genbao Shi, Kungyang Wang, Andrei Breazna, Aurora Breazna, Radu Cascaval, and Yudi Soeharyardi. Besides his own PhD students, he helped advise the theses and inspired the research of students from several countries such as Brazil, Germany and Italy. In addition Jerry has given research courses abroad with other mathematicians including Philippe Bénilan and Rainer Nagel.

In addition to his numerous research accomplishments, Jerry has also served on important national committees. He served as a member of the Advisory Committee for the Division of Mathematical Sciences at the National Science Foundation. He has served on many committees for the American Mathematical Society, including the Nominating Committee, which he chaired in 1993-1994, the Committee on Academic Freedom, Promotion and Tenure, and the Menger Prize Committee. He has served on many committees for the Mathematical Association of America; he also chaired the Mathematical Association of America's Committee on the Undergraduate Program in Mathematics for five years.

Gisèle Goldstein

Rainer Nagel

Silvia Romanelli

Hille-Yosida-Kantate

Text: Gregor Nickel
Musik: Friedrich Wille

1.Vorspiel

2. Choral

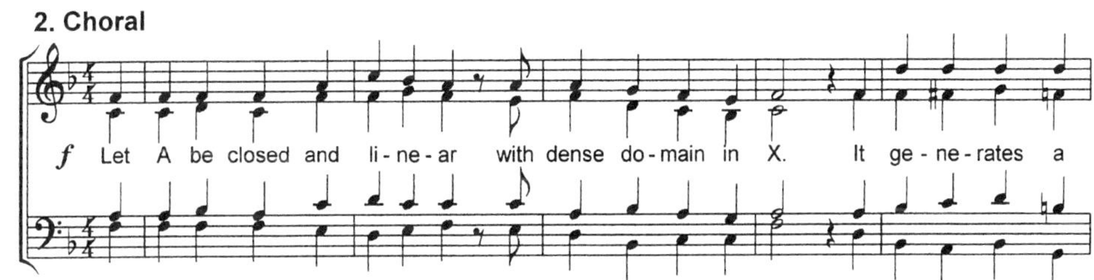

3. Rezitativ (Tenor)

Zum Be - wei - se des e - ben ge - hör - ten The - o - rems von Hil - le und auch von
p

5
Yo - si - da be - mer - ken wir zum An - fang fol - gen - des: Die E - xi - stenz der Se - mi - group

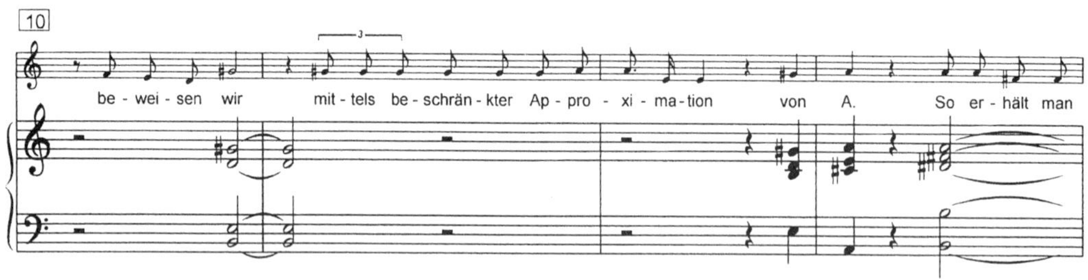
10
be - wei - sen wir mit - tels be - schrän - kter Ap - pro - xi - ma - tion von A. So er - hält man

14
schnell ne Fol - ge gleich-mä - ßig ste - ti - ger und kon - ver - gie-ren-der Halb - grup-pen die zu T, das wir su-chen, kon-ver-giern.

4. Chor

5. Arie (Mezzosopran)

9
Oh, welch ein wun-der ba-res The - o-rem! Es lässt sich täg - lich ver - wen - den.

13
Man er - ken-net da-mit so be-quem Er - zeu-ger an ih-ren Re-sol - ven - ten.

17
Auch so man-che üb-le P D E schmilzt hin bei die-sem Satz wie Früh - lings- schnee. In

21
der Phy-sik geht man-cher Trick, auch je-der Start der Welt-raum fahrt ge - lingt durch Hil-le- Yo-si-da, da-rum

24
ist der Satz so schön und wahr da-rum ist der Satz so wahr, ist so wahr, ist so wahr.
28
Oh, welch ein wun-der ba-res The - o-rem! Es lässt sich täg - lich ver-wen - den.
32
Man er - ken-net da-mit so be-quem Er - zeu-ger an ih-ren Re-sol-ven - ten.
36
39

6. Chor

Laßt uns nun lu - stig ge - ne - rie - ren, da braucht man sich gar nicht zu ge - nie - ren.
Jetzt kann man die Pro - ble - me lö - sen für die Gu - ten und auch für die Bö - sen.
Yo - si - da und Hil - le sein ge - prie - sen, dass sie uns auf die - sen Weg ge - wie - sen.
App - ro - xi - mie - ren, Nor - men um - ska - lie - ren, al - les geht uns e - le - gant jetzt von der Hand.
App - ro - xi - mie - ren, Nor - men um - ska - lie - ren, al - les ist mit ei - nem - mal jetzt voll tri - vial.

25
29
Drum stimmt ein froh und voll Ver - gnü - gen und singt mit, dass sich die Bal - ken bie - gen.
33
App - ro - xi-mie - ren, Nor - men um - ska - lie - ren, al - les geht uns e - le - gant jetzt von der Hand.
37
App - ro - xi-mie - ren, Nor - men um - ska - lie - ren, al - les ist mit ei - nem - mal jetzt voll tri - vial.
41

Matrix-Valued Generalizations of the Theorems of Borg and Hochstadt

EUGENE D. BELOKOLOS*

Institute of Magnetism, National Academy of Sciences, Ukraine
Vernadsky Str. 36B, Kiev–142, 252142, Ukraine
e-mail: bel@imag.kiev.ua

FRITZ GESZTESY*

Department of Mathematics, University of Missouri
Columbia, MO 65211, USA
e-mail: fritz@math.missouri.edu
www.math.missouri.edu/people/fgesztesy.html

KONSTANTIN A. MAKAROV

Department of Mathematics, University of Missouri
Columbia, MO 65211, USA
e-mail: makarov@math.missouri.edu
www.math.missouri.edu/people/kmakarov.html

LEV A. SAKHNOVICH

735 Crawford Ave., Brooklyn, NY 11223, USA
e-mail: Lev.Sakhnovich@verizon.net

Dedicated with great pleasure to Jerry Goldstein and Rainer Nagel on the occasion of their 60th birthdays.

ABSTRACT. We prove a generalization of the well-known theorems by Borg and Hochstadt for periodic self-adjoint Schrödinger operators without a spectral gap, respectively, one gap in their spectrum, in the matrix-valued context. Our extension of the theorems of Borg and Hochstadt replaces the periodicity condition

*Research supported in part by the CRDF grant UM1-325.

of the potential by the more general property of being reflectionless (the resulting potentials then automatically turn out to be periodic and we recover Després' matrix version of Borg's result). In addition, we assume the spectra to have uniform maximum multiplicity (a condition automatically fulfilled in the scalar context considered by Borg and Hochstadt). Moreover, the connection with the stationary matrix KdV hierarchy is established.

The methods employed in this paper rely on matrix-valued Herglotz functions, Weyl–Titchmarsh theory, pencils of matrices, and basic inverse spectral theory for matrix-valued Schrödinger operators.

1 INTRODUCTION

In a previous paper, [25], two of us constructed a class of self-adjoint $m \times m$ matrix-valued Schrödinger operators $H(\Sigma_n) = -d^2/dx^2 \mathcal{I}_m + \mathcal{Q}(\Sigma_n, \cdot)$ in $L^2(\mathbb{R})^{m\times m}$, $m \in \mathbb{N}$, with prescribed absolutely continuous finite-band spectrum Σ_n of the type

$$\Sigma_n = \left\{ \bigcup_{j=0}^{n-1} [E_{2j}, E_{2j+1}] \right\} \cup [E_{2n}, \infty), \quad n \in \mathbb{N}_0 \tag{1.1}$$

of uniform spectral multiplicity $2m$. Here

$$\{E_\ell\}_{0\le\ell\le 2n} \subseteq \mathbb{R}, \; n \in \mathbb{N}, \text{ with } E_\ell < E_{\ell+1}, \, 0 \le \ell \le 2n-1, \tag{1.2}$$

and hence $H(\Sigma_n)$ satisfies

$$\operatorname{spec}(H(\Sigma_n)) = \Sigma_n. \tag{1.3}$$

Throughout this paper all matrices will be considered over the field of complex numbers $\mathbb{C}$, and the corresponding linear space of $k \times \ell$ matrices will be denoted by $\mathbb{C}^{k\times\ell}$, $k, \ell \in \mathbb{N}$. Moreover, $\mathcal{I}_k$ denotes the identity matrix in $\mathbb{C}^{k\times k}$ for $k \in \mathbb{N}$, $\mathcal{M}^*$ the adjoint (i.e., complex conjugate transpose), $\mathcal{M}^t$ the transpose of a matrix $\mathcal{M}$, $\operatorname{diag}(m_1, \dots, m_k) \in \mathbb{C}^{k\times k}$ a diagonal $k \times k$ matrix, and $\mathrm{AC}_{\mathrm{loc}}(\mathbb{R})$ denotes the set of locally absolutely continuous functions on $\mathbb{R}$. The spectrum, point spectrum (the set of eigenvalues), essential spectrum, absolutely continuous spectrum, and singularly continuous spectrum of a self-adjoint linear operator T in a separable complex Hilbert space are denoted by $\operatorname{spec}(T)$, $\operatorname{spec}_{\mathrm{p}}(T)$, $\operatorname{spec}_{\mathrm{ess}}(T)$, $\operatorname{spec}_{\mathrm{ac}}(T)$, and $\operatorname{spec}_{\mathrm{sc}}(T)$, respectively.

The constructed matrix potentials $\mathcal{Q}_{\Sigma_n}$ (resp., $H(\Sigma_n)$) turns out to be *reflectionless* in the sense discussed in [14], [27], and [45] (cf. [15], [16], [44], [74] in the scalar context $m = 1$), that is, the half-line Weyl–Titchmarsh matrices $\mathcal{M}_\pm(\Sigma_n, z, x)$ associated with $H(\Sigma_n)$, the half-lines $[x, \pm\infty)$, and a Dirichlet boundary condition at $x \in \mathbb{R}$, satisfy

$$\lim_{\varepsilon\downarrow 0} \mathcal{M}_+(\Sigma_n, \lambda + i\varepsilon, x) = \lim_{\varepsilon\downarrow 0} \mathcal{M}_-(\Sigma_n, \lambda - i\varepsilon, x), \tag{1.4}$$

$$\lambda \in \bigcup_{j=0}^{n-1} (E_{2j}, E_{2j+1}) \cup (E_{2n}, \infty), \; x \in \mathbb{R}.$$

Especially, $\mathcal{M}_+(\Sigma_n, \cdot, x)$ is the analytic continuation of $\mathcal{M}_-(\Sigma_n, \cdot, x)$ through the set Σ_n, and vice versa. In other words, $\mathcal{M}_+(\Sigma_n, \cdot, x)$ and $\mathcal{M}_-(\Sigma_n, \cdot, x)$ are the two branches of an analytic matrix-valued function $\mathcal{M}(\Sigma_n, \cdot, x)$ on the two-sheeted Riemann surface of $\left(\prod_{\ell=0}^{2n}(z - E_\ell)\right)^{1/2}$. The reflectionless property (1.4) then implies the absolute continuity of the spectrum of H_{Σ_n} and its uniform (maximal) multiplicity $2m$.

In this sequel of paper [25], we focus on the two special cases $n = 0$ and $n = 1$ and derive matrix-valued extensions of the well-known theorems by Borg [3] and Hochstadt [33], respectively. Before describing our principal new results and the contents of each section, we briefly recall the classical results by Borg and Hochstadt in the scalar case $m = 1$.

In 1946 Borg [3] proved, among a variety of other inverse spectral theorems, the following result.

THEOREM 1.1. [3] *Let* $\Sigma_0 = [E_0, \infty)$ *for some* $E_0 \in \mathbb{R}$ *and* $q_0 \in L^1_{\text{loc}}(\mathbb{R})$ *be real-valued and periodic. Suppose that* $h_0 = -\frac{d^2}{dx^2} + q_0$ *is the associated self-adjoint Schrödinger operator in* $L^2(\mathbb{R})$ *(cf. (1.9) for* $m = 1$*) and assume that*

$$\operatorname{spec}(h_0) = \Sigma_0. \tag{1.5}$$

Then

$$q_0(x) = E_0 \text{ for a.e. } x \in \mathbb{R}. \tag{1.6}$$

Traditionally, uniqueness results such as Theorem 1.1 are called Borg-type theorems. (However, this terminology is not uniquely adopted and hence a bit unfortunate. Indeed, inverse spectral results on finite intervals recovering the potential coefficient(s) from several spectra, were also pioneered by Borg in his celebrated paper [3], and hence are also coined Borg-type theorems in the literature, see, e.g., [53; Sect. 6].) Actually, Borg assumed $q_0 \in L^2_{\text{loc}}(\mathbb{R})$, but that seems a minor detail.

REMARK 1.2.
(*i*) A closer examination of the short proof of (an extension of) Theorem 1.1 provided in [14] shows that periodicity of q_0 is not the point for the uniqueness result (1.6). The key ingredient (besides $\operatorname{spec}(h_0) = [E_0, \infty)$ and q_0 real-valued) is clearly the fact that q_0 is reflectionless in the sense of (1.4).
(*ii*) Real-valued periodic potentials are known to satisfy (1.4), but so are certain classes of real-valued quasi-periodic and almost-periodic potentials q_0 (see, e.g., [15], [16], [41], [42], [43], [44], [45], [74]). In particular, the class of real-valued algebro-geometric finite-gap potentials q_0 (a subclass of the set of real-valued quasi-periodic potentials) is a prime example satisfying (1.4) without necessarily being periodic.
(*iii*) We note that real-valuedness of q_0 is an essential assumption in Theorem 1.1. Indeed, it is well-known that $q(x) = \exp(ix)$, $x \in \mathbb{R}$, leads to the half-line spectrum $[0, \infty)$. A detailed treatment of a class of examples of this type can be found in [19], [20], [29], [61], [62]. Moreover, the example of complete exponential localization of the spectrum of a discrete Schrödinger operator with a quasi-periodic real-valued potential having two basic frequencies and no gaps in its spectrum [10] illustrates the importance of the reflectionless property of q_0 in Theorem 1.1.

Next we recall Hochstadt's theorem [33] from 1965.

THEOREM 1.3. [33] *Let* $\Sigma_1 = [E_0, E_1] \cup [E_2, \infty)$ *for some* $E_0 < E_1 < E_2$ *and* $q_1 \in L^1_{\text{loc}}(\mathbb{R})$ *be real-valued and periodic. Suppose that* $h_1 = -\frac{d^2}{dx^2} + q_1$ *is the associated self-adjoint Schrödinger operator in* $L^2(\mathbb{R})$ *(cf. (1.9) for* $m = 1$*) and assume that*

$$\operatorname{spec}(h_1) = \Sigma_1. \tag{1.7}$$

Then

$$q_1(x) = C_0 + 2\wp(x + \omega_3 + \alpha) \text{ for some } \alpha \in \mathbb{R} \text{ and a.e. } x \in \mathbb{R}. \tag{1.8}$$

Here $\wp(\cdot) = \wp(\cdot;\omega_1;\omega_3)$ denotes the elliptic Weierstrass function with half-periods $\omega_1 > 0$ and $-i\omega_3 > 0$ (cf. [1; Ch. 18]).

REMARK 1.4. Again it will turn out that periodicity of q_1 is not the point for the uniqueness result (1.8). The key ingredient (besides $\operatorname{spec}(h_1) = \Sigma_1$ and q_1 real-valued) is again the fact that q_1 is reflectionless in the sense of (1.4). Similarly, Remarks 1.2 *(ii)*, *(iii)* apply of course in the present context.

The principal results of this paper then read as follows.

THEOREM 1.5. *Let* $m \in \mathbb{N}$*, suppose* $\mathcal{Q}_\ell = \mathcal{Q}_\ell^* \in L^1_{\text{loc}}(\mathbb{R})^{m\times m}$ *and assume that the differential expressions* $-\mathcal{I}_m \frac{d^2}{dx^2} + \mathcal{Q}_\ell$*,* $\ell = 0, 1$*, are in the limit point case at* $\pm\infty$*. Define the self-adjoint Schrödinger operators* H_ℓ *in* $L^2(\mathbb{R})^{m\times m}$

$$H_\ell = -\mathcal{I}_m \frac{d^2}{dx^2} + \mathcal{Q}_\ell, \quad \ell = 0, 1, \tag{1.9}$$
$$\operatorname{dom}(H_\ell) = \{g \in L^2(\mathbb{R})^m \mid g, g' \in \mathrm{AC}_{\text{loc}}(\mathbb{R})^m;\ (-g'' + \mathcal{Q}_\ell g) \in L^2(\mathbb{R})^m\}$$

and assume that $\mathcal{Q}_\ell$ *is reflectionless (cf. (1.4)).*
(i) Let $\Sigma_0 = [E_0, \infty)$ *for some* $E_0 \in \mathbb{R}$ *and suppose that* H_0 *has spectrum*

$$\operatorname{spec}(H_0) = \Sigma_0. \tag{1.10}$$

Then

$$\mathcal{Q}_0(x) = E_0 \mathcal{I}_m \text{ for a.e. } x \in \mathbb{R}. \tag{1.11}$$

(ii) Let $\Sigma_1 = [E_0, E_1] \cup [E_2 \infty)$ *for some* $E_0 < E_1 < E_2$ *and suppose that* H_1 *has spectrum*

$$\operatorname{spec}(H_1) = \Sigma_1. \tag{1.12}$$

Then

$$\begin{aligned}\mathcal{Q}_1(x) = {} & (1/3)(E_0 + E_1 + E_2)\mathcal{I}_m \\ & + 2\mathcal{U} \operatorname{diag}(\wp(x + \omega_3 + \alpha_1), \ldots, \wp(x + \omega_3 + \alpha_m))\mathcal{U}^{-1} \\ & \text{for some } \alpha_j \in \mathbb{R},\ 1 \le j \le m \text{ and a.e. } x \in \mathbb{R},\end{aligned} \tag{1.13}$$

where $\mathcal{U}$ *is an* $m \times m$ *unitary matrix independent of* $x \in \mathbb{R}$*. In particular,* $\mathcal{Q}_1$ *satisfies the stationary KdV equation*

$$\mathcal{Q}_1''' - 3(\mathcal{Q}_1^2)' + 2(E_0 + E_1 + E_2)\mathcal{Q}_1' = 0. \tag{1.14}$$

As shown in [14], periodic Schrödinger operators in $L^2(\mathbb{R})^{m\times m}$ with spectra of uniform (maximal) multiplicity $2m$ are reflectionless in the sense that (1.4) holds for all λ in the open interior of the spectrum. Hence one obtains the following result.

THEOREM 1.6. *Let $m \in \mathbb{N}$, suppose $\mathcal{Q}_\ell = \mathcal{Q}_\ell^* \in L^1_{\text{loc}}(\mathbb{R})^{m\times m}$ is periodic and define the self-adjoint Schrödinger operators H_ℓ, $\ell = 0, 1$ in $L^2(\mathbb{R})^{m\times m}$ as in (1.9). Assume*

$$\text{spec}(H_\ell) = \Sigma_\ell, \quad \ell = 0, 1, \tag{1.15}$$

and suppose that H_ℓ, $\ell = 0, 1$ has uniform (maximal) spectral multiplicity $2m$. Then $\mathcal{Q}_\ell$, $\ell = 0, 1$, are reflectionless and hence the assertions (1.11), (1.13), and (1.14) of Theorem 1.5 hold.

REMARK 1.7.
(*i*) The assumption of uniform (maximal) spectral multiplicity $2m$ in Theorem 1.6 (*i*) is an essential one. Otherwise, one can easily construct nonconstant potentials $\mathcal{Q}$ such that the associated Schrödinger operator $H_\mathcal{Q}$ has overlapping band spectra and hence spectrum equal to a half-line. For such a construction it suffices to consider the case in which $\mathcal{Q}$ is a diagonal matrix. In the special scalar case $m = 1$, reflectionless potentials automatically give rise to maximum uniform spectral multiplicity 2 for the corresponding scalar Schrödinger operator in $L^2(\mathbb{R})$.
(*ii*) Theorem 1.6 (*i*) assuming $\mathcal{Q}_0 \in L^\infty(\mathbb{R})^{m\times m}$ to be periodic has been proved by Deprés [17] using an entirely different approach based on a detailed Floquet analysis. Deprés' result was reproved in [14] under the current general assumptions on $\mathcal{Q}_0$ using methods based on matrix-valued Herglotz functions and trace formulas.
(*iii*) For different proofs of Borg's Theorem 1.1 in the scalar case $m = 1$ we refer to [33], [37], [38], [41].
(*iv*) Without loss of generality we focus on the limit point case of the differential expression $-\mathcal{I}_m \frac{d^2}{dx^2} + \mathcal{Q}$ at $\pm\infty$ in this paper. In fact, by a result originally due to Povzner [63], scalar Schrödinger differential expressions leading to minimal operators bounded from below are in the limit point case at $\pm\infty$. Povzner's result was later also proved by Wienholtz [76] and is reproduced as Theorem 35 in [28; p. 58]. As shown in [13], Wienholtz's proof extends to the matrix case at hand. Since the spectra Σ_ℓ, $\ell = 0, 1$ are bounded from below, the limit point assumption is justified (and natural).

This paper is a modest contribution to the inverse spectral theory of matrix-valued Schrödinger (and Dirac-type) operators and part of a recent program in this area (cf. [11], [12], [13], [14], [23], [24], [25], [26], and [27]). For other relevent recent literature in this context we refer, for instance, to [5], [6], [7], [8], [9], [35], [36], [39], [54], [55], [67], [71], [72], [73]. For the applicability of this circle of ideas to the nonabelian Korteweg–de Vries hierarchy we refer to [25] and the references cited therein.

In Section 2 we recall basic facts on Weyl–Titchmarsh theory and pencils of matrices as needed in the remainder of this paper. Section 3 summarizes the principal results of paper [25]. Finally, in Section 4 we present the matrix extensions of the theorems of Borg and Hochstadt and the corresponding connection with the stationary KdV hierarchy.

2 PRELIMINARIES ON WEYL–TITCHMARSH THEORY AND PENCILS OF MATRICES

The basic assumption for this section will be the following.

HYPOTHESIS 2.1. *Fix* $m \in \mathbb{N}$, *suppose* $\mathcal{Q} = \mathcal{Q}^* \in L^1_{\mathrm{loc}}(\mathbb{R})^{m\times m}$, *introduce the differential expression*

$$\mathcal{L} = -\mathcal{I}_m \frac{d^2}{dx^2} + \mathcal{Q}, \quad x \in \mathbb{R}, \tag{2.1}$$

and suppose $\mathcal{L}$ *is in the limit point case at* $\pm\infty$.

Given Hypothesis 2.1 we consider the matrix-valued Schrödinger equation

$$-\psi''(z,x) + \mathcal{Q}(x)\psi(z,x) = z\psi(z,x) \text{ for a.e. } x \in \mathbb{R}, \tag{2.2}$$

where $z \in \mathbb{C}$ plays the role of a spectral parameter and ψ is assumed to satisfy

$$\psi(z,\cdot), \psi'(z,\cdot) \in \mathrm{AC}_{\mathrm{loc}}(\mathbb{R})^{m\times m}. \tag{2.3}$$

Throughout this paper, x-derivatives are abbreviated by a prime $'$.

Let $\Psi(z,x,x_0)$ be a $2m \times 2m$ normalized fundamental system of solutions of (2.2) at some $x_0 \in \mathbb{R}$ which we partition as

$$\Psi(z,x,x_0) = \begin{pmatrix} \theta(z,x,x_0) & \phi(z,x,x_0) \\ \theta'(z,x,x_0) & \phi'(z,x,x_0) \end{pmatrix}. \tag{2.4}$$

Here $'$ denotes d/dx, $\theta(z,x,x_0)$ and $\phi(z,x,x_0)$ are $m \times m$ matrices, entire with respect to $z \in \mathbb{C}$, and normalized according to $\Psi(z,x_0,x_0) = \mathcal{I}_{2m}$.

By Hypothesis 2.1, the $m \times m$ Weyl–Titchmarsh matrices associated with $\mathcal{L}$, the half-lines $[x, \pm\infty)$, and a Dirichlet boundary condition at x, are given by (cf. [30], [31], [32], [40], [60], [68], [69], [70])

$$\mathcal{M}_\pm(z,x) = \Psi'_\pm(z,x,x_0)\Psi_\pm(z,x,x_0)^{-1}, \quad z \in \mathbb{C}\backslash\mathbb{R}, \tag{2.5}$$

where $\Psi_\pm(z,\cdot,x_0)$ satisfy $(\mathcal{L} - z\mathcal{I}_m)\Psi_\pm(z,\cdot,x_0) = 0$ and

$$\Psi_\pm(z,\cdot,x_0) \in L^2([x_0,\pm\infty))^{m\times m}. \tag{2.6}$$

Next, we recall the definition of matrix-valued Herglotz function.

DEFINITION 2.2. A map $\mathcal{M}\colon \mathbb{C}_+ \to \mathbb{C}^{n\times n}$, $n \in \mathbb{N}$, extended to $\mathbb{C}_-$ by $\mathcal{M}(\bar{z}) = \mathcal{M}(z)^*$ for all $z \in \mathbb{C}_+$, is called an $n \times n$ Herglotz matrix if it is analytic on $\mathbb{C}_+$ and $\mathrm{Im}(\mathcal{M}(z)) \geq 0$ for all $z \in \mathbb{C}_+$.

Here we denote $\mathrm{Im}(\mathcal{M}) = (\mathcal{M} - \mathcal{M}^*)/2i$ and $\mathrm{Re}(\mathcal{M}) = (\mathcal{M} + \mathcal{M}^*)/2$.

$\pm\mathcal{M}_\pm(\cdot,x)$ are $m \times m$ Herglotz matrices of rank m and hence admit the representations

$$\pm\mathcal{M}_\pm(z,x) = \mathrm{Re}(\pm\mathcal{M}_\pm(\pm i,x)) + \int_{\mathbb{R}} d\Omega_\pm(\lambda,x)\left((\lambda - z)^{-1} - \lambda(1+\lambda^2)^{-1}\right), \tag{2.7}$$

where

$$\int_{\mathbb{R}} \|d\Omega_\pm(\lambda,x)\|_{\mathbb{C}^{m\times m}} (1+\lambda^2)^{-1} < \infty \tag{2.8}$$

and

$$\Omega_\pm((\lambda,\mu],x) = \lim_{\delta\downarrow 0}\lim_{\varepsilon\downarrow 0} \frac{1}{\pi} \int_{\lambda+\delta}^{\mu+\delta} d\nu \, \mathrm{Im}(\pm\mathcal{M}_\pm(\nu + i\varepsilon, x)). \tag{2.9}$$

Necessary and sufficient conditions for $\mathcal{M}_\pm(\cdot, x_0)$ to be the half-line $m \times m$ Weyl–Titchmarsh matrix associated with a Schrödinger operator on $[x_0, \pm\infty)$ in terms of the corresponding measures $\Omega_\pm(\cdot, x_0)$ in the Herglotz representation (2.7) of $\mathcal{M}_\pm(\cdot, x_0)$ can be derived using the matrix-valued extension of the classical inverse spectral theory approach due to Gelfand and Levitan [22], as worked out by Rofe-Beketov [64]. The following result describes sufficient conditions for a monotonically nondecreasing matrix function to be the matrix spectral function of a half-line Schrödinger operator. It extends well-known results in the scalar case $m = 1$ (cf. [50; Sects. 2.5, 2.9], [51], [59; Sect. 26.5], [75]).

THEOREM 2.3. [64] *Suppose $\Omega_+(\cdot, x_0)$ is a monotonically nondecreasing $m \times m$ matrix-valued function on $\mathbb{R}$. Then $\Omega_+(\cdot, x_0)$ is the matrix spectral function of a self-adjoint Schrödinger operator H_+ in $L^2([x_0, \infty))^m$ associated with the $m \times m$ matrix-valued differential expression $\mathcal{L}_+ = -d^2/dx^2 \mathcal{I}_m + \mathcal{Q}$, $x > x_0$, with a Dirichlet boundary condition at x_0, a self-adjoint boundary condition at ∞ (if necessary), and a self-adjoint potential matrix $\mathcal{Q}$ with $\mathcal{Q}^{(r)} \in L^1([x_0, R])^{m\times m}$ for all $R > x_0$ if and only if the following two conditions hold.*
(i) Whenever $f \in C([x_0, \infty))^{m\times 1}$ with compact support contained in $[x_0, \infty)$ and

$$\int_{\mathbb{R}} F(\lambda)^* d\Omega_+(\lambda, x_0)\, F(\lambda) = 0, \text{ then } f = 0 \text{ a.e.}, \tag{2.10}$$

where

$$F(\lambda) = \lim_{R\uparrow\infty} \int_{x_0}^{R} dx\, \frac{\sin(\lambda^{1/2}(x - x_0))}{\lambda^{1/2}} f(x), \quad \lambda \in \mathbb{R}. \tag{2.11}$$

(ii) Define

$$\widetilde{\Omega}_+(\lambda, x_0) = \begin{cases} \Omega_+(\lambda, x_0) - \frac{2}{3\pi}\lambda^{3/2}, & \lambda \geq 0, \\ \Omega_+(\lambda, x_0), & \lambda < 0 \end{cases} \tag{2.12}$$

and assume the limit

$$\lim_{R\uparrow\infty} \int_{-\infty}^{R} d\widetilde{\Omega}_+(\lambda, x_0)\, \frac{\sin(\lambda^{1/2}(x - x_0))}{\lambda^{1/2}} = \Phi(x) \tag{2.13}$$

exists with $\Phi \in L^\infty([x_0, R])^{m\times m}$ for all $R > x_0$. Moreover, suppose that for some $r \in \mathbb{N}_0$, $\Phi^{(r+1)} \in L^1([x_0, R])^{m\times m}$ for all $R > x_0$, and $\Phi(x_0) = 0$.

Assuming Hypothesis 2.1, we next introduce the self-adjoint Schrödinger operator H in $L^2(\mathbb{R})^m$ by

$$\begin{aligned} &H = -\mathcal{I}_m \frac{d^2}{dx^2} + \mathcal{Q}, \\ &\mathrm{dom}(H) = \{g \in L^2(\mathbb{R})^m \mid g, g' \in \mathrm{AC}_{\mathrm{loc}}(\mathbb{R})^m;\ (-g'' + \mathcal{Q}g) \in L^2(\mathbb{R})^m\}. \end{aligned} \tag{2.14}$$

The resolvent of H then reads

$$((H - z)^{-1} f)(x) = \int_{\mathbb{R}} dx'\, \mathcal{G}(z, x, x') f(x'), \quad z \in \mathbb{C}\backslash\mathbb{R},\ f \in L^2(\mathbb{R})^m, \tag{2.15}$$

with the Green's matrix $\mathcal{G}(z,x,x')$ of H given by

$$\mathcal{G}(z,x,x') = \Psi_{\mp}(z,x,x_0)[\mathcal{M}_-(z,x_0) - \mathcal{M}_+(z,x_0)]^{-1}\Psi_{\pm}(\overline{z},x',x_0)^*, \quad x \lesseqgtr x',\ z \in \mathbb{C}\backslash\mathbb{R}. \tag{2.16}$$

Introducing

$$\mathcal{N}_{\pm}(z,x) = \mathcal{M}_-(z,x) \pm \mathcal{M}_+(z,x), \quad z \in \mathbb{C}\backslash\mathbb{R},\ x \in \mathbb{R}, \tag{2.17}$$

the $2m \times 2m$ Weyl–Titchmarsh function $\mathcal{M}(z,x)$ associated with H on $\mathbb{R}$ is then given by

$$\begin{aligned}\mathcal{M}(z,x) &= \left(\mathcal{M}_{p,q}(z,x)\right)_{p,q=1,2} \\ &= \begin{pmatrix} \mathcal{M}_{\pm}(z,x)\mathcal{N}_-(z,x)^{-1}\mathcal{M}_{\mp}(z,x) & \mathcal{N}_-(z,x)^{-1}\mathcal{N}_+(z,x)/2 \\ \mathcal{N}_+(z,x)\mathcal{N}_-(z,x)^{-1}/2 & \mathcal{N}_-(z,x)^{-1} \end{pmatrix}, \\ &\qquad z \in \mathbb{C}\backslash\mathbb{R},\ x \in \mathbb{R}.\end{aligned} \tag{2.18}$$

Then $\mathcal{M}(z,x)$ is a $2m \times 2m$ matrix-valued Herglotz function of rank $2m$ with representations

$$\mathcal{M}(z,x) = \mathrm{Re}(\mathcal{M}(i,x)) + \int_{\mathbb{R}} d\Omega(\lambda,x)\left((\lambda - z)^{-1} - \lambda(1+\lambda^2)^{-1}\right) \tag{2.19}$$

$$= \exp\left(\mathcal{C}(x) + \int_{\mathbb{R}} d\lambda\, \Upsilon(\lambda,x)\big((\lambda - z)^{-1} - \lambda(1+\lambda^2)^{-1}\big)\right), \tag{2.20}$$

where

$$\int_{\mathbb{R}} \|d\Omega(\lambda,x)\|_{\mathbb{C}^{2m\times 2m}}\,(1+\lambda^2)^{-1} < \infty, \tag{2.21}$$

$$\mathcal{C}(x) = \mathcal{C}(x)^*, \quad 0 \le \Upsilon(\cdot,x) \le \mathcal{I}_{2m} \text{ a.e.} \tag{2.22}$$

and

$$\Omega((\lambda,\mu],x) = \lim_{\delta\downarrow 0}\lim_{\varepsilon\downarrow 0}\frac{1}{\pi}\int_{\lambda+\delta}^{\mu+\delta} d\nu\, \mathrm{Im}(\mathcal{M}(\nu + i\varepsilon,x)), \tag{2.23}$$

$$\Upsilon(\lambda,x) = \lim_{\varepsilon\downarrow 0}\pi^{-1}\mathrm{Im}(\ln(\mathcal{M}(\lambda + i\varepsilon,x))) \text{ for a.e. } \lambda \in \mathbb{R}. \tag{2.24}$$

The Herglotz, and particularly exponential Herglotz property (cf. [2], [4], [27]) of the diagonal Green's function of H,

$$\mathfrak{g}(z,x) = \mathcal{G}(z,x,x), \quad z \in \mathbb{C}\backslash\mathrm{spec}(H),\ x \in \mathbb{R}, \tag{2.25}$$

will be of particular importance in Section 4 and hence we note for subsequent purpose,

$$\mathfrak{g}(z,x) = \exp\left(\mathfrak{C}(x) + \int_{\mathbb{R}} d\lambda\, \Xi(\lambda,x)\big((\lambda - z)^{-1} - \lambda(1+\lambda^2)^{-1}\big)\right), \tag{2.26}$$

where

$$\mathfrak{C}(x) = \mathfrak{C}(x)^*, \quad 0 \le \Xi(\cdot, x) \le \mathcal{I}_m \text{ a.e.}, \tag{2.27}$$

$$\Xi(\lambda, x) = \lim_{\varepsilon \downarrow 0} \pi^{-1} \mathrm{Im}(\ln(\mathfrak{g}(\lambda + i\varepsilon, x))) \text{ for a.e. } \lambda \in \mathbb{R}. \tag{2.28}$$

We also recall the following characterization of $\mathcal{M}(z, x_0)$ to be used later. In the scalar context $m = 1$, this characterization has been used by Rofe-Beketov [65], [66] (see also [50; Sect. 7.3]).

THEOREM 2.4. [65], [66] *Assume Hypothesis 2.1, suppose that* $z \in \mathbb{C}\backslash\mathbb{R}$, $x_0 \in \mathbb{R}$, *and let* $\ell, r \in \mathbb{N}_0$. *Then the following assertions are equivalent.*
(i) $\mathcal{M}(z, x_0)$ *is the* $2m \times 2m$ *Weyl–Titchmarsh matrix associated with a Schrödinger operator* H *in* $L^2(\mathbb{R})^m$ *of the type (2.14) with an* $m \times m$ *matrix-valued potential* $\mathcal{Q} \in L^1_{\mathrm{loc}}(\mathbb{R})$ *and* $\mathcal{Q} \in C^\ell((-\infty, x_0))$ *and* $\mathcal{Q} \in C^r((x_0, \infty))$.
(ii) $\mathcal{M}(z, x_0)$ *is of the type (2.18) with* $\mathcal{M}_\pm(z, x_0)$ *being half-line* $m \times m$ *Weyl–Titchmarsh matrices on* $[x_0, \pm\infty)$ *corresponding to a Dirichlet boundary condition at* x_0 *and a self-adjoint boundary condition at* $-\infty$ *and/or* ∞ *(if any) which are associated with an* $m \times m$ *matrix-valued potential* $\mathcal{Q}$ *satisfying* $\mathcal{Q} \in C^\ell((-\infty, x_0))$ *and* $\mathcal{Q} \in C^r((x_0, \infty))$, *respectively.*
If (i) or (ii) holds, then the $2m \times 2m$ *matrix-valued spectral measure* $\Omega(\cdot, x_0)$ *associated with* $\mathcal{M}(z, x_0)$ *is determined by (2.18) and (2.23).*

Next, we consider variations of the reference point $x \in \mathbb{R}$. Since $\Psi_\pm$ satisfies the second-order linear $m \times m$ matrix-valued differential equation (2.2), $\mathcal{M}_\pm$ in (2.5) satisfies the matrix-valued Riccati-type equation (independently of any limit point assumptions at $\pm\infty$)

$$\mathcal{M}_\pm'(z, x) + \mathcal{M}_\pm(z, x)^2 = \mathcal{Q}(x) - z\mathcal{I}_m, \quad x \in \mathbb{R}, \ z \in \mathbb{C}\backslash\mathbb{R}. \tag{2.29}$$

The asymptotic high-energy behavior of $\mathcal{M}_\pm(z, x)$ as $|z| \to \infty$ has recently been determined in [11] under minimal smoothness conditions on $\mathcal{Q}$ and without assuming that $\mathcal{L}$ is in the limit point case at $\pm\infty$. Here we recall just a special case of the asymptotic expansion proved in [11] which is most suited for our discussion at hand. We denote by $C_\varepsilon \subset \mathbb{C}_+$ the open sector with vertex at zero, symmetry axis along the positive imaginary axis, and opening angle ε, with $0 < \varepsilon < \pi/2$.

THEOREM 2.5. [11] *Fix* $x_0 \in \mathbb{R}$ *and let* $x \ge x_0$. *In addition to Hypothesis 2.1 suppose that* $\mathcal{Q} \in C^\infty([x_0, \pm\infty))^{m \times m}$ *and that* $\mathcal{L}$ *is in the limit point case at* $\pm\infty$. *Let* $\mathcal{M}_\pm(z, x)$, $x \ge x_0$, *be defined as in (2.5). Then, as* $|z| \to \infty$ *in* C_ε, $\mathcal{M}_\pm(z, x)$ *has an asymptotic expansion of the form* ($\mathrm{Im}(z^{1/2}) > 0$, $z \in \mathbb{C}_+$)

$$\mathcal{M}_\pm(z, x) \underset{\substack{|z| \to \infty \\ z \in C_\varepsilon}}{=} \pm i\mathcal{I}_m z^{1/2} + \sum_{k=1}^{N} \mathcal{M}_{\pm,k}(x) z^{-k/2} + o(|z|^{-N/2}), \quad N \in \mathbb{N}. \tag{2.30}$$

The expansion (2.30) is uniform with respect to $\arg(z)$ *for* $|z| \to \infty$ *in* C_ε *and uniform in* x *as long as* x *varies in compact subsets of* $[x_0, \infty)$. *The expansion coefficients* $\mathcal{M}_{\pm,k}(x)$ *can be recursively computed from*

$$\begin{aligned}
&\mathcal{M}_{\pm,1}(x) = \mp\frac{i}{2}\mathcal{Q}(x), \quad \mathcal{M}_{\pm,2}(x) = \frac{1}{4}\mathcal{Q}'(x), \\
&\mathcal{M}_{\pm,k+1}(x) = \pm\frac{i}{2}\Big(\mathcal{M}'_{\pm,k}(x) + \sum_{\ell=1}^{k-1}\mathcal{M}_{\pm,\ell}(x)\mathcal{M}_{\pm,k-\ell}(x)\Big), \quad k \geq 2.
\end{aligned} \tag{2.31}$$

The asymptotic expansion (2.30) can be differentiated to any order with respect to x.

If one only assumes Hypothesis 2.1 (i.e., $\mathcal{Q} \in L^1([x_0, R])^{m\times m}$ *for all* $R > x_0$*), then*

$$\mathcal{M}_\pm(z,x) \underset{\substack{|z|\to\infty \\ z\in C_\varepsilon}}{=} \pm i\mathcal{I}_m z^{1/2} + o(1). \tag{2.32}$$

REMARK 2.6. Due to the recursion relation (2.31), the coefficients $\mathcal{M}_{\pm,k}$ are universal polynomials in $\mathcal{Q}$ and its x-derivatives (i.e., differential polynomials in $\mathcal{Q}$).

Finally, in addition to (2.16) (still assuming Hypothesis 2.1), one infers for the $2m \times 2m$ Weyl–Titchmarsh function $\mathcal{M}(z,x)$ associated with H on $\mathbb{R}$ in connection with arbitrary half-lines $[x, \pm\infty)$, $x \in \mathbb{R}$,

$$\mathcal{M}(z,x) = \big(\mathcal{M}_{j,j'}(z,x)\big)_{j,j'=1,2}, \quad z \in \mathbb{C}\backslash\mathbb{R}, \tag{2.33}$$

$$\begin{aligned}
\mathcal{M}_{1,1}(z,x) &= \mathcal{M}_\pm(z,x)[\mathcal{M}_-(z,x) - \mathcal{M}_+(z,x)]^{-1}\mathcal{M}_\mp(z,x) \\
&= \psi'_+(z,x,x_0)[\mathcal{M}_-(z,x_0) - \mathcal{M}_+(z,x_0)]^{-1}\psi'_-(\overline{z},x,x_0)^*,
\end{aligned} \tag{2.34}$$

$$\begin{aligned}
\mathcal{M}_{1,2}(z,x) &= 2^{-1}[\mathcal{M}_-(z,x) - \mathcal{M}_+(z,x)]^{-1}[\mathcal{M}_-(z,x) + \mathcal{M}_+(z,x)] \\
&= \psi_+(z,x,x_0)[\mathcal{M}_-(z,x_0) - \mathcal{M}_+(z,x_0)]^{-1}\psi'_-(\overline{z},x,x_0)^*,
\end{aligned} \tag{2.35}$$

$$\begin{aligned}
\mathcal{M}_{2,1}(z,x) &= 2^{-1}[\mathcal{M}_-(z,x) + \mathcal{M}_+(z,x)][\mathcal{M}_-(z,x) - \mathcal{M}_+(z,x)]^{-1} \\
&= \psi'_+(z,x,x_0)[\mathcal{M}_-(z,x_0) - \mathcal{M}_+(z,x_0)]^{-1}\psi_-(\overline{z},x,x_0)^*,
\end{aligned} \tag{2.36}$$

$$\begin{aligned}
\mathcal{M}_{2,2}(z,x) &= [\mathcal{M}_-(z,x) - \mathcal{M}_+(z,x)]^{-1} \\
&= \psi_+(z,x,x_0)[\mathcal{M}_-(z,x_0) - \mathcal{M}_+(z,x_0)]^{-1}\psi_-(\overline{z},x,x_0)^*.
\end{aligned} \tag{2.37}$$

Introducing the convenient abbreviation,

$$\mathcal{M}(z,x) = \begin{pmatrix} \mathfrak{h}(z,x) & -\mathfrak{g}_2(z,x) \\ -\mathfrak{g}_1(z,x) & \mathfrak{g}(z,x) \end{pmatrix}, \quad z \in \mathbb{C}\backslash\mathbb{R},\ x \in \mathbb{R}, \tag{2.38}$$

one then verifies from (2.33)–(2.38) and from $\mathcal{M}(\overline{z},x)^* = \mathcal{M}(z,x)$, $\mathcal{M}_\pm(\overline{z},x)^* = \mathcal{M}_\pm(z,x)$ that

$$\mathfrak{g}(\overline{z},x)^* = \mathfrak{g}(z,x), \quad \mathfrak{g}_2(\overline{z},x)^* = \mathfrak{g}_1(z,x), \quad \mathfrak{h}(\overline{z},x)^* = \mathfrak{h}(z,x), \tag{2.39}$$

$$\mathfrak{g}(z,x)\mathfrak{g}_1(z,x) = \mathfrak{g}_2(z,x)\mathfrak{g}(z,x), \tag{2.40}$$

$$\mathfrak{h}(z,x)\mathfrak{g}_2(z,x) = \mathfrak{g}_1(z,x)\mathfrak{h}(z,x), \tag{2.41}$$

$$\mathfrak{g}(z,x) = [\mathcal{M}_-(z,x) - \mathcal{M}_+(z,x)]^{-1}, \tag{2.42}$$

$$\mathfrak{g}(z,x)\mathfrak{h}(z,x) - \mathfrak{g}_2(z,x)^2 = -(1/4)\mathcal{I}_m, \tag{2.43}$$

$$\mathfrak{h}(z,x)\mathfrak{g}(z,x) - \mathfrak{g}_1(z,x)^2 = -(1/4)\mathcal{I}_m, \tag{2.44}$$

$$\mathcal{M}_\pm(z,x) = \mp(1/2)\mathfrak{g}(z,x)^{-1} - \mathfrak{g}(z,x)^{-1}\mathfrak{g}_2(z,x) \tag{2.45}$$

$$\qquad\qquad \mp (1/2)\mathfrak{g}(z,x)^{-1} - \mathfrak{g}_1(z,x)\mathfrak{g}(z,x)^{-1}, \tag{2.46}$$

assuming Hypothesis 2.1. Moreover, (2.39)–(2.46) and the Riccati-type equations (2.29) imply the following results for $z \in \mathbb{C}\backslash\mathbb{R}$ and a.e. $x \in \mathbb{R}$,

$$\mathfrak{g}' = -(\mathfrak{g}_1 + \mathfrak{g}_2), \tag{2.47}$$

$$\mathfrak{g}_1' = -(\mathcal{Q} - z\mathcal{I}_m)\mathfrak{g} - \mathfrak{h} \tag{2.48}$$

$$= (-\mathfrak{g}'' + \mathfrak{g}\mathcal{Q} - \mathcal{Q}\mathfrak{g})/2, \tag{2.49}$$

$$\mathfrak{g}_2' = -\mathfrak{g}(\mathcal{Q} - z\mathcal{I}_m) - \mathfrak{h} \tag{2.50}$$

$$= (-\mathfrak{g}'' + \mathcal{Q}\mathfrak{g} - \mathfrak{g}\mathcal{Q})/2, \tag{2.51}$$

$$\mathfrak{h}' = -\mathfrak{g}_1(\mathcal{Q} - z\mathcal{I}_m) - (\mathcal{Q} - z\mathcal{I}_m)\mathfrak{g}_2, \tag{2.52}$$

$$\mathfrak{h} = [\mathfrak{g}'' - \mathfrak{g}(\mathcal{Q} - z\mathcal{I}_m) - (\mathcal{Q} - z\mathcal{I}_m)\mathfrak{g}]/2 \tag{2.53}$$

if $\mathcal{Q} \in L^1_{\text{loc}}(\mathbb{R})^{m\times m}$, and

$$\mathfrak{g}_1'' = -2(\mathcal{Q} - z\mathcal{I}_m)\mathfrak{g}' - \mathcal{Q}'\mathfrak{g} + \mathfrak{g}_1\mathcal{Q} - \mathcal{Q}\mathfrak{g}_1, \tag{2.54}$$

$$\mathfrak{g}_2'' = -2\mathfrak{g}'(\mathcal{Q} - z\mathcal{I}_m) - \mathfrak{g}\mathcal{Q}' + \mathcal{Q}\mathfrak{g}_2 - \mathfrak{g}_2\mathcal{Q} \tag{2.55}$$

if in addition $\mathcal{Q}' \in L^1_{\text{loc}}(\mathbb{R})^{m\times m}$.

We conclude this section by recalling the definition of reflectionless matrix-valued potentials as discussed in [14], [24], [27], and [45]. We follow the corresponding notion introduced in connection with scalar Schrödinger operators and refer to [15], [16], [44], [74] for further details in this context.

DEFINITION 2.7. Assume Hypothesis 2.1 and define the self-adjoint Schrödinger operator H in $L^2(\mathbb{R})^{m\times m}$ as in (2.14). Suppose that $\text{spec}_{\text{ess}}(H) \neq \emptyset$ and let Ξ be defined by (2.28). Then $\mathcal{Q}$ is called *reflectionless* if for all $x \in \mathbb{R}$,

$$\Xi(\lambda, x) = (1/2)\mathcal{I}_m \text{ for a.e. } \lambda \in \text{spec}_{\text{ess}}(H). \tag{2.56}$$

Explicit examples of reflectionless potentials will be discussed in Section 3. If $\mathcal{Q}$ is reflectionless we will sometimes slightly abuse notation and also call the corresponding Schrödinger operator H in $L^2(\mathbb{R})^{m\times m}$ reflectionless.

3 A CLASS OF MATRIX–VALUED SCHRÖDINGER OPERATORS WITH PRESCRIBED FINITE–BAND SPECTRA

Given the preliminaries of Section 2, we now recall the construction of a class of matrix-valued Schrödinger operators with a prescribed finite-band spectrum of uniform maximum multiplicity, the principal result of [25]. Let

$$\{E_\ell\}_{0\leq\ell\leq 2n} \subseteq \mathbb{R}, \ n \in \mathbb{N}, \text{ with } E_\ell < E_{\ell+1}, \, 0 \leq \ell \leq 2n-1, \tag{3.1}$$

and introduce the polynomial

$$R_{2n+1}(z) = \prod_{\ell=0}^{2n}(z - E_\ell), \quad z \in \mathbb{C}. \tag{3.2}$$

Moreover, we define the square root of R_{2n+1} by

$$R_{2n+1}(\lambda)^{1/2} = \lim_{\varepsilon\downarrow 0} R_{2n+1}(\lambda + i\varepsilon)^{1/2}, \quad \lambda \in \mathbb{R}, \tag{3.3}$$

and

$$R_{2n+1}(\lambda)^{1/2} = |R_{2n+1}(\lambda)^{1/2}| \begin{cases} (-1)^n i & \text{for } \lambda \in (-\infty, E_0), \\ (-1)^{n+j} i & \text{for } \lambda \in (E_{2j-1}, E_{2j}),\ j = 1, \dots, n, \\ (-1)^{n+j} & \text{for } \lambda \in (E_{2j}, E_{2j+1}),\ j = 0, \dots, n-1, \\ 1 & \text{for } \lambda \in (E_{2n}, \infty), \end{cases}$$

$$\lambda \in \mathbb{R} \qquad (3.4)$$

and analytically continue $R_{2n+1}^{1/2}$ from $\mathbb{R}$ to all of $\mathbb{C}\backslash\Sigma_n$, where Σ_n is defined by

$$\Sigma_n = \left\{ \bigcup_{j=0}^{n-1} [E_{2j}, E_{2j+1}] \right\} \cup [E_{2n}, \infty). \qquad (3.5)$$

In this context we also mention the useful formula

$$\overline{R_{2n+1}(\overline{z})^{1/2}} = -R_{2n+1}(z)^{1/2}, \quad z \in \mathbb{C}_+. \qquad (3.6)$$

THEOREM 3.1. [25] *Let $z \in \mathbb{C}\backslash\Sigma_n$ and $n \in \mathbb{N}$. Define $R_{2n+1}^{1/2}$ as in (3.1)–(3.4) followed by an analytic continuation to $\mathbb{C}\backslash\Sigma_n$. Moreover, let F_n and H_{n+1} be two monic polynomials of degree n and $n+1$, respectively. Then $iR_{2n+1}(z)^{-1/2}F_n(z)$ is a Herglotz function if and only if all zeros of F_n are real and there is precisely one zero in each of the intervals $[E_{2j-1}, E_{2j}]$, $1 \leq j \leq n$. Moreover, if $iR_{2n+1}^{-1/2}F_n$ is a Herglotz function, then it can be represented in the form*

$$\frac{iF_n(z)}{R_{2n+1}(z)^{1/2}} = \frac{1}{\pi} \int_{\Sigma_n} \frac{F_n(\lambda)d\lambda}{R_{2n+1}(\lambda)^{1/2}} \frac{1}{\lambda - z}, \quad z \in \mathbb{C}\backslash\Sigma_n. \qquad (3.7)$$

Similarly, $iR_{2n+1}(z)^{-1/2}H_{n+1}(z)$ is a Herglotz function if and only if all zeros of H_{n+1} are real and there is precisely one zero in each of the intervals $(-\infty, E_0]$ and $[E_{2j-1}, E_{2j}]$, $1 \leq j \leq n$. Moreover, if $iR_{2n+1}^{-1/2}H_{n+1}$ is a Herglotz function, then it can be represented in the form

$$\begin{aligned} \frac{iH_{n+1}(z)}{R_{2n+1}(z)^{1/2}} &= \mathrm{Re}\left(\frac{iH_{n+1}(i)}{R_{2n+1}(i)^{1/2}}\right) \\ &\quad + \frac{1}{\pi} \int_{\Sigma_n} \frac{H_{n+1}(\lambda)d\lambda}{R_{2n+1}(\lambda)^{1/2}} \left(\frac{1}{\lambda - z} - \frac{\lambda}{1+\lambda^2}\right), \quad z \in \mathbb{C}\backslash\Sigma_n. \end{aligned} \qquad (3.8)$$

Actually, Theorem 3.1 can be improved by invoking ideas developed in the Appendix of [46] (cf. also [74]). Since this appears to be of independent interest we provide a brief discussion.

We start with the elementary observation that the Herglotz function

$$m(z) = \begin{cases} \frac{z-\beta}{z-\alpha}, & -\infty < \alpha < \beta < \infty, \\ z - \beta, & \alpha = -\infty,\ \beta \in \mathbb{R}, \\ \frac{-1}{z-\alpha}, & \alpha \in \mathbb{R},\ \beta = +\infty, \end{cases} \quad z \in \mathbb{C}_+, \qquad (3.9)$$

admits the (exponential) representation

$$m(z) = C(\alpha,\beta) \exp \int_{\alpha}^{\beta} d\lambda \left(\frac{1}{\lambda - z} - \frac{\lambda}{1+\lambda^2} \right), \quad z \in \mathbb{C}_+, \tag{3.10}$$

where

$$C(\alpha,\beta) = \begin{cases} \left(\frac{1+\beta^2}{1+\alpha^2}\right)^{1/2}, & -\infty < \alpha < \beta < \infty, \\ 1, & \alpha = -\infty,\ \beta \in \mathbb{R} \text{ or } \alpha \in \mathbb{R},\ \beta = +\infty. \end{cases} \tag{3.11}$$

THEOREM 3.2. *Let $z \in \mathbb{C}\backslash\Sigma_n$, $n \in \mathbb{N}$, and define $R_{2n+1}^{1/2}$ as in (3.1)–(3.4) followed by an analytic continuation to $\mathbb{C}\backslash\Sigma_n$. Suppose M is a Herglotz function such that*

$$\lim_{\varepsilon \downarrow 0} M(\lambda + i\varepsilon) \in i\mathbb{R} \quad \textit{for a.e. } \lambda \in \Sigma_n \tag{3.12}$$

and assume in addition that M is real-valued on $\mathbb{C}\backslash\Sigma$. Then M is either of the form

$$M(z) = \frac{i\widehat{F}_n(z)}{R_{2n+1}(z)^{1/2}}, \tag{3.13}$$

where $\widehat{F}_n$ is a polynomial of degree n (not necessarily monic), positive on the semi-axis (E_{2n},∞), with precisely one zero in each of the intervals $[E_{2j-1},E_{2j}]$, $1 \leq j \leq n$, or else, M is of the form

$$M(z) = \frac{i\widehat{H}_{n+1}(z)}{R_{2n+1}(z)^{1/2}}, \tag{3.14}$$

where $\widehat{H}_{n+1}$ is a polynomial of degree $n+1$ (not necessarily monic), positive on the semi-axis (E_{2n},∞), with precisely one zero in each of the intervals $(-\infty,E_0]$ and $[E_{2j-1},E_{2j}]$, $1 \leq j \leq n$.

Moreover, if $i\widehat{F}_n/R_{2n+1}^{1/2}$ is a Herglotz function, it can be represented in the form (3.7). Similarly, if $i\widehat{H}_{n+1}/R_{2n+1}^{1/2}$ is a Herglotz function, it can be represented in the form (3.8).

Proof. The Herglotz function M admits the exponential representation (cf. [2])

$$M(z) = K \exp\left(\frac{1}{2} \int_{\Sigma_n} d\lambda \left(\frac{1}{\lambda - z} - \frac{\lambda}{1+\lambda^2} \right) + \int_{\Sigma_{n,-}} d\lambda \left(\frac{1}{\lambda - z} - \frac{\lambda}{1+\lambda^2} \right) \right), \tag{3.15}$$

where $K > 0$ and

$$\Sigma_{n,-} = \{\lambda \in \mathbb{R}\backslash\Sigma_n \mid M(\lambda) < 0\}. \tag{3.16}$$

Since the Herglotz function M is strictly monotonically increasing on $\mathbb{R}\backslash\Sigma_n$, M can have at most one zero in each interval $(-\infty,E_0)$, (E_{2j-1},E_{2j}), $j = 1,\ldots,n$. Moreover, $M(E_0-0)$, $M(E_{2j}-0) \in \{+\infty,0\}$, $M(E_{2j-1}+0) \in \{-\infty,0\}$, $j =$

$1, \ldots, n$. Thus, depending on whether or not $M\big|_{(-\infty,E_0)} \geq 0$, the set $\Sigma_{n,-}$ admits one of the following two representations

$$\Sigma_{n,-} = \bigcup_{j=1}^{n} (E_{2j-1}, \mu_j) \tag{3.17}$$

or

$$\Sigma_{n,-} = (-\infty, \nu_0) \cup \bigcup_{j=1}^{n} (E_{2j-1}, \nu_j) \tag{3.18}$$

for some $\nu_0 \in (-\infty, E_0]$ and some $\mu_j, \nu_j \in [E_{2j-1}, E_{2j}]$, $1 \leq j \leq n$. Repeated use of (3.10) then proves the representation

$$\begin{aligned} M(z) &= C_1 \bigg(\prod_{j=0}^{n-1} \frac{(E_{2j+1} - z)}{(E_{2j} - z)} \frac{1}{(E_{2n} - z)} \bigg)^{1/2} \prod_{j=1}^{n} \frac{(z - \mu_j)}{(z - E_{2j-1})} \\ &= C_1 \frac{\prod_{j=1}^{n} (z - \mu_j)}{\Big(\prod_{\ell=0}^{2n} (z - E_\ell) \Big)^{1/2}} = \frac{i\widehat{F}_n(z)}{R_{2n+1}(z)^{1/2}}, \end{aligned} \tag{3.19}$$

where

$$\widehat{F}_n(z) = C_1 \prod_{j=1}^{n} (z - \mu_j) \tag{3.20}$$

for some $C_1 > 0$, whenever (3.17) holds, and

$$M(z) = C_2 (z - \nu_0) \bigg(\prod_{j=0}^{n-1} \frac{E_{2j+1} - z}{E_{2j} - z} \frac{1}{E_{2n} - z} \bigg)^{1/2} \prod_{j=1}^{n} \frac{z - \nu_j}{z - E_{2j-1}} = \frac{i\widehat{H}_{n+1}(z)}{R_{2n+1}(z)^{1/2}}, \tag{3.21}$$

where

$$\widehat{H}_{n+1}(z) = C_2 \prod_{k=0}^{n} (z - \nu_k) \tag{3.22}$$

for some $C_2 > 0$, whenever (3.18) holds. ■

Given $m \in \mathbb{N}$, we denote by

$$\mathcal{A}(z) = \sum_{k=0}^{n} \mathcal{A}_k z^k, \quad \mathcal{A}_k \in \mathbb{C}^{m \times m}, \; 1 \leq k \leq n, \; z \in \mathbb{C}, \tag{3.23}$$

a polynomial pencil of $m \times m$ matrices (in short, a pencil) in the following. $\mathcal{A}$ is called of degree $n \in \mathbb{N}_0$ if $\mathcal{A}_n \neq 0$ and monic if $\mathcal{A}_n = \mathcal{I}_m$.

DEFINITION 3.3. Let A be a pencil of the type (3.23).
(i) The pencil $\mathcal{A}$ is called *self-adjoint* if $\mathcal{A}_k = \mathcal{A}_k^*$ for all $1 \leq k \leq n$ (i.e., $\mathcal{A}(\overline{z})^* = \mathcal{A}(z)$ for all $z \in \mathbb{C}$).
(ii) A self-adjoint pencil $\mathcal{A}$ is called *weakly hyperbolic* if $\mathcal{A}_n > 0$ and for all $f \in \mathbb{C}^m \backslash \{0\}$, the roots of the polynomial $(f, \mathcal{A}(\cdot) f)_{\mathbb{C}^m}$ are real. If in addition all these

zeros are distinct, the pencil $\mathcal{A}$ is called *hyperbolic.*
(*iii*) Let $\mathcal{A}$ be a weakly hyperbolic pencil and denote by $\{p_j(\mathcal{A},f)\}_{1\leq j\leq n}$,

$$p_j(\mathcal{A},f)\leq p_{j+1}(\mathcal{A},f),\ 1\leq j\leq n-1,\ f\in\mathbb{C}^m\backslash\{0\}, \tag{3.24}$$

the roots of the polynomial $(f,\mathcal{A}(\cdot)f)_{\mathbb{C}^m}$ ordered in magnitude. The range of the roots $p_j(\mathcal{A},f)$, $f\in\mathbb{C}^m\backslash\{0\}$ is denoted by $\Delta_j(\mathcal{A})$ and called the *jth root zone* of $\mathcal{A}$.
(*iv*) A hyperbolic pencil $\mathcal{A}$ is called *strongly hyperbolic* if $\overline{\Delta_j(\mathcal{A})}$ and $\overline{\Delta_k(\mathcal{A})}$ are mutually disjoint for $j\neq k$, $1\leq j,k\leq n$.

For details on spectral theory of polynomial matrix (in fact, operator) pencils we refer, for instance, to [56], [57], [58].

COROLLARY 3.4. [25] *Let $z\in\mathbb{C}\backslash\Sigma_n$ and $m,n\in\mathbb{N}$. Define $R_{2n+1}^{1/2}$ as in (3.1)–(3.4) followed by an analytic continuation to $\mathbb{C}\backslash\Sigma_n$. Moreover let $\mathcal{F}_n$ and $\mathcal{H}_{n+1}$ be two monic $m\times m$ matrix pencils of degree n and $n+1$, respectively. Then $(i/2)R_{2n+1}^{-1/2}\mathcal{F}_n$ is a Herglotz matrix if and only if the root zones $\Delta_j(\mathcal{F}_n)$ of $\mathcal{F}_n$ satisfy*

$$\Delta_j(\mathcal{F}_n)\subseteq[E_{2j-1},E_{2j}],\quad 1\leq j\leq n. \tag{3.25}$$

Analogously, $(i/2)R_{2n+1}^{-1/2}\mathcal{H}_{n+1}$ is a Herglotz matrix if and only if the root zones $\Delta_j(\mathcal{H}_{n+1})$ of $\mathcal{H}_{n+1}$ satisfy

$$\Delta_0(\mathcal{H}_{n+1})\subset(-\infty,E_0],\quad \Delta_j(\mathcal{H}_{n+1})\subseteq[E_{2j-1},E_{2j}],\quad 1\leq j\leq n. \tag{3.26}$$

If (3.25) (resp., (3.26)) holds, then $\mathcal{F}_n$ (resp., $\mathcal{H}_{n+1}$) is a strongly hyperbolic pencil.

Next, we define the following $2m\times 2m$ matrix $\mathcal{M}_{\Sigma_n}(z,x_0)$ which will turn out to be the underlying Weyl–Titchmarsh matrix associated with the class of $m\times m$ matrix-valued Schrödinger operators with prescribed finite-band spectrum Σ_n of maximal multiplicity. We introduce, for fixed $x_0\in\mathbb{R}$,

$$\begin{aligned}\mathcal{M}_{\Sigma_n}(z,x_0)&=\big(\mathcal{M}_{\Sigma_n,p,q}(z,x_0)\big)_{1\leq p,q\leq 2} \\ &=\frac{i}{2R_{2n+1}(z)^{1/2}}\begin{pmatrix}\mathcal{H}_{n+1,\Sigma_n}(z,x_0) & -\mathcal{G}_{2,n-1,\Sigma_n}(z,x_0)\\ -\mathcal{G}_{1,n-1,\Sigma_n}(z,x_0) & \mathcal{F}_{n,\Sigma_n}(z,x_0)\end{pmatrix},\ z\in\mathbb{C}\backslash\Sigma_n.\end{aligned} \tag{3.27}$$

Here $R_{2n+1}(z)^{1/2}$ is defined as in (3.1)–(3.4) followed by analytic continuation into $\mathbb{C}\backslash\Sigma$ and the polynomial matrix pencils $\mathcal{F}_{n,\Sigma_n}$, $\mathcal{G}_{1,n-1,\Sigma_n}$, $\mathcal{G}_{2,n-1,\Sigma_n}$, and $\mathcal{H}_{n+1,\Sigma_n}$ are introduced as follows:

(*i*) $\mathcal{F}_{n,\Sigma_n}(\cdot,x_0)$ is an $m\times m$ monic matrix pencil of degree n, that is, $\mathcal{F}_{n,\Sigma}(\cdot,x_0)$ is of the type

$$\mathcal{F}_{n,\Sigma_n}(z,x_0)=\sum_{\ell=0}^{n}\mathcal{F}_{n-\ell,\Sigma_n}(x_0)z^\ell,\quad \mathcal{F}_{0,\Sigma_n}(x_0)=\mathcal{I}_m,\ z\in\mathbb{C} \tag{3.28}$$

and

$$\frac{i}{2R_{2n+1}^{1/2}}\mathcal{F}_{n,\Sigma_n}(\cdot,x_0)\text{ is assumed to be an }m\times m\text{ Herglotz matrix.} \tag{3.29}$$

Hence $\mathcal{F}_{n,\Sigma_n}(\cdot, x_0)$ is a self-adjoint (in fact, strongly hyperbolic) pencil,

$$\mathcal{F}_{n,\Sigma_n}(\overline{z}, x_0)^* = \mathcal{F}_{n,\Sigma_n}(z, x_0), \quad z \in \mathbb{C}. \tag{3.30}$$

In addition, $(i/2)R_{2n+1}^{-1/2}\mathcal{F}_{n,\Sigma_n}$ and $2iR_{2n+1}^{1/2}\mathcal{F}_{n,\Sigma_n}^{-1}$ admit the Herglotz representations

$$\frac{i}{2R_{2n+1}(z)^{1/2}}\mathcal{F}_{n,\Sigma_n}(z, x_0) = \frac{1}{2\pi}\int\limits_{\Sigma_n} \frac{d\lambda}{R_{2n+1}(\lambda)^{1/2}}\mathcal{F}_{n,\Sigma_n}(\lambda, x_0)\frac{1}{\lambda - z}, \quad z \in \mathbb{C}\backslash\Sigma_n, \tag{3.31}$$

$$\begin{aligned}
&iR_{2n+1}(z)^{1/2}\mathcal{F}_{n,\Sigma_n}(z, x_0)^{-1} \\
&= \frac{1}{\pi}\int\limits_{\Sigma_n} d\lambda\, R_{2n+1}(\lambda)^{1/2}\mathcal{F}_{n,\Sigma_n}(\lambda, x_0)^{-1}\left(\frac{1}{\lambda - z} - \frac{\lambda}{1+\lambda^2}\right) \\
&\quad + \Gamma_{\Sigma_n,0}(x_0) - \sum_{k=1}^{N}(z - \mu_k(x_0))^{-1}\Gamma_{\Sigma_n,k}(x_0), \\
&\qquad z \in \mathbb{C}\backslash\{\Sigma_n \cup \{\mu_k(x_0)\}_{1\le k\le N}\},
\end{aligned} \tag{3.32}$$

where

$$\begin{aligned}
&\Gamma_{\Sigma_n,0}(x_0) = \Gamma_{\Sigma_n,0}(x_0)^* \in \mathbb{C}^{m\times m}, \quad 0 \le \Gamma_{\Sigma_n,k}(x_0) \in \mathbb{C}^{m\times m}, \ 1 \le k \le N, \\
&\sum_{k=1}^{N} \operatorname{rank}(\Gamma_{\Sigma_n,k}(x_0)) \le mn, \quad \mu_k(x_0) \in \bigcup_{j=1}^{n}[E_{2j-1}, E_{2j}], \quad 1 \le k \le N.
\end{aligned} \tag{3.33}$$

In fact, there are precisely m numbers $\mu_k(x_0)$ in $[E_{2j-1}, E_{2j}]$ for each $1 \le j \le n$, counting multiplicity (they are the points z where $\mathcal{F}_{n,\Sigma_n}(z, x_0)$ is not invertible).

(ii) Given these facts we now define

$$\mathcal{G}_{1,n-1,\Sigma_n}(z, x_0) = \left(\sum_{k=1}^{N}\frac{\varepsilon_k(x_0)}{z - \mu_k(x_0)}\Gamma_{\Sigma_n,k}(x_0)\right)\mathcal{F}_{n,\Sigma_n}(z, x_0), \tag{3.34}$$

$$\mathcal{G}_{2,n-1,\Sigma_n}(z, x_0) = \mathcal{F}_{n,\Sigma_n}(z, x_0)\left(\sum_{k=1}^{N}\frac{\varepsilon_k(x_0)}{z - \mu_k(x_0)}\Gamma_{\Sigma_n,k}(x_0)\right), \tag{3.35}$$

$$\varepsilon_k(x_0) \in \{1, -1\}, \ 1 \le k \le N, \quad z \in \mathbb{C}\backslash\{\mu_k(x_0)\}_{1\le k\le N}, \tag{3.36}$$

and

$$\begin{aligned}
&\mathcal{H}_{n+1,\Sigma_n}(z, x_0) = R_{2n+1}(z)\mathcal{F}_{n,\Sigma_n}(z, x_0)^{-1} \\
&+ \left(\sum_{k=1}^{N}\frac{\varepsilon_k(x_0)}{z - \mu_k(x_0)}\Gamma_{\Sigma_n,k}(x_0)\right)\mathcal{F}_{n,\Sigma_n}(z, x_0)\left(\sum_{\ell=1}^{N}\frac{\varepsilon_\ell(x_0)}{z - \mu_\ell(x_0)}\Gamma_{\Sigma_n,\ell}(x_0)\right), \\
&\qquad z \in \mathbb{C}\backslash\{\mu_k(x_0)\}_{1\le k\le N}.
\end{aligned} \tag{3.37}$$

LEMMA 3.5. [25] *Let* $z \in \mathbb{C}\backslash\{\mu_k(x_0)\}_{1\le k\le N}$. $\mathcal{G}_{p,n-1,\Sigma_n}(\cdot, x_0)$, $p = 1, 2$, *are* $m\times m$ *polynomial matrix pencils of equal degree at most* $n-1$ *and* $\mathcal{H}_{n+1,\Sigma_n}(\cdot, x_0)$ *is a* *strongly hyperbolic (and hence self-adjoint) $m \times m$ monic matrix pencil of degree $n+1$. Moreover, the following identities hold.*

$$\mathcal{G}_{2,n-1,\Sigma_n}(\overline{z}, x_0)^* = \mathcal{G}_{1,n-1,\Sigma_n}(z, x_0), \tag{3.38}$$

$$\mathcal{F}_{n,\Sigma_n}(z, x_0)\mathcal{G}_{1,n-1,\Sigma_n}(z, x_0) = \mathcal{G}_{2,n-1,\Sigma_n}(z, x_0)\mathcal{F}_{n,\Sigma_n}(z, x_0), \tag{3.39}$$

$$\mathcal{H}_{n+1,\Sigma_n}(z, x_0)\mathcal{G}_{2,n-1,\Sigma_n}(z, x_0) = \mathcal{G}_{1,n-1,\Sigma_n}(z, x_0)\mathcal{H}_{n+1,\Sigma_n}(z, x_0), \tag{3.40}$$

$$\mathcal{F}_{n,\Sigma_n}(z, x_0)\mathcal{H}_{n+1,\Sigma_n}(z, x_0) - \mathcal{G}_{2,n-1,\Sigma_n}(z, x_0)^2 = R_{2n+1}(z)\mathcal{I}_m, \tag{3.41}$$

$$\mathcal{H}_{n+1,\Sigma_n}(z, x_0)\mathcal{F}_{n,\Sigma_n}(z, x_0) - \mathcal{G}_{1,n-1,\Sigma_n}(z, x_0)^2 = R_{2n+1}(z)\mathcal{I}_m. \tag{3.42}$$

Next, introducing

$$\begin{aligned} &\mathcal{M}_{\pm,\Sigma_n}(z, x_0) \\ &= \pm i R_{2n+1}(z)^{1/2}\mathcal{F}_{n,\Sigma_n}(z, x_0)^{-1} - \mathcal{G}_{1,n-1,\Sigma_n}(z, x_0)\mathcal{F}_{n,\Sigma_n}(z, x_0)^{-1} && (3.43\text{a}) \\ &= \pm i R_{2n+1}(z)^{1/2}\mathcal{F}_{n,\Sigma_n}(z, x_0)^{-1} - \mathcal{F}_{n,\Sigma_n}(z, x_0)^{-1}\mathcal{G}_{2,n-1,\Sigma_n}(z, x_0), && (3.43\text{b}) \\ &\qquad z \in \mathbb{C}\backslash\{\Sigma_n \cup \{\mu_k(x_0)\}_{1\le k\le N}\}, \end{aligned}$$

$\pm\mathcal{M}_{\pm,\Sigma_n}(\cdot, x_0)$ are $m \times m$ Herglotz matrices with representations

$$\begin{aligned} \pm\mathcal{M}_{\pm,\Sigma_n}(z, x_0) &= \frac{1}{\pi}\int_{\Sigma_n} d\lambda\, R_{2n+1}(\lambda)^{1/2}\mathcal{F}_{n,\Sigma_n}(\lambda, x_0)^{-1}\left(\frac{1}{\lambda - z} - \frac{\lambda}{1+\lambda^2}\right) \\ &\quad + \Gamma_{\Sigma_n,0}(x_0) - \sum_{k=1}^{N}\frac{1 \pm \varepsilon_k(x_0)}{z - \mu_k(x_0)}\Gamma_{\Sigma_n,k}(x_0), \\ &\qquad z \in \mathbb{C}\backslash\{\Sigma_n \cup \{\mu_k(x_0)\}_{1\le k\le N}\}. \end{aligned} \tag{3.44}$$

Applying Theorem 2.3, $\mathcal{M}_{\pm,\Sigma_n}(z, x_0)$ are seen to be the half-line Weyl–Titchmarsh matrices uniquely associated with a potential $\mathcal{Q}_{\Sigma_n} \in L^1_{\text{loc}}(\mathbb{R})^{m\times m}$. In addition, we denote by $\psi_{\pm,\Sigma_n}(z, x, x_0)$ the Weyl solutions (2.5), (2.6) associated with $\mathcal{Q}_{\Sigma_n}$,

$$\psi_{\pm,\Sigma_n}(z, x, x_0) = \theta_{\Sigma_n}(z, x, x_0) + \phi_{\Sigma_n}(z, x, x_0)\mathcal{M}_{\pm,\Sigma_n}(z, x_0), \quad z \in \mathbb{C}\backslash\Sigma_n, \tag{3.45}$$

where, in obvious notation, $\theta_{\Sigma_n}(z, x, x_0)$, $\phi_{\Sigma_n}(z, x, x_0)$ denote the fundamental system (2.4) corresponding to $\mathcal{Q}_{\Sigma_n}$. The $2m \times 2m$ Weyl–Titchmarsh matrix associated with $\mathcal{Q}_{\Sigma_n}$ on $\mathbb{R}$ is then given by

$$\begin{aligned} M_{\Sigma_n}(z, x) &= \left(M_{\Sigma,p,q}(z, x)\right)_{1\le p,q\le 2} && (3.46) \\ &= \frac{i}{2R_{2n+1}(z)^{1/2}}\begin{pmatrix} \mathcal{H}_{n+1,\Sigma_n}(z, x) & -\mathcal{G}_{2,n-1,\Sigma_n}(z, x) \\ -\mathcal{G}_{1,n-1,\Sigma_n}(z, x) & \mathcal{F}_{n,\Sigma_n}(z, x) \end{pmatrix}, \quad z \in \mathbb{C}\backslash\Sigma, \end{aligned}$$

where we abbreviated

$$\begin{aligned}
\mathcal{F}_{n,\Sigma_n}(z,x) &= \theta_{\Sigma_n}(z,x,x_0)\mathcal{F}_{n,\Sigma_n}(z,x_0)\theta_{\Sigma_n}(\overline{z},x,x_0)^* \\
&\quad + \phi_{\Sigma_n}(z,x,x_0)\mathcal{H}_{n+1,\Sigma_n}(z,x_0)\phi_{\Sigma_n}(\overline{z},x,x_0)^* \\
&\quad - \phi_{\Sigma_n}(z,x,x_0)\mathcal{G}_{1,n-1,\Sigma_n}(z,x_0)\theta_{\Sigma_n}(\overline{z},x,x_0)^* \\
&\quad - \theta_{\Sigma_n}(z,x,x_0)\mathcal{G}_{2,n-1,\Sigma_n}(z,x_0)\phi_{\Sigma_n}(\overline{z},x,x_0)^*, \qquad (3.47)\\
\mathcal{G}_{1,n-1,\Sigma_n}(z,x) &= -\theta'_{\Sigma_n}(z,x,x_0)\mathcal{F}_{n,\Sigma_n}(z,x_0)\theta_{\Sigma_n}(\overline{z},x,x_0)^* \\
&\quad - \phi'_{\Sigma_n}(z,x,x_0)\mathcal{H}_{n+1,\Sigma_n}(z,x_0)\phi_{\Sigma_n}(\overline{z},x,x_0)^* \\
&\quad + \phi'_{\Sigma_n}(z,x,x_0)\mathcal{G}_{1,n-1,\Sigma_n}(z,x_0)\theta_{\Sigma_n}(\overline{z},x,x_0)^* \\
&\quad + \theta'_{\Sigma_n}(z,x,x_0)\mathcal{G}_{2,n-1,\Sigma_n}(z,x_0)\phi_{\Sigma_n}(\overline{z},x,x_0)^*, \qquad (3.48)\\
\mathcal{G}_{2,n-1,\Sigma_n}(z,x) &= -\theta_{\Sigma_n}(z,x,x_0)\mathcal{F}_{n,\Sigma_n}(z,x_0)\theta'_{\Sigma_n}(\overline{z},x,x_0)^* \\
&\quad - \phi_{\Sigma_n}(z,x,x_0)\mathcal{H}_{n+1,\Sigma_n}(z,x_0)\phi'_{\Sigma_n}(\overline{z},x,x_0)^* \\
&\quad + \phi_{\Sigma_n}(z,x,x_0)\mathcal{G}_{1,n-1,\Sigma_n}(z,x_0)\theta'_{\Sigma_n}(\overline{z},x,x_0)^* \\
&\quad + \theta_{\Sigma_n}(z,x,x_0)\mathcal{G}_{2,n-1,\Sigma_n}(z,x_0)\phi'_{\Sigma_n}(\overline{z},x,x_0)^*, \qquad (3.49)\\
\mathcal{H}_{n+1,\Sigma_n}(z,x) &= \theta'_{\Sigma_n}(z,x,x_0)\mathcal{F}_{n,\Sigma_n}(z,x_0)\theta'_{\Sigma_n}(\overline{z},x,x_0)^* \\
&\quad + \phi'_{\Sigma_n}(z,x,x_0)\mathcal{H}_{n+1,\Sigma_n}(z,x_0)\phi'_{\Sigma_n}(\overline{z},x,x_0)^* \\
&\quad - \phi'_{\Sigma_n}(z,x,x_0)\mathcal{G}_{1,n-1,\Sigma_n}(z,x_0)\theta'_{\Sigma_n}(\overline{z},x,x_0)^* \\
&\quad - \theta'_{\Sigma_n}(z,x,x_0)\mathcal{G}_{2,n-1,\Sigma_n}(z,x_0)\phi'_{\Sigma_n}(\overline{z},x,x_0)^*. \qquad (3.50)
\end{aligned}$$

Considerations of this type can be found in [50; Sect. 8.2] in the special scalar case $m = 1$ and in the matrix context $m \in \mathbb{N}$ in [70; Sect. 9.4].

One then infers

$$\begin{aligned}
\mathcal{F}'_{n,\Sigma_n} &= -(\mathcal{G}_{1,n-1,\Sigma_n} + \mathcal{G}_{2,n-1,\Sigma_n}), \qquad (3.51)\\
\mathcal{G}'_{1,n-1,\Sigma_n} &= -(Q_{\Sigma_n} - z\mathcal{I}_m)\mathcal{F}_{n,\Sigma_n} - \mathcal{H}_{n+1,\Sigma_n} \qquad (3.52)\\
&= (-\mathcal{F}''_n + \mathcal{F}_n\mathcal{Q}_{\Sigma_n} - \mathcal{Q}_{\Sigma_n}\mathcal{F}_{n,\Sigma_n})/2, \qquad (3.53)\\
\mathcal{G}''_{1,n-1,\Sigma_n} &= -2(\mathcal{Q}_{\Sigma_n} - z\mathcal{I}_m)\mathcal{F}'_{n,\Sigma_n} - \mathcal{Q}'_{\Sigma_n}\mathcal{F}_{n,\Sigma_n} + \mathcal{G}_{1,n-1,\Sigma_n}\mathcal{Q}_{\Sigma_n} - \mathcal{Q}_{\Sigma_n}\mathcal{G}_{1,n-1,\Sigma_n}, \qquad (3.54)\\
\mathcal{G}'_{2,n-1,\Sigma_n} &= -\mathcal{F}_{n,\Sigma_n}(Q_{\Sigma_n} - z\mathcal{I}_m) - \mathcal{H}_{n+1,\Sigma_n} \qquad (3.55)\\
&= (-\mathcal{F}''_{n,\Sigma_n} + \mathcal{Q}_{\Sigma_n}\mathcal{F}_{n,\Sigma_n} - \mathcal{F}_{n,\Sigma_n}\mathcal{Q}_{\Sigma_n})/2, \qquad (3.56)\\
\mathcal{G}''_{2,n-1,\Sigma_n} &= -2\mathcal{F}'_{n,\Sigma_n}(\mathcal{Q}_{\Sigma_n} - z\mathcal{I}_m) - \mathcal{F}_{n,\Sigma_n}\mathcal{Q}'_{\Sigma_n} + \mathcal{Q}_{\Sigma_n}\mathcal{G}_{2,n-1,\Sigma_n} - \mathcal{G}_{2,n-1,\Sigma_n}\mathcal{Q}_{\Sigma_n}, \qquad (3.57)\\
\mathcal{H}'_{n+1,\Sigma_n} &= -\mathcal{G}_{1,n-1,\Sigma_n}(Q_{\Sigma_n} - z\mathcal{I}_m) - (Q_{\Sigma_n} - z\mathcal{I}_m)\mathcal{G}_{2,n-1,\Sigma_n}, \qquad (3.58)\\
\mathcal{H}_{n+1,\Sigma_n} &= [\mathcal{F}''_{n,\Sigma_n} - \mathcal{F}_{n,\Sigma_n}(\mathcal{Q}_{\Sigma_n} - z\mathcal{I}_m) - (\mathcal{Q}_{\Sigma_n} - z\mathcal{I}_m)\mathcal{F}_{n\Sigma_n}]/2 \qquad (3.59)
\end{aligned}$$

and

$$\begin{aligned}
&\mathcal{F}_{n,\Sigma_n}(\overline{z},x)^* = \mathcal{F}_{n,\Sigma_n}(z,x), \quad \mathcal{H}_{n+1,\Sigma_n}(\overline{z},x)^* = \mathcal{H}_{n+1,\Sigma_n}(z,x),\\
&\mathcal{G}_{2,n-1,\Sigma}(\overline{z},x)^* = \mathcal{G}_{1,n-1,\Sigma_n}(z,x), \qquad (3.60)\\
&\mathcal{F}_{n,\Sigma_n}(z,x)\mathcal{G}_{1,n-1,\Sigma_n}(z,x) = \mathcal{G}_{2,n-1,\Sigma_n}(z,x)\mathcal{F}_{n,\Sigma_n}(z,x), \qquad (3.61)\\
&\mathcal{H}_{n+1,\Sigma_n}(z,x)\mathcal{G}_{2,n-1,\Sigma_n}(z,x) = \mathcal{G}_{1,n-1,\Sigma_n}(z,x)\mathcal{H}_{n+1,\Sigma_n}(z,x), \qquad (3.62)\\
&\mathcal{H}_{n+1,\Sigma_n}(z,x)\mathcal{F}_{n,\Sigma_n}(z,x) - \mathcal{G}_{1,n-1,\Sigma_n}(z,x)^2 = R_{2n+1}(z)\mathcal{I}_m, \qquad (3.63)\\
&\mathcal{F}_{n,\Sigma_n}(z,x)\mathcal{H}_{n+1,\Sigma_n}(z,x) - \mathcal{G}_{2,n-1,\Sigma_n}(z,x)^2 = R_{2n+1}(z)\mathcal{I}_m. \qquad (3.64)
\end{aligned}$$

Combining (2.33)–(2.37) and (3.46) then yields

$$\begin{aligned}
&\mathcal{M}_{\pm,\Sigma_n}(z,x) \\
&= \pm i R_{2n+1}(z)^{1/2}\mathcal{F}_{n,\Sigma_n}(z,x)^{-1} - \mathcal{G}_{1,n-1,\Sigma_n}(z,x)\mathcal{F}_{n,\Sigma_n}(z,x)^{-1} \qquad (3.65a)\\
&= \pm i R_{2n+1}(z)^{1/2}\mathcal{F}_{n,\Sigma_n}(z,x)^{-1} - \mathcal{F}_{n,\Sigma_n}(z,x)^{-1}\mathcal{G}_{2,n-1,\Sigma_n}(z,x), \qquad (3.65b)\\
&\qquad z \in \mathbb{C}\backslash\mathbb{R}.
\end{aligned}$$

One observes that for each $x \in \mathbb{R}$, $\mathcal{M}_{+,\Sigma_n}(\cdot,x)$ is the analytic continuation of $\mathcal{M}_{-,\Sigma_n}(\cdot,x)$ through the set Σ_n, and vice versa,

$$\lim_{\varepsilon\downarrow 0}\mathcal{M}_{+,\Sigma_n}(\lambda+i\varepsilon,x) = \lim_{\varepsilon\downarrow 0}\mathcal{M}_{-,\Sigma_n}(\lambda-i\varepsilon,x) = \lim_{\varepsilon\downarrow 0}\mathcal{M}_{-,\Sigma_n}(\lambda+i\varepsilon,x)^*, \quad (3.66)$$
$$\lambda \in \bigcup_{j=0}^{n-1}(E_{2j},E_{2j+1}) \cup (E_{2n},\infty),\ x \in \mathbb{R}.$$

In other words, for each $x \in \mathbb{R}$, $\mathcal{M}_{+,\Sigma_n}(\cdot,x)$ and $\mathcal{M}_{-,\Sigma_n}(\cdot,x)$ are the two branches of an analytic matrix-valued function $\mathcal{M}_{\Sigma_n}(\cdot,x)$ on the two-sheeted Riemann surface of $R_{2n+1}^{1/2}$. This implies that $\mathcal{Q}_{\Sigma_n}$ is reflectionless as will be discussed in Lemma 3.7. In addition, it is worthwhile to emphasize that in the present case of reflectionless potentials, $\mathcal{F}_{n,\Sigma_n}(z,x_0)$ and $\mathcal{G}_{1,n-1,\Sigma_n}(z,x_0)$ for some fixed $x_0 \in \mathbb{R}$, uniquely determine $\mathcal{M}_{\pm,\Sigma_n}(z,x_0)$ and hence $\mathcal{Q}_{\Sigma_n}(x)$ for all $x \in \mathbb{R}$.

Introducing the open interior Σ_n^o of Σ_n defined by $\Sigma_n^o = \bigcup_{j=0}^{n-1}(E_{2j},E_{2j+1}) \cup (E_{2n},\infty)$, one obtains the following results.

THEOREM 3.6. [25] *Let $z \in \mathbb{C}\backslash\mathbb{R}$ and $x \in \mathbb{R}$. Then*
(i) $\mathcal{F}_{n,\Sigma_n}(\cdot,x)$ and $\mathcal{H}_{n+1,\Sigma_n}(\cdot,x)$ are strongly hyperbolic (and hence self-adjoint) $m\times m$ monic matrix pencils of degree n and $n+1$, respectively, and $\mathcal{G}_{p,n-1,\Sigma_n}(\cdot,x)$, $p=1,2$, are $m\times m$ matrix pencils of degree $n-1$.
(ii) The differential expression $\mathcal{L}_{\Sigma_n} = -\mathcal{I}_m\frac{d^2}{dx^2} + \mathcal{Q}_{\Sigma_n}$ is in the limit point case at $\pm\infty$.
(iii) $\mathcal{M}_{\pm,\Sigma_n}(z,\cdot)$ in (3.65) satisfy the matrix-valued Riccati-type equation

$$\mathcal{M}'_{\pm,\Sigma_n}(z,x) + \mathcal{M}_{\pm,\Sigma_n}(z,x)^2 = \mathcal{Q}_{\Sigma_n}(x) - z\mathcal{I}_m, \quad x \in \mathbb{R},\ z \in \mathbb{C}\backslash\mathbb{R}. \quad (3.67)$$

Moreover, $\mathcal{M}_{\pm,\Sigma_n}(z,x)$ in (3.65) are the $m\times m$ Weyl–Titchmarsh matrices associated with self-adjoint operators $H^D_{\pm,x,\Sigma_n}$ in $L^2([x,\pm\infty))^m$, with a Dirichlet boundary condition at the point x and an $m\times m$ matrix-valued potential $\mathcal{Q}_{\Sigma_n}$ satisfying

$$\mathcal{Q}_{\Sigma_n} = \mathcal{Q}^*_{\Sigma_n} \in C^\infty(\mathbb{R})^{m\times m}, \quad \mathcal{Q}^{(r)}_{\Sigma_n} \in L^\infty(\mathbb{R}) \text{ for all } r \in \mathbb{N}_0. \quad (3.68)$$

In addition, $\mathcal{Q}_{\Sigma_n}$ is analytic in a neighborhood of the real axis. $H^D_{\pm,x,\Sigma_n}$ is given by

$$\begin{aligned}
H^D_{\pm,x,\Sigma_n} &= -\mathcal{I}_m\frac{d^2}{dx^2} + \mathcal{Q}_{\Sigma_n}, \qquad (3.69)\\
\operatorname{dom}(H^D_{\pm,x,\Sigma_n}) &= \{g \in L^2((x,\pm\infty))^m \,|\, g,g' \in AC([x,c])^m \text{ for all } c \gtrless x;\\
&\qquad \lim_{\varepsilon\downarrow 0} g(x\pm\varepsilon) = 0;\ (-g'' + \mathcal{Q}_{\Sigma_n}g) \in L^2((x,\pm\infty))^m\}.
\end{aligned}$$

(iv) For each $x \in \mathbb{R}$, $\mathcal{M}_{\Sigma_n}(z,x)$ in (3.46) is a $2m \times 2m$ Weyl–Titchmarsh matrix associated with the self-adjoint operator H_{Σ_n} in $L^2(\mathbb{R})^m$ defined by

$$H_{\Sigma_n} = -\mathcal{I}_m \frac{d^2}{dx^2} + \mathcal{Q}_{\Sigma_n}, \tag{3.70}$$
$$\mathrm{dom}(H_{\Sigma_n}) = \{g \in L^2(\mathbb{R})^m \mid g, g' \in \mathrm{AC}_{\mathrm{loc}}(\mathbb{R})^m;\ (-g'' + \mathcal{Q}_{\Sigma_n} g) \in L^2(\mathbb{R})^m\}.$$

In particular, $\mathcal{M}_{\Sigma_n}(\cdot,x)$ is a $2m \times 2m$ Herglotz matrix of H_{Σ_n} admitting a representation of the type (2.19), with absolutely continuous measure $\Omega_{\Sigma_n}(\cdot,x)$ given by

$$d\Omega_{\Sigma_n}(\lambda,x) = \begin{cases} \frac{1}{2\pi R_{2n+1}(\lambda)^{1/2}} \begin{pmatrix} \mathcal{H}_{n+1,\Sigma_n}(\lambda,x) & -\mathcal{G}_{2,n-1,\Sigma_n}(\lambda,x) \\ -\mathcal{G}_{1,n-1,\Sigma_n}(\lambda,x) & \mathcal{F}_{n,\Sigma_n}(\lambda,x) \end{pmatrix} d\lambda, & \lambda \in \Sigma_n^o, \\ 0, & \lambda \in \mathbb{R}\backslash\Sigma_n. \end{cases} \tag{3.71}$$

(v) H_{Σ_n} has purely absolutely continuous spectrum Σ_n,

$$\mathrm{spec}(H_{\Sigma_n}) = \mathrm{spec}_{ac}(H_{\Sigma_n}) = \Sigma_n, \quad \mathrm{spec}_{\mathrm{p}}(H_{\Sigma_n}) = \mathrm{spec}_{\mathrm{sc}}(H_{\Sigma_n}) = \emptyset, \tag{3.72}$$

with $\mathrm{spec}(H_{\Sigma_n})$ *of uniform spectral multiplicity $2m$.*

It should be emphasized that the construction of $\mathcal{Q}_{\Sigma_n}$ in the scalar case $m = 1$ is due to Levitan [47] (see also [48], [49], [50; Ch. 8], [52]).

That $\mathcal{Q}_{\Sigma_n}$ is reflectionless is an elementary consequence of (3.66) as discussed next.

LEMMA 3.7. [25] *Denote by $\mathfrak{g}_{\Sigma_n}(z,x) = \mathcal{G}_{\Sigma_n}(z,x,x)$, $z \in \mathbb{C}_+$, $x \in \mathbb{R}$, the diagonal Green's function of H_{Σ_n}. Then*

$$\lim_{\varepsilon \downarrow 0} \mathfrak{g}_{\Sigma_n}(\lambda + i\varepsilon, x) = -\lim_{\varepsilon \downarrow 0} \mathfrak{g}_{\Sigma_n}(\lambda + i\varepsilon, x)^* \text{ for all } \lambda \in \Sigma_n^o \tag{3.73}$$

and hence $\mathcal{Q}_{\Sigma_n}$ is reflectionless.

Proof. Since $\mathfrak{g}(z,x) = (\mathcal{M}_-(z,x) - \mathcal{M}_+(z,x))^{-1}$, (3.66) implies (3.73). The latter implies that $\lim_{\varepsilon\downarrow 0} \mathfrak{g}(\lambda + i\varepsilon, x) = iG_{\Sigma_n}(\lambda,x)$ for all $\lambda \in \Sigma_n^o$ for some $m \times m$ matrix $G_{\Sigma_n}(\lambda,x) > 0$. This in turn implies

$$\Xi_{\Sigma_n}(\lambda,x) = \lim_{\varepsilon \downarrow 0} \pi^{-1} \mathrm{Im}(\ln(iG_{\Sigma_n}(\lambda + i\varepsilon, x))) = (1/2)\mathcal{I}_m \text{ for all } \lambda \in \Sigma^o, \tag{3.74}$$

and hence $\mathcal{Q}_{\Sigma_n}$ is reflectionless by Definition 2.7. ∎

Next, we briefly turn to the stationary matrix Korteweg–de Vries (KdV) hierarchy (cf. [18; Ch. 15], [21]) and show that the finite-band potential $\mathcal{Q}_\Sigma$ satisfies some (and hence infinitely many) equations of the stationary KdV equations.

Assuming $\mathcal{Q} = \mathcal{Q}^* \in C^\infty(\mathbb{R})^{m\times m}$, we recall the expansions (cf. Theorem 2.5)

$$\mathfrak{g}(z,x) = [\mathcal{M}_-(z,x) - \mathcal{M}_+(z,x)]^{-1} \underset{\substack{|z|\to\infty \\ z \in C_\varepsilon}}{=} \frac{i}{2z^{1/2}} \sum_{k=0}^{\infty} \widehat{\mathfrak{R}}_k(x) z^{-k} \tag{3.75}$$

for some coefficients $\widehat{\mathfrak{R}}_k$. Explicitly, one obtains

$$\widehat{\mathfrak{R}}_0 = \mathcal{I}_m, \quad \widehat{\mathfrak{R}}_1 = \tfrac{1}{2}\mathcal{Q}, \quad \widehat{\mathfrak{R}}_2 = -\tfrac{1}{8}\mathcal{Q}'' + \tfrac{3}{8}\mathcal{Q}^2, \text{ etc.} \tag{3.76}$$

The stationary KdV hierarchy is then given by

$$\text{s-KdV}_k(\mathcal{Q}) = -2\sum_{\ell=0}^{k} c_{k-\ell}\widehat{\mathfrak{R}}'_{\ell+1}(\mathcal{Q},\dots) = 0, \quad k \in \mathbb{N}_0, \tag{3.77}$$

where $c_0 = 1$ and $\{c_\ell\}_{\ell=1,\dots,k} \subset \mathbb{C}$ denotes a set of constants.
By Remark 2.6, each $\widehat{\mathfrak{R}}_\ell$ is a differential polynomial in $\mathcal{Q}$ and next we slightly abuse notation and indicate this by writing $\widehat{\mathfrak{R}}_\ell(\mathcal{Q},\dots)$ for $\widehat{\mathfrak{R}}_\ell(x)$, $\widehat{\mathfrak{R}}'_{\ell+1}(\mathcal{Q},\dots)$ for $\widehat{\mathfrak{R}}'_{\ell+1}(x)$, etc.

THEOREM 3.8. [25] *The self-adjoint finite-band potential $\mathcal{Q}_{\Sigma_n} \in C^\infty(\mathbb{R})^{m\times m}$, discussed in Theorem 3.6, is a stationary KdV solution satisfying*

$$\text{s-KdV}_n(\mathcal{Q}_{\Sigma_n}) = -2\sum_{\ell=0}^{n} c_{n-\ell}(\underline{E})\widehat{\mathfrak{R}}'_{\ell+1}(\mathcal{Q}_{\Sigma_n},\dots) = 0. \tag{3.78}$$

Here $c_\ell(\underline{E})$ are given by

$$c_0(\underline{E}) = 1,$$
$$c_k(\underline{E}) = -\sum_{\substack{j_0,\dots,j_{2n}=0\\ j_0+\cdots+j_{2n}=k}}^{k} \frac{(2j_0)!\cdots(2j_{2n})!}{2^{2k}(j_0!)^2\cdots(j_{2n}!)^2(2j_0-1)\cdots(2j_{2n}-1)} E_0^{j_0}\cdots E_{2n}^{j_{2n}},$$
$$k = 1,\dots,n. \tag{3.79}$$

4 MATRIX EXTENSIONS OF BORG'S AND HOCHSTADT'S THEOREMS

In this our principal section, we now prove Theorem 1.5, the matrix extension of Borg's and Hochstadt's theorem, Theorems 1.1 and 1.3. Our strategy of proof will be the following: First we show that the (reflectionless) Schrödinger operators H_{Σ_ℓ} constructed in Section 3 with spectrum Σ_ℓ, satisfy the conclusions (1.11) and (1.13) for $\ell = 0, 1$, respectively. Then, in a second step, we will prove that any reflectionless Schrödinger operator with spectrum given by Σ_ℓ, $\ell = 0, 1$, is precisely of the form H_{Σ_ℓ} as constructed in Section 3.

THEOREM 4.1. *Let $\ell = 0, 1$ and $\mathcal{Q}_{\Sigma_\ell}$ be the finite-band potentials constructed in Section 3, with* $\operatorname{spec}(H_{\Sigma_\ell}) = \Sigma_\ell$ *(cf. Theorem 3.6). Then*

$$\mathcal{Q}_{\Sigma_0}(x) = E_0\mathcal{I}_m \textit{ for a.e. } x \in \mathbb{R} \tag{4.1}$$

and

$$\begin{aligned}\mathcal{Q}_{\Sigma_1}(x) &= (1/3)(E_0+E_1+E_2)\mathcal{I}_m \\ &\quad + 2\mathcal{U}\operatorname{diag}(\wp(x+\omega_3+\alpha_1),\dots,\wp(x+\omega_3+\alpha_m))\mathcal{U}^{-1} \\ &\quad \textit{for some } \alpha_j \in \mathbb{R},\ 1 \le j \le m \textit{ and a.e. } x \in \mathbb{R},\end{aligned} \tag{4.2}$$

where $\mathcal{U}$ is an $m \times m$ unitary matrix independent of $x \in \mathbb{R}$. Moreover, $\mathcal{Q}_{\Sigma_0}$ satisfies the first element of the KdV hierarchy,

$$\mathcal{Q}'_{\Sigma_0} = 0, \tag{4.3}$$

and $\mathcal{Q}_{\Sigma_1}$ satisfies the stationary KdV equation

$$\mathcal{Q}'''_{\Sigma_1} - 3(\mathcal{Q}^2_{\Sigma_1})' + 2(E_0 + E_1 + E_2)\mathcal{Q}'_{\Sigma_1} = 0. \tag{4.4}$$

Proof. We consider the elementary case $\ell = 0$ first. Then the explicit expressions,

$$\mathcal{F}_{0,\Sigma_0}(z,x) = \mathcal{I}_m, \quad \mathcal{G}_{p,-1,\Sigma_0}(z,x) = 0,\ p = 1,2, \quad \mathcal{H}_{1,\Sigma_0}(z,x) = (z - E_0)\mathcal{I}_m, \tag{4.5}$$

$$\mathcal{M}_{\pm,\Sigma_0}(z,x) = \pm i(z - E_0)^{1/2}\mathcal{I}_m, \quad \mathfrak{g}_{\Sigma_0}(z,x) = (i/2)(z - E_0)^{-1/2}\mathcal{I}_m, \text{ etc.,} \tag{4.6}$$

immediately imply (4.1) and (4.3). Hence we turn to the case $\ell = 1$. In this case one obtains,

$$\mathcal{F}_{1,\Sigma_1}(z,x) = z\mathcal{I}_m + (1/2)\mathcal{Q}_{\Sigma_1}(x) + c_1\mathcal{I}_m, \tag{4.7}$$

$$\mathcal{G}_{p,0,\Sigma_1}(z,x) = -(1/4)\mathcal{Q}'_{\Sigma_1}(x),\ p = 1,2, \tag{4.8}$$

$$\mathcal{H}_{2,\Sigma_1}(z,x) = z^2\mathcal{I}_m + z(-(1/2)\mathcal{Q}_{\Sigma_1} + c_1\mathcal{I}_m) + (1/4)\mathcal{Q}''_{\Sigma_1} - (1/2)\mathcal{Q}^2_{\Sigma_1} - c_1\mathcal{Q}_{\Sigma_1}, \tag{4.9}$$

$$\mathcal{M}_{\pm,\Sigma_1}(z,x) = \pm iR_3(z)^{1/2}\mathcal{F}_{1,\Sigma_1}(z,x)^{-1} - \mathcal{G}_{1,0,\Sigma_1}(z,x)\mathcal{F}_{1,\Sigma_1}(z,x)^{-1}, \tag{4.10}$$

$$\mathfrak{g}_{\Sigma_1}(z,x) = (i/2)R_3(z)^{-1/2}(z\mathcal{I}_m + (1/2)\mathcal{Q}_{\Sigma_1}(x) + c_1\mathcal{I}_m)^{-1}, \text{ etc.,} \tag{4.11}$$

abbreviating

$$c_1 = -(1/2)(E_0 + E_1 + E_2). \tag{4.12}$$

Combining (3.61), (4.7), and (4.8), $\mathcal{Q}_{\Sigma_1}(x)$ and $\mathcal{Q}'_{\Sigma_1}(x)$ commute and hence one obtains for each $x \in \mathbb{R}$,

$$[\mathcal{Q}^{(r)}_{\Sigma_1}(x), \mathcal{Q}^{(s)}_{\Sigma_1}(x)] = 0 \text{ for all } r, s \in \mathbb{N}_0. \tag{4.13}$$

Since $\mathcal{Q}_{\Sigma_1}$ and all its derivatives are self-adjoint, one can simultaneously diagonalize the family of matrices $\{\mathcal{Q}^{(r)}_{\Sigma_1}(x_0)\}_{r\in\mathbb{N}_0}$ by a fixed unitary $m \times m$ matrix $\mathcal{U}$. By (4.7)–(4.10), this also shows that $\mathcal{F}_{1,\Sigma_1}(z,x_0)$, $\mathcal{G}_{p,0,\Sigma_1}(z,x_0)$, $p = 1,2$, $\mathcal{H}_{2,\Sigma_1}(z,x_0)$, and $\mathcal{M}_{\pm,\Sigma_1}(z,x_0)$ can all be simultaneously diagonalized by $\mathcal{U}$. In particular, the spectral measure in the Herglotz representation (3.44) of $\mathcal{M}_{\pm,\Sigma_1}(z,x_0)$ can be diagonalized by $\mathcal{U}$. After diagonalization with $\mathcal{U}$, the inverse spectral approach in [64] (i.e., the matrix-valued extension of the scalar Gelfand–Levitan method [22], [50; Ch. 2], [75]) then yields a diagonal matrix potential of the type

$$\operatorname{diag}(q_{1,\Sigma_1}(x), \dots, q_{m,\Sigma_1}(x)), \tag{4.14}$$

and hence $\mathcal{Q}_{\Sigma_1}$ itself is of the form

$$\mathcal{Q}_{\Sigma_1}(x) = \mathcal{U}\operatorname{diag}(q_{1,\Sigma_1}(x), \dots, q_{m,\Sigma_1}(x))\mathcal{U}^{-1}. \tag{4.15}$$

In order to determine the scalar potentials q_{k,Σ_1}, $1 \le k \le m$, it now suffices to solve the corresponding scalar problem ($m = 1$). But then Hochstadt's result [33] immediately yields

$$q_{k,\Sigma_1}(x) = (1/3)(E_0 + E_1 + E_2) + 2\wp(x + \omega_3 + \alpha_k), \quad 1 \le k \le m, \tag{4.16}$$

for some $\{\alpha_k\}_{1\leq k\leq m} \subset \mathbb{R}$, and hence (4.2). Finally, (4.4) is a consequence of (3.78), (3.79), taking $n = 1$. ■

To complete the proof of Theorem 1.5, we next will prove that reflectionless Schrödinger operators with spectrum equal to Σ_ℓ, $\ell = 0, 1$, in fact, coincide with some element of the family H_{Σ_ℓ}, described in Section 3.

THEOREM 4.2. *Suppose* $\mathcal{Q}_\ell$, $\ell = 0, 1$ *satisfies Hypothesis 2.1, define* H_ℓ *as in (2.14), and suppose* $\operatorname{spec}(H_\ell) = \Sigma_\ell$, $\ell = 0, 1$. *In addition, assume that* Q_ℓ, $\ell = 0, 1$, *is reflectionless. Then* H_ℓ *coincides with an element of the family of operators* H_{Σ_ℓ} *parametrized by a choice of* $\mathcal{F}_{\ell,\Sigma_\ell}(z, x_0)$, $\mathcal{G}_{1,\ell-1,\Sigma_\ell}(z, x_0)$, $\ell = 0, 1$[1].

Proof. We start with the case $\ell = 0$. Recalling $\Sigma_0 = [E_0, \infty)$, and $R_1(z)^{1/2} = (z - E_0)^{1/2}$ defined as in (3.3), (3.4), followed by an analytic continuation from $\mathbb{R}$ to $\mathbb{C}\backslash\Sigma_0$, we denote by $\mathfrak{g}(\Sigma_0, z, x)$ the diagonal Green's function of H_0 (cf. (2.25)),

$$\mathfrak{g}(\Sigma_0, z, x) = (\mathcal{M}_-(\Sigma_0, z, x) - \mathcal{M}_+(\Sigma_0, z, x))^{-1}, \tag{4.17}$$

where in obvious notation $\mathcal{M}_\pm(\Sigma_0, z, x)$ denote the Weyl-Titchmarsh matrices associated with $\mathcal{Q}_0$. Since $\mathcal{Q}_0$ is reflectionless, the matrix $\Xi(\Sigma_0, \cdot, x)$ in its associated exponential Herglotz representation (2.26) satisfies,

$$\Xi(\Sigma_0, \lambda, x) = \begin{cases} (1/2)\mathcal{I}_m & \text{for a.e. } \lambda \in (E_0, \infty), \\ 0 & \text{for a.e. } \lambda \in (-\infty, E_0). \end{cases} \tag{4.18}$$

Insertion of (4.18) into (2.26) then yields

$$\mathfrak{g}(\Sigma_0, z, x) = i(z - E_0)^{-1/2} \exp(\mathfrak{C}_0(x)), \quad z \in \mathbb{C}\backslash\Sigma_0. \tag{4.19}$$

A comparison with the high-energy asymptotics of $\mathfrak{g}$ implied by (2.32) and (4.17) yields

$$\mathfrak{g}(\Sigma_0, z, x) \underset{\substack{|z|\to\infty \\ z\in C_\varepsilon}}{=} (i/2)\mathcal{I}_m z^{-1/2} + o(1) \tag{4.20}$$

and hence $\mathfrak{C}_0(x) = -\ln(2)\mathcal{I}_m$ implying

$$\mathfrak{g}(\Sigma_0, z, x) = (i/2)(z - E_0)^{-1/2}\mathcal{I}_m, \quad z \in \mathbb{C}\backslash\Sigma_0. \tag{4.21}$$

Thus,

$$\mathcal{Q}_0(x) = \mathcal{Q}_{\Sigma_0}(x) = E_0\,\mathcal{I}_m, \quad x \in \mathbb{R}, \tag{4.22}$$

with $\mathcal{Q}_{\Sigma_0}$ constructed in Section 3 (cf. also (4.5), (4.6)).
Next we turn to the case $\ell = 1$. Recalling $\Sigma_1 = [E_0, E_1] \cup [E_2, \infty)$, and $R_3(z)^{1/2} = [(z - E_0)(z - E_1)(z - E_2)]^{1/2}$ defined as in (3.3), (3.4), followed by an analytic continuation from $\mathbb{R}$ to $\mathbb{C}\backslash\Sigma_1$, we introduce

$$\mathfrak{g}(\Sigma_1, z, x) = (\mathcal{M}_-(\Sigma_1, z, x) - \mathcal{M}_+(\Sigma_1, z, x))^{-1}, \tag{4.23}$$

[1]More precisely, a choice of $\mathcal{F}_{0,\Sigma_\ell}(z, x_0) = \mathcal{I}_m$ for $\ell = 0$ and a choice of $\mathcal{F}_{1,\Sigma_\ell}(z, x_0)$ and a set of signs $\varepsilon_k(x_0) \in \{1, -1\}$, $k = 1, \ldots, N$ (cf. (3.36)) for $\ell = 1$.

where in obvious notation $\mathcal{M}_{\pm}(\Sigma_1, z, x)$ denote the Weyl-Titchmarsh matrices associated with $\mathcal{Q}_1$, and note that $\mathfrak{g}(\Sigma_1, \lambda, x)$ is a self-adjoint matrix for $\lambda \in (-\infty, E_0) \cup (E_1, E_2)$, that is, for all $x \in \mathbb{R}$,

$$\mathfrak{g}(\Sigma_1, \lambda, x) = \mathfrak{g}(\Sigma_1, \lambda, x)^*, \ \lambda \in (-\infty, E_0) \cup (E_1, E_2). \tag{4.24}$$

Since by hypothesis $\mathcal{Q}_1$ is reflectionless, one also has for all $x \in \mathbb{R}$,

$$\lim_{\varepsilon \downarrow 0} \mathfrak{g}(\Sigma_1, \lambda + i\varepsilon, x)^* = -\lim_{\varepsilon \downarrow 0} \mathfrak{g}(\Sigma_1, \lambda + i\varepsilon, x) \text{ for a.e. } \lambda \in (E_0, E_1) \cup (E_2, \infty). \tag{4.25}$$

Combining (4.24) and (4.25) with the properties of $R_3(z)^{1/2}$ as discussed in (3.4), one infers that for all $x \in \mathbb{R}$,

$$-i \lim_{\varepsilon \downarrow 0} R_3(\lambda + i\varepsilon)^{1/2} \mathfrak{g}(\Sigma_1, \lambda + i\varepsilon, x) \text{ is self-adjoint for a.e. } \lambda \in \mathbb{R}. \tag{4.26}$$

Next, we will take a closer look at $\mathfrak{g}(\Sigma_1, \cdot)$ and show that (4.26) in fact holds for all $\lambda \in \mathbb{R}$. Since $\mathcal{Q}_1$ is reflectionless, the matrix $\Xi(\Sigma_1, \cdot, x)$ in its associated exponential Herglotz representation (2.26) satisfies,

$$\Xi(\Sigma_1, \lambda, x) = \begin{cases} (1/2)\mathcal{I}_m & \text{for a.e. } \lambda \in (E_0, E_1) \cup (E_2, \infty), \\ 0 & \text{for a.e. } \lambda \in (-\infty, E_0). \end{cases} \tag{4.27}$$

Insertion of (4.27) into (2.26) then yields

$$\begin{aligned} \mathfrak{g}(\Sigma_1, z, x) &= i\left(\frac{1 + E_0^2}{1 + E_1^2}\right)^{1/4} \left(\frac{(z - E_1)}{(z - E_0)(z - E_2)}\right)^{1/2} \\ &\times \exp\left(\mathfrak{C}_1(x) + \int_{E_1}^{E_2} d\lambda\, \Xi(\Sigma_1, \lambda, x)\big((\lambda - z)^{-1} - \lambda(1 + \lambda^2)^{-1}\big)\right), \quad z \in \mathbb{C}\backslash\Sigma_1. \end{aligned} \tag{4.28}$$

As a consequence, one obtains

$$\begin{aligned} -iR_3(z)^{1/2}\mathfrak{g}(\Sigma_1, z, x) &= \left(\frac{1 + E_0^2}{1 + E_1^2}\right)^{1/4} (z - E_1) \\ &\times \exp\left(\mathfrak{C}_1(x) + \int_{E_1}^{E_2} d\lambda\, \Xi(\Sigma_1, \lambda, x)\big((\lambda - z)^{-1} - \lambda(1 + \lambda^2)^{-1}\big)\right), \quad z \in \mathbb{C}\backslash\Sigma_1. \end{aligned} \tag{4.29}$$

By (4.29), $-i \lim_{\varepsilon \downarrow 0} R_3(\lambda + i\varepsilon)^{1/2} \mathfrak{g}(\Sigma_1, \lambda + i\varepsilon, x)$ is self-adjoint for $\lambda \in \mathbb{R}\backslash(E_1, E_2)$. However, since $\mathfrak{g}(\Sigma_1, z, x)$ is analytic in $z \in \mathbb{C}\backslash\Sigma_1$, one infers that $-i \lim_{\varepsilon \downarrow 0} R_3(\lambda + i\varepsilon)^{1/2} \mathfrak{g}(\Sigma_1, \lambda + i\varepsilon, x)$ is self-adjoint for $\lambda \in \mathbb{R}\backslash\{E_1, E_2\}$. Next, a comparison with the high-energy asymptotics of $\mathfrak{g}$ implied by (2.32) and (4.23) yields

$$\mathfrak{g}(\Sigma_1, z, x) \underset{\substack{|z| \to \infty \\ z \in C_\varepsilon}}{=} (i/2)\mathcal{I}_m z^{-1/2} + o(1) \tag{4.30}$$

and hence

$$\mathfrak{C}_1(x) = \left(\frac{1}{4}\ln\left(\frac{1 + E_1^2}{1 + E_0^2}\right) - \ln(2)\right)\mathcal{I}_m + \int_{E_1}^{E_2} d\lambda\, \Xi(\Sigma_1, \lambda, x)\frac{\lambda}{1 + \lambda^2}. \tag{4.31}$$

Moreover, since $0 \leq \Xi(\Sigma_1, z, x) \leq \mathcal{I}_m$, we obtain for $z = E_1 - \varepsilon$,

$$
\begin{aligned}
&(E_1 - z)\exp\left(\mathfrak{C}_1(x) + \int_{E_1}^{E_2} d\lambda\, \Xi(\Sigma_1, \lambda, x)\big((\lambda - z)^{-1} - \lambda(1+\lambda^2)^{-1}\big)\right) \\
&= \frac{E_2 - E_1 + \varepsilon}{2}\left(\frac{(1+E_2^2)^2}{(1+E_0^2)(1+E_1^2)}\right)^{1/4} \\
&\quad \times \exp\left(-\int_{E_1}^{E_2} d\lambda\, (\mathcal{I}_m - \Xi(\Sigma_1, \lambda, x))(\lambda - E_1 + \varepsilon)^{-1}\right)
\end{aligned} \tag{4.32}
$$

and hence (4.29) remains bounded at $z = E_1$. Thus, $-i\lim_{\varepsilon\downarrow 0} R_3(\lambda+i\varepsilon)^{1/2}\mathfrak{g}(\Sigma_1, \lambda+i\varepsilon, x)$ is self-adjoint for $\lambda \in \mathbb{R}\backslash\{E_2\}$ and by the Schwartz reflection principle, $iR_3(z)^{1/2}\mathfrak{g}(\Sigma_1, z, x)$ is analytic for $z \in \mathbb{C}\backslash\{E_2\}$. Finally, if E_2 would be a pole of $-iR_3(z)^{1/2}\mathfrak{g}(\Sigma_1, z, x)$, then $\mathfrak{g}(\Sigma_1, z, x)$ would have a $(z-E_2)^{-(3/2)}$ singularity at E_2, contradicting the Herglotz property of $\mathfrak{g}(\Sigma_1, \cdot, x)$. (Of course, the same argument applies to $z = E_1$.) Thus,

$$-iR_3(z)^{1/2}\mathfrak{g}(\Sigma_1, z, x) \text{ is entire with respect to } z. \tag{4.33}$$

By (4.29), one infers the bound

$$\|-iR_3(z)^{1/2}\mathfrak{g}(\Sigma_1, z, x)\| \leq C(x)|z| \text{ for } |z| > \max(|E_1|, |E_2|) \tag{4.34}$$

for some constant $C(x) > 0$. Thus, $\mathfrak{g}(\Sigma_1, z, x)$ is of the form

$$\mathfrak{g}(\Sigma_1, \lambda, x) = (i/2)R_3(z)^{-1/2}(\mathcal{A}(x)z + \mathcal{B}(x)), \tag{4.35}$$

for some $m \times m$ matrices $\mathcal{A}(x), \mathcal{B}(x) \in \mathbb{C}^{m\times m}$. Hence $\mathfrak{g}(\Sigma_1, z, x)$ has an asymptotic expansion to all orders as $|z| \to \infty$ and an insertion of the asymptotic expansion into the Riccati-type equation (2.29) yields (cf. also (2.31))

$$\mathfrak{g}(\Sigma_1, z, x) = (i/2)R_3(z)^{-1/2}\big(\mathcal{I}_m z + (1/2)\mathcal{Q}_1(x) + c_1\mathcal{I}_m\big), \quad z \in \mathbb{C}\backslash\Sigma_1,\ x \in \mathbb{R}, \tag{4.36}$$

with

$$c_1 = -(E_0 + E_1 + E_2)/2. \tag{4.37}$$

By (4.36), (2.35), (2.36), and (2.47), $\mathcal{Q}_1$ is locally absolutely continuous on $\mathbb{R}$. By (2.48) and (2.50), $\mathcal{Q}_1'$ is locally absolutely continuous on $\mathbb{R}$. Iterating this procedure, using (2.47)–(2.55), one infers inductively that

$$\mathcal{Q}_1 \in C^\infty(\mathbb{R}). \tag{4.38}$$

Equations (2.53) and (4.36) then yield

$$
\begin{aligned}
\mathfrak{h}(\Sigma_1, z, x) = (i/2)R_3(z)^{-1/2}\big(&\mathcal{I}_m z^2 + (-(1/2)\mathcal{Q}_1(x) + c_1\mathcal{I}_m)z \\
&+ (1/4)\mathcal{Q}_1''(x) - (1/2)\mathcal{Q}_1(x)^2 - c_1\mathcal{Q}_1(x)\big), \\
&\qquad z \in \mathbb{C}\backslash\Sigma_1,\ x \in \mathbb{R},
\end{aligned} \tag{4.39}
$$

and (4.36), (2.49), and (2.51) prove

$$\mathfrak{g}_{p,0}(\Sigma_1, z, x) = -(i/8)R_3(z)^{-1/2}\big(\mathcal{Q}_1'(x) + \mathcal{C}_p\big), \quad p = 1, 2, \tag{4.40}$$

for some constant matrices $\mathcal{C}_p \in \mathbb{C}^{m\times m}$, $p = 1, 2$. Insertion of (4.36) and (4.40) into (2.45), (2.46), taking into account the asymptotics (2.30), (2.31) of the Weyl–Titchmarsh matrices $\mathcal{M}_\pm(\Sigma_1, z, x)$ associated with $\mathcal{Q}_1$ then shows $\mathcal{C}_p = 0$, $p = 1, 2$, and hence

$$\mathfrak{g}_{p,0}(\Sigma_1, z, x) = -(i/8)R_3(z)^{-1/2}\mathcal{Q}_1'(x), \quad p = 1, 2, \ z \in \mathbb{C}\backslash\Sigma_1, \ x \in \mathbb{R}. \tag{4.41}$$

By (2.40), (4.36), and (4.41), $\mathcal{Q}_1(x)$ and $\mathcal{Q}_1'(x)$ commute and hence one obtains for each $x \in \mathbb{R}$,

$$[\mathcal{Q}_1^{(r)}(x), \mathcal{Q}_1^{(s)}(x)] = 0 \text{ for all } r, s \in \mathbb{N}_0. \tag{4.42}$$

Next, multiplying (2.44) by $R_3(z)$ and collecting the coefficients of z^k, $0 \le k \le 2$, yields

$$(1/4)\mathcal{Q}_1''(x) - (3/4)\mathcal{Q}_1(x)^2 - c_1\mathcal{Q}_1(x) + d_1\mathcal{I}_m = 0, \tag{4.43}$$

$$\begin{aligned}&((1/4)\mathcal{Q}_1''(x) - (1/2)\mathcal{Q}_1(x)^2 - c_1\mathcal{Q}_1(x))((1/2)\mathcal{Q}_1(x) + c_1\mathcal{I}_m)\\ &- (1/16)\mathcal{Q}_1'(x)^2 + E_0E_1E_2\mathcal{I}_m = 0,\end{aligned} \tag{4.44}$$

with

$$d_1 = c_1^2 - \sum_{\substack{k_1,k_2=0\\ k_1<k_2}}^{2} E_{k_1}E_{k_2}. \tag{4.45}$$

Eliminating $\mathcal{Q}_1''(x)$ in (4.43), (4.44) finally yields

$$\mathcal{Q}_1'(x)^2 = -16R_3(-(1/2)\mathcal{Q}_1(x) - c_1\mathcal{I}_m). \tag{4.46}$$

Since $\mathcal{Q}_1(x)$ is a self-adjoint $m \times m$ matrix, we may write

$$\mathcal{Q}_1(x) = \sum_{k=1}^{N} q_k(x)\mathcal{P}_k(x), \tag{4.47}$$

where $q_k(x)$ and $\mathcal{P}_k(x)$ denote the eigenvalues and corresponding self-adjoint spectral projections of $\mathcal{Q}_1(x)$, that is,

$$\mathcal{P}_k(x)\mathcal{P}_\ell(x) = \delta_{k,\ell}\mathcal{P}_\ell(x), \quad \sum_{k=1}^{N}\mathcal{P}_k(x) = \mathcal{I}_m. \tag{4.48}$$

Introducing $\mathcal{F}_1(\Sigma_1, z, x)$ by

$$\mathcal{F}_1(\Sigma_1, z, x) = z\mathcal{I}_m + (1/2)\mathcal{Q}_1(x) + c_1\mathcal{I}_m, \tag{4.49}$$

(4.47) and (4.49) imply

$$\mathcal{F}_1(\Sigma_1, z, x) = \sum_{k=1}^{N}(z - \mu_k(x))\mathcal{P}_k(x), \quad \mu_k(x) = -(1/2)q_k(x) - c_1. \tag{4.50}$$

Since by (3.4),

$$R_3(\lambda)^{1/2} = |R_3(\lambda)|^{1/2} \begin{cases} -1, & \lambda \in (E_0, E_1), \\ 1, & \lambda \in (E_2, \infty), \end{cases} \tag{4.51}$$

and

$$R_3(\lambda)^{-1/2} \mathcal{F}_1(\Sigma_1, \lambda, x) > 0, \quad \lambda \in \Sigma_1^o, \tag{4.52}$$

one concludes that for fixed $x \in \mathbb{R}$ and all $g \in \mathbb{C}^m$, $(g, \mathcal{F}_1(\Sigma_1, \lambda, x)g)_{\mathbb{C}^m}$ changes sign for $\lambda \in [E_1, E_2]$. Thus,

$$\mu_k(x) \in [E_1, E_2], \quad 1 \leq k \leq N, \tag{4.53}$$

in accordance with (3.33). A comparison with (3.32) then yields

$$\begin{aligned} \Gamma_k(\Sigma_1, x) &= -i \lim_{z \to \mu_k(x)} (z - \mu_k(x)) R_3(z)^{1/2} \mathcal{F}_1(\Sigma_1, z, x)^{-1} \\ &= -i R_3(\mu_k(x))^{1/2} \mathcal{P}_k(x). \end{aligned} \tag{4.54}$$

Hence, given a sequence of signs,

$$\varepsilon_k(x) \in \{1, -1\}, \quad 1 \leq k \leq N, \tag{4.55}$$

and temporarily assuming

$$\mu_k(x) \in (E_1, E_2), \quad 1 \leq k \leq N, \tag{4.56}$$

one computes

$$\begin{aligned} \mathcal{G}_{1,0}(\Sigma_1, z, x) &= \left(\sum_{k=1}^{N} \frac{\varepsilon_k(x)}{z - \mu_k(x)} \Gamma_k(\Sigma_1, x) \right) \mathcal{F}_1(\Sigma_1, z, x) \\ &= -i \sum_{k=1}^{n} \varepsilon_k(x) R_3(\mu_k(x))^{1/2} \Gamma_k(\Sigma_1, x) \\ &= -i R_3(-(1/2)\mathcal{Q}_1(x) - c_1 \mathcal{I}_m)^{1/2}. \end{aligned} \tag{4.57}$$

Here the choice of the matrix square root in (4.57) is a direct consequence of the choice of signs $\varepsilon_k(x)$, $1 \leq k \leq N$. By equation (4.46), one obtains

$$\mathcal{G}_{1,0}(\Sigma_1, z, x) = -(1/4)\mathcal{Q}_1'(x) \tag{4.58}$$

in accordance with (4.40). In particular, $\mathcal{F}_1(\Sigma_1, z, x)$ in (4.49) and $\mathcal{G}_{1,0}(\Sigma_1, \lambda, x)$ in (4.58) are of the form (4.7) and (4.8). Finally, the temporary restriction (4.56) can be removed by continuity. Summing up,

$$\mathcal{Q}_1(x) = \mathcal{Q}_{\Sigma_1}(x), \quad x \in \mathbb{R}, \tag{4.59}$$

with $\mathcal{Q}_{\Sigma_1}$ constructed as in Section 3 given some $\mathcal{F}_{1,\Sigma_1}(z, x_0)$ and some choice of signs $\varepsilon_k(x_0) \in \{1, -1\}$, $1 \leq k \leq N$ (cf. also (4.7)–(4.12)). ∎

We note that the case $\ell = 0$ in Theorem 4.2 was originally treated in [14] using a (matrix-valued) trace formula approach.

Combining Theorems 4.1 and 4.2 proves Theorem 1.5.

Since (self-adjoint) periodic potentials Q which lead to Schrödinger operators with uniform maximum spectral multiplicity are reflectionless in the sense of Definition 2.7 as shown in [14], Theorem 1.6 yields the proper matrix generalizations of Borg's and Hochstadt's results, Theorem 1.1 and 1.3.

ACKNOWLEDGEMENT. It is a great pleasure to dedicate this paper to Jerry Goldstein and Rainer Nagel on the occasion of their 60th birthdays. In particular, F. G. gratefully acknowledges Jerry's friendship and support throughout the years.

We thank Robert Carlson, Mark Malamud, Fedor Rofe-Beketov, and Barry Simon for many helpful discussions on the material presented in this paper.

Research leading to this paper started in 1999 when E. B. and F. G. were supported in part by the CRDF grant UM1-325.

BIBLIOGRAPHY

1. M. Abramowitz and I. A. Stegun, *Handbook of Mathematical Functions*, New York, Dover, 1972.

2. N. Aronszajn and W. F. Donoghue, *On exponential representations of analytic functions in the upper half-plane with positive imaginary part*, J. Analyse Math. **5** (1956/57), 321–388.

3. G. Borg, *Eine Umkehrung der Sturm–Liouvilleschen Eigenwertaufgabe*, Acta Math. **78** (1946), 1–96.

4. R. W. Carey, *A unitary invariant for pairs of self-adjoint operators*, J. reine angew. Math. **283** (1976), 294–312.

5. R. Carlson, *Compactness of Floquet isospectral sets for the matrix Hill's equation*, Proc. Amer. Math. Soc. **128** (2000), 2933–2941.

6. R. Carlson, *Eigenvalue estimates and trace formulas for the matrix Hill's equation*, J. Diff. Eq. **167** (2000), 211–244.

7. R. Carlson, *An inverse problem for the matrix Schrödinger equation*, J. Math. Anal. Appl. **267** (2002), 564–575.

8. R. Carlson, *A spectral transform for the matrix Hill's equation*, (preprint, 2001).

9. H.-H. Chern and C-L. Shen, *On the n-dimensional Ambarzumyan's theorem*, Inverse Probl. **13** (1997), 15–18.

10. V. A. Chulaevski and Ya. G. Sinai, *Anderson localization for the 1-D discrete Schrödinger operator with two-frequency potential*, Commun. Math. Phys. **125** (1989), 91–112.

11. S. Clark and F. Gesztesy, *Weyl–Titchmarsh M-function asymptotics for matrix-valued Schrödinger operators*, Proc. London Math. Soc. **82** (2001), 701–724.

12. S. Clark and F. Gesztesy, *Weyl–Titchmarsh M-function asymptotics, local uniqueness results, trace formulas, and Borg-type theorems for Dirac operators*, Trans. Amer. Math. Soc. **354** (2002), 3475–3534.

13. S. Clark and F. Gesztesy, *On Povzner–Wienholtz-type self-adjointness results for matrix-valued Sturm–Liouville operators*, Proc. Roy. Soc. Edinburg A, (to appear).

14. S. Clark, F. Gesztesy, H. Holden, and B. M. Levitan, *Borg-type theorems for matrix-valued Schrödinger and Dirac operators*, J. Diff. Eq. **167** (2000), 181–210.

15. W. Craig, *The trace formula for Schrödinger operators on the line*, Commun. Math. Phys. **126** (1989), 379–407.

16. P. Deift and B. Simon, *Almost periodic Schrödinger operators III. The absolutely continuous spectrum in one dimension*, Commun. Math. Phys. **90** (1983), 389–411.

17. B. Després, *The Borg theorem for the vectorial Hill's equation*, Inverse Probl. **11** (1995), 97–121.

18. L. A. Dickey, *Soliton Equations and Hamiltonian Systems*, World Scientific, Singapore, 1991.

19. M. G. Gasymov, *Spectral analysis of a class of second-order non-self-adjoint differential operators*, Funct. Anal. Appl. **14** (1980), 11–15.

20. M. G. Gasymov, *Spectral analysis of a class of ordinary differential operators with periodic coefficients*, Sov. Math. Dokl. **21** (1980), 718–721.

21. I. M. Gel'fand and L. A. Dikii, *The resolvent and Hamiltonian systems*, Funct. Anal. Appl. **11** (1970), 93–105.

22. I. M. Gelfand and B. M. Levitan, *On the determination of a differential equation from its spectral function*, Izv. Akad. Nauk SSR. Ser. Mat. **15** (1951), 309–360, 1951, (Russian); English transl.: Amer. Math. Soc. Transl. (2) **1** (1955), 253–304.

23. F. Gesztesy and H. Holden, *On trace formulas for Schrödinger-type operators*, Multiparticle Quantum Scattering with Applications to Nuclear, Atomic and Molecular Physics (New York) (D. G. Truhlar and B. Simon, eds.), Springer, 1997, pp. 121–145.

24. F. Gesztesy, A. Kiselev, and K. A. Makarov, *Uniqueness Results for Matrix-Valued Schrödinger, Jacobi, and Dirac-Type Operators*, Math. Nachr. **239–240** (2002), 103–145.

25. F. Gesztesy and L. A. Sakhnovich, *A class of matrix-valued Schrödinger operators with prescribed finite-band spectra*, Reproducing Kernel Hilbert Spaces, Positivity, System Theory and Related Topics (D. Alpay, ed.), Operator Theory: Advances and Applications, Birkhäuser, Basel, to appear.

26. F. Gesztesy and B. Simon, *On local Borg-Marchenko uniqueness results*, Commun. Math. Phys. **211** (2000), 273–287.

27. F. Gesztesy and E. Tsekanovskii, *On matrix-valued Herglotz functions*, Math. Nachr. **218** (2000), 61–138.

28. I. M. Glazman, *Direct Methods of Qualitative Spectral Analysis of Singular Differential Operators*, Israel Program for Scientific Translations, Jerusalem, 1965.

29. V. Guillemin and A. Uribe, *Hardy functions and the inverse spectral method*, Commun. PDE **8** (1983), 1455–1474.

30. D. B. Hinton and J. K. Shaw, *On Titchmarsh–Weyl $M(\lambda)$-functions for linear Hamiltonian systems*, J. Diff. Eq. **40** (1981), 316–342.

31. D. B. Hinton and J. K. Shaw, *Hamiltonian systems of limit point or limit circle type with both endpoints singular*, J. Diff. Eq. **50** (1983), 444–464.

32. D. B. Hinton and J. K. Shaw, *On boundary value problems for Hamiltonian systems with two singular points*, SIAM J. Math. Anal. **15** (1984), 272–286.

33. H. Hochstadt, *On the determination of a Hill's equation from its spectrum*, Arch. Rat. Mech. Anal. **19** (1965), 353–362.

34. A. R. Its and V. B. Matveev, *Schrödinger operators with finite-gap spectrum and N-soliton solutions of the Korteweg-de Vries equation*, Theoret. Math. Phys. **23** (1975), 343–355.

35. M. Jodeit and B. M. Levitan, *Isospectral vector-valued Sturm–Liouville problems*, Lett. Math. Phys. **43** (1998), 117–122.

36. M. Jodeit and B. M. Levitan, *The isospectrality problem for some vector-valued Sturm–Liouville boundary problems*, Russ. J. Math. Phys. **6** (1999), 375–393.

37. R. A. Johnson, *The recurrent Hill's equation*, J. Diff. Eq. **46** (1982), 165–193.

38. R. A. Johnson, *m-Functions and Floquet exponents for linear differential systems*, Ann. Mat. Pura Appl., Ser. 4, **147** (1987), 211–248.

39. R. Johnson, S. Novo, and R. Obaya, *Ergodic properties and Weyl M-functions for random linear Hamiltonian systems*, Proc. Roy. Soc. Edinburgh **130A** (2000), 1045–1079.

40. V. I. Kogan and F. S. Rofe-Beketov, *On square-integrable solutions of symmetric systems of differential equations of arbitrary order*, Proc. Roy. Soc. Edinburgh **74A** (1974), 1–40.

41. S. Kotani, *Ljapunov indices determine absolutely continuous spectra of stationary random one-dimensional Schrödinger operators*, Stochastic Analysis(K. Itǒ, ed.), North-Holland, Amsterdam, 1984, pp. 225–247.

42. S. Kotani, *One-dimensional random Schrödinger operators and Herglotz functions*, Probabilistic Methods in Mathematical Physics (K. Itǒ and N. Ikeda, eds.), Academic Press, New York, 1987, pp. 219–250.

43. S. Kotani, *Link between periodic potentials and random potentials in one-dimensional Schrödinger operators*, Differential Equations and Mathematical Physics (I. W. Knowles, ed.), Springer, Berlin, 1987, pp. 256–269.

44. S. Kotani and M. Krishna, *Almost periodicity of some random potentials*, J. Funct. Anal. **78** (1988), 390–405.

45. S. Kotani and B. Simon, *Stochastic Schrödinger operators and Jacobi matrices on the strip*, Commun. Math. Phys. **119** (1988), 403–429.

46. M. G. Krein and A. A. Nudelman, *The Markov Moment Problem and Extremal Problems*, Providence, RI, Amer. Math. Soc., 1977.

47. B. M. Levitan, *On the solvability of the inverse Sturm–Liouville problem on the entire line*, Sov. Math. Dokl. **18** (1977), 597–600.

48. B. M. Levitan, *On the solvability of the inverse Sturm–Liouville problem on the whole axis in the case of an infinite number of lacunas in the spectrum*, Sov. Math. Dokl. **18** (1977), 964–967.

49. B. M. Levitan, *An inverse problem for the Sturm–Liouville operator in the case of finitezone and infinite-zone potentials*, Trans. Moscow Math. Soc. No. **1** (1984), 1–34.

50. B. M. Levitan, *Inverse Sturm–Liouville Problems*, VNU Science Press, Utrecht, 1987.

51. B. M. Levitan and M. G. Gasymov, *Determination of a differential equation by two of its spectra*, Russ. Math. Surv. **19** (1964), 2, 1–63.

52. B. M. Levitan and A. V. Savin, *The inverse problem on the half-line for finite zone potentials*, Moscow Univ. Math. Bull. **43** (1988), no. 1, 27–34.

53. M. M. Malamud, *Similarity of Volterra operators and related questions of the theory of differential equations of fractional order*, Trans. Moscow Math. Soc. **55** (1994), 57–122.

54. M. M. Malamud, *Uniqueness questions in inverse problems for systems of ordinary differential equations on a finite interval*, Trans. Moscow Math. Soc. **60** (1999), 173–224.

55. M. M. Malamud, *Borg type theorems for first-order systems on a finite interval*, Funct. Anal. Appl. **33** (1999), 64–68.

56. A. S. Markus, *Introduction to the Spectral Theory of Polynomial Operator Pencils*, Translations of Mathematical Monographs, vol. 71, Amer. Math. Soc., Providence, RI, 1988.

57. A. S. Markus and V. I. Matsaev, *Factorization of a weakly hyperbolic bundle*, Funct. Anal. Appl. **10** (1976), 69–71.

58. A. S. Markus and V. I. Matsaev, *On the spectral factorization of holomorphic operator-functions*, Sel. Math. Sov. **4** (1985), 325–354.

59. M. A. Naimark, *Linear Differential Operators, Part II*, F. Ungar, New York, 1968.

60. S. A. Orlov, *Nested matrix disks analytically depending on a parameter, and theorems on the invariance of ranks of radii of limiting disks*, Math. USSR Izv. **10** (1976), 565–613.

61. L. A. Pastur and V. A. Tkachenko, *Spectral theory of Schrödinger operators with periodic complex-valued potentials*, Funct. Anal. Appl. **22** (1988), 156–158.

62. L. A. Pastur and V. A. Tkachenko, *An inverse problem for a class of one-dimensional Schrödinger operators with a complex periodic potential*, Math. USSR Izv. **37** (1991), 611–629.

63. A. Ya. Povzner, *The expansion of arbitrary functions in eigenfunctions of the operator* $-\Delta u + cu$, Mat. Sbornik **32** (1953), 109–156, 1953, (Russian); English transl.: Amer. Math. Soc. Transl. (2) **60** (1967), 1–49.

64. F. S. Rofe-Beketov, *Expansions in eigenfunctions of infinite systems of differential equations in the non-self-adjoint and self-adjoint cases*, Mat. Sb. **51** (1960), 293–342, (Russian).

65. F. S. Rofe-Beketov, *The spectral matrix and the inverse Sturm–Liouville problem on the axis* $(-\infty, \infty)$, Teor. Funktsiĭ Funkts. Analiz Prilozh. **4** (1967), 189–197, (Russian).

66. F. S. Rofe-Beketov, *The inverse Sturm–Liouville problem for the spectral matrix on the whole axis and associated problems*, Integral Equations and Inverse Problems (V. Petkov and R. Lazarov, eds.), Longman, New York 1991, pp. 234–238.

67. F. S. Rofe-Beketov and A. M. Kholkin, *Spectral Analysis of Differential Operators. Connection between Spectral and Oscillatory Properties*, Mariupol, 2001.

68. A. L. Sakhnovich, *Spectral functions of a canonical system of order* $2n$, Math. USSR Sbornik **71** (1992), 355–369.

69. L. A. Sakhnovich, *Method of operator identities and problems of analysis*, St. Petersburg Math. J. **5** (1994), 1–69.

70. L. A. Sakhnovich, *Spectral Theory of Canonical Differential Systems. Method of Operator Identities*, Operator Theory: Advances and Applications, vol. 107, Birkhäuser, Basel, 1999.

71. C.-L. Shen, *Some eigenvalue problems for the vectorial Hill's equation*, Inverse Probl. **16** (2000), 749–783.

72. C.-L. Shen, *Some inverse spectral problems for vectorial Sturm–Liouville equations*, Inverse Probl. **17** (2001), 1253–1294.

73. C.-L. Shen and C.-T. Shieh, *Two inverse eigenvalue problems for vectorial Sturm–Liouville equations*, Inverse Probl. **14** (1998), 1331–1343.

74. M. Sodin and P. Yuditskii, *Almost periodic Sturm–Liouville operators with Cantor homogeneous spectrum*, Comment. Math. Helv. **70** (1995), 639–658.

75. C. Thurlow, *A generalisation of the inverse spectral theorem of Levitan and Gasymov*, Proc. Roy. Soc. Edinburgh **84A** (1979), 185–196.

76. E. Wienholtz, *Halbbeschränkte partielle Differentialoperatoren zweiter Ordnung vom eliptischen Typus*, Math. Ann. **135** (1958), 50–80.

Local and global well-posedness results for generalized BBM-type equations

JERRY L. BONA

Department of Mathematics, Statistics and Computer Science
University of Illinois at Chicago, Chicago, IL 60607
e-mail: bona@math.uic.edu

HONGQIU CHEN

Department of Mathematical Sciences
The University of Memphis, Memphis, TN 38152
e-mail: hchen1@memphis.edu

The authors dedicate this paper to Jerome Goldstein, friend and mentor, on the occasion of his 60th birthday.

ABSTRACT. The BBM or regularized long-wave equation was originally proposed as an alternative to the Korteweg-de Vries equation. It was shown in the paper of Benjamin *et al.* (1972) to be globally well-posed in $H^1(\mathbb{R})$, the class of square-integrable-functions whose derivative is also square-integrable. Recently, Bona and Tzvetkov (2002) have shown that the initial-value problem

$$\left.\begin{aligned} &u_t + u_x + uu_x - u_{xxt} = 0, && x \in \mathbb{R}, t > 0, \\ &u(x,0) = u_0(x), && x \in \mathbb{R}, \end{aligned}\right\} \tag{0.1}$$

is globally well posed in $H^s(\mathbb{R})$ for any $s \geq 0$.

It is our purpose here to extend this well-posedness theory in weak spaces to some members of a more general class of evolution equations of the form

$$u_t + u_x + g(u)_x + Lu_t = 0 \tag{0.2}$$

where L is a Fourier-multiplier related to the linearized dispersion relation and g is a smooth, real-valued function of a real variable. Results are established analogous to those for equation (0.1), for (0.2) posed on the entire real axis. In addition, local and global well-posedness theory is established for bore-like or kink-like initial data, wherein u_0 has different limits as x tends to $\pm\infty$.

1 INTRODUCTION

The regularized long-wave equation or BBM-equation

$$u_t + u_x + uu_x - u_{xxt} = 0 \tag{1.1}$$

was put forward by Peregrine (1964, 1967) and Benjamin *et al.* (1972) as an alternative model to the Korteweg-de Vries equation for small-amplitude, long wavelength surface water waves. In their paper, Benjamin and his co-workers discussed not only (1.1), but a class of evolution equations of the more general form

$$u_t + u_x + g(u)_x + Lu_t = 0 \tag{1.2}$$

where $g : \mathbb{R} \to \mathbb{R}$ is a smooth function (typically a polynomial in applications) and L is a Fourier multiplier operator with symbol α, say, so that

$$\widehat{Lv}(\xi) = \alpha(\xi)\widehat{v}(\xi)$$

for all wave numbers ξ, where the circumflex connotes the Fourier transform (with respect to the spatial variable x) of the function it surmounts and the symbol α of L is related to the dispersion suffered by infinitesimal waves. Thus, if $\omega(\xi)$ connotes the frequency corresponding to wavenumber ξ in the linearized theory, then at least for small values of wave number ξ (long waves)

$$\omega(\xi) = \frac{\xi}{1 + \alpha(\xi)}.$$

As pointed out by Benjamin *et al.* (1972) and many others, model equations of the form (1.2) arise in the description of waves in quite a number of physical situations.

In the analysis following the derivation of (1.1), Benjamin *et al.* (1972) showed (1.1) to be globally well posed in the Sobolev class $H^1(\mathbb{R})$ and in spaces such as $C_b^k(\mathbb{R}) \cap H^r(\mathbb{R})$ provided $r \geq 1$. In the Appendix to their paper, they sketched theory relating to the more general class of equations (1.2).

It is our purpose here to establish local and global well-posedness results for (1.2) in weaker L_p-based spaces for appropriate values of p. In this endeavor, we generalize some of the recent work of Bona & Tzvetkov (2002) concerned with (1.1). We also countenance bore-like initial data as well as L_p-based data.

The plan for the remainder of the paper is the following. In Section 2, the pure initial-value problem is converted into an integral equation. Local existence is then established for this integral equation by an application of the contraction-mapping principle in appropriate L_p-spaces. For a restricted class of the equations possessing a local well-posedness theory, an *a priori* bound is derived that leads to global well posedness. Section 3 follows a similar development relative to propagation of bore-like disturbances.

2 NOTATION AND LOCAL WELL-POSEDNESS

We begin with a brief synopsis of our notational conventions and function-space designations.

2.1 Notation

For $1 \leq r < \infty$, $L_r = L_r(\mathbb{R})$ connotes the rth-power Lebesgue-integrable functions with the usual modification for the case $r = \infty$. The norm of a function $f \in L_r$ is written $|f|_r$. The Sobolev class $H^s(\mathbb{R})$ for $s \geq 0$ is the class of $L_2(\mathbb{R})$-functions whose Fourier transform $\widehat{f}$ has the property

$$\|f\|_s^2 = \frac{1}{2\pi} \int_{-\infty}^{\infty} (1 + \xi^2)^s |\widehat{f}(\xi)|^2 \, d\xi < +\infty$$

where $\widehat{f}(\xi) = \int_{-\infty}^{\infty} f(x) e^{-ix\xi} \, d\xi$. Note that $\|f\|_0 = |f|_2$ and we will in fact write the L_2-norm of f unadorned as simply $\|f\|$. If X is any Banach space and $T > 0$ given, $C(0,T;X)$ is the class of continuous functions from $[0,T]$ into X with its usual norm

$$\|u\|_{C(0,T;X)} = \max_{0 \leq t \leq T} \|u(t)\|_X.$$

If $S \subset X$ is a subset, then $C(0,T;S)$ is the collection of elements u in $C(0,T;X)$ such that $u(t) \in S$ for $0 \leq t \leq T$. When $T = \infty$, $C(0,\infty;X)$ is a Fréchet space with defining set of semi-norms

$$p_n(u) = \max_{0 \leq t \leq n} \|u(t)\|_X$$

for $n = 1, 2, \cdots$. The subspace $C_b(0,\infty;X)$ of elements of $C(0,\infty;X)$ which are uniformly bounded is a Banach space with norm

$$\|u\|_{C(0,\infty;X)} = \sup_{t \geq 0} \|u(t)\|_X.$$

The Banach space $C^1(0,T;X)$ is the subspace of $C(0,T;X)$ for which the limit

$$u'(t) = \lim_{h \to 0} \frac{u(t+h) - u(t)}{h}$$

exists in $C(0,T;X)$. It is equipped with the obvious norm. Inductively, one defines $C^k(0,T;X)$ and, by analogy, $C^k(0,\infty;X)$ and $C_b^k(0,\infty;X)$.

2.2 Associated Integral Equations

The theory begins by converting the original initial-value problem into an associated integral equation. For this, we operate formally and consider afterward the issue of whether or not solutions of the integral equation are solutions of the initial-value problem.

Write the evolution equation (1.2) posed on all of $\mathbb{R}$ in the form

$$(I + L)u_t = -\big(u + g(u)\big)_x \tag{2.1}$$

and take the Fourier transform with respect to the spatial variable x. Writing the Fourier transform of u with respect to x as $\widehat{u}$, there appears the formal relation

$$\big(1 + \alpha(\xi)\big)\widehat{u}_t = -i\xi\big(\widehat{u} + \widehat{g(u)}\big).$$

Dividing by $1+\alpha$ and taking the inverse Fourier transform leads to the integral equation

$$u_t = K * \big(u + g(u)\big) \tag{2.2}$$

where the kernel K is the inverse Fourier transform of the function

$$\widehat{K}(\xi) = -i\xi/\big(1+\alpha(\xi)\big).$$

Of course the convolution may have to be interpreted in the sense of tempered distributions. A formal integration in the temporal variable then leads to the BBM-type integral equation (Benjamin *et al.* 1972)

$$u(x,t) = u_0(x) + \int_0^t \int_{-\infty}^{\infty} K(x-y)\Big(u(y,s) + g(u(y,s))\Big)\, dy\, ds \tag{2.3}$$

where $u_0(x) = u(x,0)$ is the initial data. For classes of functions v defined on $\mathbb{R} \times [0,T]$ to be discussed presently, let $w = A(v)$ be the function obtained from v by replacing u by v on the right-hand side of (2.3). The equation (2.3) then takes the form

$$u = A(u). \tag{2.4}$$

In terms of the integral equation, a solution is thus seen to comprise a fixed point of the nonlinear operator A.

2.3 Local well-posedness with non-smooth initial data

To use (2.3) or (2.4) in a precise way, assumptions about the nonlinearity g and the dispersion operator L must be made and appropriate function classes put forward. As mentioned earlier, our goal is to work in relatively large function spaces.

Assumptions on g and the symbol α of L are now delineated.

(H1) The function $g : \mathbb{R} \to \mathbb{R}$ is C^1 and has the property that there is a $p > 1$ and a constant C_0 such that

$$\big|1 + g'(z)\big| \le C_0(1 + |z|^{p-1})$$

for all $z \in \mathbb{R}$. This assumption will be referred to as the assumption of polynomial growth. Without loss of generality we take it that $g(0) = g'(0) = 0$.

(H2) The symbol α of L has the property that the tempered distribution K whose Fourier transform is $-i\xi/\big(1+\alpha(\xi)\big)$ is given by a measurable function lying in $L_1(\mathbb{R}) \cap L_r(\mathbb{R})$ for some $r > 1$.

Examples: If $L = -\partial_x^2$, then

$$K(x) = \frac{1}{2}\,\mathrm{sgn}(x)e^{-|x|}$$

as one ascertains by a direct calculation using the Residue Theorem (see Benjamin *et al.* 1972). Clearly, this version of K satisfies (H2) for any positive value of r.

If $L = D^s$ with $s > 1$ where $\widehat{D^s h}(\xi) = |\xi|^s \widehat{h}(\xi)$, then

$$K(x) = \mathcal{F}^{-1}\Big\{\frac{i\xi}{1+|\xi|^s}\Big\} = \frac{1}{2\pi}\int_{-\infty}^{\infty} \frac{i\xi e^{ix\xi}}{1+|\xi|^s}\,d\xi = -\frac{1}{\pi}\int_0^{\infty} \frac{\xi\sin(x\xi)}{1+\xi^s}\,d\xi$$

where $\mathcal{F}$ connotes the Fourier transform in the spatial variable x and $\mathcal{F}^{-1}$ is its inverse. It thus follows that K is odd and in particular $K(0) = 0$. For any $x > 0$, integration by parts twice yields

$$\begin{aligned} K(x) &= -\frac{1}{\pi x}\int_0^{\infty} \frac{1-(s-1)\xi^s}{(1+\xi^s)^2}\cos(x\xi)\,d\xi \\ &= -\frac{1}{\pi x^2}\int_0^{\infty} \frac{s(s+1)\xi^{s-1} - s(s-1)\xi^{2s-1}}{(1+\xi^s)^3}\sin(x\xi)\,d\xi. \end{aligned}$$

It is thereby concluded that $K(x) = O(\frac{1}{x^2})$ as $x \to \infty$. On the other hand, for $x > 0$, $K(x)$ may also be represented in the form

$$\begin{aligned} K(x) &= -\frac{1}{\pi}\int_0^{\frac{1}{x}} \frac{\xi\sin(\xi x)}{1+\xi^s}\,d\xi - \frac{1}{\pi}\int_{\frac{1}{x}}^{\infty} \frac{\xi\sin(\xi x)}{1+\xi^s}\,d\xi \\ &= -\frac{1}{\pi}\int_0^{\frac{1}{x}} \frac{\xi\sin(\xi x)}{1+\xi^s}\,d\xi - \frac{x^s\cos 1}{\pi x^2(x^s+1)} - \frac{1}{\pi x}\int_{\frac{1}{x}}^{\infty} \frac{1-(s-1)\xi^s}{(1+\xi^s)^2}\cos(x\xi)\,d\xi \\ &= -\frac{1}{\pi}\int_0^{\frac{1}{x}} \frac{\xi\sin(\xi x)}{1+\xi^s}\,d\xi - \frac{x^s\cos 1}{\pi x^2(x^s+1)} - \frac{x^s}{\pi x^2}\int_1^{\infty} \frac{x^s-(s-1)y^s}{(x^s+y^s)^2}\cos y\,dy. \end{aligned}$$

It follows immediately that

$$|K(x)| \le \frac{1}{\pi}\int_0^{\frac{1}{x}} \frac{\xi^2 x}{1+\xi^s}\,d\xi + x^{s-2}(x^s+1).$$

Since the integrand $x\xi^2/(1+\xi^s)$ is bounded by $x\xi^{2-s}$, it follows that

$$|K(x)| = O(x^{s-2})$$

as $x \to 0$. These considerations imply $K \in L_1 \cap L_r$ for any $r < 1/(2-s)$ if $s < 2$, and for $r = \infty$ if $s \ge 2$.

As for the nonlinearity, if p is an integer, $p \ge 2$, and $g(z) = \sum_{j=2}^{p} c_j z^j$ is a polynomial of degree p, then

$$|1 + g'(z)| = |1 + \sum_{j=2}^{p} j c_j z^{j-1}| \le C_0(1 + |z|^{p-1})$$

for a suitable constant C_0 depending only on the coefficients $c_2, \cdots, c_p$.

Here is a local existence theory based on (H1) and (H2).

THEOREM 2.1. *Consider the integral equation (2.3) and suppose the nonlinear function g and the integral kernel K satisfy hypotheses (H1) and (H2), respectively. Let q be such that*

$$q \geq \max\{p, (pr - r)/(r - 1)\}$$

where p and r, specified in the hypotheses (H1) and (H2), represent the properties of the nonlinear function g and the integral kernel K, respectively. Then for any initial data $u_0 \in L_q$, there is a positive number $T = T(|u_0|_q)$ such that (2.3) has a unique solution u lying in the space $C(0,T;L_q)$. Moreover, the mapping $u_0 \mapsto u$ from the space L_q to $C(0,T;L_q)$ is continuous.

Recall that, for any $r_1, r_2 \geq 1$, if $u \in L_{r_1}$ and $v \in L_{r_2}$, then $u * v \in L_r$ and $|u*v|_r \leq |u|_{r_1}|v|_{r_2}$, where r is determined by $\frac{1}{r} = \frac{1}{r_1} + \frac{1}{r_2} - 1$. With this inequality in mind, the tool for proving the proposition is the standard version of the contraction mapping theorem.

Proof. The remark just made implies that if $u \in L_q$, then

$$K * (g(u) + u) \in L_q$$

because

$$|g(u) + u| = \left| \int_0^1 \Big(1 + g'(su)\Big)\, ds\, u \right| \leq C_0\Big(|u| + |u|^p\Big),$$

so,

$$|K * (g(u) + u)|_q \leq C_0\big(|K|_1|u|_q + |K|_{q/(q-p+1)}|u|_q^p\big). \tag{2.5}$$

Hence, the integral operator A may be considered as a map of $C(0,\infty;L_q)$ to itself. For any $\beta > 0$, let $B_\beta = \{u \in L_q : |u|_q \leq \beta\}$, and for any $T > 0$, let $X = X_{T,\beta} = C(0,T;B_\beta)$. It is asserted that if $\beta > 0$ is chosen properly and $T > 0$ is sufficiently small, then A maps X to itself and is a contraction mapping. In fact, for any $u \in X$,

$$||Au||_X \leq |u_0|_q + TC_0\Big(|K|_1||u||_X + |K|_{q/(q-p+1)}||u||_X^p\Big).$$

To estimate the norm of difference $Au - Av$ in the space X, the following sublemma will be useful.

SUBLEMMA 2.2. *If $u, v \in L_q$ and $|u|_q, |v|_q \leq \beta$, then*

$$\begin{aligned}
&\Big|K * \big(u - v + g(u) - g(v)\big)\Big|_q \\
&= \left|K * \int_0^1 \big(1 + g'(v + s(u-v))\big)\, ds(u - v)\right|_q \\
&\leq \left||K| * \int_0^1 C_0(1 + |v + s(u-v)|^{p-1})\, ds|u - v|\right|_q \\
&\leq C_0 \int_0^1 \Big\{|K|_1|u - v|_q + |K|_{q/(q-p+1)}|v + s(u-v)|_q^{p-1}|u - v|_q\Big\}\, ds \\
&\leq C_0\Big\{|K|_1 + |K|_{q/(q-p+1)}\beta^{p-1}\Big\}|u - v|_q
\end{aligned} \tag{2.6}$$

since $|v + s(u-v)|_q \leq \beta$ *because* u *and* v *both lie in the ball of radius* β *about* 0 *in* L_q.

Hence, it is straightforward to adduce the relation

$$\begin{aligned}
&\left|Au(\cdot,t) - Av(\cdot,t)\right|_q \\
&= \left|\int_0^t K * \Big(u(\cdot,\tau) - v(\cdot,\tau) + g(u(\cdot,\tau)) - g(v(\cdot,\tau))\Big)\, d\tau\right|_q \\
&\leq \int_0^t \left|K * \Big(u(\cdot,\tau) - v(\cdot,\tau) + g(u(\cdot,\tau)) - g(v(\cdot,\tau))\Big)(\cdot,\tau)\right|_q d\tau \\
&\leq C_0 \int_0^t \Big\{|K|_1 + |K|_{q/(q-p+1)}\beta_q^{p-1}\Big\}|u(\cdot,\tau) - v(\cdot,\tau)|_q\, d\tau.
\end{aligned}$$

Taking the maximum in this inequality for $t \in [0,T]$ yields

$$||Au - Av||_X \leq C_0 T(|K|_1 + |K|_{q/(q-p+1)}\beta^{p-1})||u - v||_X. \tag{2.7}$$

It follows readily that if we choose

$$\beta = 2|u_0|_q \quad \text{and} \quad T = \frac{1}{2C_0(|K|_1 + |K|_{q/(q-p+1)}\beta^{p-1})}, \tag{2.8}$$

then the operator A maps X to X and is contractive. Since X is a complete metric space, the contraction mapping theorem completes the proof. ∎

REMARK. In fact, the result of 2.1 is true if the nonlinear function g is a C^α-function for some number $\alpha \in (0,1]$ and there is a number $C_1 > 0$ such that for any $x, y \in \mathbb{R}$,

$$|g(x) - g(y)| \leq C_1(|x|^{1-\alpha} + |y|^{1-\alpha} + |x|^{p-1} + |y|^{p-1})|x - y|^\alpha. \tag{2.9}$$

Furthermore, if there is a positive integer k such that $g \in C^{k+\alpha}$, and $g^{(k)}$ satisfies an inequality like (2.9) in which p is replaced by $p - k$, then the mapping $u_0 \mapsto u$ from the space L_q to $C(0,T;L_q)$ is k^{th}-order differentiable. Here we only outline the proof for $k = 1$. This case amounts to the assertion that the mapping taking u_0 to the associated solution u is differentiable. Write the solution with initial data u_0 as $u = u_{u_0}$. Fix initial data ϕ and a perturbation h in L_q and let $\delta \neq 0$ be an arbitrary real number. Clearly,

$$\frac{u_{\phi+\delta h} - u_\phi}{\delta} = h + \int_0^t K * \Big(\frac{u_{\phi+\delta h} - u_\phi}{\delta} + \frac{g(u_{\phi+\delta h}) - g(u_\phi)}{\delta}\Big)\, d\tau$$

and so, by the sublemma, it is thus implied that for any $T' > 0$ such that $u_{\phi+\delta h}$ and u_ϕ are well defined on $C(0,T';L_q)$,

$$\left\|\frac{u_{\phi+\delta h} - u_\phi}{\delta}\right\|_{C(0,T';L_q)} \leq |h|_q + C'(\delta,\phi)T'\left\|\frac{u_{\phi+\delta h} - u_\phi}{\delta}\right\|_{C(0,T';L_q)}$$

where $C'(\delta,\phi) = C_0(|K|_1 + |K|_{q/(q-p+1)})(\|u_\phi\|^{p-1}_{C(0,T';L_q)} + o(\delta))$ and the little o denotes terms for which $\lim_{\delta\to 0} o(\delta) = 0$. Notice that $C'T' \le \frac{1}{2}$ when $\delta \ne 0$ and T' are chosen sufficiently small. In these circumstances, it is not hard to verify that the difference quotient on the left-hand side of the last inequality is also Cauchy as $\delta \to 0$, and therefore one infers existence of the limit

$$\lim_{\delta\to 0} \frac{u_{\phi+\delta h} - u_\phi}{\delta}$$

in $C(0,T';L_q)$. As differentiability is a local issue and the time interval is compact, this completes the proof of the remark. The argument for this result is similar in case $k > 1$. If g is analytic (e.g. if g is a polynomial), then the mapping is also analytic.

The L_q-results just established may be generalized in various directions. Here is an aspect of further regularity corresponding to enhanced smoothness of the initial data.

THEOREM 2.3. *(Regularity 1) Let $u \in C(0,T;L_q)$ be the solution of (2.3) described in Theorem 2.1. In addition, suppose $g \in C^k(\mathbb{R})$ for some $k \ge 1$ and $g^{(k)}$ is bounded by a polynomial with degree less than or equal to $p-k$. Then it follows that*

$$\frac{\partial u}{\partial t}, \cdots, \frac{\partial^k u}{\partial t^k} \in C(0,T;L_q).$$

Proof. Because of Theorem 2.1, u is given locally in time as the fixed point of the operator A as in (2.3). The fixed point of this contraction mapping may be obtained as the limit of the sequence $\{u_n\}_{n=1}^\infty$ generated by the iteration

$$u_1 = A(\theta), \cdots, u_{n+1} = A(u_n), \cdots, \tag{2.10}$$

commencing from the starting point θ which is the constant function equal to 0 everywhere. Thus, this iterated sequence $\{u_n\}_{n=1}^\infty$ is Cauchy in $C(0,T;L_q)$. Applying Sublemma 2.2, it follows that for any $n, m \ge 0$,

$$\begin{aligned}\Big|\frac{\partial u_{n+m+1}(\cdot,t)}{\partial t} - \frac{\partial u_{n+1}(\cdot,t)}{\partial t}\Big|_q &= |K * \big(u_{n+m} - u_n + g(u_{n+m}) - g(u_n)\big)(\cdot,t)|_q \\ \le C_0\{|K|_1 + |K|_{q/(q-p+1)} &\max_{s\in[0,1]} \{|u_n + s(u_{n+m} - u_n)|_q^{p-1}\}|u_{n+m}(\cdot,t) - u_n(\cdot,t)|_q.\end{aligned} \tag{2.11}$$

Since $\{u_n\}_{n\ge 1}$ converges to u in $C(0,T;L_q)$, $\max_{s\in[0,1]}\{|u_n + s(u_{n+m} - u_n)|_q^{p-1}\}$ is bounded by a constant $C = C(\|u\|_{C(0,T;L_q)})$ dependent only on u. Taking supremum of this inequality for $t \in [0,T]$ yields

$$\Big\|\frac{\partial u_{n+m+1}}{\partial t} - \frac{\partial u_{n+1}}{\partial t}\Big\|_{C(0,T;L_q)} \le C\|u_{n+m} - u_n\|_{C(0,T;L_q)}.$$

In consequence, the sequence $\{\frac{\partial u_n}{\partial t}\}_{n\ge 1} \subset C(0,T;L_q)$ is also a Cauchy sequence. Inductively, for $j = 0, 1, \cdots, k-1$ (where it is presumed that $\partial^0 u/\partial t^0 = u$), it is adduced that

$$\frac{\partial^{j+1} u_{n+1}}{\partial t^{j+1}} = K * \frac{\partial^j u_n}{\partial t^j} + K * \Big(g'(u_n)\frac{\partial^j u_n}{\partial t^j} + \cdots + g^{(j)}(u_n)(\frac{\partial u_n}{\partial t})^j\Big). \tag{2.12}$$

Arguing as above, the sequence $\{\frac{\partial^{j+1}u_n}{\partial t^{j+1}}\}_{n\geq 1}$ is also Cauchy in $C(0,T;L_q)$. The completeness of $C(0,T;L_q)$ implies that the sequence $\{\frac{\partial^{j+1}u_n}{\partial t^{j+1}}\}_{n\geq 1}$ is convergent to some function $w_{j+1} \in C(0,T;L_q)$, say, as $n \to \infty$. To prove $w_{j+1} = \partial^{j+1}u/\partial t^{j+1}$, simply let $n \to \infty$ in (2.12) to reach the conclusion

$$w_{j+1} = K * w_j + K * \big(g'(u)w_j + \cdots + g^{(j)}(u)w_1^j\big).$$

The uniqueness of the solution implies that $w_1 = \frac{\partial u}{\partial t}$, and inductively, it is seen that $w_{j+1} = \frac{\partial^{j+1}u}{\partial t^{j+1}}$. ■

REMARKS: If a function f has the property that $\widehat{f} \in L_p$ for some $p \in [1,2]$, then $f \in L_q$ and $|f|_q \leq |\widehat{f}|_p$, where $1/p + 1/q = 1$. For the integral equation (2.3), if the initial data u_0 has the property that $\widehat{u}_0 \in L_{q/(q-1)}$, which guarantees that $u_0 \in L_q$, then the results of Theorem 2.1 hold true, and furthermore, at least in the case where g is a polynomial, the Fourier transform $\widehat{u}$ of the solution u with respect to the spatial variable lies in the space $C(0,T;L_{q/(q-1)})$. The proof is made by considering the Fourier transform of (2.3), namely

$$\widehat{u}(\xi,t) = \widehat{u_0}(\xi) + \frac{i\xi}{1+\alpha(\xi)} \int_0^t \Big(\widehat{u}(\xi,\tau) + \widehat{g(u)}(\xi,\tau)\Big)\, d\tau.$$

One shows that the mapping $\widetilde{A}$ defined by the right-hand side of the last formula is a contraction in $C(0,T;\widehat{B_\beta})$ for T and β chosen appropriately, where $\widehat{B_\beta}$ is the ball of radius β about 0 in $L_{q/(q-1)}$. For this step, we use the fact that g is a polynomial so $\widehat{g(u)}$ is a finite sum of the form ,

$$\sum_{k=2}^{N} a_k \widehat{u} * \widehat{u} * \cdots \widehat{u}$$

where the kth term features a k-fold convolution of $\widehat{u}$ with itself. One thus infers existence of a solution of the above integral equation whose Fourier transform satisfies (2.3) and lies in $C(0,T;L_q)$.

Another point worth mention is the smoothing associated with a temporal derivative. Indeed, since $\widehat{u_t} = \frac{i\xi}{1+\alpha(\xi)}\big(\widehat{u} + \widehat{g(u)}\big)$, u_t is smoother than $u + g(u)$ if α grows super-linearly at infinity. For simplicity, let the nonlinear function g be homogeneous, say $g(z) = z^p$. For the dispersion α, suppose there is a positive number $s > 1 + \frac{p-1}{q}$ such that

$$\liminf_{|\xi|\to\infty} \frac{\alpha(\xi)}{|\xi|^s} > 0.$$

Then, for any ϵ in the range $[0, s - 1 - \frac{p-1}{q})$,

$$\begin{aligned}\left|(1+\xi^2)^{\frac{\epsilon}{2}}\widehat{u_t}(\xi,t)\right|_{q/(q-1)} &= \left|\frac{i\xi(1+\xi^2)^{\frac{\epsilon}{2}}}{1+\alpha(\xi)}\Big(\widehat{u}(\xi,t) + \widehat{u^p}(\xi,y)\Big)\right|_{q/(q-1)} \\ &\leq \gamma_1 |\widehat{u}(\cdot,t)|_{q/(q-1)} + \gamma_2 |\widehat{u^p}(\cdot,t)|_{q/(q-p)} \\ &\leq \gamma_1 |\widehat{u}(\cdot,t)|_{q/(q-1)} + \gamma_2 |\widehat{u}(\cdot,t)|^p_{q/(q-1)}\end{aligned}$$

where the numbers γ_1 and γ_2 are determined to be

$$\gamma_1 = \sup_{\xi \in \mathbb{R}} \frac{|\xi|(1+\xi^2)^{\frac{\epsilon}{2}}}{1+\alpha(\xi)} \quad \text{and} \quad \gamma_2^{\frac{q}{p-1}} = \int_{\mathbb{R}} \left(\frac{|\xi|(1+\xi^2)^{\frac{\epsilon}{2}}}{1+\alpha(\xi)} \right)^{\frac{q}{p-1}} d\xi.$$

Thus, $u_t \in C(0,T;W^{\epsilon,q})$ where $W^{\epsilon,q} = \{u \in L_q : (1+\xi^2)^{\frac{\epsilon}{2}}\widehat{u} \in L_q\}$. In particular, for the original BBM-equation where $s=2$ and $p=2$, if the initial data $u_0 \in L_2$, then the solution $u \in C(0,\infty;L_2)$ as proved by Bona and Tzvetkov 2001. The above ruminations and a bootstrap argument imply that the time derivative u_t lies in $C(0,\infty;H^1)$ and so is spatially smoother than u.

THEOREM 2.4. *(Regularity 2) Let $u \in C(0,T;L_q)$ be the solution whose existence is guaranteed in Theorem 2.1. Suppose in addition that $u_0 \in C_b^k$ and $g \in C^{k+1}$ for some $k \geq 0$ and $g^{(k+1)}$ is bounded by a polynomial of degree $p-k$. Then $u \in C(0,T;C_b^k \cap W_q^k)$.*

Proof. It is sufficient to prove that the sequence $\{\partial^j u_n/\partial x^j\}_{n\geq 1}$ for $j = 0,1,\cdots,k$ is Cauchy in $C(0,T_0;C_b \cap L_q)$ for $T_0 > 0$ sufficiently small, where $\{u_n\}_{n\geq 1}$ is the iterated sequence defined in (2.10). A straightforward analysis based on Gronwall's inequality allows one to extend the result to any time interval $[0,T]$ for which the solution u of (2.3) is known to lie in $C(0,T;L_q)$.

Since $u_0 \in C_b^k$ and $k \geq 0$, define functions $u_j = u_j(x,t)$ in $C(0,\infty;C_b)$ as follows:

$$u_1 = u_0 + \int_0^t K * \Big(u_0 + g(u_0)\Big)(\cdot,\tau)\,d\tau,$$

and inductively, for $n = 1,2\cdots$,

$$u_{n+1} = u_0 + \int_0^t K * \Big(u_n + g(u_n)\Big)(\cdot,\tau)\,d\tau. \tag{2.13}$$

Naturally,

$$\begin{aligned} &u_{n+1}(\cdot,t) - u_n(\cdot,t) \\ &= \int_0^t K * \Big(\int_0^1 (1+g'(u_{n-1}+s(u_n-u_{n-1}))(\cdot,\tau)\,ds(u_n-u_{n-1})\Big)\,d\tau \end{aligned}$$

and, as in Sublemma 2.2, it follows that

$$\begin{aligned} &\Big|u_{n+1}(\cdot,t) - u_n(\cdot,t)\Big|_\infty \\ &\leq \int_0^t C_0\Big(|K|_1 + |K|_{q/(q-p+1)} \max_{s\in[0,1]}\{|u_{n-1}+s(u_n-u_{n-1})|_q^{p-1}\}\Big) \\ &\qquad\qquad \Big|u_n(\cdot,\tau) - u_{n-1}(\cdot,\tau)\Big|_\infty\,d\tau. \end{aligned}$$

Since $\{u_n\}_{n\geq 1}$ is Cauchy in $C(0,T;L_q)$, it follows that for any $t \in [0,T]$ and $s \in [0,1]$,

$$\limsup_{n\to\infty} |u_{n-1} + s(u_n - u_{n-1})(\cdot,t)|_q \leq |u(\cdot,t)|_q.$$

In consequence, the sequence $\{u_n\}_{n\geq 1} \subset C(0,T;C_b)$ is Cauchy. To prove that the limit is the solution u of (2.3), let $w = \lim_{n\to\infty} u_n$ and consider the limit as $n \to \infty$ in (2.13). There obtains the relation

$$w(\cdot,t) = u_0 + \int_0^t K * (w + g(w))(\cdot,\tau)\, d\tau,$$

whence,

$$\begin{aligned}
&|w(\cdot,t) - u(\cdot,t)|_\infty \\
&= \left| \int_0^t K * (w - u + g(w) - g(u))(\cdot,\tau)\, d\tau \right|_\infty \\
&\leq C_0 \int_0^t \left(|K|_1 + |K|_{q/(q-p+1)} \left(|w|_q + |u|_q \right)^{p-1} \right) |w(\cdot,\tau) - u(\cdot,\tau)|_\infty \, d\tau.
\end{aligned}$$

It then follows from Gronwall's inequality that $w = u \in C(0,T;C_b \cap L_q)$.

Next consider the sequence $\{\partial u_n/\partial x\}_{n\geq 1}$. It is obvious that, for any $t \geq 0$,

$$\frac{\partial u_1}{\partial x} = u_0' + \int_0^t K * \left((1 + g'(u_0)) u_0' \right) d\tau$$

and the right-hand side of the last equation is in $C(0,\infty;C_b \cap L_q)$ because

$$\left| \frac{\partial u_1(\cdot,t)}{\partial x} \right|_\infty \leq |u_0'|_\infty + C_0 \int_0^t \left(|K|_1 + |K|_{q/(q-p+1)} |u_0|_q^{p-1} \right) |u_0'|_\infty \, d\tau$$

and

$$\left| \frac{\partial u_1(\cdot,t)}{\partial x} \right|_q \leq |u_0'|_q + C_0 \int_0^t \left(|K|_1 + |K|_{q/(q-p+1)} |u_0|_q^{p-1} \right) |u_0'|_q \, d\tau.$$

Induction on n yields that

$$\frac{\partial u_{n+1}}{\partial x} = u_0' + \int_0^t K * \left((1 + g'(u_n)) \frac{\partial u_n}{\partial x} \right)(\cdot,\tau)\, d\tau \in C(0,\infty;C_b \cap L_q) \qquad (2.14)$$

because

$$\left| \frac{\partial u_{n+1}(\cdot,t)}{\partial x} \right|_\infty \leq |u_0'|_\infty + C_0 \int_0^t \left(|K|_1 + |K|_{q/(q-p+1)} |u_n|_q^{p-1}(\cdot,\tau) \right) \left| \frac{\partial u_n(\cdot,\tau)}{\partial x} \right|_\infty d\tau$$

and

$$\left| \frac{\partial u_{n+1}(\cdot,t)}{\partial x} \right|_q \leq |u_0'|_q + C_0 \int_0^t \left(|K|_1 + |K|_{q/(q-p+1)} |u_n|_q^{p-1}(\cdot,\tau) \right) \left| \frac{\partial u_n(\cdot,\tau)}{\partial x} \right|_q d\tau.$$

Therefore, for t restricted to the interval $[0, T_0]$, where

$$T_0 = \min\Big\{T, \frac{1}{2C_0(|K|_1 + |K|_{q/(q-p+1)}|u|^{p-1}_{C(0,T;L_q)})}\Big\},$$

the sequence $\{\partial u_{n+1}/\partial x\}_{n\geq 1}$ is bounded in $C(0, T_0; C_b \cap L_q)$, and in fact for $t \in [0, T_0]$,

$$\Big|\frac{\partial u_{n+1}(\cdot,t)}{\partial x}\Big|_\infty \leq 2|u_0'|_\infty \quad \text{and} \quad \Big|\frac{\partial u_{n+1}(\cdot,t)}{\partial x}\Big|_q \leq 2|u_0'|_q.$$

Furthermore, for any $n \geq 0$

$$\begin{aligned}
&\frac{\partial u_{n+1}(\cdot,t)}{\partial x} - \frac{\partial u_n(\cdot,t)}{\partial x} \\
&= \int_0^t K * \Big(g'(u_n)\frac{\partial u_n}{\partial x} - g'(u_{n-1})\frac{\partial u_{n-1}}{\partial x} + \frac{\partial u_n}{\partial x} - \frac{\partial u_{n-1}}{\partial x}\Big)(\cdot,\tau)\, d\tau \\
&= \int_0^t K * \Big(\Big[g'(u_n) - g'(u_{n-1})\Big]\frac{\partial u_n}{\partial x} + (1 + g'(u_{n-1}))\Big[\frac{\partial u_n}{\partial x} - \frac{\partial u_{n-1}}{\partial x}\Big]\Big)(\cdot,\tau)\, d\tau.
\end{aligned}$$

Notice that

$$g'(u_n) - g'(u_{n-1}) = \int_0^1 g''(u_{n-1} + s(u_n - u_{n-1}))\, ds(u_n - u_{n-1}),$$

and g'' is bounded by a polynomial of degree $p-2$, then there is a positive number C such that $|g''(x)| \leq C(1 + |x|^{p-2})$. Thus applying Sublemma 2.2 yields

$$|K * (g'(u_n) - g'(u_{n-1}))| \leq C|K| * \int_0^1 (1 + |u_{n-1} + s(u_n - u_{n-1})|^{p-2})\, ds|u_n - u_{n-1}|,$$

so,

$$\begin{aligned}
&|K * (g'(u_n(\cdot,t)) - g'(u_{n-1}(\cdot,t))|_\infty \\
\leq& C(|K|_1 + |K|_{q/(q-p+2)})\int_0^1 |u_{n-1} + s(u_n - u_{n-1})|_q^{p-2}\, ds|u_n(\cdot,t) - u_{n-1}(\cdot,t)|_\infty
\end{aligned}$$

and

$$\begin{aligned}
&\Big|K * (1 + g'(u_{n-1}))\Big(\frac{\partial u_n}{\partial x} - \frac{\partial u_{n-1}}{\partial x}\Big)(\cdot,t)\Big|_\infty \\
\leq& C_0(|K|_1 + |K|_{q/(q-p+1)}|u_{n-1}(\cdot,t)|_q^{p-1})\Big|\frac{\partial u_n(\cdot,t)}{\partial x} - \frac{\partial u_{n-1}(\cdot,t)}{\partial x}\Big|_\infty.
\end{aligned}$$

Hence, there is a number C' dependent only on the solution u such that

$$\Big|\frac{\partial u_{n+1}(\cdot,t)}{\partial x} - \frac{\partial u_n(\cdot,t)}{\partial x}\Big|_\infty \leq C'\int_0^t \Big|\frac{\partial u_n(\cdot,\tau)}{\partial x} - \frac{\partial u_{n-1}(\cdot,\tau)}{\partial x}\Big|_\infty\, d\tau.$$

In a same fashion, it follows that

$$\left|\frac{\partial u_{n+1}(\cdot,t)}{\partial x} - \frac{\partial u_n(\cdot,t)}{\partial x}\right|_q \leq C' \int_0^t \left|\frac{\partial u_n(\cdot,\tau)}{\partial x} - \frac{\partial u_{n-1}(\cdot,\tau)}{\partial x}\right|_q d\tau.$$

Thus, for $T_1 > 0$ chosen sufficiently small, $\{\frac{\partial u_n}{\partial x}\}_{n\geq 1}$ is Cauchy in $C(0,T_1;C_b \cap L_q)$. To prove that the limit of this sequence as $n \to \infty$ is equal to u_x for t restricted to $[0,T_1]$, denote by $w_1 = \lim_{n\to\infty} \frac{\partial u_n}{\partial x}$, and let $n \to \infty$ in (2.14) to obtain

$$w_1 = u_0' + \int_0^t K * \Big((1+g'(u))w_1\Big)\, d\tau.$$

On the other hand, taking the derivative with respect to x in (2.3) leads to

$$u_x = u_0' + \int_0^t K * \Big((1+g'(u))u_x\Big)\, d\tau. \tag{2.15}$$

Forming the difference and estimating yields

$$|w_1(\cdot,t) - u_x(\cdot,t)|_\infty \leq C_0|K|_1 \int_0^t (1 + |u|^{p-1}_{C(0,T;C_b)})|w_1(\cdot,\tau) - u_x(\cdot,\tau)|_\infty\, d\tau.$$

Gronwall's inequality then implies $u_x = w_1$, which is to say, the solution u of (2.3) lies in $C(0,T_1;C_b^1 \cap W_q^1)$. Furthermore, (2.15) implies that

$$|u_x|_\infty \leq |u_0'|_\infty + C_0 \int_0^t (|K|_1 + |K|_{q/(q-p+1)}|u(\cdot,\tau)|_q^{p-1})|u_x(\cdot,\tau)|_\infty\, d\tau$$

and

$$|u_x|_q \leq |u_0'|_q + C_0 \int_0^t (|K|_1 + |K|_{q/(q-p+1)}|u(\cdot,\tau)|_q^{p-1})|u_x(\cdot,\tau)|_q\, d\tau.$$

Gronwall's inequality shows that the time interval over which the solution u can be extended is in fact the interval $[0,T]$ on which $u(\cdot,t)$ is known to lie in L_q, which is to say, $u \in C(0,T;C_b^1 \cap W_q^1)$. By induction, it can be shown that $u \in C(0,T;C_b^k \cap W_q^k)$. ∎

THEOREM 2.5. *In Theorem 2.1, suppose the relationship between r, the index appearing in (H2), and p, which governs the growth of the nonlinearity g in (H1) are further restricted by the relation*

$$r \geq \frac{p+1}{2}.$$

Then the integral equation (2.3) is locally well posed in $L_2 \cap L_{p+1}$, so, if the initial data $u_0 \in L_2 \cap L_{p+1}$, then there is a $T > 0$ such that (2.3) has an unique solution *u lying in* $C(0,T;L_2 \cap L_{p+1})$*, and the mapping* $u_0 \to u$ *is Lipschitz from the space* $L_2 \cap L_{p+1}$ *to* $C(0,T;L_2 \cap L_{p+1})$*. Moreover, its* L_2*-norm is bounded by*

$$||u(\cdot,t)||^2 \leq e^{Ct}||u_0||^2 + 2\int_0^t e^{C(t-\tau)}|u(\cdot,\tau)|_{p+1}^{p+1}\,d\tau,$$

where $C = 2C_0|K|_1$.

Proof. The existence of the unique solution $u \in C(0,T;L_{p+1})$ of (2.3) for some $T > 0$ is a direct result of Theorem 2.1. Again, it is known that the sequence $\{u_n\}_{n\geq 1}$ defined in (2.10) lies in $C(0,T;L_{p+1})$ and converges to u. We know further that

$$\begin{aligned}||u_1(\cdot,t)|| &\leq ||u_0|| + \int_0^t ||K * \big(u_0 + g(u_0)\big)||\,d\tau \\ &\leq ||u_0|| + C_0\int_0^t \big(|K|_1||u_0|| + |K|_{(2p+2)/(p+3)}|u_0|_{p+1}^{p}\big)\,d\tau,\end{aligned}$$

so, $u_1 \in L_2 \cap L_{p+1}$. Inductively, for $n \geq 1$,

$$\begin{aligned}||u_{n+1}(\cdot,t)|| &\leq ||u_0|| + \int_0^t ||K * u_n(\cdot,\tau) + K * g(u_n(\cdot,\tau))||\,d\tau \\ &\leq ||u_0|| + C_0\int_0^t \big(|K|_1||u_n(\cdot,\tau)|| + |K|_{(2p+2)/(p+3)}|u_n(\cdot,\tau)|_{p+1}^{p}\big)\,d\tau,\end{aligned}$$

which means $\{u_n(\cdot,t)\}_{n\geq 1} \subset L_2 \cap L_{p+1}$ at least for $t \in [0,T]$. Moreover, for any $n > 0$,

$$||u_{n+1}(\cdot,t) - u_n(\cdot,t)|| = \Big\|\int_0^t K * \Big(u_n - u_{n-1} + g(u_n) - g(u_{n-1})\Big)\Big\|,$$

since u_n converges to $u \in C(0,T;L_{p+1})$ as $n \to \infty$, for sufficiently large value of n, $||u_n||_{C(0,T;L_{p+1})} \leq 2||u||_{C(0,T;L_{p+1})}$, and therefore, by Sublemma 2.2,

$$\begin{aligned}&||u_{n+1}(\cdot,t) - u_n(\cdot,t)|| \\ \leq &C_0\int_0^t \Big(|K|_1 + 2^{p-1}|K|_{(p+1)/2}||u||_{C(0,T;L_p+1)}^{p-1}\Big)||u_n(\cdot,\tau) - u_{n-1}(\cdot,\tau)||\,d\tau.\end{aligned}$$

Thus, for $T_0 > 0$ chosen small, the sequence is Cauchy in $C(0,T_0;L_2)$, whence,

$$u \in C(0,T_0;L_2 \cap L_{p+1}).$$

Moreover, the solution u is continuously dependent on the initial data u_0. The regularity result of Theorem 2.3 allows us to formally multiply the equation (2.2) by $2u$ and integrate over $\mathbb{R}$ to reach the relations

$$\frac{d}{dt}\|u\|^2 = \int_{-\infty}^{\infty} 2uK * \big(u + g(u)\big)\, dx \le 2C_0|K|_1\|u\|^2 + 2C_0|K|_1|u|_{p+1}^{p+1}.$$

Gronwall's inequality gives

$$\|u(\cdot,t)\|^2 \le e^{Ct}\|u_0\|^2 + 2\int_0^t e^{C(t-\tau)}|u(\cdot,\tau)|_{p+1}^{p+1}\, d\tau$$

for any $t \in [0, T_0]$. The continuous dependence result allows us to conclude this relation holds for rough data. It is thereby implied that the time interval $[0, T_0]$ where $|u(\cdot,t)|_{(p+1)/2}$ stays finite can be extended to $[0, T]$. The theorem is complete. ∎

LEMMA 2.6. *Let $u \in C(0,T; L_2 \cap L_{p+1})$ be a solution of (2.3). The functional*

$$\int_{-\infty}^{\infty} \big(F(u) + \frac{1}{2}u^2\big)\, dx$$

is invariant with respect to the time variable t, where F is the anti-derivative $F(z) = \int_0^z g(z)\, dz$ of g.

Proof. For smooth solutions, the following calculation is decisive:

$$\begin{aligned}
&\frac{d}{dt}\int_{-\infty}^{\infty} \big(F(u(x,t)) + \frac{1}{2}u^2(x,t)\big)\, dx \\
&= \int_{-\infty}^{\infty} (g(u) + u)u_t\, dx \\
&= -\int_{-\infty}^{\infty} (g(u) + u)(I + L)^{-1}\partial_x\big(g(u) + u\big)\, dx.
\end{aligned}$$

As $(I + L)^{-1}\partial_x$ is skew-adjoint, the right-hand side is obviously zero. For solutions in the advertised class, the result follows from the regularity theory, the continuous dependence of solutions on the initial data and density of, say, $\mathcal{D}(\mathbb{R})$ in $L_2 \cap L_{p+1}$. ∎

COROLLARY 2.7. *In Lemma 2.6, if there is a positive number $\underline{\gamma}$ such that the function F satisfies $2F(x) + x^2 > \underline{\gamma}(x^2 + |x|^{p+1})$ for any $x \in \mathbb{R}$, then it follows that, for all $t > 0$ for which the solution u of (2.3) exists,*

$$\int_{-\infty}^{\infty} (u^2 + |u|^{p+1})\, dx \le \frac{1}{\underline{\gamma}}\int_{-\infty}^{\infty} (2F(u) + u^2)\, dx.$$

In consequence of this a priori deduced estimate, it follows that the local existence result can be iterated to produce a solution u *of (2.3) which lies in* $C(0,\infty;L_2 \cap L_{p+1})$.

The next result is a special case of Theorem 2.1 and Corollary 2.7.

COROLLARY 2.8. *Let* $p \geq 1$ *be any integer. The generalized BBM-equation*

$$u_t + u_x + u^{p-1}u_x - u_{xxt} = 0, \quad x \in \mathbb{R},\ t > 0,$$

is locally well-posed in L_q *for any* $q \geq p$. *That is, if the initial data* $u(\cdot,0) = u_0 \in L_q$, *then there exists a positive number* $T = T(|u_0|_q)$ *such that the above equation has an unique solution* $u \in C(0,T;L_q)$ *which is continuously dependent on* u_0. *If* $p \geq 3$ *is an odd integer and the initial data* $u_0 \in L_2 \cap L_{p+1}$, *then the solution* u *lies in* $C_b(0,\infty;L_2 \cap L_{p+1})$ *and so is globally defined.*

REMARK. Unfortunately, except $p = 2$, we don't have a global result in $L_2 \cap L_{p+1}$ for p an even integer greater than 1.

3 BORE-LIKE INITIAL DATA

The theory developed in Section 2 has concentrated on initial profiles that decay to zero at $\pm\infty$, at least in a weak sense. Attention is turned now to initial data that possesses different asymptotic states at $+\infty$ and $-\infty$. In the water wave context, this corresponds to bore propagation in field situations (see Peregrine 1964, 1967) and hydraulic surges in laboratory configurations. In other physical systems, such data is generated when a signal corresponding to a surge moves into an undisturbed stretch of the medium of propagation. Theoretical work on the bore problem in the context of the BBM-equation was initiated by Bona and Bryant (1973) (see also the paper of Bona, Rajopadhye and Schonbek 1994, where further theory was developed for both BBM and the Korteweg-de Vries equations).

In the present contribution, the assumptions on the initial data is weakened and the theory extended to the broader class of models featured in (1.2).

The mathematical problem amounts to being confronted with the prospect of solutions $u = u(x,t)$ satisfying the boundary conditions

$$\lim_{x\to-\infty} u(x,t) = l, \qquad \lim_{x\to\infty} u(x,t) = 0, \tag{3.1}$$

where $l > 0$ is a constant. The question is, if the initial disturbance is bore-shaped, will the wave evolve in a bore-like pattern? If so, how long will this pattern last? Bona, Rajopadhye and Schonbek (1994) showed that the BBM-equation with bore-like initial data as in (3.1) is globally well posed and that the solution maintains the boundary behavior (3.1) for all time. In this section, the generalized BBM-type model equations (1.2) will be discussed in the bore context.

Consider the initial-value problem

$$\left.\begin{aligned} &u_t + u_x + g(u)_x + Lu_t = 0, \\ &u(x,0) = u_0(x), \end{aligned}\right\} \tag{3.2}$$

where the operator L and nonlinear function g are as described in Section 2 and the initial data u_0 satisfies the bore condition (3.1). Following the technique used by Bona, Rajopadhye and Schonbek (1994), u_0 can be decomposed into the sum of two parts v_0 and ϕ, say, where $\phi \in C^\infty(\mathbb{R})$ satisfies the bore condition (3.1) and its derivative ϕ' lies in H^∞, and v_0 is a measurable function on $\mathbb{R}$ whose smoothness is determined by the smoothness of u_0.

Introduce a new variable $v = v(x,t)$ by $u(x,t) = v(x,t) + \phi(x)$. Upon substitution of this form into (3.2), there follows the initial-value problem

$$\left.\begin{aligned}&(I+L)v_t + v_x + \big(g(v+\phi) - g(\phi)\big)_x = -\big(1+g'(\phi)\big)\phi'\\ &v(x,0) = v_0\end{aligned}\right\} \tag{3.3}$$

for v. Inverting the operator $I+L$ and then integrating with respect to t over $[0,t]$, there appears the integral equation

$$\begin{aligned} v =& v_0 + \int_0^t K * \{v + g(v+\phi) - g(\phi)\}(\cdot,\tau)\,d\tau \\ &+ tM * \big(1+g'(\phi)\big)\phi' \end{aligned} \tag{3.4}$$

where the integral kernels K and M are determined via their Fourier symbols, *viz.*

$$\widehat{K}(\xi) = -i\xi/(1+\alpha(\xi)) \quad \text{and} \quad \widehat{M}(\xi) = -1/(1+\alpha(\xi)),$$

respectively. The following result is the analog in the bore context of Theorem 2.1.

THEOREM 3.1. *Suppose the nonlinear function g and the integral kernel K satisfy hypotheses (H1) and (H2), respectively. Moreover, suppose that*

$$\inf_{\xi\in\mathbb{R}} \alpha(\xi) > -1 \quad \text{and} \quad \liminf_{|\xi|\to\infty} \frac{\alpha(\xi)}{|\xi|} > 0.$$

(This is true for most cases encountered in practice). For any q such that

$$q \geq \frac{pr - r}{r-1},$$

if $v_0 \in L_q$, then there is a positive number $T = T(|\phi|_\infty, |\phi'|_q) > 0$ such that the integral equation (3.4) has an unique solution $v \in C(0,T;L_q)$ and, moreover, the mapping $v_0 \mapsto v$ is continuous from L_q to $C(0,T;L_q)$.

Proof. For any $v \in C(0,\infty;L_q)$, modify the definition of the operator A in Section 2 as follows:

$$Av = v_0 + tM * \Big(\big(1+g'(\phi)\big)\phi'\Big) + \int_0^t K * \Big\{v + g(v+\phi) - g(\phi)\Big\}\,d\tau. \tag{3.5}$$

It is sufficient to prove that A has a fixed point in $C(0,T;L_q)$ for some $T>0$. Note as before that for any $v \in L_q$,

$$v + g(v+\phi) - g(\phi) = \int_0^1 \Big(1 + g'(\phi + sv)\Big)\,ds\,v.$$

In consequence, it follows that

$$|K*(v+g(v+\phi)-g(\phi))|_q \le C_0\Big||K|*\Big(\big(1+(|\phi|+|v|)^{p-1}\big)|v|\Big)\Big|_q$$
$$\le C_0|K|_1|v|_q + C_0\sum_{j=0}^{p-1}\binom{p-1}{j}|\phi|_\infty^j|K|_{q/(q+1-p+j)}|v|_q^{p-j}.$$

Hence, it is seen that

$$K*\big(v+g(v+\phi)-g(\phi)\big)\in L_q.$$

Since $\phi\in C_b^\infty$, $\phi'\in H^\infty$, g is a C^1-function and the operator M is defined by its Fourier symbol $\frac{1}{1+\alpha(\xi)}$ where α has the decay property just described, it follows that

$$M*\Big(\big(1+g'(\phi)\big)\phi'\Big)\in H^1\subset L_q$$

because

$$\int_{-\infty}^{\infty}(1+\xi^2)\Big|\mathcal{F}\Big(M*((1+g'(\phi))\phi')\Big)(\xi)\Big|^2\,d\xi$$
$$=\int_{-\infty}^{\infty}\frac{1+\xi^2}{(1+\alpha(\xi))^2}\Big|\mathcal{F}((1+g'(\phi))\phi')(\xi)\Big|^2\,d\xi<\infty.$$

So, A maps $C(0,\infty;L_q)$ to itself. Let B_β be, as before, the closed ball of radius $\beta>0$ centered at the origin in L_q. For any $v,w\in C(0,\infty;L_q)$,

$$Av(\cdot,t)-Aw(\cdot,t)=\int_0^t K*\{v-w+g(v+\phi)-g(w+\phi)\}(\cdot,\tau)\,d\tau,$$

hence, if $v,w\in C(0,\infty;B_\beta)$, then applying Sublemma 2.2 yields

$$|Av(\cdot,t)-Aw(\cdot,t)|_q\le C\int_0^t\big(1+(|\phi|_\infty+\beta)^{p-1}\big)|v(\cdot,\tau)-w(\cdot,\tau)|_q\,d\tau \tag{3.6}$$

where the constant C may be taken to be

$$C=C_0\max_{0\le j\le p-1}\{|K|_{q/(q-j)}\}.$$

Following the line of argument laid down in the proof of Theorem 2.1, choose

$$\beta=2|v_0|_q+2\Big|M*\Big(\big(1+g'(\phi)\big)\phi'\Big)\Big|_q$$

and

$$T=\min\big\{1,1/\big(2C(|\phi|_\infty+\beta)^{p-1}\big)\big\}.$$

The operator A is then contractive on $C(0,T;B_\beta)$ and the stated results follow directly. ∎

THEOREM 3.2. *(Regularity 3) Let $v \in C(0,T;L_q)$ be the solution in Theorem 3.1. In addition, suppose for some $k \geq 1$, the nonlinear function $g \in C^k$ and $g^{(k)}$ is bounded by a polynomial of degree less than or equal to $p-k$. Then for $j = 1, \cdots, k$,*

$$\frac{\partial^j v}{\partial t^j} \in C(0,T;L_q).$$

Proof. Define a sequence $\{v_n\}_{n\geq 1}$ iteratively by

$$\begin{aligned} v_1 &= A\theta \\ &= v_0 + tM * \left((1+g'(\phi))\phi'\right) \end{aligned} \tag{3.7}$$

where θ is, as before, the zero-function and for $n \geq 1$,

$$\begin{aligned} v_{n+1} &= Av_n \\ &= v_0 + tM * \left((1+g'(\phi))\phi'\right) \\ &+ \int_0^t K * \{v_n + g(v_n+\phi) - g(\phi)\}(\cdot,\tau)\, d\tau. \end{aligned} \tag{3.8}$$

The solution v in Theorem 3.1 can be obtained as the limit of the sequence $\{v_n\}_{n\geq 1}$, so in particular, the sequence is Cauchy in $C(0,T;L_q)$. By the Fundamental Theorem of Calculus,

$$\begin{aligned} \partial v_1/\partial t &= M * \left((1+g'(\phi))\phi'\right) \\ &\vdots \\ \partial v_{n+1}/\partial t &= M * \left((1+g'(\phi))\phi'\right) + K * \{v_n + g(v_n+\phi) - g(\phi)\}. \end{aligned}$$

It is straightforward to see that $\{\partial v_n/\partial t\}_{n\geq 1} \subset C(0,T;L_q)$. As before, for $n > 1$,

$$\partial v_{n+1}/\partial t - \partial v_n/\partial t = K * \Big(v_n - v_{n-1} + g(\phi+v_n) - g(\phi+v_{n-1})\Big),$$

and thus, the Sublemma implies again that for sufficiently large values of n,

$$\begin{aligned} &|\partial v_{n+1}(\cdot,t)/\partial t - \partial v_n(\cdot,t)/\partial t|_q \\ \leq & C_1\Big(1 + \left[|\phi|_\infty + 2\|v\|_{C(0,T;L_q)}\right]^{p-1}\Big)|v_n(\cdot,t) - v_{n-1}(\cdot,t)|_q \end{aligned}$$

where $C_1 = C_0 \max_{0\leq j\leq p-1}\{|K|_{q/(q-j)}\}$. Since the sequence $\{v_n\}_{n\geq 1}$ is Cauchy in $C(0,T;L_q)$, so is $\{\partial v_n/\partial t\}_{n\geq 1}$. Denote its limit as $n \to \infty$ by w_1. The function w_1 satisfies the equation

$$w_1 = M * \left((1+g'(\phi))\phi'\right) + K * \{v + g(v+\phi) - g(\phi)\},$$

and by a uniqueness argument as expounded previously, $w_1 = v_t$. By induction on j, it follows that for any j with $1 \leq j \leq k$, $\{\partial^j v_n/\partial t^j\}_{n\geq 1}$ is Cauchy in $C(0,T;L_q)$ and the limit as $n \to \infty$ is equal to $\partial^j v/\partial t^j$. ∎

The following further regularity result is the analog of Theorem 2.4. As the proof is entirely similar, it is omitted.

THEOREM 3.3. *(Regularity 4) Let $v \in C(0,T;L_q)$ be the solution of (3.2) obtained in Theorem 3.1. Suppose in addition that for some $k \geq 1$, $v_0 \in C_b^{k-1}$ and $g \in C^k$ and its kth derivative $g^{(k)}$ is bounded by a polynomial of degree less than or equal to $p-k$. Then $v \in C(0,T;C_b^{k-1} \cap W_q^{k-1})$.*

PROPOSITION 3.4. *In the above Theorem, if $p = 2n-1 > 1$ is an odd number and there are two positive numbers γ_1 and γ_2 such that the nonlinear function $g(z) \geq (\gamma_1 - 1)z + 2n\gamma_2 z^{2n-1}$ for all $z \geq 0$, then the equation (3.4) is well-posed in $L_2 \cap L_{2n}$ globally in time, in the sense that for any initial data $v_0 \in L_2 \cap L_{2n}$ and $\bar{T} > 0$, the solution v lies in $C(0,\bar{T};L_2 \cap L_{2n})$.*

Proof. Theorem 3.1 guarantees that there is $T > 0$ such that (3.4) has a unique solution $v \in C(0,T;L_{2n})$. As in the proof of Theorem 2.4, it can be shown that v also lies in $C(0,T;L_2)$. It is sufficient to show that the solution can be extended to times that are arbitrarily large.

Define a function F on $\mathbb{R}$ by $F(z) = \int_0^z g(z)\,dz$. Because of hypothesis (H2) and the restriction on g, it is easily deduced that $F(z) \geq \frac{1}{2}(\gamma_1 - 1)z^2 + \gamma_2 z^{2n}$ for some positive constants γ_1 and γ_2. Define a functional I by

$$I(v) = \int_{-\infty}^{\infty} \big(\frac{1}{2}v^2 + F(v)\big)\,dx.$$

If v is a solution of (3.4), then formally

$$\begin{aligned}
\frac{d}{dt}I(v) &= \int_{-\infty}^{\infty} (v + g(v))u_t\,dx \\
&= \int_{-\infty}^{\infty} (v+g(v))\big[K * (v+g(v)) + M * (1+g'(\phi))\phi'\big]\,dx \\
&= \int_{-\infty}^{\infty} (v+g(v))\big(M * (1+g'(\phi))\phi'\big)\,dx \\
&\leq C_0\|M * (1+g'(\phi))\phi'\|\|v\| + C_0|M * (1+g'(\phi))\phi'|_{p+1}|v|_{p+1}^p \\
&\leq C_1(\|v\|^2 + |v|_{2n}^p) + C_2 \\
&\leq \bar{\gamma}I(v) + C_2,
\end{aligned}$$

where C_1 and C_2 are constants dependent only on the quantities $|\phi|_\infty$ and $\|\phi'\|_1$ and $\bar{\gamma} = C_1/\min\{\frac{1}{2}\gamma_1, \gamma_2\}$. As before, this formal calculation is justified by the regularity theory combined with the continuous dependence result. A Gronwall-type inequality shows that for any $t > 0$,

$$I(v(\cdot,t)) \leq I(u_0)e^{\gamma_1 t} + \frac{C_2}{\gamma_1}(e^{\gamma_1 t} - 1).$$

This means that on any time interval $[0,\bar{T}]$, the L_2- and L_{2n}- norm of the solution v is finite. The standard extension argument then completes the proof. ∎

COROLLARY 3.5. *For the generalized BBM-equation*

$$u_t + u_x + u^{2n-2}u_x - u_{xxt} = 0,$$

where $n \geq 2$*, if the initial data* $u_0 = v_0 + \phi$ *where* ϕ *is an infinitely smooth bore and* $v_0 \in L_2 \cap L_{2n}$*, then there is a unique solution* $u = v + \phi$ *where* $v \in C(0, \infty; L_2 \cap L_{2n})$ *which depends continuously on* v_0*.*

ACKNOWLEDGEMENT. This work was partially supported by the US National Science Foundation. The manuscript was finished while both authors were visiting the Institute of Mathematical Sciences at the Chinese University of Hong Kong. The authors are grateful for the excellent working environment and hospitality.

BIBLIOGRAPHY

1. T. B. Benjamin, *Lectures on nonlinear wave motion*, Lect. Appl. Math., vol. 15 American Math. Soc., Providence, 1974, pp. 3–74.

2. T. B. Benjamin, J. L. Bona, and J. J. Mahony, *Model equations for long waves in nonlinear dispersive systems*, Philos. Trans. Royal Soc. London A **272** (1972), 47–78.

3. J. L. Bona, S. Rajopadhye, and M. E. Schonbeck, *Models for propagation of bores I. Two-dimensional theory*, Differential & Int. Eq. **7** (1994), 699–734.

4. J. L. Bona and M. E. Schonbeck, *Traveling-wave solutions of the Koreteweg-de Vries-Burgers equations*, Proc. Royal Soc. Edinburgh A **101** (1985), 207–226.

5. J. L. Bona and M. Tzvetkov, *Sharp well-posedness results for BBM-type equations*, (preprint, 2002).

6. D. H. Peregrine, *Calculations of the development of an undular bore*, J. Fluid Mech. **25** (1964), 321–330.

7. D. H. Peregrine, *Long waves on a beach*, J. Fluid Mech. **27** (1967), 815–827.

Variable Coefficient KdV Equations and Waves in Elastic Tubes

RADU C. CASCAVAL

Department of Mathematics, University of Missouri, Columbia
MO 65211, USA
e-mail: radu@math.missouri.edu

Dedicated to Jerry Goldstein and Rainer Nagel on the occasion of their 60th birthdays

ABSTRACT. We present a simplified one-dimensional model for pulse wave propagation through fluid-filled tubes with elastic walls, which takes into account the elasticity of the wall as well as the tapering effect. The spatial dynamics in this model is governed by a variable coefficient KdV equation with conditions given at the inflow site. We discuss an existence theory for the associated evolution equation, based on a semilinear Hille-Yosida theory, which was previously developed for the classical KdV equation.

1 INTRODUCTION

The study of pulse wave propagation in blood vessels constitutes a major component in the effort to better understand the dynamics of the circulatory system, both in normal and pathological conditions. The blood is a suspension of cells and other particles in plasma. Due to the complexity of blood rheology, a mathematical description of blood itself has not yet been completely formulated. Nevertheless, there have been many attempts of describing the dynamics in the circulatory system through mathematical models built on various simplifying assumptions, see [10], [16], [17], [19], [20], [21]. In the systemic circulation, the large vessels are approximated by tubes with thin, elastic walls, while the blood filling the vessels is considered as a continuum, incompressible fluid. For smaller vessels, the continuum assumption is no longer valid. The walls of the smaller arteries and arterioles become less elastic and consist of a muscular tissue with important role in controlling the arterial pressure.

The pressure wave, initiated in the ascending aorta by the heart, is propagated along the arterial tree until it reaches the smallest sites of the circulation, the capillaries. A major factor in the wave propagation is the elasticity of the wall. If the vessel walls were rigid, the wave motion would be in bulk, even in the pulsatile regime, with any disturbance at one end of the tube propagating with infinite speed along the tube. This is not the case when the vessels walls are compliant, as in the sistemic portion of the circulation. The compliance of the walls is reflected by the presence of a circumferential stress corresponding to a given amount of wall displacement from its initial configuration. This pressure is similar to the hydrostatic pressure which appears in the models of the water waves in shallow channels.

The heart plays the role of a wave maker in the systemic circulation. As an analogy, there are mathematical models of wave generation in water channels, described by Korteweg-de Vries equations in a quarter plane. The initial-boundary value problem has been studied extensively, including well-posedness and regularity of solutions (see, e.g., [1], [2], [3].) As noted in [2], laboratory coordinates are preferred for studying the initial-boundary value problem, since it restricts the (t, x) in the quarter plane. Nevertheless, in case of a variable depth channel, other coordinates were used, such as in [13] and [14]. It has been proven that with a particular choice of a moving frame, the equation describing the wave propagation over an uneven bottom is a variable coefficients Korteweg-de Vries equation. The evolution is described with respect to the spatial variable x rather than the physical time t. The advantage of such a description is that the inflow boundary condition becomes an initial condition of the evolution. This is similar to the case of wave propagation in fiber optics, where the nonlinear Schrödinger equation describes the spatial dynamics of the optical signal rather than its temporal evolution.

2 THE GOVERNING EQUATIONS

Throughout this paper we will assume that our fluid-filled tube has compliant walls and circular cross-sections, at rest and during deformations. We assume that the fluid is incompressible and inviscid, and therefore its motion is governed by the Euler equations. We also assume that the motion is axisymmetric, therefore $\bar{u} = \bar{u}(\bar{t}, \bar{x}, \bar{r})$ and same for $\bar{v}$,

$\bar{p}$, while the circumferential component of the velocity is identically zero. Written in cylindrical coordinates, the equations of the fluid motion are

$$\bar{u}_{\bar{t}} + \bar{u}\bar{u}_{\bar{x}} + \bar{v}\bar{u}_{\bar{r}} + \frac{1}{\rho}\bar{p}_{\bar{x}} = 0, \tag{2.1}$$

$$\bar{v}_{\bar{t}} + \bar{u}\bar{v}_{\bar{x}} + \bar{v}\bar{v}_{\bar{r}} + \frac{1}{\rho}\bar{p}_{\bar{r}} = 0, \tag{2.2}$$

$$\bar{u}_{\bar{x}} + \bar{v}_{\bar{r}} + \frac{1}{\bar{r}}\bar{v} = 0, \tag{2.3}$$

whenever $0 \le \bar{r} \le \bar{r}^w(\bar{t}, \bar{x}) = \bar{r}_o(\bar{x}) + \bar{\eta}(\bar{t}, \bar{x})$. Here $\bar{r}^w = \bar{r}^w(\bar{t}, \bar{x})$ is the inner radius of the tube, with $\bar{r}_o = \bar{r}_o(\bar{x})$ is the radius of the unstressed tube and $\bar{\eta} = \bar{\eta}(\bar{t}, \bar{x})$ the wall displacement from the unstressed position.

To obtain the boundary conditions at the wall, we assume that the fluid velocity at the wall equals the velocity of the wall itself, (no-slip condition), i.e.

$$\begin{aligned}\bar{v}(\bar{t},\bar{x},\bar{r})\big|_{\bar{r}=\bar{r}^w(\bar{t},\bar{x})} &= \frac{d}{d\bar{t}}\,\bar{r}^w(\bar{t},\bar{x}) \\ &= \bar{\eta}_{\bar{t}}(\bar{t},\bar{x}) + (\bar{r}_o(\bar{x}) + \bar{\eta}(\bar{t},\bar{x}))_{\bar{x}}\,\bar{u}(\bar{t},\bar{x},\bar{r})\big|_{\bar{r}=\bar{r}^w},\end{aligned}$$

or, ommiting the independent variables,

$$\bar{v} = \bar{\eta}_{\bar{t}} + (\bar{r}_{o\bar{x}} + \bar{\eta}_{\bar{x}})\,\bar{u}, \tag{2.4}$$

whenever $\bar{r} = \bar{r}^w(\bar{t},\bar{x}) = \bar{r}_o(\bar{x}) + \bar{\eta}(\bar{t},\bar{x})$. Note that, under the incompressibility assumption (2.7), the equation (2.4) is equivalent to

$$\frac{\partial \bar{A}}{\partial \bar{t}} + \frac{\partial \bar{Q}}{\partial \bar{x}} = 0, \tag{2.5}$$

where $\bar{A} = \pi(\bar{r}_o(\bar{x}) + \bar{\eta}(\bar{t},\bar{x}))^2$ is the cross-sectional area and Q is the flux at site x and time t.

$$\bar{Q}(\bar{t},\bar{x}) = \pi \int_0^{\bar{r}^w(\bar{t},\bar{x})} su(\bar{t},\bar{x},s)ds. \tag{2.6}$$

Note that (2.5) is precisely the equation of conservation of total mass of the fluid.

The wall motion is determined by the transmural pressure $\bar{p}^w = \bar{p} - \bar{p}_o$ (difference between pressure exerted by the fluid particles "pushing" the wall and the atmospheric pressure) and the circumpherential stress, $S_{\theta\theta}$, which has the form

$$S_{\theta\theta} = \frac{\bar{E}_\sigma h}{\bar{r}_o^2}\bar{\eta}.$$

Here $\bar{E}_\sigma = \frac{\bar{E}}{1-\sigma^2}$, $\bar{E}$ is the Young modulus of elasticity (may vary along the length of the tube), σ is the Poisson ratio of the elastic wall (usually one takes $\sigma = 1/2$, since the wall is considered incompressible), and h is the thickness of the tube. In reality the strain-stress relation for the arterial wall is nonlinear, which means that the Young modulus is dependent on the wall displacement as well as on the pulse frequency. Nevertheless, in this paper we neglect such effects and consider only linear elastic properties of the wall. We also neglect any gravitational effects.

If ρ^w denotes the density of the tube wall, then the equation of the wall motion is

$$\rho^w h\,\bar{\eta}_{\bar{t}\bar{t}} = \bar{p}^w - \frac{\bar{E}_\sigma(\bar{x})h}{\bar{r}_o(\bar{x})^2}\bar{\eta}. \tag{2.7}$$

Note that we neglect the wall motion in the axial direction due to stretching.

Finally, the last restriction imposed, that the radial velocity vanishes in the center of the tube,

$$\bar{v}\big|_{\bar{r}=0} = 0, \tag{2.8}$$

follows from the axisymmetry of the physical system.

The first step in analyzing the system of equations (2.1)–(2.8) is to rewrite it in terms of non-dimensional variables. To this end, consider new variables (t, x, r), η as follows;

$$\bar{x} = \bar{\Lambda}\, x, \qquad \bar{r} = \bar{R}\, r, \qquad \bar{\eta} = \bar{A}\, \eta, \tag{2.9}$$

$$\bar{t} = \frac{\bar{\Lambda}}{\bar{c}_{\mathrm{MK}}}\, t, \tag{2.10}$$

where $\bar{\Lambda}$ is a typical wave length of the waves propagating in the tube, $\bar{R}$ is a typical radius of a cross-section of the tube, and $\bar{A}$ is a typical amplitude of the wall displacement from the unstressed position. The quantity $\bar{c}_{\mathrm{MK}} = \sqrt{\frac{\bar{E}h}{2\bar{R}\rho}}$ is the Moens-Korteweg velocity of a wave propagating along an elastic tube when all non-linear terms are neglected. Here ρ is the density of the fluid and $\bar{E}$ is the Young modulus of elasticity of the vessel wall.

Let $\varepsilon = \frac{\bar{A}}{\bar{R}}$ and $\delta = \frac{\bar{R}}{\bar{\Lambda}}$. In the sequel we assume that the following long-wave hypothesis holds true:

$$\varepsilon << 1, \quad \delta^2 = k\varepsilon, \quad k = O(1). \tag{2.11}$$

In vivo, the ratios ε and δ vary considerably, depending on the vessel type; thus the subsequent analysis and the model derived herein are valid only on scales compatible with those satisfying (2.11). As a typical example of an artery under consideration is the brachial artery, with typical radius $\bar{R} = 0.3$ mm and ratio $\varepsilon = 0.1$ and $\delta = 0.4$.

Rescaling the axial and radial velocity and the pressure,

$$\bar{u} = \varepsilon \bar{c}_{\mathrm{MK}} u, \qquad \bar{v} = \varepsilon \bar{c}_{\mathrm{MK}} \delta v, \tag{2.12}$$

$$\bar{p} - \bar{p}_{\mathrm{yo}} = \varepsilon \rho \bar{c}_{\mathrm{MK}}^{2}\, p, \tag{2.13}$$

then the non-dimensional variables $u = u(t, x, r)$, $v = v(t, x, r)$ and $p = p(t, x, r)$ satisfy the system

$$u_t + \varepsilon u\, u_x + \varepsilon v\, u_r + p_x = 0, \tag{2.14}$$

$$\delta^2 \left[v_t + \varepsilon u\, v_x + \varepsilon v\, v_r \right] + p_r = 0, \tag{2.15}$$

$$u_x + v_r + \frac{1}{r} v = 0, \tag{2.16}$$

in the region $0 \leq r \leq r^w(t, x) = r_{\mathrm{o}}(x) + \varepsilon\eta(t, x)$, with the boundary conditions

$$v = \eta_t + r_{\mathrm{o}x}(x) u + \varepsilon \eta_x u, \tag{2.17}$$

$$\frac{\rho^w h}{\rho R} \delta^2\, \eta_{tt} = p^w - 2\frac{E(x)}{r_{\mathrm{o}}(x)^2}\eta, \tag{2.18}$$

whenever $r = r^w(t, x) = r_{\mathrm{o}}(x) + \varepsilon\eta(t, x)$. Here $p^w(t, x) = p(t, x, r^w(t, x))$ and $E(x) = \frac{\bar{E}_\sigma(x)}{\bar{E}}$ is non-dimensional quantity depending on the axial position x. In the case of a tube with same elasticity throughout its length, one can choose $E(x) \equiv 1$. Finally, the axisymmetry assumption translates into

$$v = 0, \qquad \text{whenever } r = 0. \tag{2.19}$$

Next, consider new independent variables τ and ξ, similar to those used in [13], to account for the variable "landscape" (radius and elasticity) in which the wave propagation occurs, namely

$$\tau = \varepsilon x, \quad \xi = G(x; \varepsilon) - t. \tag{2.20}$$

The choice for $G(x;\varepsilon)$ will be indicated below. The change of variables implies

$$\partial_t = -\partial_\xi, \tag{2.21}$$
$$\partial_x = \varepsilon\partial_\tau + g(\tau)\partial_\xi, \tag{2.22}$$

where $g = \frac{dG}{dx}$. Thus, the new set of equations satisfied by the velocity, pressure and displacement in terms of the new variables, $u = u(\tau, \xi, r)$, $v = v(\tau, \xi, r)$, $p = p(\tau, \xi, r)$ and $\eta = \eta(\tau, \xi)$, is

$$-u_\xi + \varepsilon u\left(\varepsilon u_\tau + g(\tau)u_\xi\right) + \varepsilon v u_r + \varepsilon p_\tau + g(\tau)p_\xi = 0, \tag{2.23}$$
$$k\varepsilon\left[-v_\xi + \varepsilon u\left(\varepsilon v_\tau + g(\tau)v_\xi\right) + \varepsilon v v_r\right] + p_r = 0, \tag{2.24}$$
$$\varepsilon u_\tau + g(\tau)u_\xi + v_r + \frac{1}{r}v = 0, \tag{2.25}$$

whenever $0 \le r \le r_o(\tau) + \varepsilon\eta(\tau, \xi)$, together with

$$v = -\eta_\xi + \varepsilon r_{o\tau} u + \varepsilon\left(\varepsilon\eta_\tau + g(\tau)\eta_\xi\right)u, \tag{2.26}$$
$$\gamma k\varepsilon\eta_{\xi\xi} = p^w - 2\frac{E(\tau)}{r_o(\tau)^2}\eta. \tag{2.27}$$

whenever $r = r_o(\tau) + \varepsilon\eta(\tau, \xi)$. For convenience, the notation $\gamma = \frac{\rho^w h}{\rho R}$ was introduced. In addition,

$$v = 0, \text{ when } r = 0. \tag{2.28}$$

A standard method of solving the system (2.23)–(2.28) is to look for solutions η, u and v (similarly for p) in the form

$$\eta(\tau, \xi; \varepsilon) = \eta_0(\tau, \xi) + \varepsilon\eta_1(\tau, \xi) + o(\varepsilon), \tag{2.29}$$
$$u(\tau, \xi, r; \varepsilon) = u_0(\tau, \xi, r) + \varepsilon u_1(\tau, \xi, r) + o(\varepsilon), \tag{2.30}$$
$$v(\tau, \xi, r; \varepsilon) = v_0(\tau, \xi, r) + \varepsilon v_1(\tau, \xi, r) + o(\varepsilon). \tag{2.31}$$

The zero-order terms must satisfy the system

$$\begin{cases} -u_{0\xi} + g(\tau)p_{0\xi} = 0 \\ p_{0r} = 0 & \quad 0 \le r \le r_o(\tau) \\ g(\tau)u_{0\xi} + v_{0r} + \frac{1}{r}v_0 = 0 \end{cases} \tag{2.32}$$

and

$$\begin{cases} v_0 = -\eta_{0\xi} & \quad r = r_o(\tau) \\ 0 = p_0 - 2\frac{E(\tau)}{r_o(\tau)^2}\eta_0. \end{cases} \tag{2.33}$$

In addition, $v_0 = 0$ when $r = 0$. From (2.32) it follows that p_0 is independent of r,

$$p_0(\tau,\xi,r) = p_0^w(\tau,\xi) = 2\frac{E(\tau)}{r_o(\tau)^2}\eta_0, \qquad \text{for all } 0 \le r \le r_o(\tau). \tag{2.34}$$

After a few computations, one obtains

$$v_0 = -rg(\tau)^2\frac{E(\tau)}{r_o(\tau)^2}\eta_{0\xi}, \qquad \text{for all } 0 \le r \le r_o(\tau). \tag{2.35}$$

Note that v_0 depends linearly on r. To satisfy the condition (2.33) at $r = r_o(\tau)$, the following must hold

$$g(\tau) = \frac{r_o(\tau)^{1/2}}{E(\tau)^{1/2}}, \tag{2.36}$$

which is equivalent to choosing (in (2.20))

$$G(x;\varepsilon) = \int_0^x \frac{r_o(\varepsilon x')^{1/2}}{E(\varepsilon x')^{1/2}}dx' = \frac{1}{\varepsilon}\int_0^{\varepsilon x}\frac{r_o(y)^{1/2}}{E(y)^{1/2}}dy. \tag{2.37}$$

It is worth mentioning that in the case of an uniform tube (same radius and elasticity along its length), then $G(x;\varepsilon) = gx$ for some constant g, which means that our choice of the variable ξ in (2.20) represents the moving frame (to the right) with constant speed $c = 1/g$.

Thus far, the following relations hold true for the zero-order terms;

$$v_0 = -\frac{r}{r_o(\tau)}\eta_{0\xi}, \tag{2.38}$$

$$p_0 = 2\frac{E(\tau)}{r_o(\tau)^2}\eta_0, \tag{2.39}$$

$$u_{0\xi} = 2\frac{E(\tau)^{1/2}}{r_o(\tau)^{3/2}}\eta_{0\xi}. \tag{2.40}$$

We note that $u_{0\xi}$ is also independent of r, although u_0 is not.

The first-order $O(\varepsilon)$ terms, evaluated for $r = r_o(\tau)$, obey the system of equations

$$\begin{cases} -u_{1\xi} + g(\tau)u_0u_{0\xi} + v_0u_{0r} + p_{0\tau} + g(\tau)p_{1\xi} = 0, \\ -kv_{0\xi} + p_{1r} = 0, \\ u_{0\tau} + g(\tau)u_{1\xi} + v_{1r} + \frac{1}{r}v_1 = 0 \end{cases} \qquad 0 \le r \le r_o(\tau) \tag{2.41}$$

and

$$\begin{cases} \eta_0v_{0r} + v_1 = -\eta_{1\xi} + r_{o\tau}(\tau)u_0 + g(\tau)u_0\eta_{0\xi}, & r = r_o(\tau) \\ \gamma k\eta_{0\xi\xi} = p_1^o - 2\frac{E(\tau)}{r_o(\tau)^2}\eta_1. \end{cases} \tag{2.42}$$

Additionally, $v_1 = 0$ when $r = 0$.

Our goal is to derive, from this system, equations for η_0 and u_0. First, (2.41) yields

$$p_{1r} = kv_{0\xi} = -k\frac{r}{r_o(\tau)}\eta_{0\xi\xi},$$

thus

$$p_1(\tau,\xi,r) = -k\frac{r^2}{2r_o(\tau)}\eta_{0\xi\xi} + p_1(\tau,\xi,0), \tag{2.43}$$

which implies (setting $r = r_o(\tau)$)

$$p_1^o(\tau,\xi) = -k\frac{r_o(\tau)}{2}\eta_{0\xi\xi} + p_1(\tau,\xi,0).$$

From (2.42) we can write

$$-\eta_1 = \frac{r_o(\tau)^2}{2E(\tau)}\left[\gamma + \frac{r_o(\tau)}{2}\right]k\eta_{0\xi\xi} - \frac{r_o(\tau)^2}{2E(\tau)}p_1\big|_{r=0}.$$

and

$$\begin{aligned} v_1^o =& \frac{1}{r_o(\tau)}\eta_0\eta_{0\xi} + \frac{r_o(\tau)^2}{2E(\tau)}\left[\gamma + \frac{r_o(\tau)}{2}\right]k\eta_{0\xi\xi\xi} \\ &- \frac{r_o(\tau)^2}{2E(\tau)}p_{1\xi}\big|_{r=0} + r_{o\tau}(\tau)u_0^o + \frac{r_o(\tau)^{1/2}}{E(\tau)^{1/2}}u_0^o\eta_{0\xi}. \end{aligned} \tag{2.44}$$

We now turn to (2.41). Eliminating $u_{1\xi}$ yields

$$\begin{aligned} \frac{1}{r}(rv_1)_r =& -u_{0\tau} - \frac{r_o(\tau)}{E(\tau)}u_0u_{0\xi} + \frac{r}{r_o(\tau)^{1/2}E(\tau)^{1/2}}u_{0r}\eta_{0\xi} \\ &- 2\frac{r_o(\tau)^{1/2}}{E(\tau)^{1/2}}\left[\frac{E(\tau)}{r_o(\tau)^2}\eta_0\right]_\tau - \frac{r_o(\tau)}{E(\tau)}p_{1\xi}. \end{aligned} \tag{2.45}$$

Let

$$Q_0 = Q_0(\tau,\xi) = \int_0^{r_o(\tau)} ru_0(\tau,\xi,r)dr \tag{2.46}$$

be the "zero-order" flux at "time" τ and "position" ξ. Solving (2.45) for v_1 and substituting $r = r_o(\tau)$, ($v_1^o = v_1\big|_{r=r_o(\tau)}$), we obtain

$$\begin{aligned} v_1^o =& -\frac{1}{r_o(\tau)}Q_{0\tau} + r_{o\tau}(\tau)u_0^o - \frac{4}{r_o(\tau)^{3/2}E(\tau)^{1/2}}Q_0\eta_{0\xi} \\ &+ \frac{r_o(\tau)^{1/2}}{E(\tau)^{1/2}}u_0^o\eta_{0\xi} - \frac{r_o(\tau)^{3/2}}{E(\tau)^{1/2}}\left[\frac{E(\tau)}{r_o(\tau)^2}\eta_0\right]_\tau \\ &+ k\frac{r_o(\tau)^3}{8E(\tau)}\eta_{0\xi\xi\xi} - \frac{r_o(\tau)^2}{2E(\tau)}p_{1\xi}\big|_{r=0}. \end{aligned} \tag{2.47}$$

Comparing this with the expression (2.44) we obtain

$$\begin{aligned} &\frac{E(\tau)^{1/2}}{r_o(\tau)^{1/2}}\eta_{0\tau} + \frac{r_o(\tau)^{3/2}}{E(\tau)^{1/2}}\left[\frac{E}{r_o(\tau)^2}\right]_\tau\eta_0 + \frac{1}{r_o(\tau)}\eta_0\eta_{0\xi} + k\frac{r_o(\tau)^2}{2E(\tau)}\left[\gamma + \frac{r_o(\tau)}{4}\right]\eta_{0\xi\xi\xi} \\ &\quad + \frac{1}{r_o(\tau)}Q_{0\tau} + \frac{4}{r_o(\tau)^{3/2}E(\tau)^{1/2}}Q_0\eta_{0\xi} = 0. \end{aligned} \tag{2.48}$$

Now, since $u_{0\xi}$ is independent of r (see (2.40)), we have

$$Q_{0\xi} = \int_0^{r_o(\tau)} r dr\, u_{0\xi} = E(\tau)^{1/2} r_o(\tau)^{1/2} \eta_{0\xi},$$

therefore

$$Q_0(\tau,\xi) = E(\tau)^{1/2} r_o(\tau)^{1/2} \eta_0(\tau,\xi) + f(\tau), \tag{2.49}$$

where $f(\tau)$ is independent of ξ. This is an important observation; it allows us to determine $f(\tau)$ from the initial state of the tube, by setting $t = 0$ or, equivalently, $\xi = G(x;\varepsilon)$. Once f is known, the equation governing the τ-evolution of the wall displacement reads

$$\eta_{0\tau} + \frac{3}{4}\left[\frac{E_\tau(\tau)}{E(\tau)} - \frac{r_{o\tau}(\tau)}{r_o(\tau)}\right]\eta_0 + \frac{5}{2}\frac{1}{r_o(\tau)^{1/2}E(\tau)^{1/2}}\eta_0\eta_{0\xi}$$
$$+ \frac{k}{2}\frac{r_o(\tau)^{5/2}}{E(\tau)^{3/2}}\left(\gamma + \frac{r_o}{4}\right)\eta_{0\xi\xi\xi} + \frac{1}{2r_o(\tau)^{1/2}E(\tau)^{1/2}}f'(\tau) + \frac{2}{r_o(\tau)E(\tau)}f(\tau)\eta_{0\xi} = 0. \tag{2.50}$$

This is a variable coefficient Korteweg-de Vries equation with a forcing term (both in wave amplitude and in wave velocity) which depends on the initial state of the tube.

REMARK 2.1. (i) As noted earlier, if one considers that the tube is initially in a quiescent state (zero flux), then in the above equation one has $f(\tau) = 0$, for all τ.

(ii) In the case of an uniformly elastic tube, when the elasticity does not change along the tube one can take without loss of generality $E(\tau) \equiv 1$ and thus obtain

$$\eta_{0\tau} - \frac{3}{4}\frac{r_{o\tau}(\tau)}{r_o(\tau)}\eta_0 + \frac{5}{2r_o(\tau)^{1/2}}\eta_0\eta_{0\xi} + \frac{k}{2}r_o(\tau)^{5/2}\left(\gamma + \frac{r_o(\tau)}{4}\right)\eta_{0\xi\xi\xi} = 0. \tag{2.51}$$

If, in addition, the radius of the unstressed tube remains constant, then $r_o(\tau) \equiv r_o$ and the equation reads

$$\eta_{0\tau} + \sigma\eta_0\eta_{0\xi} + \kappa\eta_{0\xi\xi\xi} = 0, \tag{2.52}$$

with $\sigma = \frac{5}{2{r_o}^{1/2}}$, $\kappa = \frac{k}{2}{r_o}^{5/2}\left(\gamma + \frac{r_o}{4}\right)$. This is the classical (constant coeficient) KdV equation, derived in the context of elastic tubes in [7], [19], [20].

(iii) Another special case worth mentioning is when the elasticity of the tube is proportional to the radius of the cross-section (without loss of generality, taking $E(\tau) = r_o(\tau)$.)

3 INITIAL VALUE PROBLEM

In this section we will restrict our attention to the equation (2.50) with the forcing term $f(\tau) = 0$, and show that under appropriate conditions, the corresponding initial value problem is well-posed. We will consider that the walls are thin compared to the radius of the tube and thus we can take $\gamma = 0$. (this is only for simplicity of the formulas.)

Let q be defined by

$$\eta_0(\tau,\xi) = \frac{r_o(\tau)^{3/4}}{E(\tau)^{3/4}} q(\tau,\xi), \tag{3.1}$$

so that q satisfies the equation

$$q_\tau + \frac{5}{2}\frac{r_o(\tau)^{1/4}}{E(\tau)^{5/4}} q q_\xi + \frac{k}{8}\frac{r_o(\tau)^{7/2}}{E(\tau)^{3/2}} q_{\xi\xi\xi} = 0. \tag{3.2}$$

Introducing the new "time" variable τ' satisfying

$$\frac{d\tau'}{d\tau} = \frac{5}{2}\frac{r_o(\tau)^{1/4}}{E(\tau)^{5/4}}, \tag{3.3}$$

we finally obtain,

$$q_{\tau'} + q q_\xi + \frac{k}{20}\frac{r_o(\tau')^{13/4}}{E(\tau')^{1/4}} q_{\xi\xi\xi} = 0. \tag{3.4}$$

Denote the time dependent coefficient, appearing in the equation above, by $\alpha(\tau') = \frac{k}{20}\frac{r_o(\tau')^{13/4}}{E(\tau')^{1/4}}$. Then one has the variable coefficient KdV equation (we write, for simplicity, τ instead of τ')

$$q_\tau + q q_\xi + \alpha(\tau) q_{\xi\xi\xi} = 0. \tag{3.5}$$

Often "constant coefficient" problems are easier to solve, but spatially dependent coefficients appear more often in models coming from the physical world, when the phenomenon under consideration occurs on a variable landscape. The coefficient $\alpha(\tau)$ in (3.5) looks like a real valued function of time alone, but it is actually spatially dependent, since τ is a variable that comes from rescaling x. This way of writing the problem leads to significant simplifications.

We assume that the original system is periodic in time (i.e. laboratory time t) with period T, that is we assume the system is in basal condition. Therefore the same is true for the system described in the new coordinate ξ. Under this assumption, all solutions of the equation (3.5) are $\xi-$periodic

$$q(\tau,\xi) = q(\tau,\xi+T), \text{ for all } \tau,\xi. \tag{3.6}$$

The initial data for (3.5) corresponds to $\tau = 0$, which, in laboratory coordinates, means $x = 0$, i.e. the inflow condition.

$$q(\tau=0,\xi) = q_0(\xi)\ [= q_0(-t, x=0)]. \tag{3.7}$$

One can verify that the following functional is invariant under the flow governed by equation (3.5).

$$\varphi_0(\tau,q) = \frac{1}{2}\int_{\mathbb{T}} q(\tau,\xi)^2 d\xi \tag{3.8}$$

Consider also the functionals

$$\varphi_1(\tau,q) = \frac{1}{2}\int_{\mathbb{T}} \left[\alpha(\tau) q_\xi(\tau,\xi)^2 - \frac{1}{3}q(\tau,\xi)^3\right] d\xi \tag{3.9}$$

and

$$\varphi_2(\tau, q) = \frac{1}{2} \int_{\mathbb{T}} \left[\alpha(\tau) q_{\xi\xi}(\tau, \xi)^2 + \beta(\tau) q(\tau, \xi)^2 q_{\xi\xi}(\tau, \xi) + \frac{\beta(\tau)}{6\alpha(\tau)} q(\tau, \xi)^4 \right] d\xi. \quad (3.10)$$

For our purposes, the choice of the funtion $\beta(t)$ will be

$$\beta(\tau) = \frac{\alpha(\tau) + 4}{\alpha(\tau) + 5}.$$

A formal computation (assuming that $q = q(\tau, \xi)$ is a solution of the initial-value problem (3.5)–(3.7) and all derivatives of q involved exist), yields

$$\frac{d}{d\tau} \varphi_1(\tau, q) = \frac{1}{2} \alpha'(\tau) \int_{\mathbb{T}} q_\xi^2 \quad (3.11)$$

and, similarly,

$$\frac{d}{d\tau} \varphi_2(\tau, q) = \frac{1}{2} \alpha'(\tau) \int_{\mathbb{T}} q_{\xi\xi}^2 + \gamma'(\tau) \int_{\mathbb{T}} q^2 q_{\xi\xi} + \left(\frac{\gamma(\tau)}{6\alpha(\tau)} \right)' \int_{\mathbb{T}} q^4. \quad (3.12)$$

Thus a sufficient condition for controlling the growth of the functionals φ_1 and φ_2 is

$$\alpha'(\tau) \le 0. \quad (3.13)$$

The main result of this paper is a generalization of the results of [5] and [6], to the nonautonomous case.

THEOREM 3.1. *Assume that* $\alpha = \alpha(\tau) \in W^{1,\infty}$ *is defined on some (possibly infinite) interval* $\tau \in [0, L]$. *Then the initial value problem (3.5)–(3.7) is well-posed in the space* $H^2(\mathbb{T})$. *If, in addition,* $q_0 \in H^3(\mathbb{T})$, *then* $q(\tau, \xi)$ *is a classical solution. Moreover, if* $\alpha(\tau)$ *is positive, nonincreasing function, then the solution exists globally in time* τ.

The semilinear Hille-Yosida theory presented in [5] and [6] finds a stongly continuous semigroup solution to the autonomous problem

$$\frac{dq}{d\tau} = \mathcal{A}q, \quad q(0) = f, \quad (3.14)$$

in circumstances in which the Crandall-Liggett theory (and various others as well) does not apply. For the nonautonomous case,

$$\frac{dq}{d\tau} = \mathcal{A}(\tau)u, \quad q(s) = f, \quad (3.15)$$

$\mathcal{A}(\tau)$ can be approximated on $[s, \tau]$ by $\mathcal{A}_\pi(\tau) = \mathcal{A}(\tilde{\tau}_i)$ on $[\tau_{i-1}, \tau_i)$, for a given partition π, $s = \tau_0 < \tau_1 < \ldots < \tau_n = \tau$, and a choice of $\tilde{\tau}_i \in [\tau_{i-1}, \tau_i)$. Thus

$$q_\pi(\tau) = \prod_{j=1}^{n} T(\tau_j - \tau_{j-1}; A(\tilde{\tau}_j))\, f, \quad (3.16)$$

where $T(s; \mathcal{A}(\tau))$ is the semigroup generated by $\mathcal{A}(\tau)$ and initial time s, and the product is ordered so that $\prod_{j=1}^{n} S_j f$ means $S_n S_{n-1} \dots S_2 S_1 f$. Finally one shows that the solution of the IVP in the nonautonomous case is

$$q(t) = \lim q_\pi(t)$$

where the limit is taken as the mesh of π $(= \max_j(\tau_j - \tau_{j-1}))$ tends to zero.

The proof of our theorem relies upon writing the IVP as the abstract Cauchy problem (on the Hilbert space $\mathcal{H} = L^2(\mathbb{T})$)

$$\frac{dq}{d\tau}(\tau) = \alpha(\tau)Aq(\tau) + B(q(\tau)) \ [=: \mathcal{A}(\tau)q]\,, \tag{3.17}$$

$$q(0) = q_0 \in \mathcal{H}, \tag{3.18}$$

where $A = -\partial_\xi^3$ generates a C_0 semigroup on $\mathcal{H}$ and $B(u) = -u\partial_\xi u$ is a nonlinear operator satisfying a local quasi-dissipative condition. "Local" here means that the quasi-dissipative constant depends on the level set of the functionals φ_0, φ_1 and φ_2. The estimates on the growth of these functionals along solutions follow from (3.11) and (3.12), using Holder inequalities. More precisely, for all τ,

$$\varphi_0(\tau, q) = \varphi_0(0, q_0), \tag{3.19}$$

$$\varphi_1(\tau, q) \leq \varphi_1(0, q_0), \tag{3.20}$$

$$\varphi_2(\tau, q) \leq e^{\omega\tau}\varphi_2(0, q_0), \tag{3.21}$$

(3.22)

where $\omega = \omega(||q_0||_{H^1(\mathbb{T})})$. The last inequality allows us to extend solution for all "times" τ, thus obtaining global solutions. The solution itself is obtained as a limit of approximate solutions obtained by discretizing the "time". Thus it will be sufficient to solve (for q) the resolvent equation

$$q - \lambda\alpha Aq = \lambda B(q) + p \tag{3.23}$$

for every $p \in \mathcal{H}$, and for sufficiently small $\lambda > 0$. This can be done via a fixed point argument similar to the one presented in [5].

ACKNOWLEDGEMENT. The author would like to thank the referee for many useful remarks, which helped improve the paper considerably.

BIBLIOGRAPHY

1. J. L. Bona and L. Luo, *A generalized korteweg-de vries equation in a quarter plane*, Cont. Math. **221** (1999), 59–125.

2. J. L. Bona, S. M. Sun, and B. Y. Zhang, *A non-homogeneous boundary-value problem for the Korteweg-de Vries equation in a quarter plane*, Trans. AMS **354** (2001), 427–490.

3. J. L. Bona and R. Winther, *The Korteweg-de Vries equation, posed in a quarter-plane*, SIAM J. Math. Anal. **14** (1983), 1056–1106.

4. R. Camassa, D. Holm, and C. D. Levermore, *Long-time shallow water equations with a varying bottom*, J. Fluid. Mech. **349** (1997), 173–189.

5. R. C. Cascaval, *Global well-posedness for a class of dispersive equations*, Ph.D. thesis, The University of Memphis, Memphis, TN, 2000.

6. R. C. Cascaval and J. A. Goldstein, *A semigroup approach to dispersive waves*, Evolution Equations and Their Applications in Physical and Life Sciences (New York), Marcel Dekker, 2001, pp. 225–233.

7. H. Demiray, *Solitary waves in prestressed elastic tubes*, Bull. Math. Biol. **58** (1996), 939–955.

8. M. Epstein and C. Johnston, *Improved solution for solitary waves in arteries*, J. Math. Biol. **39** (1999), 1–18.

9. L. Formaggia, J. F. Gerbeau, F. Nobile, and A. Quarteroni, *On the coupling of 3D and 1D Navier-Stokes equations for flow problems in compliant vessels*, Comp. Methods in Appl. Mech. Eng. **191** (2001), 561–582.

10. L. Formaggia, F. Nobile, A. Quarteroni, and A. Veneziani, *Multiscale modelling of the circulatory system: a preliminary analysis*, Comput. Visual Sci. **2** (1999), 75–83.

11. J. A. Goldstein, *Semigroups of linear operators and applications*, Oxford University Press, Oxford, New York, 1985.

12. J. A. Goldstein, *The KdV equation via semigroups*, Theory and Applications of Non-linear Operators of Accretive and Monotone Type (New York) (A. Kartsatos, ed.), Marcel Dekker, 1996, pp. 107–114.

13. R. S. Johnson, *On the developement of a solitary wave moving over an uneven bottom*, Proc. Camb. Phil. Soc. **73** (1973), 183–203.

14. A. C. Newell, *Solitons in mathematics and physics*, SIAM, Philadelphia, 1985.

15. M. Olufsen, *A structured tree outflow condition for blood flow in the larger systemic arteries*, Am. J. Physiol. **276** (1999), 257–268.

16. M. Olufsen, *A one-dimensional fluid dynamic model of the systemic arteries*, Stud. Health Technol. Inform. **71** (2000), 79–97.

17. A. Quarteroni, M. Tuveri, and A. Veneziani, *Computational vascular fluid dynamics: problems, models and methods*, Comp. Visual Sci. **2** (2000), 163–197.

18. S. Tsutsui, K. Suzuyama, and H. Ohki, *Model equations of nonlinear dispersive waves in shallow water and an application of its simplified version to wave evolution on the step-type reef*, Coastal Eng. J. **40** (1998), 41–60.

19. A. N. Volobuev, *Fluid flow in tubes with elastic walls*, Physics Uspekhi **38** (1995), 169–185.

20. S. Yomosa, *Solitary waves in large blood vessels*, J. Phys. Soc. Japan **56** (1987), 506–520.

21. M. Zamir, *The physics of pulsatile flow*, Springer, New York, 2000.

Infinitely many solutions for a superlinear Neumann problem in tileable regions

ALFONSO CASTRO

Department of Applied Mathematics
The University of Texas at San Antonio
San Antonio TX 78249-0664
e-mail: castro@math.utsa.edu

ABSTRACT. We establish the existence of infinitely many solutions to a superlinear Neumann problem on regions that can be tiled by regular prisms. Our proofs take advantage of recent results on the existence of sign-changing solutions for superlinear problems. The nonlinearities under consideration are not perturbations of odd nonlinearities where Liusternik-Schnirelman arguments may be applied.

1 INTRODUCTION.

Let Δ the Laplacian operator, Ω a bounded region in $\mathbf{R}^N$, and $f \in C^1(\mathbf{R}, \mathbf{R})$ such that $f(0) = 0$. We consider the Neumann problem

$$\begin{cases} \Delta u + f(u) = 0 & \text{in } \Omega \\ \frac{\partial u}{\partial \eta} = 0 & \text{in } \partial\Omega. \end{cases} \tag{1.1}$$

We assume that there exist constants $A > 0$ and $p \in (1, \frac{N+2}{N-2})$ such that $|f'(u)| \leq A(|u|^{p-1} + 1)$ for all $u \in \mathbf{R}$. Hence f is subcritical, i.e., there exists $B > 0$ such that $|f(u)| \leq B(|u|^p + 1)$. Also, we assume that there exists $m \in (0, 1)$ and $\rho > 0$ such that

$$muf(u) \geq 2F(u) \tag{1.2}$$

for $|u| > \rho$, where $F(u) = \int_0^u f(s)\, ds$. Finally, we make the assumption that f satisfies

$$f'(u) > \frac{f(u)}{u} \quad \text{for } u \neq 0, \quad \text{and} \quad \lim_{|u| \to \infty} \frac{f(u)}{u} = \infty \quad (f \text{ is superlinear}). \tag{1.3}$$

The reader can easily verify that if f is defined by $f(u) = |u|^{p-1}u$ for $u \geq 0$ and $f(u) = |u|^{q-1}u$ for $u < 0$, then f satisfies the above assumptions even when $p \neq q$.
Our main result is result:

THEOREM 1.1. *If Ω is a bounded region in $\mathbf{R}^N$ that can be tiled by prims congruent to*

$$\begin{aligned} \{(x_1, \ldots, x_n);\ & x_i \geq 0 \textit{ for } i = 1, \ldots, n,\ x_n \leq x_i, \\ & x_n \leq a - x_i \textit{ for } i = 1, \ldots, n-1\}, \end{aligned} \tag{1.4}$$

with $a > 0$ arbitrarily small, then the problem (1.1) *has infinitely many sign-changing solutions that satisfy*

$$\frac{\partial u}{\partial x_i} \not\equiv 0 \textit{ for } i = 1, \ldots, n. \tag{1.5}$$

Examples of regions Ω satisfying the hypothesis of Theorem 1.1 include regions that can be tiled by a hypercube or regions in $\mathbf{R}^2$ that can be tiled by an isosceles right triangle. In fact, a hypercube can be tiled by arbitrarily small hypercubes which in turn can tiled by a prims. Also, any isosceles right triangles can also be tiled by arbitrarily small isosceles right triangles.

Our proofs take advantage of recent results on the existence of sign-changing solutions for superlinear problems. Let $0 = \lambda_1 < \lambda_2 \leq \lambda_3 \leq \cdots$ be the eigenvalues of $-\Delta$ with zero Neumann boundary condition in Ω. Obvious modifications of Theorem 1.1 of [4] imply that if $f'(0) < 0$ then in any bounded region Ω with Lipschitzian boundary the problem (1.1) has a solution that changes sign at least once. Similarly, from Theorem 1.1 of [6] it follows that if $f'(0) \in [\lambda_k, \lambda_{k+1})$ the problem (1.1) has a solution that changes sign and has no more that $k+1$ nodal regions. For future reference we summarize the above discussion in the following statement.

LEMMA 1.2. *If Ω is a bounded region with Lipschitzian boundary the problem (1.1) has a sign-changing solution.*

Unlike results derived from the study of perturbed even functionals (see [2]) our nonlinearities need not be sublinear perturbations of superlinear odd functionals. For historical remarks concerning the existence of sign changing solutions to semilinear elliptic boundary value problems we refer the reader to [5]. See also [10].

ACKNOWLEDGEMENT. The author thanks his colleagues of the Applied Mathematics Department at the University of Texas in San Antonio for their valuable suggestions.

2 PROOF OF MAIN THEOREM.

Since (1.1) is invariant under rotations and translations we may assume that there exists $a > 0$ such Ω can be tiled by

$$\begin{aligned} W_1 = \{(x_1, \ldots, x_n); x_i \geq 0 \text{ for } i = 1, \ldots, n,\ \ x_n \leq x_i, \\ x_n \leq a - x_i \text{ for } i = 1, \ldots, n-1\} \subset \Omega. \end{aligned}$$

Let u_1 be a sign-changing solution to (1.1) in W_1. Since symmetric extensions of u_1 to reflections of W_1 across its faces also satisfy (1.1), and Ω is the union of finitely many of these reflections we see that u_1 can be extended to a solution to (1) on Ω. We denote such an extension also by u_1.

Let us see that $u_1 \equiv u$ satisfies (1.5). We prove this by mathematical induction on n.

Let $n = 2$ and suppose that $\partial u/\partial x_1 \equiv 0$. Hence $u(x_1, x_2) \equiv v(x_2)$. Since $0 = (\partial u/\partial \eta)(t,t) = \sqrt{2}[\partial u/\partial x_1)(t,t) - (\partial u/\partial x_2)(t,t)] = \sqrt{2}v'(t)$ for all $t \in [0, a/2]$, we see that v is a constant function. Hence u is also a constant function. Since this contradict that u is a sign changing function we conclude that $\partial u/\partial x_1 \not\equiv 0$. Similar argument shows that $\partial u/\partial x_2 \equiv 0$. This proof the result for $n = 2$

Let $n > 2$. Let (1.5) hold for prisms of dimension less than n. Assuming that $\partial u/\partial x_i \equiv 0$ for some $i = 1, \ldots, n-1$, the function

$$v(x_1, \ldots, x_{n-1}) \equiv u(x_1, \ldots, x_{i-1}, a/2, x_i, \ldots, x_{n-1}) \tag{2.1}$$

satisfies the equation (1.1) in a prism W_2 of dimension $n - 1$. The prism W_2 is congruent to the intersection of W_1 with the hyperplane $x_i = a/2$. By induction hypothesis there exists $x \in W_2$ such that $(\partial v/\partial x_{n-1})(x) \neq 0$ Let $\hat{x} \in W_2$ be the element resulting from inserting in x the number $a/2$ in the ith-component. Let $\tilde{x}$ be the element of W_1 resulting from replacing in $\hat{x}$ the i-th component by its n-th component. Hence $\hat{x} \in \partial W_1$ and

$$\frac{\partial u}{\partial \eta}(\tilde{x}) = \frac{\partial u}{\partial x_n}(\tilde{x}) - \frac{\partial u}{\partial x_i}(\tilde{x}) = \frac{\partial u}{\partial x_n}(\tilde{x}) = \frac{\partial v}{\partial x_n}(\tilde{x}) \neq 0. \tag{2.2}$$

Since this contradicts the boundary condition in (1.1) we conclude that $\partial u/\partial x_i \not\equiv 0$, for any $i = 1, \ldots, n-1$.

Suppose now that $\partial u/\partial x_n \equiv 0$. Hence there exits $v : [0,a]^{n-1} \to \mathbf{R}$ such that $v(x_1, \ldots, x_{n-1}) = u(x_1, \ldots, x_{n-1}, x_n)$. For $\epsilon \in (0, a/2)$ let $R_\epsilon = [(a/2) - \epsilon, (a/2) + \epsilon]^{n-1}$. For $x_2, x_{n-1} \in ((a/2) - \epsilon, (a/2) + \epsilon)$, $\alpha \equiv ((a/2) \pm \epsilon, x_2, \ldots, x_{n-1}, (a/2) \pm \epsilon) \in \partial W_1$. Hence

$$\begin{aligned} 0 &= (\partial u/\partial \eta)(\alpha) = (\partial u/\partial x_n)(\alpha) - (\partial u/\partial x_1)(\alpha) \\ &= (\partial v/\partial x_1)(((a/2) \pm \epsilon, x_2, \ldots, x_{n-1})). \end{aligned} \tag{2.3}$$

Similarly

$$(\partial v/\partial x_i)(x) = 0, \text{ for any } x \in \partial R_\epsilon. \tag{2.4}$$

Since u is independent of its n-th variable, $\partial^2 u/\partial x_n^2 = 0$. Thus v satisfies the Neumann problem

$$\begin{cases} \Delta v + f(v) = 0 & \text{in } R_\epsilon \\ \frac{\partial v}{\partial \eta} = 0 & \text{in } \partial R_\epsilon. \end{cases} \tag{2.5}$$

From (2.4) it follow that v is constant along any line of the form

$$\{((a/2), \ldots, (a/2), t, (a/2), \ldots, (a/2)); t \in (0, a)\}. \tag{2.6}$$

Let $z = (a/2, \ldots, (a/2))$. Hence $(\partial^2 v/\partial x_i^2)(x)$ for all $i = 1, \ldots, n-1$. Thus, by (2.5), $f(v(z)) = 0$. Let $M = \max\{f'(v(x)); \|x - z\| \leq .5\}$. By Poincarés inequality, for ϵ sufficiently small

$$\int_{R_\epsilon} (v - v(z))^2 dy \leq (.25/M) \int_{R_\epsilon} \|\nabla(v - v(z))\|^2 dy. \tag{2.7}$$

On the other hand multiplying in (2.5) by $v - v(z)$ and using that $f(v(z)) = 0$ we have

$$\begin{aligned}\int_{R_\epsilon} \|\nabla(v - v(z))\|^2 dy &= \int_{R_\epsilon} (f(v(y)) - f(v(z)))(v - v(z))dy \\ &\leq M \int_{R_\epsilon} (v - v(z))^2 dy.\end{aligned} \tag{2.8}$$

From (2.7) and (2.8) we conclude that $v \equiv v(z)$ on R_ϵ. Hence by unique continuation properties of solutions to linear elliptic equations (see [8]) we have $v \equiv v(z)$ in W_1, which contradicts that v changes sign. Hence v satisfies (1.5).

Since u_1 changes sign and Ω can be tiled by arbitrarily small prisms, there exists a prim W_2 that tiles Ω and such that $W_2 \subset \{x \in W_1; u(x) > 0\}$. By Lemma 1.2, the problem (1.1) has a solution u_2 in W_2 that changes sign. Extending u_2 to Ω by successive reflections we obtain a solution to (1.1) that changes sign. For the same reasons that u_1 satisfies (1.5), u_2 also satisfies (1.5). That $u_2 \not\equiv u_1$ follows from the fact that u_1 is positive in W_2 while u_2 changes sign in W_2. Iterating this process we obtain a sequence $\{u_j\}_j$ of solutions to (1.1) which change sign and satisfy (1.1). This proves Theorem 1.1.

BIBLIOGRAPHY

1. R. Adams, *Sobolev spaces*, Academic Press, New York, 1975.

2. A. Bahri and P. L. Lions, *Morse index of some min-max critical points. I. Applications to multiplicity results*, Comm. Pure Appl. Math. **41** (1988), 1027–1037.

3. A. Castro, J. Cossio, and J. M. Neuberger, *On multiple solutions of a nonlinear Dirichlet problem*, Nonlinear Analysis TMA **30** (1997), no. 6, 3657–3662.

4. A. Castro, J. Cossio, and J. M. Neuberger, *A sign-changing solution for a superlinear Dirichlet problem*, Rocky Mountain J. of Math. **27** (1997), no. 4, 1041–1053.

5. A. Castro, J. Cossio, and J. M. Neuberger, *A minmax principle, index of the critical point, and existence of sign-changing solutions to elliptic boundary value problems*, Electronic J. of Diff. Eq. **1998** (1998), no. 2, 1–18.

6. A. Castro, P. Drabek, and J. M. Neuberger, *A sign-changing solution for a superlinear Dirichlet problems, II.*, (preprint).

7. D. Gilbarg and N. Trudinger, *Elliptic partial differential equations of second order*, Springer-Verlag, 1983.

8. D. Jerison and C. Kenig, *Unique continuation in the absence of positive eigenvalues for schrodinger operators*, Ann. of Math **12** (1985), no. 2, 463–494.

9. J. M. Neuberger, *A sign-changing solution for a superlinear dirichlet problem with a reaction term nonzero at zero*, Nonlinear Analysis **33** (1998), 427–441.

10. Z. Q. Wang, *On a superlinear elliptic equation*, Ann. Inst. H. Poincare Analyse Non Lineaire **8** (1991), 43–57.

On applications of maximal regularity to inverse problems for integrodifferential equations of parabolic type

FABRIZIO COLOMBO

Dipartimento di Matematica, Politecnico di Milano
Via Bonardo n.9, 20133 Milano, Italy
e-mail: fabcol@mate.polimi.it

DAVIDE GUIDETTI

Dipartimento di Matematica, Università
Piazza di Porta S. Donato 5, 40126 Bologna (Italy)
e-mail: guidetti@dm.unibo.it

ALFREDO LORENZI

Dipartimento di Matematica, Università degli Studi di Milano
Via Cesare Saldini 50, 20133 Milan (Italy)
e-mail: Alfredo.Lorenzi@mat.unimi.it

Dedicated to Jerome Goldstein and Rainer Nagel on occasion of their 60th birthday

0 INTRODUCTION

The study of maximal regularity for linear parabolic mixed problems systematically based on real interpolation theory started with the fundamental papers of Grisvard and Da Prato in the sixties and seventies (see, for example, [11] and [7]). For a more recent introduction to the subject, see also the book [20].

It is well known that maximal regularity results establish linear and topological isomorphisms between certain Banach spaces of functions and are very useful to treat nonlinear problems (see, for example, [8]). As inverse problems, even if concerning linear equations, are essentially nonlinear in nature, it is not surprising that maximal regularity can be very effective to study them.

It is the aim of this paper to illustrate some applications of maximal regularity to inverse problems for integrodifferential equations of parabolic type. So we start with some specific problems of the form that we want to consider.

In the linear theory of heat flow in a rigid isotropic body consisting of material with thermal memory the generalized heat equation

$$D_t u(t,x) = \mathcal{A}u(t,x) + \int_0^t m(t-s)\mathcal{A}u(s,x)ds + f(t,x) \tag{0.1}$$

holds in the cylinder $[0,T]\times O$, where O is a bounded domain in $\mathbf{R}^n$ with suitably regular boundary and $\mathcal{A}$ is a second order strongly elliptic operator (see [14], [15], [24]). Here u is the temperature of the body. Further, m is the memory kernel of heat flux and f the heat supply.

An analogous parabolic integrodifferential equation occurs in the theory of flow in porous media, where u now denotes the pressure and f the mass source density (see, for instance, [16]).

Besides equation (0.1), the function u is requested to satisfy the initial and boundary conditions

$$u(0,x) = u_0(x) \tag{0.2}$$

and

$$\mathcal{B}u(t,x) = g(t,x) \quad \text{on} \quad [0,T]\times\partial O. \tag{0.3}$$

Here $\mathcal{B}$ is a differential operator of order not exceeding one such that, for $p \in (1,+\infty)$, setting

$$D(A) := \{u \in W^{2,p}(O) : \mathcal{B}u_{|\partial O} = 0\}, Au = \mathcal{A}u \quad \forall u \in D(A),$$

A is the infinitesimal generator of an analytic semigroup in $L^p(O)$ (see [22], 3.7, 3.8).

In practise, it is very difficult to measure the memory kernel m. So we have an inverse problem if we want to determine, not only the unknown function u, but also the kernel m, which, in a traditional setting, is usually assumed to be one of the data of the problem. Of course, we need some further information. This is given by prescribing the value $\Psi[u(t)]$ for every $t \in [0,T]$, where Ψ is a suitable functional. For example, if u is a temperature, we can imagine to insert some sensor in a neighborhood of a point y inside O and to measure $u(t,y)$ for every $t \in [0,T]$. So we add to (0.1) – (0.3) the condition

$$\Psi[u(t,.)] = g(t), \quad t \in [0,T]. \tag{0.4}$$

The organization of the paper is the following: in the first section we consider linear problems of the form (0.1)-(0.4) together with some abstract generalizations of them. We examine also some cases where the memory kernel m may depend also on some of the space variables. We state first a very general theorem of maximal regularity, which will include all maximal regularity results used in this paper.

In the second section we consider some quite general nonlinear problems.

We conclude this introduction by pointing out some notation we shall use in the sequel.

If X and Y are Banach spaces, we shall indicate with $\mathcal{L}(X, Y)$ the Banach space of linear and bounded operators from X to Y. If $X = Y$, we shall simply write $\mathcal{L}(X)$. If Y coincides with the scalar field, we shall write X' instead of $\mathcal{L}(X, Y)$.

If A is a linear operator in the Banach space X, we shall indicate with $D(A)$ its domain and with $\rho(A)$ its resolvent set.

The symbol $*$ will be used to indicate the convolution in a bounded interval $[0, T]$.

If $A \subset \mathbf{R}^n$, we shall use the notation ∂A to denote its boundary.

In the following we shall mention also the Besov spaces $B^{\alpha}_{p,q}(O)$. For their definition and properties we refer to [23].

1 LINEAR PROBLEMS

We want to state the general maximal regularity result which will be used throughout the paper. We start with some well known definitions.

Let D and X be Banach spaces with norms $\|\cdot\|_D$ and $\|\cdot\|$ respectively, $D \subseteq X$ with continuous embedding, $A \in \mathcal{L}(D, X)$. We shall mainly think of A as an unbounded linear operator in X. Assume the following:

h1) there exist $R_0 > 0$ and $M > 0$ such that $\Sigma := \{\lambda \in \mathbf{C} : \ |\lambda| \geq R_0, |Arg\lambda| \leq \frac{\pi}{2}\} \subseteq \rho(A)$; moreover, for every $\lambda \in \Sigma$

$$||(\lambda - A)^{-1}||_{\mathcal{L}(X)} \leq M|\lambda|^{-1}.$$

It is well known (see, for example, [20]), that, if condition $h1)$ is satisfied, then A is the infinitesimal generator of an analytic semigroup $\{S(t)\}_{t \geq 0}$, possibly not strongly continuous at 0.

Let now $\alpha \in (0, 1)$ and $p \in [1, +\infty]$. We set

$$D_A(\alpha, p) := \{x \in X : t \to v(t) = ||t^{1-\alpha-1/p} AS(t)x|| \in L^p((0, 1))\} \tag{1.1}$$

and, if $x \in D_A(\alpha, p)$,

$$[x]_{D_A(\alpha,p)} := ||v||_{L^p((0,1))}, \tag{1.2}$$

$$||x||_{D_A(\alpha,p)} := ||x|| + [x]_{D_A(\alpha,p)}. \tag{1.3}$$

$D_A(\alpha, p)$ is a Banach space with the norm $||\cdot||_{D_A(\alpha,p)}$ and coincides, up to equivalent norms, with the real interpolation space $(X, D)_{\alpha,p}$ (see [20], proposition 2.2.2).

We introduce now, for $-\infty < a < b < +\infty$, $\beta \in (0, 1), p \in [1, +\infty]$, the fractional Sobolev spaces $W^{\beta,p}(a, b; X)$. We set

$$W^{\beta,p}(a, b; X) := \begin{cases} \{f \in L^p(a, b; X) : \ \int_a^b (\int_a^t \frac{||f(t)-f(s)||^p}{(t-s)^{1+\beta p}} ds) dt < +\infty\} \\ \qquad \text{if } 1 \leq p < +\infty, \\ \\ \sup\{\frac{||f(t)-f(s)||}{(t-s)^{\beta}} : \quad a \leq s < t \leq b\} < +\infty \\ \qquad \text{if } p = +\infty. \end{cases} \tag{1.4}$$

Observe that the space $W^{\beta,\infty}(a,b;X)$ coincides with the space $C^{\beta}([a,b];X)$ of Hölder continuous functions. We set also, for $k \in \mathbf{N}$,

$$\begin{aligned} &W^{k+\beta,p}(a,b;X) \\ &:= \{f \in W^{k,p}(a,b;X) | f^{(j)} \in W^{\beta,p}(a,b;X) \ \ \forall j \in \{1,...,k\}\}. \end{aligned} \tag{1.5}$$

Of course, here the derivatives are intended in the sense of vector valued distributions.

If $\beta > \frac{1}{p}$, then, by Sobolev's embedding theorem (see [1] 7.57), every element of $W^{\beta,p}(a,b;X)$ can be identified with a continuous function.

Now we can state the promised maximal regularity theorem.

THEOREM 1.1. *Consider the problem*

$$\begin{cases} u'(t) = Au(t) + f(t), & t \in [a,b], \\ u(a) = u_0, \end{cases} \tag{1.6}$$

under assumption h1) on A. Then the following conditions are necessary and sufficient in order that the mild solution

$$u(t) := S(t-a)u_0 + \int_a^t S(t-s)f(s)ds \tag{1.7}$$

of problem (1.6) belong to $W^{1+\beta,p}(a,b;X) \cap W^{\beta,p}(a,b;D)$, *for* $1 < p \leq +\infty$ *and* $\beta \in (0,1) \setminus \{\frac{1}{p}\}$:

(I) $f \in W^{\beta,p}(a,b;X)$;

(II) if $0 < \beta < \frac{1}{p}$, $u_0 \in D_A(1+\beta-\frac{1}{p},p)$; *if* $\frac{1}{p} < \beta < 1$, $u_0 \in D$;

(III) if $\frac{1}{p} < \beta < 1$, $Au_0 + f(a) \in D_A(\beta - \frac{1}{p},p)$.

For a proof, see, e.g., [2], appendix. A strictly related paper related to the L^p-theory, $p \in (1,+\infty)$, for the above mentioned Cauchy problem is [6]. The case $p = +\infty$ was proved by E. Sinestrari (see [21]). The usefulness of fractional Sobolev spaces with respect to spaces of Hölder continuous functions lies of course in the fact that they allow a higher degree of irregularity in the data of the problem.

The first paper in which maximal regularity is used to treat a problem of the form (0.1)-(0.4) is (to our knowledge) [18], but its role is more evident in [17]. In this paper the following abstract inverse problem is considered (see in particular Theorem 1.3):

Let A be a closed operator satisfying condition h1) in the Banach space X, let $B : D(B) \subseteq X \to X$ *be another linear operator, such that* $D \subseteq D(B)$ *and let* $B(\lambda - A)^{-1} \in \mathcal{L}(X)$ $\forall \lambda \in \rho(A)$, $p \in]1,+\infty[$ $\sigma \in]0,1[\setminus\{\frac{1}{p}\}$. *Search for a couple of functions* $u \in W^{1,p}(0,T;X) \cap L^p([0,T];D)$ *and* $m \in L^1(0,T)$ *such that*

$$\begin{cases} u'(t) = Au(t) + \int_0^t m(t-s)Bu(s)ds + f(t) & t \in [0,T], \\ u(0) = u_0, \\ \phi(u(t)) = g(t), & t \in [0,T], \end{cases} \tag{1.8}$$

with $f \in L^p(0,T;X)$, $u_0 \in D$, $\phi \in X'$, g *being a scalar-valued function with domain* $[0,T]$.

We present here the main result in [17] in a slightly simplified form.

THEOREM 1.2. *Let* $\sigma \in (0,1) \setminus \{\frac{1}{p}\}$, $T_0 > 0$. *Assume that*

(I) $u_0 \in D$;
(II) $f \in W^{1+\sigma}(0,T_0;X)$;
(III) $Au_0 + f(0) \in D_A(\sigma+1-\frac{1}{p},p)$ *if* $\sigma < \frac{1}{p}$, $u_0 \in D$ *if* $\sigma > \frac{1}{p}$;
(IV) $g \in W^{2+\sigma,p}(0,T_0)$;
(V) $\phi(u_0) = g(0)$, $\phi(Au_0 + f(0)) = g'(0)$;
(VI) $\chi := \phi(Bu_0) \neq 0$;
(VII) if $\sigma > \frac{1}{p}$,

$$A[Au_0 + f(0)] + f'(0) + \mathcal{H}Bu_0 \in D_A(\beta - \frac{1}{p}, p),$$

where

$$\mathcal{H} := \chi^{-1}\{g''(0) - \Phi[A(Au_0 + f(0)) + f'(0)]\}.$$

(VIII) $Bu_0 \in D_A(\theta,p)$ *for some* $\theta \in (0,1)$ *if* $\sigma \in (0,\frac{1}{p})$, $Bu_0 \in D_A(\sigma - \frac{1}{p},p)$ *if* $\sigma > \frac{1}{p}$.

Then there exists $T > 0$ *such that problem (1.8) has a unique solution* (u,h) *of domain* $[0,\tau]$, *with the following properties:*
(a) $u \in W^{2+\sigma,p}(0,T;X) \cap W^{1+\sigma,p}(0,T;D)$;
(b) $m \in W^{\sigma,p}(0,T)$.

This abstract result is then applied to problem (0.1-0.4) in case $X = L^p(\Omega)$, $D = W^{2,p}(\Omega) \cap W_0^{1,p}(\Omega)$, $Au = \mathcal{A}u$, where $\mathcal{A}$ is a strongly second-order elliptic operator.

To illustrate the basic idea of the proof, we start by observing that the main difficulty lies on the fact that, if we consider (0.1) as an equation in the unknown m, we get a Volterra integral equation of the first kind, which is a typical ill-posed problem (see [12] 2.2). However, if it is possible to differentiate, we obtain a Volterra integral equation of the second kind, which is usually quite amenable. The assumptions of Theorem 1.2 are in fact directed to guarantee this possibility. So, if we put $v := D_t u$, we obtain

$$\begin{cases} D_t v(t) = Av(t) + m(t)Bu_0 + \int_0^t m(t-s)Bv(s)ds + f'(t) & t \in [0,T], \\ v(0) = Au_0 + f(0), & \\ \phi(v(t)) = g'(t), & t \in [0,T], \end{cases} \tag{1.9}$$

We stress that functions mBu_0 and f' are assumed to belong to the same space $W^{\sigma,p}(0,T;X)$ (cf. (II)). This fact guarantees the maximal regularity for the direct problem.

Applying ϕ to the first equation in (1.9) and using the fact that m is scalar valued, we get

$$m(t)\phi(Bu_0) + \phi(Av(t) + \int_0^t m(t-s)Bv(s)ds + f'(t)) = g''(t). \tag{1.10}$$

Recalling now that A generates the analytic semigroup $(S(t))_{t\geq 0}$, we easily derive the operator equation

$$\begin{array}{c} v(t) = S(t)(Au_0 + f(0)) + \int_0^t S(t-s)[m(s)Bu_0 + \\ + f'(s) + \int_0^s m(s-\sigma)Bv(\sigma)d\sigma)]ds \end{array} \tag{1.11}$$

Next, we set $w(t) := Av(t)$. Assume for simplicity that A is invertible and use assumption (VI). Then we obtain the following system for the unknowns m and w:

$$\left\{ \begin{array}{c} w(t) = A\int_0^t S(t-s)[\int_0^s m(s-\sigma)BA^{-1}w(\sigma)d\sigma]ds \\ +A\int_0^t S(t-s)m(s)Bu_0 ds + \\ +A\int_0^t S(t-s)f'(s)ds + AS(t)[Au_0 + f(0)], \\ m(t) = -\chi^{-1}\phi(\int_0^t m(t-s)w(s)ds + \\ +A\int_0^t S(t-s)(\int_0^t m(s-\sigma)w(\sigma)d\sigma)ds) + \chi^{-1}(g''(t) - f'(t)). \end{array} \right. \tag{1.12}$$

(1.12) can be viewed as a system of Volterra integral equations of the second kind in the space $W^{\sigma,p}(0,T;X) \times W^{\sigma,p}(0,T)$, which can be solved by simple fixed-point techniques, at least in a small interval $[0,T]$. We stress the fact that Theorem 1.1 is crucial to get the estimates allowing to apply the contraction mapping theorem. Just to get some flavour, assume for simplicity that $\sigma < \frac{1}{p}$. Then, if $m \in W^{\sigma,p}(0,T)$, $f \in W^{1+\sigma,p}(0,T;X)$ and $Au_0 + f(0) \in D_A(\sigma + 1 - \frac{1}{p}, p)$, we obtain from Theorem 1.1 that

$$t \to A\int_0^t S(t-s)m(s)Bu_0 ds + A\int_0^t S(t-s)f'(s)ds + AS(t)[Au_0 + f(0)]$$

belongs to $W^{\sigma,p}(0,T;X)$ and continuously depends on m.

We have illustrated in some length the basic idea of the proof of the main result in [17] because the method is of general application in problems of this kind.

Since the problem is nonlinear, Theorem 1.2 gives a solution only in a small interval $[0,T]$. The problem of existence of a global solution provided we know an a-priori estimate of u on a maximal interval $[0,T)$ is treated in [19]. In more recent papers the global existence can be proved using suitable weighted spaces (see, for example, [4]).

The assumptions of Theorem 1.2 may seem quite heavy, but one has to consider that our problem is nonlinear and (in nature) ill-posed, so that it is conceivable that the existence of a solution can be assured only under quite burdensome and restrictive conditions.

Up to now we have considered only the case that the kernel m depends only on t.

In [5] the authors face for the first time the case that m may depend, not only on time, but also on some of the space variables. The dependence of m on time only was used before in the passage from (1.9) to (1.10), when we wrote $m(t)\phi(Bu_0)$ instead of $\phi(m(t)Bu_0)$.

In [5] the authors consider a problem of the form

$$\begin{cases} D_t u(t,x,y) = \mathcal{A}u(t,x,y) + \int_0^t m(t-s,x)\mathcal{A}u(s,x,y)ds + f(t,x,y), \\ \qquad (t,x,y) \in [0,T] \times O, \\ u(t,x,y) = 0, (t,x,y) \in [0,T] \times O, \\ u(0,x,y) = u_0(x,y), (x,y) \in O. \end{cases} \tag{1.13}$$

under the further assumptions:

i) the variable can be written in the form (x, y), with $x \in \mathbf{R}$ and $y \in \mathbf{R}^n$;

ii) $O = O_1 \times O_2$, where O_1 is an interval in $\mathbf{R}$, while O_2 is an open bounded subset of $\mathbf{R}^n$ with smooth boundary;

iii) $\mathcal{A}$ is of the form

$$\mathcal{A} = \mathcal{B}_1(x, \partial_x) + \mathcal{B}_2(y, \partial_y). \tag{1.14}$$

Moreover, they replace condition (0.4) with a condition of the form

$$\Phi[u(t,x,.)] = \int_{O_2} \phi(y)u(t,x,y)dy = g(t,x). \tag{1.15}$$

The key fact is that m and Φ commute: more explicitly we have:

$$\Phi[m(t,x)\mathcal{A}u_0(s,x,.)] = m(t,x)\Phi[\mathcal{A}u_0(s,x,.)]. \tag{1.16}$$

Under suitable assumptions of regularity and compatibility conditions concerning the data f, u_0 and g, the authors are able to find a unique solution (u, m) in a certain space $\mathcal{U}^{2+\sigma,p}(B_2, B_1 + B_2) \times W^{\sigma,p}(0,T;\mathcal{K})$, for $\sigma \in (0,1) \setminus \{\frac{1}{p}\}$ (see Theorem 4.1 in the first of the two papers in [5]). B_1 and B_2 are operators related to $\mathcal{B}_1$ and $\mathcal{B}_2$ whose resolvent operators commute, due to the fact that $\mathcal{B}_1$ and $\mathcal{B}_2$ depend on different variables. From this point of view, the authors are influenced by the theory of sums of operators developed in [7].

Problem (1.13) with a supplementary condition of the form

$$\int_{O_x} \phi(x,y)u(t,x,y)dy = g(t,x), \tag{1.17}$$

where $O_x := \{y \in \mathbf{R} : (x,y) \in O\}$, has been instead considered in [4] in the case when O is a curvilinear polygon in the two-dimensional space, without assuming for $\mathcal{A}$ the particular structure (1.14). Here the authors have employed certain recent results concerning the elliptic Dirichlet problem in a plane curvilinear polygon in spaces of continuous or measurable and bounded functions contained in [3]. It is worth mentioning the fact that one can show the existence of a unique global solution of the inverse problem. This is a consequence of the following abstract result (see Theorem 2.1 in [4]):

THEOREM 1.3. *Consider the abstract problem*

$$\begin{cases} u'(t) = Au(t) + k * (Bu + r)(t) + f(t), & t \in [0,T], \\ u(0) = u_0, & \\ \Phi[u(t)] = g(t), & t \in [0,T]. \end{cases} \tag{1.18}$$

under the following hypothesis:

(H1) X, Y, Z are Banach spaces, $(k, y) \to ky$ is a bilinear and continuous mapping from $Z \times Y$ to X;
(H2) $A : D(A) \subseteq X \to X$ is a closed linear operator satisfying condition h1);
(H3) $B \in \mathcal{L}(D(A); Y)$;
(H4) $\Phi \in \mathcal{L}(X, Z)$;
(H5) $u_0 \in D(A)$ and the linear continuous mapping of Z into itself

$$k \to \Phi[k(Bu_0 + r(0))]$$

is one to one and onto whose inverse mapping is denoted by M;
(H6) there exist $\gamma \in (0, 1)$, a closed linear operator A_1 in Z with domain $D(A_1)$ and $\Psi \in \mathcal{L}(D_A(\gamma, \infty); Z)$ such that, for every $u \in D(A)$,

$$\Phi(Au) = A_1\Phi[u] + \Psi u; \tag{1.19}$$

(H7) concerning A_1 we also assume the following: let

$$h \in C^1([0, T]; Z) \cap C([0, T]; D(A_1))$$

be such that the Cauchy problem

$$\begin{cases} h'(t) = A_1 h(t) & \forall t \in [0, T], \\ h(0) = 0. \end{cases}$$

admits the null solution, only;
(H8) $r \in C^{1+\beta}([0, T]; Y)$, for some $\beta \in (0, 1)$.

Then the following conditions are necessary and sufficient in order that problem (1.18) have a solution $(u, k) \in [C^{2+\beta}([0, T]; X) \cap C^{1+\beta}([0, T]; D(A))] \times C^{\beta}([0, T]; Z)$:
(a) $v_0 := Au_0 + f(0) \in D(A)$;
(b) $f \in C^{1+\beta}([0, T]; X)$;
(c) $g \in C^{2+\beta}([0, T]; Z) \cap C^{1+\beta}([0, T]; D(A_1))$;
(d) $\Phi[u_0] = g(0)$, $\Phi[v_0] = g'(0)$;
(e) $Av_0 + f'(0) + k_0(Bu_0 + r(0)) \in D_A(\beta, \infty)$, where k_0 is defined as

$$k_0 := M(g''(0) - \Phi[Av_0 + f'(0)]). \tag{1.20}$$

(1.19) is the key assumption and is used in the application to a problem of the form (1.13) together with (1.17) in the following way: assume that $\mathcal{A}$ is of the form

$$\mathcal{A} = a(x)D_x^2 + 2b(x, y)D_{xy}^2 + c(x, y)D_y^2 + \text{ lower order terms},$$

with a, b, c suitably regular. Let A_1 be the operator $a(x)D_x^2$ with Dirichlet boundary conditions in the interval which is the projection of O onto the first coordinate axis (recall that we are working in dimension 2). Then it is easily seen, employing Green's formula, that, if $u \in D(A)$ (so that u vanishes in ∂O),

$$\int_{O_x} \phi(x, y)\mathcal{A}u(t, x, y)dy = A_1(x \to \int_{O_x} \phi(x, y)u(t, x, y)dy)$$

$$+\text{terms depending only on } u,\ D_x u \text{ and } D_y u.$$

2 NONLINEAR PROBLEMS

In this section we shall consider some nonlinear generalizations of problem (0.1)-(0.4). To start with an application, we consider the well-known Lotka-Volterra model with diffusion, arising in population dynamics (for other results concerning systems of this form, see also [9]). The nonlinearity in the integral term appearing in the first equation in the sequel, is analogous to that of the Karmack-McKendrik system arising in the theory of spread of infections.

Let O be an open bounded set in $\mathbf{R}^n$, lying on one side of ∂O, which is a submanifold of $\mathbf{R}^n$ of class C^2. The problem is:

determine two functions $u : [0,T] \times O \to \mathbf{R}$ *and* $m : [0,T] \to \mathbf{R}$ *satisfying the integro-differential system:*

$$\left\{ \begin{array}{c} D_t u(t,x) = d(u(t,x))\Delta u(t,x) \\ +bu(t,x) \int_0^t m(t-s)u(s,x)ds + f(u(t,x)), (t,x) \in (0,\tau) \times O, \\ u(0,x) = u_0(x), \quad x \in O \\ u(t,x') = 0, \quad (t,x') \in (0,\tau) \times \partial O \\ \int_O \phi(x)u(t,x)\,dx = g(t), \quad t \in (0,\tau). \end{array} \right. \tag{2.1}$$

where d is real valued and positive and b is a real number different from 0.

We introduce a proper functional setting. We put

$$X := L^p(O), \tag{2.2}$$

with

$$2p > n, \tag{2.3}$$

$$D = W^{2,p}(O) \cap W_0^{1,p}(O). \tag{2.4}$$

Observe that, owing to condition (2.3), $D \subseteq C(\overline{O})$. Set

$$\left\{ \begin{array}{l} F : D \times D \to X, \\ F(u_1,u_2) := bu_1u_2, \quad u_1, u_2 \in D, \end{array} \right. \tag{2.5}$$

$$\left\{ \begin{array}{c} G : D \to X, \\ G(u) := d(u)\Delta u + f(u), \quad u \in D. \end{array} \right. \tag{2.6}$$

and

$$\Phi(u) := \int_O \phi(x)u(x)dx, \tag{2.7}$$

under the condition that

$$\phi \in L^{p'}(O). \tag{2.8}$$

Then we can write the problem (2.1) in the abstract form

$$\begin{cases} u'(t) = G(u(t)) + \int_0^t m(t-s)F(u(t),u(s))ds, & t \in [0,\tau], \\ u(0) = u_0, & \\ \Phi(u(t)) = g(t), & t \in (0,\tau) \end{cases} \tag{2.9}$$

A slightly simplified version of the main result in [2] is the following

THEOREM 2.1. *Consider problem (2.9) under the following conditions k1)-k6):*

k1) X and D are Banach space, D is continuously imbedded into X and dense in it;

k2) $F \in C^2(D \times D; X)$ and F'' - the second Frechet derivative of F - is uniformly Lipschitz continuous from $D \times D$ to $\mathcal{L}(D \times D; \mathcal{L}(D \times D; X))$ in every bounded subset of $D \times D$;

k3) $G \in C^2(D; X)$ and G'' is uniformly Lipschitz continuous from D to $\mathcal{L}(D; \mathcal{L}(D; X))$ in every bounded subset of D;

k4) $u_0 \in D$;

k5) if we set $A := G'(u_0)$, A, considered as an unbounded operator in X, satisfies assumption h1);

k6) $\Phi \in X'$, the dual space of X;

Fix $p \in (1,+\infty]$, $\beta \in (0,1) \setminus \{\frac{1}{p}\}$, $T > 0$. Assume, moreover, that:

(I) if $\beta < \frac{1}{p}$, $G(u_0) \in D_A(1+\beta-\frac{1}{p},p)$, if $\beta > \frac{1}{p}$, $G(u_0) \in D$;

(II) $g \in W^{2+\beta,p}(0,T)$;

(III) $\Phi(u_0) = g(0)$, $\Phi(G(u_0)) = g'(0)$;

(IV) $\chi := \Phi(F(u_0,u_0)) \neq 0$;

(V) if $\beta > \frac{1}{p}$,

$$A[G(u_0)] + \mathcal{H}F(u_0,u_0) \in D_A(\beta - \frac{1}{p}, p),$$

where $\mathcal{H}$ is defined as

$$\mathcal{H} := \chi^{-1}\{g''(0) - \Phi[A(G(u_0))]\}. \tag{2.10}$$

Then there exists $\tau \in [0,T]$ such that problem (2.9) has a unique solution (u,h) belonging to $(W^{2+\beta,p}(0,\tau;X) \cap W^{1+\beta,p}(0,\tau;D)) \times W^{\beta,p}(0,\tau)$.

To apply Theorem 2.1 to the problem (2.1), we introduce the following further conditions:

h_{11}) d and f belong to $C^3(\mathbf{R})$;

h_{12}) $g \in W^{2+\beta,p}(0,T)$, for some $\beta \in (0,1)$, $\beta \notin \{\frac{1}{p}, \frac{3}{2p}-1, \frac{3}{2p}\}$.

h_{13}) $u_0 \in W^{2,p}(O) \cap W_0^{1,p}(O)$;

h_{14}) if $G(u_0) := d(u_0)\Delta u_0 + f(u_0)$,

$$G(u_0) \in \begin{cases} B_{p,p}^{2(1+\beta-\frac{1}{p})}(O) & \text{if} \quad \beta < \frac{3}{2p}-1, \\ \{u \in B_{p,p}^{2(1+\beta-\frac{1}{p})}(O) : u_{|\partial O} = 0\} & \text{if} \quad \frac{3}{2p}-1 < \beta < \frac{1}{p}, \\ W^{2,p}(O) \cap W_0^{1,p}(O) & \text{if} \quad \beta > \frac{1}{p}; \end{cases}$$

h_{15}) $\int_O \phi(x)u_0(x)dx = g(0), \int_O \phi(x)[d(u_0(x))\Delta u_0(x) + f(u_0(x))]dx = g'(0)$;

h_{16}) $\int_O \phi(x)u_0(x)^2dx \neq 0$;

h_{17})

$$d'(u_0)G(u_0)\Delta u_0 + d(u_0)\Delta G(u_0) + f'(u_0)G(u_0) + \mathcal{H}bu_0^2$$

$$\in \begin{cases} B_{p,p}^{2(\beta-\frac{1}{p})}(O) & \text{if} \quad \frac{1}{p} < \beta < \frac{3}{2p}, \\ \{u \in B_{p,p}^{2(\beta-\frac{1}{p})}(O) : u_{|\partial O} = 0\} & \text{if} \quad \frac{3}{2p} < \beta, \end{cases}$$

with

$$\mathcal{H} := \chi^{-1}[g''(0) - \int_O \phi\{d(u_0)\Delta G(u_0) + [d'(u_0)\Delta u_0 + f'(u_0)]G(u_0)\}dx].$$

It is easily seen that F satisfies $k2)$ and G satisfies $k3)$. $k4)$ is exactly assumption $h_{13})$. We have also, for every $v \in D$,

$$G'(u_0)v = d(u_0)\Delta v + [d'(u_0)\Delta u_0 + f'(u_0)]v,$$

so that $k5)$ is satisfied by 3.8 in [22].

$k6)$ follows from (2.8), setting

$$\Phi(u) := \int_O \phi(x)u(x)dx. \tag{2.11}$$

Concerning assumptions $(I) - (V)$ in Theorem 2.1, it suffices to note that, if $\theta \in (0,1) \setminus \{\frac{1}{2p}\}$, we have

$$D_A(\theta, p) = \begin{cases} B_{p,p}^{2\theta}(O) & \text{if} \quad \theta < \frac{1}{2p}, \\ \{u \in B_{p,p}^{2\theta}(O) : u_{|\partial O} = 0\} & \text{if} \quad \frac{1}{2p} < \theta \end{cases} \tag{2.12}$$

(see [13]). Observe finally that D is dense in X.

The proof of Theorem 2.1 follows the scheme of the proof of Theorem 1.2 already outlined in Section 1: one differentiates the first equation in (2.9) and introduces the new unknown function $v := u'$. Finally one is reduced to study the system

$$\begin{cases} v(t) = S(t)[G(u_0) + f(0)] + \int_0^t S(t-s)f'(s)ds \\ + \int_0^t S(t-s)m(s)F(u_0, u_0)ds + \int_0^t S(t-s)\mathcal{R}(v,h)(s)ds, \\ m(t) = \chi^{-1}[g''(t) - \Phi(Av(t) + \mathcal{R}(v,h)(t) + f'(t))], \end{cases} \tag{2.13}$$

where $\mathcal{R}(v, h)$ is "small" in a proper sense. Here again, of course, Theorem 1.1 is crucial to get the proper estimates allowing to apply (in a short interval) the contraction mapping theorem.

We conclude quoting a paper by M. Grasselli and A. Lorenzi ([10]). They study the following problem, which is more general than (2.9):

$$\begin{cases} u'(t) + (h * u')(t) = G(u(t)) + \\ + \int_0^t m(t-s)F(u(t), u(s))ds + f(t), & t \in [0, \tau], \\ u(0) = u_0, \\ \Phi(u(t)) = g(t), & t \in (0, \tau) \\ M(u(t)) + (m * N(u))(t) = \phi(t), & t \in (0, \tau). \end{cases} \tag{2.14}$$

The main novelty is the introduction of the second unknown kernel h (the case $f \not\equiv 0$ was considered also in [2]). Such introduction is balanced by the condition

$$M(u(t)) + (m * N(u))(t) = \phi(t), \quad t \in (0, \tau). \tag{2.15}$$

Here M and N are not necessarily linear functionals of domain D coinciding with the domain of G. The authors obtain results of existence and uniqueness of a local solution (u, h, m). They consider a functional setting which is partially similar to that of Theorem 2.1, with $\beta \in (0, \frac{1}{p})$ only, but their results, when applied to the less general situation previously considered, are not completely comparable with those of Theorem 2.1, although they are not optimal, in the sense that the assumptions of [10] are only sufficient to get their conclusions. On the contrary, the assumptions of Theorem 2.1, under certain basic structural conditions, are also necessary (see for this Lemma 5.1 in [2]). Here the authors apply again their abstract results to a problem of heat conduction with memory, where h can be interpreted as the so-called "internal energy relaxation" function.

BIBLIOGRAPHY

1. R. Adams, *Sobolev spaces*, Pure Appl. Math. 65, Academic Press, 1975.

2. F. Colombo and D. Guidetti, *A unified approach to nonlinear integrodifferential inverse problems of parabolic type*, Zeitschrift für Analysis und ihre Anwendungen **21** (2002) 2, 431–464.

3. F. Colombo, D. Guidetti, and A. Lorenzi,*Elliptic equations in non-smooth plane domains with an application to a parabolic problem*, Diff. Int. Eq. **7** (2002), 695–716.

4. F. Colombo, D. Guidetti, and A. Lorenzi, *Integrodifferential identification problems for thermal materials with memory in non-smooth plane domains*, preprint, 2001.

5. F. Colombo and A. Lorenzi, *Identification of time and space dependent relaxation kernels for materials with memory related to cylindrical domains I-II*, J. Math. Anal. Appl. **213** (1997), 32–62, 63–90.

6. G. Di Blasio, *Linear parabolic evolution equations in L^p-spaces*, Ann. Mat. Pura Appl. **8** (1984), 55–104.

7. G. Da Prato and P. Grisvard, *Sommes d'opérateurs linéaires et équations différentielles opérationelles*, J. Math. Pures Appliqueés **54** (1975), 305–387.

8. G. Da Prato and P. Grisvard, *Equations d'évolution abstraites non linéaires de type parabolique*, Ann. Mat. Pura Appl. (IV) **120** (1979), 329–396.

9. S. A. Gourley and N. F. Britton, *On a modified Volterra population equation with diffusion* , Nonlinear Analysis, Theory, Methods and Application **21** (1993), 389–395.

10. M. Grasselli and A. Lorenzi, *An inverse problem for an abstract nonlinear parabolic integrodifferential equation*, Diff. Int. Eq. **6** (1993), 63–81.

11. P. Grisvard, *Équations differentielles abstraites*, Ann. Scient. Éc. Norm. Sup., 4e série 2 (1969), 311–395.

12. C. W. Groetsch, *Inverse problems in the mathematical sciences*, Vieweg Mathematics for Scientists and Engineers, 1993.

13. D. Guidetti, *On interpolation with boundary conditions*, Math. Z. **207** (1991), 439–460.

14. J. Janno and L. v. Wolfersdorf, *Inverse problems for identification of memory kernels in heat flow*, J. Inv. Ill-Posed Probl. **4** (1996), 39–66.

15. J. Janno and L. v. Wolfersdorf, *Identification of weakly singular memory kernels in heat conduction*, Z. Angew. Math. Mech. **77** (1997), 243–257.

16. J. Janno, *An inverse problem arising in compression of a poro-viscoelastic medium*, Proc. Estonian Acad. Sci. Phys. Math. **49** (2000), 75–89.

17. A. Lorenzi and E. Paparoni, *Direct and inverse problems in the theory of materials with memory*, Rend. Sem. Univ. Padova **87** (1992), 105–138.

18. A. Lorenzi and E. Sinestrari, *An inverse problem in the theory of materials with memory*, Nonlinear Anal. **12** (1988), no. 12, 1317–1335.

19. A. Lorenzi and E. Sinestrari, *Stability results for a partial integrodifferential inverse problem*, Pitman Research Notes in Math. **190** (1989), 271–294.

20. A. Lunardi, *Analytic semigroups and optimal regularity in parabolic problems*, Progress in Nonlinear Differential Equations and Their Applications vol. 16, Birkhäuser, 1995.

21. E. Sinestrari, *On the abstract Cauchy problem in spaces of continuous functions*, J. Math. Anal. Appl. **107** (1985), 16-66.

22. H. Tanabe, *Equations of evolution*, Pitman, 1979.

23. H. Triebel, *Interpolation theory, function spaces, differential operators*, North-Holland Mathematical Library, vol. 18, 1978.

24. L. v. Wolfersdorf, *Inverse problems for memory kernels in heat flow and viscoelasticity*, J. Inv. Ill-Posed Probl. **4** (1996), 341–354.

A semilinear integrodifferential inverse problem

FABRIZIO COLOMBO
Dipartimento di Matematica, Politecnico di Milano
Via Bonardi n. 9, 20133, Milano, Italy
e-mail: fabcol@mate.polimi.it

VINCENZO VESPRI
Dipartimento di Matematica, Università degli Studi di Firenze
Viale Morgagni 67/a, 50134, Firenze, Italy
e-mail: vespri@math.unifi.it

ABSTRACT. We prove an existence and uniqueness theorem for the abstract version of a semilinear integrodifferential inverse problem. We apply such result to the Kermack-McKendrick model with diffusion. Our main tools are: the analytic semigroups theory, optimal regularity results and fixed point arguments.

1 INTRODUCTION

We study an identification problem for a semilinear integrodifferential system arising in the theory of the spread of infections. More precisely, we investigate the Kermack-McKendrick model with diffusion. For more details about this model see [10].
In this section we state our problem then, we formulate it in an abstract setting and finally we point out the novelty of our results.

We will use the spaces $\mathcal{B}([0,T]; D_A(\theta,\infty))$ of bounded functions with values in the interpolation spaces $D_A(\theta,\infty)$. For precise definitions see the beginning of Section 2.

Let Ω be an open bounded set in $\mathbf{R}^3$ and let T be a positive constant. If u and v denote the susceptible and the infective populations densities, respectively their evolution equations are

$$u_t(t,x) = K_1\Delta u(t,x) - a_1 u(t,x)$$

$$-b_1 u(t,x)\int_0^t h(t-s)v(s,x)\,ds, \qquad (t,x) \in (0,T] \times \Omega, \tag{1.1}$$

$$v_t(t,x) = K_2\Delta v(t,x) - a_2 v(t,x)$$

$$+b_2 u(t,x)\int_0^t h(t-s)v(s,x)\,ds, \qquad (t,x)\in(0,T]\times\Omega, \tag{1.2}$$

with the related initial–boundary conditions

$$u(0,x) = u_0(x), \qquad v(0,x) = v_0(x), \qquad x\in\Omega, \tag{1.3}$$

$$D_\nu u(t,x) = D_\nu v(t,x) = 0, \qquad (t,x)\in[0,T]\times\partial\Omega, \tag{1.4}$$

where D_ν denotes the external normal derivative, a_i and b_i, $i=1,2$ are called the constant rates, $K_i>0$ are the diffusion coefficients, $u_0, v_0 : \Omega\to\mathbf{R}$ are given functions. The kernel $h(t)$ is considered unknown. The addition of the terms $K_1\Delta u$ and $K_2\Delta v$ to the original Kermack-McKendrick model allows the migration of both infective and susceptible populations, and the integral term in (1.1) and (1.2) represents a transfer mechanism of infection.

Because of the interpretation of the model we assume $a_i \le 0$ and $b_i > 0$ for $i=1,2$.

The fact that the kernel h is not directly measurable leads us to consider the inverse problem, i.e. u, v and h have to be determined simultaneously. To obtain our purpose we need additional information on u, which can be analytically represented in integral form as follows

$$\int_\Omega \phi(x)u(t,x)\,dx = g(t), \tag{1.5}$$

where $\phi:\Omega\to\mathbf{R}$ and $g:[0,T]\to\mathbf{R}$ are given functions whose regularity will be specified in the sequel. More precisely, condition (1.5) represents additional measurements on the susceptible population u in some parts of Ω, so ϕ is a suitable compact support function that depends on the type of device used for the measure, while g stands for the results of the measurements on u.

We are in position to formulate our **Inverse Problem** (IP).
Determine three continuous functions: $u:[0,T]\times\Omega\to\mathbf{R}$, $v:[0,T]\times\Omega\to\mathbf{R}$ *and* $h:[0,T]\to\mathbf{R}$ *satisfying system (1.1)-(1.5).*

We now formulate our problem in a more general setting relating it to a Banach algebra X.

Let X be a Banach algebra and assume that $A:D(A)\subset X\to X$ and $B:D(B)\subset X\to X$ are two linear and closed operators with domains $D(A)$, and $D(B)$, respectively. We denote by $D_A(\theta,\infty)$ the interpolation spaces, for their definition see Section 2, with the inclusions $D(A)\subset D_A(\theta,\infty)\subset X$. Analogously we define $D_B(\theta,\infty)$.

Consider now three functions $u:[0,T]\to D_A(\theta,\infty)$, $v:[0,T]\to D_B(\theta,\infty)$ and $h:[0,T]\to\mathbf{R}$ satisfying the system

$$u'(t) = Au(t) - a_1u(t) - b_1u(t)\int_0^t h(t-s)v(s)\,ds, \tag{1.6}$$

$$v'(t) = Bv(t) - a_2 v(t) + b_2 u(t) \int_0^t h(t-s)v(s)\,ds, \tag{1.7}$$

$$u(0) = u_0, \qquad v(0) = v_0, \tag{1.8}$$

$$\Phi[u(t)] = g(t), \quad t \in [0,T], \tag{1.9}$$

where Φ is a given bounded linear functional on $D_A(\theta,\infty)$, $g : [0,T] \to \mathbf{R}$, $u_0 : \overline{\Omega} \to \mathbf{R}$ and $v_0 : \overline{\Omega} \to \mathbf{R}$ are given functions and $a_i \le 0$ and $b_i > 0$ for $i = 1, 2$.

We can formulate our **Abstract Inverse Problem** (AIP).
Determine three functions $u : [0,T] \to D_A(\theta,\infty)$, $v : [0,T] \to D_B(\theta,\infty)$ *and* $h : [0,T] \to \mathbf{R}$ *satisfying (1.6)-(1.9).*

Optimal regularity results and the analytic semigroup theory are fundamental tools in the study of direct and inverse parabolic problems. Our strategy is to formulate the abstract version of the inverse problem in terms of a system of equivalent fixed point equations. Several optimal regularity results are at our disposal for such equivalent formulation.

Consider the following optimal regularity results for the Cauchy Problem (CP):

$$\begin{cases} u'(t) = Au(t) + f(t), & t \in [0,T], \\ u(0) = u_0. \end{cases}$$

Let $A : D(A) \subset X \to X$ be the infinitesimal generator of the analytic semigroup e^{tA} and $\beta \in (0,1)$. In [9] [12] we can find the proofs of the following results:

THEOREM 1.1. **(Strict solution in Hölder spaces $C^\beta([0,T];X)$)** *For any* $f \in C^\beta([0,T];X)$, $u_0 \in D(A)$ *with* $Au_0 + f(0) \in D_A(\beta,\infty)$ *the Cauchy problem (CP) admits a unique solution* $u \in C^{1+\beta}([0,T];X) \cap C^\beta([0,T];D(A))$.

THEOREM 1.2. **(Strict solution in spaces $\mathcal{B}([0,T];D_A(\beta,\infty))$)** *For any* $f \in C([0,T];X) \cap \mathcal{B}([0,T];D_A(\beta,\infty))$, $u_0 \in D_A(\beta+1,\infty)$ *the Cauchy problem (CP) admits a unique solution* $u \in C^1([0,T];X) \cap C([0,T];$
$D(A)) \cap \mathcal{B}([0,T];D_A(\beta+1,\infty))$

What we prove in this paper is an existence and uniqueness theorem for the (AIP) and we apply it to the (IP) in the case X is the space of continuous functions $C(\overline{\Omega})$. Using the generation results in [6] we could also prove a similar result if we set $X = C^1(\overline{\Omega})$ if A and B are second order differential operators.

On the optimal regularity Theorem 1.1 are based many results that we can find in the recent literature also for the Phase-Field models, for the theory of materials with memory and for the population dynamic models see [1] [4] [5]. In [7], [2] and in [3] the authors study the above model requiring a different set of conditions on the data without investigating the $\mathcal{B}([0,T];D_A(\theta,\infty))$ regularity for some $\theta \in \mathbf{R}^+$.

The novelty of this paper is that we apply Theorem 1.2, (and not Theorem 1.1) which is particularly suitable for the semilinear model introduced above.

We recall that in the paper [8], for the first time, the analytic semigroup theory and Theorem 1.1 have been used to face integrodifferential parabolic inverse problems for the heat equation with memory.

The plan of the paper is the following.

In section 2 we define the function spaces and we introduce some preliminary material related to the analytic semigroups theory.
In section 3 we state our main result for the abstract problem (1.6)-(1.9) (see Theorem 3.1) and we give an application to a concrete case (see Theorem 3.2).
In section 4 we prove, that under suitable conditions the integrodifferential system (1.6)-(1.9) is equivalent to a fixed point system of three Volterra integral equations of the second kind (see Theorem 4.1).
Finally in section 5 we prove Theorem 3.1.

2 PRELIMINARY MATERIAL

The results that we are going to recall in this section hold in the case X is a Banach space only. For our purpose, since we apply our results to the (IP) we consider X to be a Banach algebra with norm $||\cdot||$. Let $T>0$, we denote by $C([0,T];X)$ the usual space of continuous functions with values in X, while we denote by $\mathcal{B}([0,T];X)$ the space of bounded functions with values in X. Equipped both with the sup–norm:

$$||u||_{C([0,T];X)} = ||u||_{\mathcal{B}([0,T];X)} := \sup_{0\le t\le T} ||u(t)|| \tag{2.1}$$

they become Banach spaces. For $\beta \in (0,1)$ we define

$$C^{\beta}([0,T];X) = \{u \in C([0,T];X) : |u|_{C^{\beta}([0,T];X)} = \sup_{0\le s<t\le T} \frac{||u(t)-u(s)||}{(t-s)^{\beta}} < \infty\} \tag{2.2}$$

and we endow it with the norm

$$||u||_{C^{\beta}([0,T];X)} = ||u||_{C([0,T];X)} + |u|_{C^{\beta}([0,T];X)}. \tag{2.3}$$

Let Ω a bounded open set in $\mathbf{R}^n$, $n \in \mathbf{N}$, for $\beta \in (0,1)$ we define

$$C^{\beta}(\Omega) = \{u \in C(\Omega) : |u|_{C^{\beta}(\Omega)} = \sup_{x,y\in\Omega,\ x\neq y} \frac{|u(x)-u(y)|}{|x-y|^{\beta}} < \infty\} \tag{2.4}$$

and we endow it with the norm

$$||u||_{C^{\beta}(\Omega)} = ||u||_{C(\Omega)} + |u|_{C^{\beta}(\Omega)}, \tag{2.5}$$

where

$$||u||_{C(\Omega)} := \sup_{x\in\Omega} |u(x)|. \tag{2.6}$$

By $\mathcal{L}(X)$ we denote the space of all bounded linear operators from X into itself equipped with the sup–norm, while $\mathcal{L}(X;\mathbf{R})$ is the space of all bounded linear functionals on X considered with the natural norm.

DEFINITION 2.1. Let $A : D(A) \subset X \to X$, be a linear operator, possibly with $\overline{D(A)} \neq X$. A is said to be sectorial if it satisfies the following assumptions:

(H1) there exists $\theta \in (\pi/2, \pi)$ such that any $\lambda \in \mathbf{C}\backslash\{0\}$ with $|\arg\lambda| \le \theta$ and $\lambda = 0$ belong to the resolvent set of A;

(H2) there exists $M>0$ such that $||\lambda(\lambda I - A)^{-1}||_{\mathcal{L}(X)} \le M$ for any $\lambda \in \mathbf{C}\backslash\{0\}$ with $|\arg\lambda| \le \theta$.

The fact that the resolvent set of A is not void implies that A is closed, so that $D(A)$ endowed with the graph norm becomes a Banach space.
According to assumptions *H1, H2*, it is possible to define the semigroup $\{e^{tA}\}_{t\geq 0}$, of bounded linear operators in $\mathcal{L}(X)$, so that $t \to e^{tA}$ is an analytic function from $(0,\infty)$ to $\mathcal{L}(X)$ satisfying for $k \in \mathbf{N}$ the relations

$$\frac{d^k}{dt^k} e^{tA} = A^k e^{tA}, \quad t > 0, \tag{2.7}$$

and $Ae^{tA}x = e^{tA}Ax$, for all $x \in D(A)$ and $t \geq 0$. Moreover there exist positive constants M_k, for $k \in \mathbf{N}_0$ such that

$$\|t^k A^k e^{tA}\|_{\mathcal{L}(X)} \leq M_k, \qquad t > 0. \tag{2.8}$$

For more details see for example [9] [11]. Let us define the family of interpolation spaces ([11] or [13]) $D_A(\beta,\infty)$, $\beta \in (0,1)$, between $D(A)$ and X by

$$D_A(\beta,\infty) = \left\{x \in X : \ |x|_{D_A(\beta,\infty)} := \sup_{t>0} t^{1-\beta}\|Ae^{tA}x\| < \infty\right\} \tag{2.9}$$

with the norm (2.11) for $j = 0$ only. We also set

$$\mathcal{D}_A(1+\beta,\infty) - \{x \in \mathcal{D}(A) : Ax \in \mathcal{D}_A(\beta,\infty)\}, \tag{2.10}$$

$D_A(1+\beta,\infty)$ turn out to be Banach spaces when equipped with the norms

$$\|x\|_{D_A(1+\beta,\infty)} = \sum_{j=0}^{1} \|A^j x\| + |Ax|_{D_A(\beta,\infty)}. \tag{2.11}$$

We reconsider Theorem 1.1 with the related fundamental estimates, related to the Cauchy problem

$$u'(t) = Au(t) + f(t), \qquad t \in [0,T], \tag{2.12}$$

$$u(0) = u_0. \tag{2.13}$$

THEOREM 2.2. *Let $A : D(A) \subset X \to X$ be the generator of the analytic semigroups e^{tA}. Then for any $f \in C([0,T];X)\cap\mathcal{B}([0,T];D_A(\beta,\infty))$, $u_0 \in D_A(\beta+1,\infty)$ problem (2.12)-(2.13) admits a unique solution $u \in C^1([0,T];X)\cap C([0,T];D(A))\cap \mathcal{B}([0,T];D_A(\beta+1,\infty))$ represented by the formula*

$$u(t) = e^{tA}u_0 + \int_0^t e^{(t-s)A} f(s)\,ds := e^{tA}u_0 + (e^{tA} * f)(t), \tag{2.14}$$

and u', $Au \in C([0,T];X) \cap \mathcal{B}([0,T];D_A(\beta,\infty))$ and $Au \in C^\beta([0,T];X)$. Moreover, the following estimate holds

$$\|e^{tA} * f\|_{C([0,T];X)\cap\mathcal{B}([0,T];D_A(\beta,\infty))} \leq TM_0\|f\|_{C([0,T];X)\cap\mathcal{B}([0,T];D_A(\beta,\infty))}, \tag{2.15}$$

$$\|e^{tA}u_0\|_{C([0,T];X)\cap\mathcal{B}([0,T];D_A(\beta,\infty))} \leq C\|u_0\|_{D_A(\beta+1,\infty)}. \tag{2.16}$$

Proof. One can argue as in Corollary 4.3.9 iii) of [9]. To prove estimate (2.15) we note that

$$||e^{tA} * f||_{C([0,T];X)} \leq TM_0||f||_{C([0,T];X)}, \tag{2.17}$$

$$||e^{tA} * f||_{\mathcal{B}([0,T];D_A(\beta,\infty))} \leq TM_0||f||_{\mathcal{B}([0,T];D_A(\beta,\infty))}. \tag{2.18}$$

Adding (2.17) and (2.18) we get (2.15). As in Corollary 4.3.9 iii) [9] we get estimate (2.16). ■

THEOREM 2.3. *Let $h \in C([0,T];\mathbf{R})$ and $u \in C([0,T];X) \cap \mathcal{B}([0,T];D_A(\beta,\infty))$. Define the convolution operator*

$$h * u(t) := \int_0^t h(t-s)u(s)ds. \tag{2.19}$$

Then $*$ maps $C([0,T];\mathbf{R}) \times [C([0,T];X) \cap \mathcal{B}([0,T];D_A(\beta,\infty))]$ into $C([0,T];X)$ $\cap\mathcal{B}([0,T];D_A(\beta,\infty))$ and the following estimate holds:

$$||h * u||_{C([0,T];X)\cap\mathcal{B}([0,T];D_A(\beta,\infty))} \leq T||h||_{C([0,T];\mathbf{R})}||u||_{C([0,T];X)\cap\mathcal{B}([0,T];D_A(\beta,\infty))}. \tag{2.20}$$

Proof. Consider the estimates

$$||h * u||_{C([0,T];X)} \leq T||h||_{C([0,T];\mathbf{R})}||u||_{C([0,T];X)} \tag{2.21}$$

and

$$|h * u(t)|_{D_A(\beta,\infty)} = \sup_{\tau>0} \tau^{1-\beta}||Ae^{\tau A}\int_0^t h(t-s)v(s)ds||$$

$$\leq \sup_{\tau>0} \int_0^t h(t-s)\tau^{1-\beta}||Ae^{\tau A}u(s)||ds \leq T||h||_{C([0,T];\mathbf{R})}||u||_{\mathcal{B}([0,T];D_A(\beta,\infty))} \tag{2.22}$$

so that

$$||h * u||_{\mathcal{B}([0,T];D_A(\beta,\infty)} \leq T||h||_{C([0,T];\mathbf{R})}||u||_{\mathcal{B}([0,T];D_A(\beta,\infty))}. \tag{2.23}$$

From (2.21) and (2.23) we get the statement. ■

3 THE MAIN RESULTS

In this section we state our main abstract result and then we apply it to the case of the continuous functions space.

Suppose X be a Banach algebra, $\beta \in (0,1)$. Let us introduce the following conditions

Regularity Conditions

R_1 $u_0 \in D(A)$, $v_0 \in D(B)$;

R_2 $Au_0 - a_1u_0 \in D_A(\beta+1,\infty)$, $Bv_0 - a_2v_0 \in D_B(\beta+1,\infty)$;

R_3 $g \in C^2([0,T])$;

R_4 $\Phi \in \mathcal{L}(D_A(\beta+1,\infty);\mathbf{R})$;

R_5 $\Phi[Au] = \Psi[u]$, with $\Psi \in \mathcal{L}(D_A(\beta,\infty);\mathbf{R})$.

Compatibility Conditions

C_1 $\Phi[u_0] = g(0)$;

C_2 $\Phi[Au_0 - a_1 u_0] = g'(0)$;

C_3 $\Phi[v_0 u_0] \neq 0$ *and* $b_1 \neq 0$.

Our main abstract result is:

THEOREM 3.1. *Let A and B be sectorial operators (see Definition 2.1). Under assumptions R_1-R_5 and C_1-C_3 there exists $T^* \in (0,T]$ such that for any $\tau \in (0,T^*)$ problem (1.6)-(1.9) has a unique solution $(u,v,h) \in [C^1([0,\tau];X) \cap C([0,\tau];D(A)) \cap \mathcal{B}([0,\tau];D_A(\beta+1,\infty)] \times [C^1([0,\tau];X) \cap C([0,\tau];D(B)) \cap \mathcal{B}([0,\tau];D_B(\beta+1,\infty)] \times C([0,\tau];\mathbf{R})$.*

Proof. It is in Section 5. ∎

We are now in the position to apply Theorem 3.1 when we choose as reference space:

$$X = C(\bar{\Omega}). \tag{3.1}$$

We suppose that $D(A) = D(B)$ with the Neumann homogeneous boundary conditions, where

$$D(A) = \{u \in C(\bar{\Omega}) : Au \in C(\bar{\Omega}), \quad D_\nu u = 0 \ \ on\ \partial\Omega\}, \tag{3.2}$$

and

$$A := K_1\Delta, \quad B := K_2\Delta. \tag{3.3}$$

Moreover, in [9] it is proved that the operators defined in (3.3) whose domains are defined in (3.2) are sectorial in $C(\bar{\Omega})$. Then we recall the following characterizations concerning the interpolation spaces related to A (see [9]), for $\beta \neq 1/2$:

$$D_A(\beta,\infty) = C^{2\beta}(\overline{\Omega}), \quad if \ \ \beta \in (0,1/2), \tag{3.4}$$

$$D_A(\beta,\infty) = C^{2\beta}_\nu(\overline{\Omega}), \quad if \ \ \beta \in (1/2,1), \tag{3.5}$$

where

$$C^{2\beta}_\nu(\overline{\Omega}) = \{u \in C^{2\beta}(\overline{\Omega}) : \ D_\nu u = 0 \ \ on \ \ \partial\Omega\ \}. \tag{3.6}$$

Finally, we can define the set of admissible data consisting of all those functions u_0, v_0, ϕ, g, satisfying the following assumptions, for $\beta \in (0,1)/\{1/2\}$:

Regularity Conditions

$\mathcal{R}_1$ $u_0, v_0 \in D(A)$;

$\mathcal{R}_2$ $K_1\Delta u_0 - a_1 u_0,\ K_2\Delta v_0 - a_2 v_0 \in C^{2(\beta+1)}(\overline{\Omega})$, $\quad if \ \ \beta \in (0,1/2)$;

$\mathcal{R}_2'$ $K_1\Delta u_0 - a_1 u_0,\ K_2\Delta v_0 - a_2 v_0 \in C^{2(\beta+1)}_\nu(\overline{\Omega})$, $\quad if \ \ \beta \in (1/2,1)$;

$\mathcal{R}_3$ $g \in C^2([0,T])$;

$\mathcal{R}_4$ $\Phi \in \mathcal{L}(C^{2(\beta+1)}(\overline{\Omega});\mathbf{R})$, $\quad if \ \ \beta \in (0,1/2)$;

$\mathcal{R}_4'$ $\Phi \in \mathcal{L}(C^{2(\beta+1)}_\nu(\overline{\Omega});\mathbf{R})$, $\quad if \ \ \beta \in (1/2,1)$;

$\mathcal{R}_5$ $\int_\Omega \phi(x) K_1 \Delta u(t,x)\,dx = \Psi[u(t,\cdot)]$ with $D_\nu\phi = 0$ on $\partial\Omega$;

$\mathcal{R}_6$ $\Psi \in \mathcal{L}(C^{2\beta}(\overline{\Omega});\mathbf{R})$ $\quad if \ \ \beta \in (0,1/2)$;

$\mathcal{R}_6'$ $\Psi \in \mathcal{L}(C^{2\beta}_\nu(\overline{\Omega});\mathbf{R})$ $\quad if \ \ \beta \in (1/2,1)$;

where

$$\Psi[u(t,\cdot)] = \int_\Omega \psi(x)u(t,x)\,dx, \quad \psi(x) := K_1\Delta\phi(x) \tag{3.7}$$

Compatibility Conditions
$\mathcal{C}_1$ $\int_\Omega u_0(x)\,dx = g(0)$;
$\mathcal{C}_2$ $\int_\Omega [K_1\Delta u_0(x)\,dx - a_1 u_0(x)]\,dx = g'(0)$;
$\mathcal{C}_3$ $\int_\Omega v_0(x)u_0(x)\,dx \neq 0$ *and* $b_1 \neq 0$;

THEOREM 3.2. *Let A and B be sectorial operators (see Definition 2.1). Under assumptions $\mathcal{R}_1$-$\mathcal{R}_5$ and $\mathcal{C}_1$-$\mathcal{C}_3$ there exists $T^* \in (0,T]$ such that for any $\tau \in (0,T^*)$ problem (1.6)–(1.9) has a unique solution $(u,v,h) \in [C^1([0,\tau];C(\bar\Omega)) \cap \mathcal{B}([0,\tau];C^{2(\beta+1)}(\overline\Omega)] \times [C^1([0,\tau];C(\bar\Omega)) \cap \mathcal{B}([0,\tau];C^{2(\beta+1)}(\overline\Omega)] \times C([0,\tau];\mathbf{R})$ when $\beta \in (0,1/2)$ and $(u,v,h) \in [C^1([0,\tau];C(\bar\Omega)) \cap \mathcal{B}([0,\tau];C_\nu^{2(\beta+1)}(\overline\Omega)] \times [C^1([0,\tau]; C(\bar\Omega)) \cap \mathcal{B}([0,\tau];C_\nu^{2(\beta+1)}(\overline\Omega)] \times C([0,\tau];\mathbf{R})$ when $\beta \in (1/2,1)$.*
Moreover $u, v \in C([0,\tau];D(A))$.

Proof. It is an application of the Theorem 3.1, while condition (3.7) follows from the Green formulae for the Laplace operator and the homogeneous Neumann boundary conditions. ∎

4 AN EQUIVALENT FIXED POINT SYSTEM

In this section we formulate the abstract inverse problem (1.6)-(1.9) in terms of an equivalent non–linear fixed point system. The main result of this section is the equivalence Theorem 4.1, which can be considered the heart of the paper since the inverse problem (1.6)-(1.9), without additional conditions on the data is, in general, not well posed. The equivalence theorem gives a set of hypotheses on the data such that the inverse problem becomes well posed, and starting from the equivalence fixed point system we can obtain existence and uniqueness results for system (1.6)-(1.9). A theorem of continuous dependence on the data can also be proved starting from Theorem 4.1. We omit the statement and the proof for the sake of brevity. Here we prove:

THEOREM 4.1. *Let A and B be sectorial operators. Let us assume the data g, u_0, v_0, Φ and Ψ satisfy the regularity conditions R_1-R_5 and the compatibility conditions C_1-C_3. Suppose $\beta \in (0,1)$ and let $(u,v,h) \in [C^1([0,T];X) \cap C([0,T];D(A)) \cap \mathcal{B}([0,T];D_A(\beta+1,\infty))] \times [C^1([0,T];X) \cap C([0,T];D(B)) \cap \mathcal{B}([0,T];D_B(\beta+1,\infty))] \times C([0,T];\mathbf{R})$ be a solution of the problem (1.6)–(1.9). Then the triplet (w,z,h), where $w = u'$, $z = v'$, belongs to*

$$[C([0,T];X)\cap\mathcal{B}([0,T];D_A(\beta,\infty))]\times[C([0,T];X)\cap\mathcal{B}([0,T];D_B(\beta,\infty))]\times C([0,T];\mathbf{R})$$

and solves problem (4.15)(defined in the sequel). Conversely, if $(w,z,h) \in [C([0,T]; X) \cap \mathcal{B}([0,T];D_A(\beta,\infty))] \times [C([0,T];X) \cap \mathcal{B}([0,T];D_B(\beta,\infty))] \times C([0,T];\mathbf{R})$ is a solution of the problem (4.15), then the triplet (u,v,h), where $u = u_0 + 1 * w$, $v = v_0 + 1 * z$, belongs to $[C^1([0,T];X) \cap C([0,T];D(A)) \cap \mathcal{B}([0,T];D_A(\beta+1,\infty))] \times [C^1([0,T];X) \cap C([0,T];D(B)) \cap \mathcal{B}([0,T];D_B(\beta+1,\infty))] \times C([0,T];\mathbf{R})$ and solves problem (1.6)–(1.9).

Proof. Let $(u, v, h) \in [C^1([0,T];X) \cap C([0,T];D(A)) \cap \mathcal{B}([0,T];D_A(\beta+1,\infty))] \times [C^1([0,T];X) \cap C([0,T];D(B)) \cap \mathcal{B}([0,T];D_B(\beta+1,\infty))] \times C([0,T];\mathbf{R})$ be a solution of the problem (1.6)–(1.9). With (u, v, h) we associate the triplet (w, v, h) where the first two components are defined by

$$w(t) = u'(t), \qquad u(t) = u_0 + 1 * w(t), \tag{4.1}$$

$$z(t) = v'(t), \qquad v(t) = u_0 + 1 * z(t). \tag{4.2}$$

Differentiating with respect to the time (1.6) and (1.7) we get

$$\begin{cases} u''(t) = Au'(t) - a_1 u'(t) - b_1[u'(t)[h * v(t)] \\ +u(t)h(t)v_0 - u(t)[h * v'(t)]], \\ v''(t) = Bv'(t) - a_2 v'(t) + b_2[u'(t)[h * v(t)] \\ +u(t)h(t)v_0 + u(t)[h * v'(t)]], \\ u'(0) = Au_0 - a_1 u_0, \quad v'(0) = Bv_0 - a_2 v_0, \\ \Phi[u'(t)] = g'(t), \quad t \in [0,T], \end{cases} \tag{4.3}$$

and thanks to positions (4.1) and (4.2) we obtain

$$\begin{cases} w'(t) = Aw(t) - a_1 w(t) - b_1[w(t)[h * [u_0 + 1 * z(t)]] \\ +[u_0 + 1 * w(t)][h(t)v_0 + h * z(t)]], \\ z'(t) = Bz(t) - a_2 z(t) + b_2[z(t)[h * [u_0 + 1 * z(s)]] \\ +[u_0 + 1 * w(t)][h(t)v_0 + h * z(t)]], \\ w(0) = Au_0 - a_1 u_0, \quad z(0) = Bv_0 - a_2 v_0, \\ \Phi[w(t)] = g'(t), \quad t \in [0,T], \end{cases} \tag{4.4}$$

thanks to the regularity conditions R_1-R_3 and compatibility condition C_1 and C_2 we have that the (AIP) is equivalent to (4.4). Apply the functional Φ to both hands sides of the first equation of system (4.4) we get

$$\begin{cases} w'(t) = Aw(t) - a_1 w(t) - b_1[w(t)[h * [u_0 + 1 * z(t)]] \\ +[u_0 + 1 * w(t)][h(t)v_0 + h * z(t)]], \\ z'(t) = Bz(t) - a_2 z(t) + b_2[z(t)[h * [u_0 + 1 * z(s)]] \\ +[u_0 + 1 * w(t)][h(t)v_0 + h * z(t)]], \\ w(0) = Au_0 - a_1 u_0, \quad z(0) = Bv_0 - a_2 v_0, \\ g''(t) = \Phi[Aw](t) - a_1 g'(t) - b_1 h(t)\Phi[u_0 v_0] \\ -b_1 \Phi\Big([w(t)[h * [v_0 + 1 * z(t)]] \\ +u_0[h * z(t)] + [1 * w(t)][h(t)v_0 + h * z(t)]]\Big), \end{cases} \tag{4.5}$$

where we have taken into account the regularity conditions R_4. Set

$$\chi^{-1} := b_1 \Phi[u_0 v_0] \neq 0 \tag{4.6}$$

and supposed C_3 holds. Thanks to the fundamental condition R_5, that is $\Phi[Aw] =$

$\Psi[w]$, we obtain the system

$$\begin{cases} w'(t) = Aw(t) - a_1 w(t) - b_1[w(t)[h * [u_0 + 1 * z(t)] \\ +[u_0 + 1 * w(t)][h(t)v_0 + h * z(t)]], \\ z'(t) = Bz(t) - a_2 z(t) + b_2[v'(t)[h * [u_0 + 1 * z(s)]] \\ +[u_0 + 1 * w(t)][h(t)v_0 + h * z(t)], \\ w(0) = Au_0 - a_1 u_0, \qquad z(0) = Bv_0 - a_2 v_0, \\ h(t) = \chi\Big\{\Psi[w](t) - a_1 g'(t) - g''(t) - b_1 \Phi\Big([w(t)[h * [v_0 + 1 * z(t)] \\ +u_0[h * z(t)] + [1 * w(t)][h(t)v_0 + h * z(t)]]\Big)\Big\}. \end{cases} \tag{4.7}$$

Using the constant variation formula (see (2.14)), we represent in integral form the first two equations in (4.7) and we get the equivalent system

$$\begin{cases} w(t) = e^{tA}[Au_0 - a_1 u_0] + e^{tA} * \Big[- a_1 w(t) - b_1[w(t)[h * [v_0 + 1 * z(t)]] \\ +[u_0 + 1 * w(t)][h(t)v_0 + h * z(t)]]\Big], \\ z(t) = e^{tB}[Bv_0 - a_2 v_0] + e^{tB} * \Big[- a_2 z(t) + b_2[z(t)[h * [u_0 + 1 * z(t)]] \\ +[u_0 + 1 * w(t)][h(t)v_0 + h * z(t)]]\Big], \\ h(t) = \chi\Big\{\Psi[w](t) - a_1 g'(t) - g''(t) - b_1 \Phi\Big(w(t)[h * [v_0 + 1 * z(t)] \\ +u_0[h * z(t)] + [1 * w(t)][h(t)v_0 + h * z(t)]]\Big)\Big\}, \end{cases} \tag{4.8}$$

replacing the first equation into the term $\Psi[w]$ in the last equation in system (4.8), which is the only term which is not contractive, and define the nonlinear operators

$$\Gamma_1(w, z, h) := e^{tA} * \Big[- a_1 w(t) - b_1[w(t)[h * [v_0 + 1 * z(t)]] \\ +[u_0 + 1 * w(t)][h(t)v_0 + h * z(t)]]\Big], \tag{4.9}$$

$$\Gamma_2(w, z, h) := e^{tB} * \Big[- a_2 z(t) + b_2[z(t)[h * [u_0 + 1 * z(t)]] \\ +[u_0 + 1 * w(t)][h(t)v_0 + h * z(t)]]\Big], \tag{4.10}$$

we get

$$\begin{cases} w(t) = e^{tA}[Au_0 - a_1 u_0] + \Gamma_1(w, z, h), \\ z(t) = e^{tB}[Bv_0 - a_2 v_0] + \Gamma_2(w, z, h), \\ h(t) = \chi\Big\{\Psi\big[e^{tA}[Au_0 - a_1 u_0]\big] - a_1 g'(t) - g''(t) \\ +\Psi\big[\Gamma_1(w, z, h)\big](t) - b_1 \Phi\Big([w(t)[h * [v_0 + 1 * z(t)]] \\ +u_0[h * z(t)] + [1 * w(t)][h(t)v_0 + h * z(t)]]\Big)\Big\}, \end{cases} \tag{4.11}$$

with the positions

$$w_0(t) := e^{tA}[Au_0 - a_1 u_0], \qquad z_0(t) := e^{tB}[Bv_0 - a_2 v_0], \tag{4.12}$$

$$h_0(t) := \chi\Big\{\Psi[e^{tA}[Au_0 - a_1 u_0]] - a_1 g'(t) - g''(t)\Big\}, \tag{4.13}$$

$$\Gamma_3(w,z,h) := \chi\Big\{\Psi\big[\Gamma_1(w,z,h)\big](t) - b_1\Phi\Big([w(t)[h*[v_0+1*z(t)]]$$

$$+u_0[h*z(t)] + [1*w(t)][h(t)v_0 + h*z(t)]\Big)\Big\}, \tag{4.14}$$

we finally get the system (for $t \in [0,T]$)):

$$\begin{cases} w(t) = w_0(t) + \Gamma_1(w,z,h)(t), \\ z(t) = z_0(t) + \Gamma_2(w,z,h)(t), \\ h(t) = h_0(t) + \Gamma_3(w,z,h)(t), \end{cases} \tag{4.15}$$

and this completes the proof. ∎

5 PROOF OF THEOREM 3.1

In order to apply the Contraction Principle to system (4.15) we begin by defining the complete metric space

$$Y_m(\beta) := \{(w,z,h) \in Y \ : \|(w,z,h)\|_Y \le 2m\}, \quad m \in \mathbf{R}_+, \quad \beta \in (0,1) \tag{5.1}$$

where

$$Y := [C([0,T];X) \cap \mathcal{B}([0,T];D_A(\beta,\infty))]$$

$$\times[C([0,T];X) \cap \mathcal{B}([0,T];D_B(\beta,\infty))] \times C([0,T];\mathbf{R}). \tag{5.2}$$

Y is a Banach space when endowed with the norm

$$\|(w,z,h)\|_Y = \|w\|_{C([0,T];X)\cap\mathcal{B}([0,T];D_A(\beta,\infty))}$$

$$+\|z\|_{C([0,T];X)\cap\mathcal{B}([0,T];D_B(\beta,\infty))} + \|h\|_{C([0,T];\mathbf{R})}. \tag{5.3}$$

Moreover, we choose m to satisfy the inequality

$$\|w_0\|_{C([0,T];X)\cap\mathcal{B}([0,T];D_A(\beta,\infty))} + \|z_0\|_{C([0,T];X)\cap\mathcal{B}([0,T];D_B(\beta,\infty))}$$

$$+\|h_0\|_{C([0,T];\mathbf{R})} \le m, \tag{5.4}$$

where w_0, z_0, h_0 are defined in (4.12), (4.13). By virtue of Theorems 2.2 and 2.3 and the regularity conditions the vector function (w_0, z_0, h_0) belongs to Y. We consider the nonlinear vector operator

$$\Gamma(w,z,h) := (w_0 + \Gamma_1(w,z,h), z_0 + \Gamma_2(w,z,h), h_0 + \Gamma_3(w,z,h)), \tag{5.5}$$

$$\Gamma: \quad Y_m(\beta) \to Y_m(\beta). \tag{5.5'}$$

Then we proceed to estimate the nonlinear operators Γ_i $(i = 1,2,3)$ defined in (4.9), (4.10) and (4.14), respectively. In the following we denote by $C_i(m,T)$, $i = 1,...,6$ positive constants continuously depending on the arguments pointed out. We first consider the estimates

$$\|\Gamma_1(w,z,h)\|_{C([0,T];X)\cap\mathcal{B}([0,T];D_A(\beta,\infty))}$$

$$\le a_1\|e^{tA} * w\|_{C([0,T];X)\cap\mathcal{B}([0,T];D_A(\beta,\infty))}$$

$$+b_1\|e^{tA} * [w[h*[v_0 + 1*z]]]\|_{C([0,T];X)\cap\mathcal{B}([0,T];D_A(\beta,\infty))}$$

$$+b_1\|e^{tA} * \big([u_0 + 1 * w][hv_0 + h * z]\big)\|_{C([0,T];X)\cap\mathcal{B}([0,T];D_A(\beta,\infty))}$$
$$\leq TC_1(m,T), \tag{5.6}$$
$$\|\Gamma_2(w,z,h)\|_{C([0,T];X)\cap\mathcal{B}([0,T];D_B(\beta,\infty))}$$
$$\leq a_2\|e^{tB} * z\|_{C([0,T];X)\cap\mathcal{B}([0,T];D_B(\beta,\infty))}$$
$$+b_2\|e^{tB} * [z[h * [u_0 + 1 * z]]]\|_{C([0,T];X)\cap\mathcal{B}([0,T];D_B(\beta,\infty))}$$
$$\|e^{tB} * \big([u_0 + 1 * w(t)][h(t)v_0 + h * z(t)]\big)\|_{C([0,T];X)\cap\mathcal{B}([0,T];D_B(\beta,\infty))}$$
$$\leq TC_2(m,T) \tag{5.7}$$

and finally

$$\|\Gamma_3(w,z,h)\|_{C([0,T];\mathbf{R})} \leq$$
$$\chi \max\Big\{\|\Psi\|_{\mathcal{L}(D_A(\beta,\infty);\mathbf{R})}; \|\Phi\|_{\mathcal{L}(D_A(\beta+1,\infty);\mathbf{R})}\Big\}$$
$$\times\Big\{\|\Gamma_1(w,z,h)\|_{C([0,T];X)\cap\mathcal{B}([0,T];D_A(\beta,\infty))}$$
$$+b_1\|w(t)[h * [v_0 + 1 * z(t)]]\|_{C([0,T];X)\cap\mathcal{B}([0,T];D_A(\beta,\infty))}$$
$$+b_1\|u_0[h * z(t)]\|_{C([0,T];X)\cap\mathcal{B}([0,T];D_A(\beta,\infty))}$$
$$+b_1\|[1 * w(t)][h(t)v_0 + h * z(t)]\|_{C([0,T];X)\cap\mathcal{B}([0,T];D_A(\beta,\infty))}\Big\}$$
$$\leq TC_3(m,T). \tag{5.8}$$

Adding (5.6)-(5.8) we have

$$\|\Gamma(w,z,h)\|_{Y_m} \leq T\sum_{j=1}^{3} C_j(m,T)$$
$$:= T\delta_1(m,T), \tag{5.9}$$

for a suitable T^* such that

$$T^*\delta_1(m,T^*) \leq m \tag{5.10}$$

operator $\Gamma(w,z,h)$ maps Y_m into itself. Analogously, by Theorems 2.2 and 2.3, we can prove the estimates

$$\|\Gamma_1(w_2,z_2,h_2) - \Gamma_1(w_1,z_1,h_1)\|_{C([0,T];X)\cap\mathcal{B}([0,T];D_A(\beta,\infty))} \leq TC_4(m,T)$$
$$\times\Big[\|w_2 - w_1\|_{C([0,T];X)\cap\mathcal{B}([0,T];D_A(\beta,\infty))}$$
$$+\|z_2 - z_1\|_{C([0,T];X)\cap\mathcal{B}([0,T];D_B(\beta,\infty))} + \|h_2 - h_1\|_{C([0,T];\mathbf{R})}\Big], \tag{5.11}$$

$$\|\Gamma_2(w_2,z_2,h_2) - \Gamma_2(w_1,z_1,h_1)\|_{C([0,T];X)\cap\mathcal{B}([0,T];D_B(\beta,\infty))} \leq TC_5(m,T)\}$$
$$\times\Big[\|w_2 - w_1\|_{C([0,T];X)\cap\mathcal{B}([0,T];D_A(\beta,\infty))}$$
$$+\|z_2 - z_1\|_{C([0,T];X)\cap\mathcal{B}([0,T];D_B(\beta,\infty))} + \|h_2 - h_1\|_{C([0,T];\mathbf{R})}\Big], \tag{5.12}$$

and therefore

$$||\Gamma_3(w_2, z_2, h_2) - \Gamma_3(w_1, z_1, h_1)||_{C([0,T];\mathbf{R})} \leq TC_6(m, T)$$

$$\times \Big[||w_2 - w_1||_{C([0,T];X)\cap \mathcal{B}([0,T];D_A(\beta,\infty))}$$

$$+ ||z_2 - z_1||_{C([0,T];X)\cap \mathcal{B}([0,T];D_B(\beta,\infty))} + ||h_2 - h_1||_{C([0,T];\mathbf{R})} \Big]. \qquad (5.13)$$

Lastly adding (5.11)-(5.13) we get

$$||\Gamma(w_2, z_2, h_2) - \Gamma(w_1, z_1, h_1)||_{Y_m} \leq T\delta_2(m, T)$$

$$\times \Big[||w_2 - w_1||_{C([0,T];X)\cap \mathcal{B}([0,T];D_A(\beta,\infty))}$$

$$+ ||z_2 - z_1||_{C([0,T];X)\cap \mathcal{B}([0,T];D_B(\beta,\infty))} + ||h_2 - h_1||_{C([0,T];\mathbf{R})} \Big] \qquad (5.14)$$

where we have set

$$\delta_2(m, T) := \sum_{j=4}^{6} C_j(m, T). \qquad (5.15)$$

If we now choose T^+ such that Γ is a contraction operator in Y_m

$$(T^+)\delta_2(m, T^+) < 1, \qquad (5.16)$$

for

$$T_0 := \min\{T^*, T^+\} \qquad (5.17)$$

we have that Γ, has a unique fixed point in Y_m. Hence we get the statement thanks to the equivalence Theorem 4.1 and the existence of a unique fixed point of Γ . ■

BIBLIOGRAPHY

1. F. Colombo, *Direct and inverse problems for a phase field model with memory,* J. Math. Anal. Appl. **260** (2001), 517–545.

2. F. Colombo, *An inverse problem for a generalized Kermack-McKendrick model,* J. Inv. Ill Pos. Prob. **10** (2002), 221–241.

3. F. Colombo and D. Guidetti, *A unified approach to nonlinear integrodifferential inverse problems of parabolic type*, Zeit. Anal. Anwen. **21** (2002), no. 2, 431–464.

4. F. Colombo and A. Lorenzi, *An inverse problem in the theory of combustion of materials with memory,* Adv. Diff. Equ. **3** (1998), 133–154.

5. F. Colombo and A. Lorenzi, *Identification of time and space dependent relaxation kernels in the theory of materials with memory I,* J. Math. Anal. Appl. **213** (1997), 32–62.

6. F. Colombo and V. Vespri, *Generation of analytic semigroups in $W^{k,p}(\Omega)$ and $C^k(\bar{\Omega})$*, Diff. Int. Equ. **6** (1996), 421–436.

7. M. Grasselli, *An inverse problem in population dynamics,* Num. Funct. Anal. Optim. **18** (1997), no. 3-4, 311–323.

8. A. Lorenzi and E. Sinestrari, *An inverse problem in the theory of materials with memory,* Nonlinear Anal. T. M. A. **12** (1988), 1317–1335.

9. A. Lunardi, *Analytic Semigroups and Optimal Regularity in Parabolic Problems.* Birkhäuser Verlag, Basel, 1995.

10. C. V. Pao, *Nonlinear Parabolic and Elliptic Equations*, Plenium Press, New York, 1992.

11. A. Pazy, *Semigroups of linear operators and applications to partial differential equations*, Springer Verlag, New York, Berlin, Heidelberg, Tokyo, 1983.

12. E. Sinestrari, *On the Cauchy problem in spaces of continuous functions,* J. Math. Anal. Appl. **107** (1985), 16–66.

13. H. Triebel, *Interpolation theory, function spaces, differential operators,* North Holland, Amsterdam, New York, Oxford, 1978.

Gearhart-Prüss Theorem in stability for wave equations: a survey

DAVID CRAMER

Department of Mathematics, University of Missouri, Columbia
MO 65211, USA
e-mail: mathgr6@math.missouri.edu,

YURI LATUSHKIN

Department of Mathematics, University of Missouri, Columbia
MO 65211, USA
e-mail: yuri@math.missouri.edu

To Jerry Goldstein and Rainer Nagel on the occasion of their 60th birthdays.

ABSTRACT. We give a brief survey of applications of the Gearhart-Prüss spectral mapping theorem for abstract strongly continuous semigroups on Hilbert spaces to the study of stability of solitary waves for a large class of Hamiltonian partial differential equations of mathematical physics including Klein-Gordon, nonlinear Schrödinger, Boussinesq, Benjamin-Bona-Mahoney (regularized long-wave), Korteweg-deVries, and Green-Naghdi.

1 INTRODUCTION

The main goal of this survey is to attract the attention of analysts working in the asymptotic theory of strongly continuous semigroups and abstract evolution equations to applications arising in the theory of linearized stability for distinguished solutions (solitary waves, standing waves, etc.) of a number of equations of mathematical physics. This exposition is concentrated around applications of one particularly important abstract result, the *Gearhart-Prüss Theorem.* This result is well known to and widely used by many experts in nonlinear waves. Their beautiful and important work on stability for nonlinear equations appears to be unfamiliar to the semigroup community.

Let A be the generator of a strongly continuous semigroup $\{e^{tA}\}_{t\geq 0}$ on a Hilbert space $\mathcal{H}$. The position of the spectrum $\sigma(e^{tA})$ of the semigroup is responsible for its stability: if $\sigma(e^{tA}) \subset \mathbb{D} := \{z \in \mathbb{C} : |z| < 1\}$, $t \neq 0$, then the semigroup is uniformly asymptotically stable. However, in any actual problem the generator A (and hopefully, its spectrum $\sigma(A)$) is given, not the semigroup $\{e^{tA}\}_{t\geq 0}$. The classical Lyapunov Theorem takes care of this problem: for a wide range of semigroups if $\sigma(A) \subset \mathbb{C}_- := \{z \in \mathbb{C} : \operatorname{Re} z < 0\}$, then $\sigma(e^{tA}) \subset \mathbb{D}, t \neq 0$. This class of semigroups includes analytic semigroups, most frequently arising in applications due to their connections to parabolic problems for PDE's.

Examples of strongly continuous semigroups that do not have the spectral mapping property $\sigma(e^{tA})\backslash\{0\} = \overline{\exp(t\sigma(A))}$, $t \neq 0$, are well known and thoroughly studied by analysts (see [13, 51]). M. Renardy gave an example of the generator A (for which this property does not hold) that appears if one rewrites as an abstract Cauchy problem the following first-order perturbation of the two-dimensional wave equation: $\partial_t^2 u = \Delta u + e^{ix_2}\partial_{x_1} u$, $u = u(t, x_1, x_2)$, $x_1, x_2 \in [0, 2\pi]$ (see [42]). This example is especially exciting, since for the generators arising in one-dimensional hyperbolic PDE's at least the spectral bound $s(A) = \sup\{\operatorname{Re} z : z \in \sigma(A)\}$ is equal to the growth bound $\omega(A) = t^{-1} \log \sup\{|z| : z \in \sigma(e^{tA})\}$, $t \neq 0$ (see [34]).

As Renardy's example shows, one has to be very careful applying results of the spectral analysis of A to stability of $\{e^{tA}\}_{t\geq 0}$ when dealing with "hyperbolic" problems. Quite often the generators of this type appear in the linear stability analysis of some "distinguished" solutions (traveling waves, trains, solitary waves, bound states, etc.) for nonlinear equations of fluid mechanics, nonlinear optics and other applications. We provide a list of the equations and the distinguished solutions as well as the corresponding linearizations in the next section. Since the aforementioned Lyapunov Theorem does not generally work, one needs another tool to derive information about the linear stability of the solution from the spectral information about the generator given by the linearized equation. This is where the following Gearhart-Prüss Theorem is used.

GEARHART-PRÜSS THEOREM. *For a strongly continuous semigroup on a Hilbert space* $\omega(A) < 0$ *if and only if* $s(A) < 0$ *and* $\sup\{\|(z-A)^{-1}\| : \operatorname{Re} z > 0\} < \infty$.

This is a fundamental and well-documented result in the literature on strongly continuous semigroups. The original papers are [15, 40] and detailed discussions may be found in [10, 13, 51]. This result is also well known to the control theory community as the Huang Theorem (see [21] and [12]). It finds numerous applications in "abstract" control (see e.g. [2, 22, 41, 52]) as well as applications in more specific models (see e.g. [11, 31, 32]).We will not review all applications of the Gearhart-Prüss Theorem here, but will concentrate on stability of steady-states for nonlinear equations (see [16, 27, 30, 33, 38, 43, 55]).

Typically, the Gearhart-Prüss Theorem is applied as follows. Consider a nonlinear equation $\partial_t u = L(\partial_x)u + \mathcal{N}(u)$ from the list given in the next section. Here $L(\cdot)$ is a polynomial and $\mathcal{N}$ is the nonlinear term. Take a "distinguished" solution, e.g., a traveling wave of the form $u(x, t) = Q(x - ct)$, $x \in \mathbb{R}, t \geq 0$, where c is a constant (the speed of the wave). In the moving frame $\xi = x - ct$ the original equation can be recast as $\partial_t u = L(\partial_\xi)u + c\partial_\xi u + \mathcal{N}(u)$, such that Q is its steady state. The linearization of the last equation about the steady state gives the operator $\mathcal{L} = L(\partial_\xi) + c\partial_\xi + D\mathcal{N}(Q)$. The stability of the semigroup generated by $\mathcal{L}$ corresponds to the "stability" of Q.

Depending on the type of "stability" of Q that we are interested in, we choose an appropriate space where we want to study $\sigma(\mathcal{L})$. For instance, for the study of orbital asymptotic stability it is natural to consider $\mathcal{L}$ on a space of exponentially weighted functions (see Section 2 for more information).

The spectral analysis of the semigroup $\{e^{t\mathcal{L}}\}_{t\geq 0}$ consists of two parts which are outlined in Sections 3 and 4, respectively. First, the point spectrum $\sigma_p(\mathcal{L})$ is discussed. The main tool of this analysis is the *Evans function*. This is a Wronskian-type function whose zeros correspond to the eigenvalues of $\mathcal{L}$, see excellent reviews [14, 43] and the literature therein. Note that the definition of the Evans function is based on the notion of exponential dichotomy and Palmer-Ben-Artzi-Gohberg type results, well known in the field of abstract evolution equations (see, e.g. [13; p.480] or [10]). Second, resolvent estimates for $\mathcal{L}$ are done to make sure that the resolvent is bounded in the right half plane and to therefore derive stability from the Gearhart-Prüss Theorem.

Besides the stability analysis, the Gearhart-Prüss Theorem is used to establish the existence of local invariant manifolds in a neighborhood of a steady-state solution for a nonlinear equation (see [16, 27, 43]). General results on the existence of the manifolds (see, e. g., [4]) are proved under the assumptions that $\sigma(e^{tA})$ has "spectral gaps"; here A is the operator obtained by the linearization about a steady-state. Since only the information on $\sigma(A)$ is available, one needs to make sure that the spectral mapping property $\sigma(e^{tA})\backslash\{0\} = \exp(t\sigma(A))$ holds for the generator in hand. For this, it is often convenient to use the Gearhart-Prüss Theorem in the following form: $\sigma(e^{tA}) \setminus \{0\}$ consists of all points $e^{\lambda t}$ such that either $\lambda_k = \lambda + 2\pi ik/t \in \sigma(A)$ for some $k \in \mathbb{Z}$, or the sequence $\{\|(\lambda_k - A)^{-1}\|\}_{k\in\mathbb{Z}}$ is unbounded.

Finally, we remark that there are recent generalizations of the Gearhart-Prüss Theorem (for the case of the semigroups acting on Banach spaces) given in terms of Fourier multipliers properties of the resolvent of A (see [1, 20, 28, 29]). A typical result in this direction says that a strongly continuous semigroup $\{e^{tA}\}_{t\geq 0}$ on a *Banach* space X is uniformly exponentially stable if and only if the resolvent of the generator is bounded on $\mathbb{C}_+ = \{z \in \mathbb{C} : \operatorname{Re} z > 0\}$ and, in addition, the function $s \mapsto (is - A)^{-1}$ is an $L_p(\mathbb{R}; X)$–Fourier multiplier, $1 \leq p < \infty$. We do not know if the generalizations can be applied to questions of stability for solutions of nonlinear equations.

2 EQUATIONS AND LINEARIZATIONS

First, we will briefly discuss a general setup for the linearization of nonlinear equations about traveling waves, see [43]. Consider a partial differential equation of the form

$$\partial_t u = L(\partial_x)u + \mathcal{N}(u), \quad u = u(x,t), \quad x \in \mathbb{R}, \quad t \geq 0. \tag{2.1}$$

Here $L(\cdot)$ is a polynomial, u may be a vector-valued function, and $\mathcal{N}$ is a nonlinear term. In the moving coordinate frame $\xi = x - ct$ this equation reads:

$$\partial_t v = L(\partial_\xi)v + c\partial_\xi v + \mathcal{N}(v), \quad v = v(\xi,t), \quad \xi \in \mathbb{R}, \quad t \geq 0. \tag{2.2}$$

This means that if v and u are related via

$$v(\xi,t) = u(\xi + ct, t), \quad \xi = x - ct, \tag{2.3}$$

then $u = u(x,t)$ satisfies (2.1) if and only if $v = v(\xi,t)$ satisfies (2.2). A solution $u(x,t) = Q_c(x - ct)$ of (2.1), or the function Q_c, is called a *traveling wave* for (2.1). In other words, Q_c is a traveling wave for (2.1) if and only $v(\xi,t) = Q_c(\xi)$, related to $u(x,t) = Q_c(x - ct)$ via (2.3), is a time-independent solution of (2.2):

$$0 = L(\partial_\xi)Q_c + c\partial_\xi Q_c + \mathcal{N}(Q_c). \tag{2.4}$$

The linearization of (2.2) about the steady-state Q_c is given by

$$\partial_t u = \mathcal{L}u, \quad \text{where} \quad \mathcal{L}u = L(\partial_\xi)u + c\partial_\xi u + D\mathcal{N}(Q_c)u, \tag{2.5}$$

and $D\mathcal{N}$ is the derivative of the nonlinear term. Thus, $\mathcal{L}$ is a variable coefficient differential operator since $Q_c = Q_c(\xi)$ is ξ-dependent.

In many cases, $\mathcal{L}$ is "asymptotically autonomous", that is, the ξ-dependent coefficients of $\mathcal{L}$ have limits as $|\xi| \to \infty$. The corresponding constant coefficient differential operator $\mathcal{L}^\infty$ often has spectrum in $i\mathbb{R}$, and therefore the essential spectrum of $\mathcal{L}$ is located on $i\mathbb{R}$, since $\mathcal{L}$ is a relatively compact perturbation of $\mathcal{L}^\infty$. In addition, $\mathcal{L}$ might have point spectrum in $\mathbb{C}_+$. These "modes" are called unstable and cause instabilities for the nonlinear equation. In some cases $\mathcal{L}$ might have essential spectrum in $\mathbb{C}_+$ and this also causes instability (see an important general result on instability for nonlinear equations in [46]).

Quite often a nonlinear equation has a two-parametric family

$$\mathcal{M} = \{Q_c(\cdot - ct + \gamma) : c > 0, \gamma \in \mathbb{R}\} \tag{2.6}$$

of solitary waves. It is called *orbitally stable* if a solution of the nonlinear equation that is initially close to a solitary wave $Q_c(\cdot - ct)$ in some (say, Sobolev $H^1(\mathbb{R})$)norm will remain close to the set $\mathcal{M}$ of translates $\{Q_c(\cdot - ct + \gamma)\}$ of the wave, that is if $\inf_\gamma \|u(\cdot,t) - Q_c(\cdot - ct + \gamma)\|$ is small for all $t > 0$ provided it is small for $t = 0$ (see [8, 9, 37, 53]). A more refined stability notion is that of *orbital asymptotic stability* (or *convective stability*), see [37, 38, 55]. In this case one wants to show that if u at $t = 0$ is a small perturbation of a given solitary wave $Q_c(\cdot - ct + \gamma)$, then $u(\cdot,t) - Q_{c_+}(\cdot - c_+t + \gamma_+) \to 0$ as $t \to +\infty$ for some $c_+ \approx c$ and $\gamma_+ \approx \gamma$. Note that convergence here is *not* expected in the sense of a translation-invariant norm (like $L^2(\mathbb{R})$ on $H^1(\mathbb{R})$). This happens because in the coordinate frame moving together with a "large" (and therefore fast) solution Q_c the smaller solitary waves do not decay in the translation—invariant norm, but are being outrun by the large wave (see a very clear explanation of this point and more references in [37], and Fig.1).

Therefore, stability is studied in the spaces of exponentially weighted functions

$$L^2_a = \{u : e^{(\cdot)a}u(\cdot) \in L^2(\mathbb{R})\} \quad \text{and} \quad H^1_a = \{u : e^{(\cdot)a}u(\cdot) \in H^1(\mathbb{R})\}. \tag{2.7}$$

Note that the transition from H^1 to H^1_a and L^2 to L^2_a "shifts" the essential spectrum of $\mathcal{L}$ from $i\mathbb{R}$ into $\mathbb{C}_-$ [30, 33, 37, 55].

We cite a typical "orbital asymptotic stability" result from [37]. Let $Q_c(x - ct + \gamma)$ be a solitary wave solution of the KdV equation (see the equation below). Fix $a \in (0, \sqrt{c/3})$ and $b \in (0, a(c - a^2))$. For constants $C > 0$ and $\epsilon > 0$ sufficiently small consider the initial value problem with $u(x,0) = Q_c(x + \gamma) + v_0(x)$ where $v_0 \in H^2 \cap H^1_a$ and $\|v_0\|_{H^1} + \|v_0\|_{H^1_a} < \epsilon$. Then there exist $c_+ > 0$ and $\gamma_+ \in \mathbb{R}$ such that $|c - c_+| < C\epsilon$, $|\gamma - \gamma_+| < C\epsilon$ and for all $t \geq 0$ we have:

$$\|u(\cdot,t) - Q_{c_+}(\cdot - c_+t + \gamma_+)\|_{H^1} \leq C\epsilon \quad \text{and} \quad \|u(\cdot,t) - Q_{c_+}(\cdot - c_+t + \gamma_+)\|_{H^1_a} \leq C\epsilon e^{-bt}.$$

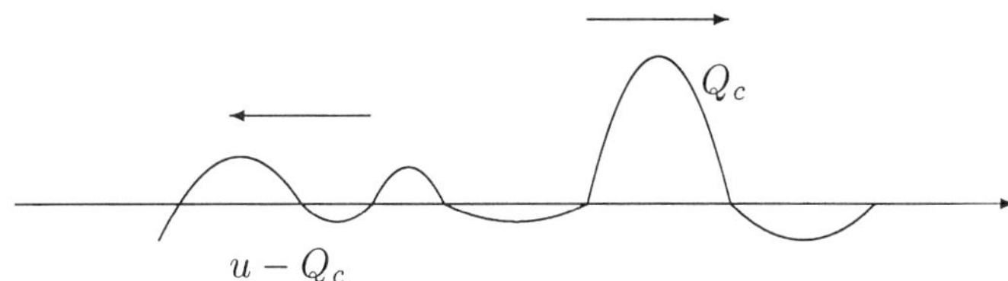

Figure 1.1: Smaller waves move to the left relative to the large wave Q_c. Their L^2 and H^2 norms do not change, but L^2_a and H^2_a norms decay exponentially, if the frame moving with Q_c is considered.

More advanced results of this type for the Boussinesq, BBM, and Green-Naghdi equations are obtained in [33, 37] and [30].

Next, we will give a brief inventory of equations of interest. For each equation, where appropriate, we will include: (a) the general form of the equation, (b) a formula for the nonlinear term most commonly used, (c) the equation written in the "moving frame" $\xi = x - ct$, (d) a formula for the "distinguished" solution (solitary wave or standing wave), (e) the equation linearized about the "distinguished" solution, and (f) the formula for the linearized operator $\mathcal{L}$. In what follows, we assume that $u = u(x,t) \in \mathbb{R}$, $x \in \mathbb{R}$, $t \geq 0$, unless explicitly mentioned otherwise. We cite sources containing detailed historical remarks and bibliography.

The generalized Korteweg-deVries (KdV) equation (see [8, 9, 36]):

$$\begin{aligned}
&\text{(a)} \quad \partial_t u + \partial_x f(u) + \partial_x^3 u = 0; \quad \text{(b)} \quad f(u) = u^{p+1}/(p+1);\\
&\text{(c)} \quad \partial_t v + v^p \partial_\xi v + \partial_\xi^3 v - c\partial_\xi v = 0;\\
&\text{(d)} \quad Q(\xi) = [c(p+1)(p+2)/2]^{\frac{1}{p}} \operatorname{sech}^{\frac{2}{p}}(p\sqrt{c}\xi/2);\\
&\text{(e)} \quad \partial_t u + [Q^p \partial_\xi + \partial_\xi Q^p]u + \partial_\xi^3 u - c\partial_\xi u = 0;\\
&\text{(f)} \quad \mathcal{L} = -[Q^p \partial_\xi + \partial_\xi Q^p] - \partial_\xi^3 + c\partial_\xi.
\end{aligned}$$

The generalized Benjamin-Bona-Mahoney (BBM) equation (and regularized long-wave (RLW) equation, see [36] and [33, 55]):

$$\begin{aligned}
&\text{(a)} \quad \partial_t u + \partial_x u + \partial_x f(u) - \partial_x^2 \partial_t u = 0; \quad \text{(b)} \quad f(u) = u^{p+1}/(p+1);\\
&\text{(c)} \quad \partial_t v + \partial_\xi v + v^p \partial_\xi v - c\partial_\xi v - \partial_\xi^2 \partial_t v + c\partial_\xi^3 v = 0;\\
&\text{(d)} \quad Q(\xi) = \alpha \operatorname{sech}^{\frac{2}{p}}(\beta x),\ \alpha = [(c-1)(p+1)(p+2)/2]^{\frac{1}{p}},\ \beta = p\Big(\frac{c-1}{c}\Big)^{\frac{1}{2}}/2;\\
&\text{(e)} \quad (I - \partial_\xi^2)\partial_t u = \partial_\xi(-c\partial_\xi^2 + (c-1) - Q^p)u;\\
&\text{(f)} \quad \mathcal{L} := (I - \partial_\xi^2)^{-1}(\partial_\xi(-c\partial_\xi^2 + (c-1) - Q^p)).
\end{aligned}$$

The generalized regularized Boussinesq equation (see [36, 38]):

$$\text{(a)}\quad \partial_t^2 u - \partial_x^2 u - \partial_x^2 f(u) - \partial_x^2\partial_t^2 u = 0; \quad \text{(b)}\quad f(u) = u^{p+1}/(p+1);$$
$$\text{(c)}\quad (\partial_t - c\partial_\xi)^2 (I - \partial_\xi^2) v - \partial_\xi^2 v - \partial_\xi (v^p \partial_\xi v) = 0;$$
$$\text{(d)}\quad Q(\xi) = \alpha\, \operatorname{sech}^{\frac{2}{p}}(\beta x),\ \alpha = \left[(c^2-1)(p+1)(p+2)/2\right]^{\frac{1}{p}},\ \beta = p(\frac{c^2-1}{c^2})^{\frac{1}{2}}/2;$$
$$\text{(e)}\quad (\partial_t - c\partial_\xi)^2 (I - \partial_\xi^2) u = \partial_\xi^2 (u + Q^p u).$$

One can re-write ([38; p.359]) the second order in t equation as a first order system and obtain:

$$\text{(f)}\quad \mathcal{L} = \begin{bmatrix} c\partial_\xi & S \\ S(I+Q^p) & c\partial_\xi \end{bmatrix}, \qquad S = \partial_\xi (I - \partial_\xi^2)^{-\frac{1}{2}}.$$

Green-Naghdi system of equations (see [30]):

$$\text{(a)}\quad \partial_t \eta + \partial_x w = 0, \quad \partial_t w = -\partial_x\left(w^2/\eta\right) - \eta\partial_x\eta + \frac{1}{3}\partial_x\left(\eta^2 \frac{d}{dt}\left(\eta\partial_x\left(w/\eta\right)\right)\right),$$
$$w = u\eta, \quad u = u(x,t), \quad \eta = \eta(x,t), \quad \frac{d}{dt} = \partial_t + u\partial_x;$$
$$\text{(d)}\quad w_c = c(c^2-1)\operatorname{sech}^2((3(c^2-1))^{\frac{1}{2}}(x-ct)/(2c)), \quad \eta_c = 1 + w/c;$$
$$\text{(f)}\quad \mathcal{L}\begin{bmatrix} w \\ \eta \end{bmatrix} = \begin{bmatrix} (J_1 w + J_2 w' + J_3 w'' + J_4\eta + J_5\eta' + J_6\eta'')' \\ c\eta' - w' \end{bmatrix},$$
$$\text{where}\quad ' = \partial_\xi, \quad \xi = x - ct,$$
$$J_1 = c - 2w_c/\eta_c - 2\eta_c' w_c'/3 - 2w_c\eta_c''/3 + 2\eta_c w_c''/3 + 2w_c(\eta_c')^2/(3\eta_c)$$
$$J_2 = c\eta_c\eta_c'/3 - 2\eta_c' w_c/3, \quad J_3 = 2\eta_c w_c/3 - c\eta_c^2/3,$$
$$J_4 = -2c\eta_c w_c''/3 + w_c^2/\eta_c^2 - \eta_c + c\eta_c' w_c'/3 + 2w_c w_c''/3 - w_c^2(\eta_c')^2/(3\eta_c^2),$$
$$J_5 = c\eta_c w_c'/3 - 2w_c w_c'/3 + 2w_c^2\eta_c'/(3\eta_c), \quad J_6 = -w_c^2/3.$$

Klein-Gordon equation (see [4, 5, 7, 26, 44, 45, 49]):

$$\text{(a)}\quad \partial_t^2 u = \Delta u + f(u), \quad u = u(x,t), \quad x \in \mathbb{R}^n, \quad n \geq 3,$$
$$\text{(b)}\quad f(u) = |u|^\gamma u - m^2 u, \quad \gamma > 0; \quad \text{or} \quad f(u) = u(|u|^2 - |u|^4 - 1) \quad [44];$$
$$\text{(d)}\quad Q(\cdot) = \text{any radially symmetric stationary solution } [7, 26, 49].$$

As usual, one can re-write the equation as a first order system as follows:

$$\text{(a)}\quad \partial_t \begin{bmatrix} u \\ v \end{bmatrix} = \mathcal{L}\begin{bmatrix} u \\ v \end{bmatrix} + \begin{bmatrix} 0 \\ f(u+Q) - f(Q) - f'(Q)u \end{bmatrix}, \quad \partial_t u = v$$
$$\text{(f)}\quad \mathcal{L} = \begin{bmatrix} 0 & I \\ \Delta + f'(Q) & 0 \end{bmatrix}.$$

Nonlinear Schrödinger (NLS) equation (see [16–19, 45, 47, 48, 54]:

$$\text{(a)}\quad iu_t = -\Delta u - f(x,|u|^2)u + \beta u, \quad u = u(x,t) \in \mathbb{C}, \quad x \in \mathbb{R}^n, \quad \beta \in \mathbb{R},$$
$$\text{(b)}\quad f(x,|u|^2) = 4|u|^2,$$
$$\text{(d)}\quad Q(x) = \sqrt{\beta/2}\,\operatorname{sech}(\sqrt{\beta}x).$$

$\partial_t u = JDE(u)$ $\qquad\qquad$ $\partial_t u = -DE(u)$

Figure 1.2: Hamiltonian versus gradient systems (dotted lines are the level surfaces of the energy functional E).

Writing $u = v + iw$, one can re-write this equation as a system as follows:

$$\text{(a)}\quad \partial_t \begin{bmatrix} v \\ w \end{bmatrix} = \begin{bmatrix} \Delta w + f(x, v^2 + w^2)w + \beta w \\ -\Delta v - f(x, v^2 + w^2)v - \beta v \end{bmatrix}, \quad v = \operatorname{Re} u, \quad w = \operatorname{Im} u$$

$$\text{(f)}\quad \mathcal{L} = \begin{bmatrix} 0 & -L_R \\ L_I & 0 \end{bmatrix}, \quad L_R = -\Delta - \beta + \partial_2 f(x, Q^2(x)),$$

$$L_I = -\Delta - \beta - f(x, Q^2(x)) - 2\partial_2 f(x, Q^2(x))Q^2(x).$$

We conclude this section by the following important remark. All equations above have a Hamiltonian structure, and can be written as $\partial_t u = JDE(u)$, where E is an energy functional on an appropriate function space and J is a skew-symmetric linear operator. This fact is the basic fact for the stability/instability analysis, see [17–19, 45, 50, 54, 55] and the literature therein. The energy E is a conserved quantity. Thus, the solution $u(\cdot, t)$ moves along the level surfaces of E. Note the contrast with gradient systems of the form $\partial_t u = -DE(u)$, where solutions cross the level surfaces (see Fig. 2).

For many Hamiltonian equations of interest the energy E is composed from two conserved quantities, the Hamiltonian H and charge (or impulse) N. The distinguished solutions (solitary waves or standing waves) are critical points for E. Their stability or instability is, formally, determined by the fact that the Hessian D^2E is sign definite. Since there are two conserved quantities present, the solitary or standing wave Q_c could be a constrained minimum of E. For example, for the generalized KdV we have ([37]):

$$E(u) = H(u) + cN(u), \quad H(u) = \int_{\mathbb{R}} \left[\frac{1}{2}(\partial_x u)^2 - F(u)\right] dx, \quad N(u) = \frac{1}{2}\|u\|_{L^2}^2,$$

where $F'(u) = f(u)$, $F(0) = 0$, and for the NLS $E(u) = H(u) - \beta N(u)$ ([53, 55]). A general abstract theory for stability of critical points in Hamiltonian equations is developed in [19].

3 POINT SPECTRUM AND EVANS FUNCTION

We now return to the general discussion in the beginning of Section 2 and again follow [43]. We want to study the spectral problem $\mathcal{L}w = \lambda w$ for the operator $\mathcal{L}$ as in (2.5), where $\lambda \in \mathbb{C}$ and $w = w(\xi)$, or equivalently, solutions of $\partial_t u = \mathcal{L}u$ of the form $e^{t\lambda}w(\xi)$. Note that the steady state equation (2.4) is an ODE that contains higher order derivatives in ξ and thus can be written as a first order system $\partial_\xi V = f(V)$, where $V = V(\xi)$ and f (depending also on c) is a vector field whose dimension depends on the order of $L(\cdot)$. The corresponding eigenvalue problem $\mathcal{L}w = \lambda w$ for $\mathcal{L}$ is also an ODE with higher order derivatives in ξ and can also be recast as $\partial_\xi Y = (Df(V) + \lambda B)Y$ for an appropriate matrix $B = B(\xi)$ and vector-function $Y = Y(\xi)$.

As a result, the linearized stability of a traveling wave for (2.1) is related to the study of the nonautonomous linear equation $\partial_\xi Y = A(\xi;\lambda)Y$ or the spectral properties of the operator $\Gamma(\lambda) := -\partial_\xi + A(\cdot;\lambda)$ on a space of vector-functions $Y = Y(\xi)$ (say, $L^2(\mathbb{R};\mathbb{C}^n)$). We will assume that $A(\xi,\lambda) = \tilde{A}(\xi) + \lambda B(\xi)$, where $\tilde{A}, B \in C^\infty(\mathbb{R};\mathbb{R}^{n\times n})$ (see [43]).

Fix $\lambda \in \mathbb{C}$ and let $U(\xi,\zeta)$ denote the evolution family (propagator) for $\partial_\xi Y = A(\xi;\lambda)Y$. Recall ([10, 13, 43]) that $\partial_\xi Y = A(\xi;\lambda)Y$ is said to have exponential dichotomy on $I(=\mathbb{R}^+, \mathbb{R}^-$ or $\mathbb{R})$ if there exists a continuous projection-valued function P and constants $M \geq 1$, $\alpha > 0$ such that, for $\xi,\zeta \in I$, we have $U(\xi,\zeta)P(\zeta) = P(\xi)U(\xi,\zeta)$ and

$$||U^s(\xi,\zeta)|| \leq Me^{-\alpha(\xi-\zeta)}, \quad \xi \geq \zeta; \qquad ||U^u(\xi,\zeta)|| \leq Me^{\alpha(\xi-\zeta)}, \quad \xi \leq \zeta.$$

Here and below $\mathbf{R}(\cdot)$ and $\mathbf{N}(\cdot)$ denote the range and kernel of an operator, and $U^s(\xi,\zeta) = U(\xi,\zeta) \mid \mathbf{R}(P(\zeta)), U^u(\xi,\zeta) = U(\xi,\zeta) \mid \mathbf{N}(P(\zeta))$ are the corresponding restrictions. The ξ-independent $\dim \mathbf{N}(P(\xi))$ is called the *Morse index*. If $\partial_\xi Y = A(\xi;\lambda)Y$ has exponential dichotomy on $\mathbb{R}_-$ and $\mathbb{R}_+$ with projections P_- and P_+, then we denote $i_\pm(\lambda) = \dim \mathbf{N}(P_\pm(0))$.

A well-known result in [3, 6, 35] relates the Fredholm properties of the operator $\Gamma(\lambda)$ on $L^2(\mathbb{R};\mathbb{C}^n)$ and dichotomy of $\partial_\xi Y = A(\xi;\lambda)Y$ as follows: $\Gamma(\lambda)$ is invertible if and only if there is exponential dichotomy on $\mathbb{R}$; also, $\Gamma(\lambda)$ is Fredholm if and only if there are dichotomies on $\mathbb{R}_-$ and $\mathbb{R}_+$. If this is the case then the spaces $\mathbf{N}(P_-(0)) \cap \mathbf{R}(P_+(0))$ and $\mathbf{N}(\Gamma(\lambda))$ are isomorphic via $Y(0) \mapsto Y(\cdot)$, and the index $\operatorname{ind}\Gamma(\lambda) = i_-(\lambda) - i_+(\lambda)$.

Suppose $\Omega \subset \mathbb{C}$ is a simply-connected open set such that for each $\lambda \in \Omega$ the operator $\Gamma(\lambda)$ is Fredholm with index zero. Then $P_+ = P_{+,\lambda}$ and $P_- = P_{-,\lambda}$, the dichotomy projections on $\mathbb{R}_+$ and $\mathbb{R}_-$ for $\partial_\xi Y = A(\xi,\lambda)Y$, exist and can be chosen analytically in $\lambda \in \Omega$ (see [39]). Also, $k := \dim \mathbf{N}(P_{-,\lambda}(0)) = \dim \mathbf{N}(P_{+,\lambda}(0))$ is constant for $\lambda \in \Omega$. Choose (analytically in λ) bases $Y_1(\lambda), \ldots, Y_k(\lambda)$ of $\mathbf{N}(P_{-,\lambda}(0))$ and $Y_{k+1}(\lambda), \ldots, Y_n(\lambda)$ of $\mathbf{R}(P_{+,\lambda}(0))$ and define the *Evans function* by

$$E(\lambda) = \det[Y_1(\lambda), \ldots, Y_n(\lambda)], \quad \lambda \in \Omega. \tag{3.1}$$

Remark that $E(\lambda) = 0$ if and only if $\mathbf{N}(P_{-,\lambda}(0)) \cap \mathbf{R}(P_{+,\lambda}(0)) \neq \{0\}$ or, equivalently, $\mathbf{N}(\Gamma(\lambda)) \neq \{0\}$. Going back to the spectral problem $\mathcal{L}w = \lambda w$, we conclude that $\lambda \in \sigma_p(\mathcal{L})$ if and only if $E(\lambda) = 0$ ([14; Thm.3.1]).

Let us consider three specific examples of the applications of the Evans function.

First, we shall follow an excellent exposition in [14] to sketch the proof of the instability of the scalar reaction-diffusion equation linearized about a pulse solution. The reaction-diffusion equation

$$\partial_t u = \partial_x^2 u - u + 2u^3, \quad (x,t) \in \mathbb{R} \times \mathbb{R}^+, \tag{3.2}$$

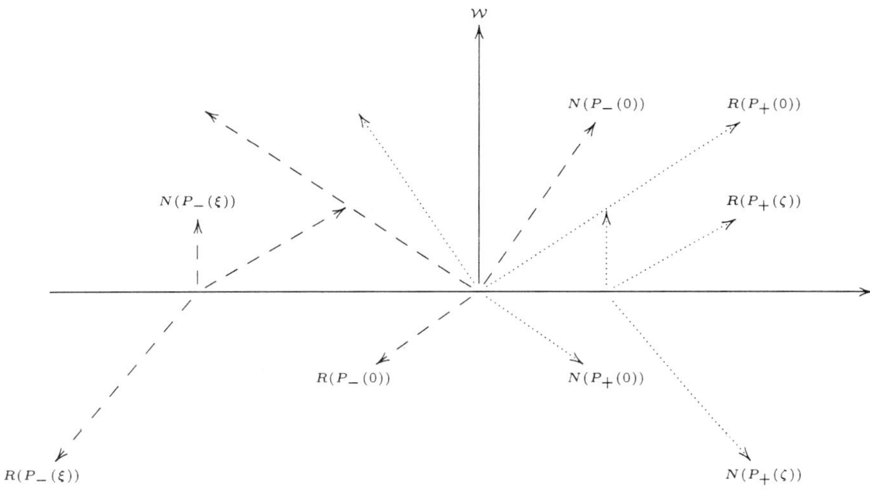

Figure 1.3: Dichotomy: $\mathcal{W} = N(P_-(0)) \cap R(P_+(0))$

supports a pulse solution of the form $Q(x) = \text{sech}x$. The eigenvalue problem for the linearization about Q is then:

$$w'' - (1 - 6Q^2(x))w = \lambda w, \quad ' = \partial_x \tag{3.3}$$

We shall indicate how the Evans function may be used to prove the existence of an unstable mode and thus to show that the linearization is unstable. Let the linearized equation be recast as a first-order system for $Y = [w, w']^T$:

$$Y' = (M(\lambda) + R(x))Y, \quad M(\lambda) = \begin{bmatrix} 0 & 1 \\ 1+\lambda & 0 \end{bmatrix}, \quad R(x) = \begin{bmatrix} 0 & 0 \\ -6Q^2(x) & 0 \end{bmatrix}. \tag{3.4}$$

Now assume that $\text{Re}\,\lambda > -1$. The eigenvalues of $M(\lambda)$ are $\mu^{\pm}(\lambda) = \pm\sqrt{1+\lambda}$ with eigenvectors $\eta^{\pm}(\lambda) = [1, \mu^{\pm}(\lambda)]^T$. Consider solutions $Y^+(\lambda, x)$ and $Y^-(\lambda, x)$ to (3.4) such that

$$\lim_{x\to+\infty} Y^+(\lambda, x)e^{-\mu^-(\lambda)x} = \eta^-(\lambda), \quad \lim_{x\to-\infty} Y^-(\lambda, x)e^{-\mu^+(\lambda)x} = \eta^+(\lambda). \tag{3.5}$$

Note that $\lim_{x\to+\infty}|Y^+(\lambda, x)| = 0$ and $\lim_{x\to-\infty}|Y^-(\lambda, x)| = 0$. The Evans function is then defined to be: $E(\lambda) = \det[Y^-(\lambda, x), Y^+(\lambda, x)]$ and does not depend upon x (see [14]).

By the remark after equation (3.1), we know that $E(\lambda) = 0$ if and only if λ is an eigenvalue of (3.3). Note that $\lambda = 0$ is an eigenvalue of (3.3) with eigenfunction $Q'(x)$. We then know that $E(0) = 0$. It can be shown that $E'(0) > 0$ and that $E(\lambda)$ has a negative limit for large λ (see [14]). These facts then prove the existence of a positive root for $E(\lambda) = 0$. This gives the existence of a positive eigenvalue which then implies the instability of the linearized equation for (3.2).

Secondly, the Evans function may also be used to show that there does not exist a positive real root to $E(\lambda) = 0$ and thus assures the asymptotic stability of the linearized system. Consider the KdV equation as discussed in Section 2. For $p = 1, 2$ the Evans function takes the form (see [37])

$$E(\lambda) = \left(\frac{\mu_1(\lambda) + \sqrt{c}}{\mu_1(\lambda) - \sqrt{c}} \right)^2, \tag{3.6}$$

where $\mu_1(\lambda)$ is the root of $\mu^3 - c\mu + \lambda = 0$ with smallest real part. Then, if $E(\lambda) = 0$ we have that $\mu_1(\lambda) = -\sqrt{c}$ which in turn gives $\lambda = 0$. So for the cases $p = 1$ or $p = 2$, $E(\lambda) = 0$ only for $\lambda = 0$.

Lastly, a similar stability result holds for the RLW equation under certain scaling conditions (see [55]). The scaled Evans function for the RLW equation converges to the Evans function (3.6) for the KdV equation. For $(c-1)/c$ positive and sufficiently small, $\lambda = 0$ is the only eigenvalue of the linearized RLW problem for sufficiently small $|\lambda|$. For larger values of $|\lambda|$, one can show that $E(\lambda) \sim 1$. The reader is encouraged to seek details in [38, 55].

4 RESOLVENT ESTIMATES FOR GEARHART-PRÜSS THEOREM AND CONSEQUENCES

We now return to the general discussion in the beginning of Section 2 and sketch the abstract scheme that lead to the application of the Gearhart-Prüss Theorem. We follow an excellent exposition in [55], see also [30,33,37,38] for detailed analyses of the KdV, BBM, Boussinesq and GN equations.

Suppose that equation (2.2) has a two-parametric family $\mathcal{M}$ of fixed points $Q(\cdot, \vec{p})$, where $\vec{p}$ is a parameter (it could be the pair (γ, c) as in (2.6)). We represent the solution v of (2.2) in the form $v(\xi, t) = Q(\xi, \vec{p}(t)) + w(\xi, t)$. The evolution of $\vec{p}$ describes the dynamics along $\mathcal{M}$ while w describes the dynamics in the "transversal" to $\mathcal{M}$ direction. Fix a reference fixed point $Q_c \in \mathcal{M}$ of (2.2). Linearizing (2.2) about Q_c, and taking in account certain symmetries in the equation (e.g. translation invariance in space or phase), after some substantial work ([33, 55]) we arrive to the following equations:

$$\partial_t w = \mathcal{L} w + \mathcal{F}(w, \vec{p}), \quad \partial_t \vec{p} = \mathcal{S}(w, \vec{p}(t)).$$

Here $\mathcal{L}$ is given in (2.5). For instance, for the regularized long-wave equation (BBM with $p = 1$) we have ([33, 55]):

$$\mathcal{L} = (I - \partial_\xi^2)^{-1} \partial_\xi L_{Q_c}, \quad \text{where } L_{Q_c} = -c\partial_\xi^2 + c - 1 - Q_c, \quad c > 1.$$

Based on the heuristic argument in Section 2 regarding the *asymptotic* orbital stability, we do not expect that $\omega(\mathcal{L}) < 0$ on a function space with a translation-invariant norm. Indeed, the Hamiltonian structure of the system usually implies that the L^2-spectrum of $\mathcal{L}$ belongs to $i\mathbb{R}$. It is natural, however, to consider $\mathcal{L}$ on *weighted* spaces (with an exponential weight for the KdV, RLW, Boussinesq and GN equations, and power weight for the NLS). Speaking about the exponentially weighted spaces, we want to study $\mathcal{L}$ on H^1_a, $a > 0$, see (2.7), or equivalently, we want to study on $H^1(\mathbb{R})$ the operator $\mathcal{L}_a = e^{(\cdot)a} \mathcal{L} e^{-(\cdot)a}$. Note that under the transformation $\mathcal{L} \mapsto \mathcal{L}_a$ the spectrum of the constant coefficient "asymptotic" operator $\mathcal{L}_a^\infty$ will shift from $i\mathbb{R}$ to form a curve in $\mathbb{C}_-$. Using the Evans function, one can give a (highly nontrivial) argument to show that $\lambda = 0$ is the only eigenvalue for $\mathcal{L}_a$ in $\overline{\mathbb{C}_+}$ with two-dimensional eigenspace $\ker \mathcal{P}$ for the spectral projection $\mathcal{P}$ ([33, 55]).

Finally, one is ready to apply the Gearhart-Prüss Theorem, and to derive the required estimate

$$||e^{\mathcal{L}_a t}\mathcal{P}w||_{H^1} \leq Ce^{-bt}||w||_{H^1}, \quad t \geq 0.$$

For this, one has to show that the resolvent estimate $\sup\{||(\lambda - \mathcal{L}_a\mathcal{P})^{-1}|| : \operatorname{Re}\lambda > -b\} < \infty$ holds for some $b > 0$. The proof of the resolvent estimate is based on the identity

$$\lambda - \mathcal{L}_a = (\lambda - \mathcal{L}_a^{\infty})(I + K(\lambda)), \quad \text{where } K(\lambda) = (\lambda - \mathcal{L}_a^{\infty})^{-1}(\mathcal{L}_a - \mathcal{L}_a^{\infty}) \qquad (4.1)$$

is a compact operator. Since $\mathcal{L}_a^{\infty}$ is a constant coefficient differential operator, its resolvent $(\lambda - \mathcal{L}_a^{\infty})^{-1}$ is given explicitly, and is proved to be bounded for $\operatorname{Re}\lambda > -b$. Again, using the Evans function, one can prove that $||K(\lambda)|| < 1$ for large $|\lambda|$. This proves the resolvent estimate, and therefore the desired exponential decay is established.

Note that an approach similar to (4.1) was used in [16] to prove the spectral mapping theorem for the linearized NLS equation.

BIBLIOGRAPHY

1. W. Arendt and S. Bu, *The operator-valued Marcinkiewicz multiplier theorem and maximal regularity*, Ulmer Seminare **5** (2000), 36–58.

2. V. Barbu, *Optimal control of linear periodic resonant systems in Hilbert spaces*, SIAM J. Control Optim. **35** (1997), 2137–2156.

3. H. Bart, I. Gohberg, and M. A. Kaashoek, *Wiener-Hopf factorization, inverse Fourier transform and exponentially dichotomous operators*, J. Funct. Anal. **68** (1986), 1–42.

4. P. W. Bates and C. K. R. T. Jones, *Invariant manifolds for semilinear partial differential equations*, Dynamics Reported **2** (1989), 1–38.

5. P. Bates and C. Jones, *The solution of the nonlinear Klein-Gordon equation near a steady state*, Advance topics in the theory of dynamical systems (G. Fusco, M. Iannelli, and L. Salvadori, eds.), pp. 1–9.

6. A. Ben-Artzi, I. Gohberg, and M. A. Kaashoek, *Invertibility and dichotomy of differential operators on a half line*, J. Dynam. Diff. Eq. **5** (1993), 1–36.

7. H. Berestycki and P. L. Lions, *Nonlinear scalar field equations, II. Existence of infinitely many solutions*, Arch. Rat. Mech. Anal. **82** (1983), 347–376.

8. J. Bona, *On the stability theory of solitary waves*, Proc. Roy. Soc. London Ser. A **344** (1975), 363-374.

9. J. Bona, P. E. Souganidis, and W. Strauss, *Stability and instability of solitary waves of Korteweg-de Vries type*, Proc. Roy. Soc. London Ser. A **411** (1987), 395–412.

10. C. Chicone and Y. Latushkin, *Evolution semigroups in dynamical systems and differential equations*, Math. Surv. Monogr. vol. 70, AMS, Providence, 1999.

11. S. P. Chen, K. S. Liu and Z. Y. Liu, *Spectrum and stability for elastic systems with global or local Kelvin-Voigt damping*, SIAM J. Appl. Math. **59** (1998), 651–668.

12. R. Curtain and H. J. Zwart, *An introduction to infinite-dimensional linear control systems theory*, Springer-Verlag, New York, 1995.

13. K.-J. Engel and R. Nagel, *One-parameter semigroups for linear evolution equations*, Springer-Verlag, New York, 2000.

14. R. Gardner, C. Jones, and T. Kapitula, *Stability of fronts, pulses, and wave-trains*, (preprint, 2001).

15. L. Gearhart, *Spectral theory for contraction semigroups on Hilbert spaces*, Trans. Amer. Math. Soc. **236** (1978), 385-394.

16. F. Gesztesy, C. Jones, Y. Latushkin, and M. Stanislavova, *A spectral mapping theorem and invariant manifolds for nonlinear Schrödinger equations*, Indiana University Math. Journal **49** (2000), 221–243.

17. M. Grillakis, *Linearized instability for nonlinear Schrödinger and Klein-Gordon equations*, Commun. Pure Appl. Math. **41** (1988), 747–774.

18. M. Grillakis, *Analysis of the linearization around a critical point of an infinite dimensional Hamiltonian system*, Commun. Pure Appl. Math. **43** (1990), 299–333.

19. M. Grillakis, J. Shatah, and W. Strauss, *Stability theory of solitary waves in the presence of symmetry, I, II*, J. Funct. Anal. **74** (1987), 160–197, **94** (1990), 308–348.

20. M. Hieber, *A characterization of the growth bound of a semigroup via Fourier multipliers*, In: Evolution equations and their applications in physical and life

sciences (Bad Herrenalb, 1998), Lect. Notes Pure Appl. Math., vol. 215, pp 121–124, Dekker, New York, 2001.

21. F. Huang, *Characteristic conditions for exponential stability of linear dynamical systems in Hilbert spaces*, Ann. Diff. Eq. **1** (1985), 45–53.

22. B. Jacob and H. Zwart, *Equivalent conditions for stabilizability of infinite-dimensional systems with admissable control operators*, SIAM J. Control Optim. **37** (1999), 1419–1455.

23. C. Jones, *Instability of standing waves for non-linear Schrödinger equations*, Ergod. Th. and Dynam. Sys. **8** (1988), 119–138.

24. C. Jones, *An instability mechanism for radially symmetric standing waves of a nonlinear Schrödinger equation*, J. Diff. Eq. **71** (1988), 34–62.

25. C. Jones, *Standing waves of nonlinear Schrödinger equations: existence and stability*, Material instabilities in continuum mechanics (Edinburgh, 1985–1986), Oxford Sci. Publ., Oxford University Press, New York, 1988, pp. 197–216.

26. C. Jones and T. Küpper, *On the infinitely many solutions of a semilinear elliptic equation*, SIAM J. Math. Anal. **17** (1986), 803–835.

27. T. Kapitula and B. Sandstede, *Stability of bright solitary-wave solutions to perturbed nonlinear Schrödinger equations*, Physica D **124** (1998), 58–103.

28. Y. Latushkin and F. Räbiger, *Fourier multipliers in stability and control theory*, (preprint, 2001).

29. Y. Latushkin and R. Shvydkoy, *Hyperbolicity of semigroups and Fourier multipliers*, Systems, Approximation, Singular Integral Operators, and Related Topics, International Workhop on Operator Theory and Applications, IWOTA 2000, (Alexander A. Borichev and Nikolai K. Nikolski, eds.), Oper. Theory Adv. Appl., vol. 129, 2001, pp. 341–364.

30. Yi Li, *Linear stability of solitary waves of the Green-Naghdi equations*, Comm. Pure Appl. Math. **54** (2001), 501–536.

31. K. S. Liu and Z. Y. Liu, *Boundary stabilization of a nonhomogeneous beam with rotary inertia at the tip*, J. Comput. Appl. Math. **114** (2000), 1–10.

32. Z. Liu, S. A. Trogdon and J. M. Yong, *Modeling and analysis of a laminated beam*, Math. Comput. Modelling **30** (1999), 149–167.

33. J. Miller and M. Weinstein, *Asymptotic stability of solitary waves for the regularized long-wave equation*, Comm. Pure Appl. Math. **49** (1996), 399-441.

34. A. F. Neves, H. S. Ribeiro, and D. Lopes, *On the spectrum of evolution operators generated by hyperbolic systems*, J. Funct. Anal. **67** (1986), 320–344.

35. K. Palmer, *Exponential dichotomy and Fredholm operators*, Proc. Amer. Math. Soc. **104** (1988), 149–156.

36. R. Pego and M. Weinstein, *Eigenvalues and instabilities of solitary waves*, Phil. Trans. R. Soc. Lond. **340** (1992), 47–94.

37. R. Pego and M. Weinstein, *Asymptotic stability of solitary waves*, Comm. Math. Phys. **164** (1994), 305–349.

38. R. Pego and M. Weinstein, *Convective linear stability of solitary waves for Boussinesq equations*, Stud. Appl. Math. **99** (1997), 311–375.

39. D. Peterhof, B. Sandstede, and A. Scheel, *Exponential dichotomies for solitary-wave solutions of semilinear equations on infinite cylinders*, J. Diff. Eq. **140** (1997), 266–308.

40. J. Prüss, *On the spectrum of C_0-semigroups*, Trans. Amer. Math. Soc. **284** (1984), no. 2, 847–857.

41. R. Rebarber and S. Townley, *Robustness with respect to delays for exponential stability of distributed parameter systems*, SIAM J. Control Optim. **37** (1998), 230–244.

42. M. Renardy, *On the linear stability of hyperbolic PDEs and viscoelastic flows*, Z. Angew Math. Phys. **45** (1994), 854–865.

43. B. Sandstede, *Stability of travelling waves*, Handbook of Dynamical Systems/Towards Applications (B. Fiedler, ed.), Elsevier, at press.

44. J. Shatah, *Stable standing waves of nonlinear Klein-Gordon equations*, Comm. Math. Phys. **91** (1983), 313–327.

45. J. Shatah and W. Strauss, *Instability of nonlinear bound states*, Comm. Math. Phys. **100** (1985), 173–190.

46. J. Shatah and W. Strauss, *Spectral condition for instability. Nonlinear PDE's, dynamics and continuum physics (South Hadley, MA, 1998)*, Contemp. Math., **255** (2000), 189–198.

47. A. Soffer and M. Weinstein, *Multichannel nonlinear scattering for nonintegrable equations*, Comm. Math. Phys. **133** (1990), 119–146.

48. A. Soffer and M. Weinstein, *Multichannel nonlinear scattering for nonintegrable equations. II. The case of anisotropic potentials and data*, J. Diff. Eq. **98** (1992), 376–390.

49. W. A. Strauss, *Existence of solitary waves in higher dimensions*, Comm. Math. Phys. **55** (1977), 149–162.

50. C. Sulem and P.'Sulem, *The Nonlinear Schrödinger Equation*, Springer-Verlag, New York, 1999.

51. J. M. A. M. van Neerven, *The asymptotic behavior of semigroups of lincar operators*, Oper. Theory Adv. Appl., vol. 88, Birkhäuser, 1996.

52. G. Weiss and R. Rebarber, *Optimizability and estimability for infinite-dimensional linear systems*, SIAM J. Control Optim. **39** (2000), 1204–1232.

53. *M. Weinstein, Lyapunov stability of ground states of nonlinear dispersive evolution equations*, Comm. Pure Appl. Math. **39** (1986), 51–67.

54. M. Weinstein, *Modulational stability of ground states of nonlinear Schrödinger equations*, SIAM J. Math. Anal. **16** (1985), 472-491.

55. M. Weinstein, *Asymptotic stability of nonlinear bound states in conservative dispersive systems*, Mathematical problems in the theory of water waves. (Luminy, 1995), Contemp. Math., vol. 200, Amer. Math. Soc., Providence, RI, 1996, pp. 223–235.

A Note on Generalized Maximum Principles for Elliptic and Parabolic PDE

M. G. CRANDALL
University of California
Santa Barbara, CA 93106, U.S.A.
e-mail: crandall@math.ucsb.edu

A. ŚWIĘCH
Georgia Institute of Technology
Atlanta, GA 30332, U.S.A.
e-mail: swiech@math.gatech.edu

Dedicated to Jerry Goldstein on the occasion of his 60th birthday.

INTRODUCTION

This brief note concerns estimates on maxima of subsolutions of scalar fully nonlinear nondegenerate elliptic equations of the form

$$F[u] = F(x, u, Du, D^2u) = f(x)$$

in an open domain Ω in $\mathbb{R}^n$, together with parabolic generalizations. In the elliptic case, the estimates at issue have the form

$$\max_{\overline{\Omega}} u \leq \max_{\partial\Omega} u^+ + C\left(\int_{\Omega} (f^+)^p\right)^{\frac{1}{p}} \tag{0.1}$$

where $u^+ = \max(u, 0)$, etc. There are two parameters in (0.1), the constant C and the exponent p. When (0.1) holds for all subsolutions of $F[u] = f$, these parameters depend on structure assumptions on F, the dimension n and the domain Ω; in addition C will vary with p. Our contribution has to do with the range of p for which (0.1) holds (with some C). We show that this range can be taken to be independent of the Lipschitz constant γ we impose on the dependence of F in Du as well as Ω; in fact, the range of p's for which (0.1) is valid for F includes the range for which it is valid for an associated Pucci extremal operator with no gradient dependence as explained below. Previous works provided a range depending on the product of γ and the diameter of Ω.

We explain in more detail. When $p = n$ in (0.1), it is a maximum principle of Aleksandrov-Bakelman-Pucci type. When $p < n$ we refer to (0.1) as a generalized maximum principle or GMP. Since the paper of Fabes and Stroock [8] it has been known that given a linear non-divergence form uniformly elliptic equation

$$-\sum_{i,j=1}^{n} a_{ij}(x)\frac{\partial^2 u}{\partial x_i \partial x_j} = f(x) \quad \text{in } \Omega \subset \mathbb{R}^n,$$

where $0 < \lambda I \leq (a_{ij}) \leq \Lambda I$, there exists an exponent $p = p(n, \Lambda/\lambda) < n$ such that any solution $u \in W^{2,p}(\Omega) \cap C(\overline{\Omega})$ can be estimated in the form

$$\max_{\overline{\Omega}} u \leq \max_{\partial\Omega} u + C(d(\Omega))^{2-\frac{n}{p}} \left(\int_{\Omega} (f^+)^p \right)^{\frac{1}{p}}, \tag{0.2}$$

where $C = C(n, \Lambda/\lambda)$ depends only on Λ/λ. If $p = n$, (0.2) follows from the classical Alexandrov-Bakelman-Pucci inequality. Parabolic variants of the ABP estimate are due to Krylov and Tso and in this case $p = n+1$. See Escauriaza [7] concerning the validity of a parabolic version of the result of Fabes and Stroock for some $p < n+1$.

This circle of results was subsequently extended to the weak notion of "L^p-viscosity" subsolutions of fully nonlinear elliptic and parabolic equations of the forms

$$F(x, u, Du, D^2u) = f(x) \quad \text{in } \Omega,$$
$$u_t + F(x, t, u, Du, D^2u) = f(x, t) \quad \text{in } Q.$$

Here Ω is a bounded open domain in $\mathbb{R}^n$, $T > 0$, $Q = \Omega \times (-T, 0]$, $\mathcal{S}(n)$ is the set of real symmetric $n \times n$ matrices and $F\colon\ Q \times \mathbb{R} \times \mathbb{R}^n \times \mathcal{S}(n) \to \mathbb{R}$ satisfies the structure condition

$$\begin{cases} F(x,t,r,p,X) & \text{is nondecreasing in } r,\ F(x,t,0,0,0) = 0, \text{ and} \\ & \mathcal{P}^-(X-Y) - \gamma|p-q| - \omega(|s-r|) \\ & \quad \leq F(x,t,r,p,X) - F(x,t,s,q,Y) \\ & \qquad \leq \mathcal{P}^+(X-Y) + \gamma|p-q| + \omega(|r-s|) \end{cases} \tag{SC}$$

for $X, Y \in \mathcal{S}(n)$, $p, q \in \mathbb{R}^n$ and $r, s \in \mathbb{R}$ for some $\lambda, \Lambda > 0, \gamma \geq 0$ and modulus of continuity ω (see [4, 5, 6]). The "Pucci extremal" operators $\mathcal{P}^+, \mathcal{P}^-$ are given by

$$\begin{cases} \mathcal{P}^-(X) = -\Lambda\,\mathrm{trace}(X^+) + \lambda\,\mathrm{trace}(X^-), \\ \mathcal{P}^+(X) = \Lambda\,\mathrm{trace}(X^-) - \lambda\,\mathrm{trace}(X^+) \end{cases}$$

for $X \in \mathcal{S}(n)$, and where $\mathrm{trace}(X^+)$ (respectively, $\mathrm{trace}(X^-)$) is the sum of the positive eigenvalues of X (respectively, $-X$). Note that there are no conditions on the x, t dependence of $F(x,t,r,p,X)$; F can be merely measurable in the independent variables.

Full proofs of the GMP for L^p-viscosity subsolutions are given in [9] for the elliptic case, and in [5] for the parabolic case. The arguments there generated an exponent p_0 defining the range $p_0 < p$ for which GMP was shown to hold which depended not only on n, λ, Λ, but also on the product of γ and the diameter of the domain. This type of dependence also appears in Cabré [2] for linear equations. The presence of the dependence on γ and the diameter makes some of the statements of results of [4, 5, 6] rather cumbersome. In this note we show that p_0 can be chosen to depend only on n and Λ/λ. This is a simple statement but its proof relies on highly nontrivial results, namely it depends on the strong solvability of extremal equations and suitable estimates on L^q norms of derivatives of the solutions and an argument that allows one to "dispose of the gradient" term at a cost of introducing a new right hand side function that however has a better integrability. A similar argument was used recently in Koike and Takahashi [10] to prove the Harnack inequality for equations with almost quadratic growth in Du.

1 THE RESULTS

We begin with the GMP for L^p-strong solutions of the problems

$$\begin{aligned} &\text{(i) } \mathcal{P}^-(D^2u) \leq f(x), \\ &\text{(ii) } \mathcal{P}^-(D^2u) - \gamma|Du| \leq f(x) \end{aligned} \tag{1.1}$$

in the elliptic case and

$$\begin{aligned} &\text{(i) } u_t + \mathcal{P}^-(D^2u) \leq f(x,t), \\ &\text{(ii) } u_t + \mathcal{P}^-(D^2u) - \gamma|Du| \leq f(x,t) \end{aligned} \tag{1.2}$$

in the parabolic case. An L^p-strong solution of (1.1) (i), (ii) in Ω (respectively, (1.2) (i), (ii) in Q) is a continuous function in $W^{2,p}_{\text{loc}}(\Omega)$ (respectively, $W^{2,1,p}_{\text{loc}}(Q)$) which satisfies the relation pointwise a.e. Of course, $\gamma \geq 0$ thoughout.

The parabolic distance between (x,t) and (y,s) in $\mathbb{R}^n \times \mathbb{R}$ is

$$d((x,t),(y,s)) = (|x-y|^2 + |t-s|)^{1/2}.$$

This is the metric in which the diameter diam(Q) is measured. We will use the abbreviations $d(Q)$ for this parabolic diameter diam(Q) and $d(\Omega)$ for the standard Euclidean diameter diam(Ω).

We take it as known (see [2, 5, 7, 8, 9]) that there exists a $p_0 = p_0(n, \Lambda/\lambda)$ satisfying $n/2 \leq p_0 < n$ such that for $p > p_0$ there is a constant $C = C(n,\lambda,\Lambda,p)$ such that if $f \in L^p(\Omega)$ and $u \in W^{2,p}(\Omega) \cap C(\overline{\Omega})$ is a strong solution of the $\gamma = 0$ problem (1.1) (i) then

$$\max_{\overline{\Omega}} u \leq \max_{\partial\Omega} u + C(d(\Omega))^{2-\frac{n}{p}} \left(\int_\Omega (f^+)^p \right)^{\frac{1}{p}}. \tag{1.3}$$

Similarly, there exists an exponent $p_0 = p_0(n, \Lambda/\lambda)$ satisfying $(n+2)/2 \leq p_0 < n+1$ with the following property: for $p > p_0$ there is a constant $C = C(n,\lambda,\Lambda,p)$ such that if $f \in L^p(Q)$ and $u \in W^{2,1,p}(Q) \cap C(\overline{Q})$ is a strong solution of (1.2) (i) then

$$\max_{\overline{Q}} u \leq \max_{\partial_p Q} u + C(d(Q))^{2-\frac{n+2}{p}} \left(\int_Q (f^+)^p \right)^{\frac{1}{p}} \tag{1.4}$$

where $\partial_p Q$ is the parabolic boundary of Q.

Theorem 1.1 below states the GMP for strong solutions (1.1) (ii) and (1.2) (ii) with the improved information on p_0. It is followed by a corollary extending the results to L^p-viscosity subsolutions of more general equations.

THEOREM 1.1. *(i) Let $p_0 = p_0(n, \Lambda/\lambda)$ be the exponent above such that (1.3) holds for L^p-strong solutions of (1.1) (i) when $p_0 < p$. Then there is another constant $C = C(n, \Lambda, \lambda, p, \gamma \mathrm{d}(\Omega))$ such that (1.3) holds for strong solutions $u \in W^{2,p}(\Omega) \cap C(\overline{\Omega})$ of (1.1) (ii) when $p_0 < p$.*

(ii) Let $p_0 = p_0(n, \Lambda/\lambda)$ be the exponent above such that (1.4) holds for L^p-strong solutions of (1.2) (i) when $p_0 < p$. Then there is another constant

$$C = C(n, \Lambda, \lambda, p, \gamma \mathrm{d}(Q))$$

such that (1.4) holds for strong solutions $u \in W^{2,p}(Q) \cap C(Q)$ of (1.2) (ii) when $p_0 < p$.

COROLLARY 1.2. *Let F satisfy (SC).*

(i) Let p_0 be from Theorem 1.1 (i). Then there exists a constant

$$C = C(n, \Lambda, \lambda, p, \gamma \mathrm{d}(\Omega))$$

such that if $u \in C(\overline{\Omega})$ is an L^p-viscosity solution of $F(x, u, Du, D^2u) \leq f(x)$ in Ω then

$$\max_{\overline{\Omega}} u \leq \max_{\partial\Omega} u^+ + C(d(\Omega))^{2-\frac{n}{p}} \left(\int_\Omega (f^+)^p \right)^{\frac{1}{p}}.$$

(ii) Let p_0 be from Theorem 1.1 (ii). Then there exists a constant

$$C = C(n, \Lambda, \lambda, p, \gamma \mathrm{d}(Q))$$

such that if $u \in C(\overline{Q})$ is an L^p-viscosity solutions of $u_t + F(x, t, u, Du, D^2u) \leq f(x, t)$ in Q then

$$\max_{\overline{Q}} u \leq \max_{\partial_p Q} u^+ + C(d(Q))^{2-\frac{n+2}{p}} \left(\int_Q (f^+)^p \right)^{\frac{1}{p}}.$$

The first step in deducing the corollary from the theorem is to note that owing to the montonicity of F in u the function u^+ is an L^p viscosity solution of $F(x, 0, Du^+, D^2u^+) \leq f^+(x)$ in case (i) and of $u_t^+ + F(x, t, 0, Du^+, D^2u^+) \leq f^+(x, t)$ in case (ii). Then (SC) implies that u^+ is an L^p-viscosity solution of (1.1) (ii) in the elliptic case and of (1.2) (ii) in the parabolic case. Thus the main point is that the estimates of Theorem 1.1 for strong solutions remain valid for L^p-viscosity subsolutions of the same equations. We refer to Section 2 of [6] for details on how this goes and say no more here.

Proof of Theorem 1.1. We will only prove (ii) as the proof of (i) is similar. Let p_0 be as in the statement. If u is a strong solution of (1.2) we can choose $u^j \in C^2(Q)$ such that $u^j \to u$ in $W^{2,1,p}_{\text{loc}}(Q) \cap C(Q)$ and then

$$u_t^j + \mathcal{P}^-(D^2u^j) - \gamma|Du^j| \to u_t + \mathcal{P}^-(D^2u) - \gamma|Du|$$

in $L^p_{\text{loc}}(Q)$. Therefore, considering first slightly smaller domains we may assume that $u \in C^2(\overline{Q})$ and $f \in C(\overline{Q})$ and then obtain (1.4) in the limit.

It is enough to take $\text{d}(Q) = 1$ as the general case then follows by scaling. Hereafter B_r denotes the ball of radius r and center 0 in $\mathbb{R}^n$, while $Q_r = B_r \times (-r^2, 0]$. Assuming that $0 \in \Omega$, we have

$$Q \subset B_{\text{d}(Q)} \times (-\text{d}^2(Q), 0] = Q_{\text{d}(Q)} = Q_1.$$

From now on, we assume that $Q \subset Q_1$.

Define

$$\tilde{f}(x,t) = \begin{cases} f^+(x,t) & (x,t) \in Q, \\ 0 & (x,t) \in Q_2 \setminus Q. \end{cases}$$

As we reduced to the case of continuous f, the function $\tilde{f}$ is bounded and therefore (see [5]) there exists a strong solution $v \in W^{2,1,n+1}_{\text{loc}}(Q_2) \cap C(\overline{Q})$ (actually $v \in W^{2,1,q}_{\text{loc}}(Q_2)$ for all $q < \infty$) of

$$\begin{cases} v_t + \mathcal{P}^-(D^2v) = \tilde{f}(x,t) & \text{in } Q_2 \\ v = 0 \quad \text{on } \partial_p Q_2. \end{cases}$$

Then for $p_0 < p < n+2$

$$0 \le v \le C(\lambda, \Lambda, n, p)\|\tilde{f}\|_{L^p(Q_2)} = C(\lambda, \Lambda, n, p)\|f^+\|_{L^p(Q)} \tag{1.5}$$

where the upper-bound is from the assumptions and the lower from $0 \le \tilde{f}$, $u = 0$ on $\partial_p Q$ (see, e.g., [5]). Moreover

$$\|v\|_{W^{2,1,p}(Q_{3/2})} \le C(n, \lambda, \Lambda, p)\|f^+\|_{L^p(Q)}.$$

Using the imbedding theorem for anisotropic Sobolev spaces [1], Theorem 10.2, and the fact that if $n + 2 > p > (n+2)/2$ then $n + 1 < (n+2)p/(n+2-p)$ we thus obtain

$$\|Dv\|_{L^{n+1}(Q)} \le C(n,p)\|v\|_{W^{2,1,p}(Q_{3/2})} \le C(\lambda, \Lambda, n, p)\|f^+\|_{L^p(Q)}. \tag{1.6}$$

Of course, this inequality persists for $n + 2 \le p$.

As $\mathcal{P}^-$ is superadditive, $\mathcal{P}^-(D^2u) - \mathcal{P}^-(D^2v) \ge \mathcal{P}^-(D^2(u-v))$, so the function $w = u - v$ is a strong solution of

$$w_t + \mathcal{P}^-(D^2w) - \gamma|Dw| \le \gamma|Dv(x,t)| \quad \text{in } Q.$$

Moreover the right hand side of the above inequality is in $L^{n+1}(Q)$. Therefore by the classical Aleksandrov-Bakelman-Pucci-Krylov-Tso maximum principle, the relation $0 \le v$ from (1.5) and (1.6) we have

$$\begin{aligned} \max_{\overline{Q}} w &\le \max_{\partial_p Q} w + C(n, \lambda, \Lambda, \gamma)\gamma\|Dv\|_{L^{n+1}(Q)} \\ &\le \max_{\partial_p Q} u + C(n, p, \lambda, \Lambda, \gamma)\|f^+\|_{L^p(Q)} \end{aligned}$$

where C varied from line to line. From this, using $u = w + v$ and (1.5) again, we finally obtain

$$\max_{\overline{Q}} u \leq \max_{\partial_p Q} u + C(n, p, \lambda, \Lambda, \gamma)\|f^+\|_{L^p(Q)}$$

and the proof is complete. ∎

ACKNOWLEDGEMENT. A. Święch's research was partially supported by NSF grant DMS 0098565.

BIBLIOGRAPHY

1. O. V. Besov, V. P. Il'in, and S.M. Nikolskii, *Integral representations of functions and embedding theorems*, Scripta Series in Mathematics, vol. 1, John Wiley & Sons, 1978.

2. X. Cabré, *On the Alexandroff-Bakelman-Pucci estimate and the reversed Hölder inequality for solutions of elliptic and parabolic equations*, Comm. Pure Appl. Math. **48** (1995), 539–570.

3. L. A. Caffarelli and X. Cabré, *Fully nonlinear elliptic equations*, American Mathematical Society, Providence, 1995.

4. L. Caffarelli, M. G. Crandall, M. Kocan, and A. Święch, *On viscosity solutions of fully nonlinear equations with measurable ingredients*, Comm. Pure Appl. Math. **49** (1996), 365–397.

5. M. G. Crandall, M. Kocan, K. Fok, and A. Święch, *Remarks on nonlinear uniformly parabolic equations*, Indiana Univ. Math. J. **47** (1998), 1293–1326.

6. M. G. Crandall, M. Kocan, and A. Święch, *L^p-Theory for fully nonlinear uniformly parabolic equations*, Comm. Partial Differential Equations **25** (11& 12) (2000), 1997–2053.

7. L. Escauriaza, *$W^{2,n}$ a priori estimates for solutions to fully non-linear equations*, Indiana Univ. Math. J. **42** (1993), 413–423.

8. E. Fabes and D. Stroock, *The L^p integrability of Green's functions and fundamental solutions for elliptic and parabolic equations*, Duke J. Math. **51** (1984), 997–1016.

9. P. Fok, *A nonlinear Fabes-Stroock result*, Comm. Partial Differential Equations **23** (5& 6) (1998), 967–983.

10. S. Koike and T. Takahashi, *Remarks on viscosity solutions for fully nonlinear uniformly elliptic equations with measurable ingredients*, Adv. Differential Equations **7** no. 4, (2002), 493–512.

Finite dimensional convex gradient systems perturbed by noise

G. DA PRATO
Scuola Normale Superiore di Pisa
Piazza dei Cavalieri 7, 56126, Pisa, Italy
e-mail: DaPrato@sns.it

1 INTRODUCTION AND SETTING OF THE PROBLEM

We fix $N \in \mathbb{N}$, and set in all the paper $H = \mathbb{R}^N$. We denote by $\{e_1, ..., e_N\}$ the canonical basis of H. We set $x_h = \langle x, e_h \rangle$, and denote by D_h the partial derivative with respect to x_h for all $h = 1, ..., N$.

We are concerned with the following linear differential operator

$$N_0\varphi(x) = \frac{1}{2}\,\Delta\varphi(x) - \langle DU(x), D\varphi(x)\rangle, \quad \varphi \in C_b^\infty(H), \tag{1.1}$$

where $U : H \to \mathbb{R}$ is such that

HYPOTHESIS 1.1. *(i) U is convex and of class C^1. Moreover*

$$\lim_{|x|\to+\infty} U(x) = +\infty, \tag{1.2}$$

and

$$\int_H (1 + |x|^2)e^{-2U(x)}dx < +\infty. \tag{1.3}$$

(ii) We have

$$\int_H |DU(x)|^2 e^{-2U(x)}dx < +\infty. \tag{1.4}$$

By $C_b^\infty(H)$ we mean the space of all real functions on H of class C^∞ which are bounded together with all their derivatives of any order.

Let us consider the following probability measure on H,

$$\nu(dx) = Z^{-1}e^{-2U(x)}dx := \rho(x)dx,$$

where $Z = \int_H e^{-2U(x)}dx$. Notice that, by (1.4) we have

$$D \log \rho = -2DU \in L^2(H, \nu; H). \tag{1.5}$$

EXAMPLE 1.2. *Let* $U(x) = |x|^{2m}$, $m \in \mathbb{N}$, *or* $U(x) = e^{|x|^2}$, $x \in H$. *Then* U *fulfills Hypothesis 1.1.*

It is well known, see e.g. [8], that N_0 is symmetric in $L^2(H, \nu)$ and that the measure ν is infinitesimally invariant for the operator N_0, that is

$$\int_H N_0\varphi(x)\nu(dx) = 0, \quad \varphi \in C_b^\infty(H), \tag{1.6}$$

see Proposition 1.3 below. Since N_0 is symmetric, it is closable. We shall denote by N_2 its closure.

Let us describe the content of the paper. In section 2 we will show that N_2 is self–adjoint. This result was proved under more general assumptions in [6]; however we present here a (different) proof to make the paper self–consistent.

Section 3 contains the main results of the paper. It is devoted to the characterization of the domain of N_2, first when U is C^2 and then (under a slightly stronger assumption) when it is C^1. In both cases we prove that

$$D(N_2) = \left\{\varphi \in W^{2,2}(H, \nu) : \langle DU, D\varphi\rangle \in L^2(H, \nu)\right\}. \tag{1.7}$$

Such a characterization was known when DU is linear (see [11], [3]). The general case was announced, without complete proofs in [5], see also [6].

This problem has been also studied in $L^p(H, \nu)$ for $p \neq 2$. We quote [2] and [12] for the case when U is quadratic and [7] under different assumptions.

We notice that (1.7) can be easily extended under the weaker assumption that U is semiconvex.

Finally, in section 4 we extend the previous results to a continuous convex function U. We consider the differential operator (1.1) where $DU(x)$ represents now the element of minimal norm of the subdifferential of U at x. We still prove that N_0 is symmetric in $L^2(H, \nu)$ and also that the domain of its closure N_2 is given by (1.7).

Let us give some preliminaries.

PROPOSITION 1.3. *Assume that Hypothesis* 1.1 *holds. Then the following statements hold:*

(i) The measure ν *is infinitesimally invariant for the operator* N_0.

(ii) For any φ, $\psi \in C_b^\infty(H)$ *we have*

$$\int_H N_0\varphi\psi d\nu = -\frac{1}{2}\int_H \langle D\varphi, D\psi\rangle d\nu. \tag{1.8}$$

(iii) N_0 *is symmetric in* $L^2(H, \nu)$.

Proof. First notice that N_0 is well defined thanks to Hypothesis 1.1–(ii).

Let us prove (i). Taking into account (1.2) we have,

$$\frac{1}{2}\int_H \Delta\varphi d\nu = \frac{1}{2}\int_H \Delta\varphi\ \rho dx = \int_H \langle D\varphi, DU\rangle d\nu,$$

and (1.6) follows.

Let us prove (ii). Let $\varphi,\ \psi \in C_b^\infty(H)$, then

$$\frac{1}{2}\int_H \Delta\varphi\ \psi\ d\nu = \frac{1}{2}\int_H \Delta\varphi\ \psi\ \rho\ dx$$

$$= -\frac{1}{2}\int_H \langle D\varphi, D\psi\rangle\ \rho\ dx - \frac{1}{2}\int_H \langle D\varphi, D\log\rho\rangle\psi\ d\nu$$

$$= -\frac{1}{2}\int_H \langle D\varphi, D\psi\rangle\ d\nu + \int_H \langle D\varphi, DU\rangle\psi\ d\nu,$$

and (1.8) follows. Finally, (iii) is an immediate consequence of (1.8). ■

We end this section with the definition of the Sobolev spaces $W^{1,2}(H,\nu)$ and $W^{2,2}(H,\nu)$. For this we will use the following *integration by parts formula*, which follows easily taking into account (1.2). Let $\varphi,\psi \in C_b^\infty(H)$ and $h = 1,...,N$. Then we have

$$\int_H D_h\varphi\ \psi\ d\nu = -\int_H \varphi\ D_h\psi\ d\nu + 2\int_H \varphi\ \psi\ D_hU\ d\nu. \tag{1.9}$$

PROPOSITION 1.4. *Assume that Hypothesis* 1.1 *holds. Then for any* $h = 1,...,N$ *the linear operator* D_h *defined as*

$$D_h : C_b^\infty(H) \subset L^2(H,\nu) \to L^2(H,\nu),\ \varphi \to D_h\varphi.$$

is closable in $L^2(H,\nu)$.

We shall denote by $\overline{D_h}$ the closure of D_h.

Proof. Let $h \in H$, $g \in L^2(H,\nu)$ and let $\{\varphi_n\}$ be a sequence in $C_b^\infty(H)$ such that

$$\varphi_n \to 0,\ \ D\varphi_n \to g \quad \text{in } L^2(H,\nu).$$

We have to show that $g = 0$. Let $\psi \in C_b^\infty(H)$. Then by (1.9) we have

$$\int_H D_h\varphi_n\ \psi\ d\nu = -\int_H \varphi_n\ D_h\psi\ d\nu + 2\int_H D_hU\ \varphi_n\ \psi\ d\nu.$$

Clearly

$$\lim_{n\to\infty}\int_H D_h\varphi_n\ \psi\ d\nu = \int_H g\ \psi\ d\nu,$$

$$\lim_{n\to\infty}\int_H \varphi_n\ D_h\psi\ d\nu = 2\lim_{n\to\infty}\int_H D_hU\ \varphi_n\ \psi\ d\nu = 0.$$

Consequently

$$\int_H g\psi d\nu = 0,$$

so that $g = 0$ from the arbitrariness of ψ and the density of $C_b^\infty(H)$ in $L^2(H,\nu)$. ∎

COROLLARY 1.5. *The linear operator*

$$D : C_b^\infty(H) \subset L^2(H,\nu) \to L^2(H,\nu;H),\ \varphi \to (D_1\varphi, ..., D_N\varphi),$$

is closable.

We shall denote by $\overline{D}$ the closure of D on $L^2(H,\nu)$. Clearly, if φ belongs to the domain of $\overline{D}$, it belongs to the domain of $\overline{D_h}$, $h = 1, ..., N$ and we have

$$\overline{D}\varphi = (\overline{D_1}\varphi, ..., \overline{D_N}\varphi).$$

Now we define $W^{1,2}(H,\nu)$ as the linear space of all functions $\varphi \in L^2(H,\nu)$ which belong to the domain of $\overline{D}$. It is easy to check that $W^{1,2}(H,\nu)$, endowed with the inner product,

$$\langle\varphi,\psi\rangle_{W^{1,2}(H,\nu)} = \langle\varphi,\psi\rangle_{L^2(H,\nu)} + \sum_{h=1}^N \int_H \overline{D_h}\varphi\ \overline{D_h}\psi\, d\nu,$$

is a Hilbert space.

PROPOSITION 1.6. *Assume that* $\varphi \in W^{1,2}(H,\nu)$ *and that* $\overline{D}\varphi = 0$. *Then* φ *is constant.*

Proof. Let $\{\varphi_n\} \subset W^{1,2}(H,\nu)$ be such that

$$\varphi_n \to \varphi, \quad D\varphi_n \to 0 \quad \text{in } L^2(H,\nu;H).$$

Then it is easy to see that, for any $R > 0$, we have

$$\varphi_n \to \varphi, \quad D_h\varphi_n \to 0 \quad h = 1, ..., N, \text{ in } L^2(B_R,\lambda),$$

where λ is the Lebesgue measure and B_R is the ball with center 0 and radius R. Consequently φ is constant on B_R and the conclusion follows. ∎

Let us finally define the space $W^{2,2}(H,\nu)$. We set

$$\gamma(\varphi) = (D\varphi, D^2\varphi), \quad \varphi \in C_b^\infty(H),$$

where

$$D\varphi = (D_1\varphi, ..., D_N\varphi),$$

and

$$D^2\varphi = (D_iD_j\varphi)_{i,j=1,...,N}.$$

PROPOSITION 1.7. γ *is closable.*

Proof. Let $g_h, g_{h,k} \in H,\ ,\ h,k = 1,...,N$ and let $\{\varphi_n\} \subset C_b^\infty(H)$ be such that

$$\lim_{n\to\infty} \varphi_n = 0 \quad \text{in } L^2(H,\nu),$$

$$\lim_{n\to\infty} D_h\varphi_n = g_h \quad \text{in } L^2(H,\nu),\ h = 1,...,N,$$

$$\lim_{n\to\infty} D_hD_k\varphi_n = g_{h,k} \quad \text{in } L^2(H,\nu),\ h,k = 1,...,N.$$

Then $g_h = 0$ since D_h is closable, $h = 1,...,N$. Set

$$D_k\varphi_n = \varphi_n^k, \quad k = 1,...,N,\ n \in \mathbb{N}.$$

Then we have

$$\lim_{n\to\infty} \varphi_n^k = 0, \quad k = 1,...,N,\ n \in \mathbb{N},$$

$$\lim_{n\to\infty} D_h\varphi_n^k = g_{h,k} \quad \text{in } L^2(H,\nu),\ h,k = 1,...,N.$$

Again by the closure of D_h, we deduce that $g_{h,k} = 0,\ h,k = 1,...,N$, so that γ is closable. ■

We shall denote by $\overline{\gamma}$ the closure of γ, and by $W^{2,2}(H,\nu)$ the domain of $\overline{\gamma}$. We shall set moreover

$$\overline{\gamma}(\varphi) = (\overline{D}\varphi, \overline{D^2}\varphi), \quad \varphi \in W^{2,2}(H,\nu),$$

and

$$\overline{D^2}\varphi = (\overline{D_iD_j}\varphi)_{i,j=1,...,N}.$$

It is easy to check that $W^{2,2}(H,\nu)$, endowed with the inner product,

$$\begin{aligned}\langle\varphi,\psi\rangle_{W^{2,2}(H,\nu)} &= \langle\varphi,\psi\rangle_{W^{1,2}(H,\nu)} + \sum_{h,k=1}^N \int_H \overline{D_hD_k}\varphi\, \overline{D_hD_k}\psi\, d\nu \\ &= \langle\varphi,\psi\rangle_{W^{1,2}(H,\nu)} + \int_H \mathrm{Tr}\,[\overline{D^2}\varphi\, \overline{D^2}\psi]\, d\nu,\end{aligned}$$

is a Hilbert space. Obviously $C_b^\infty(H)$ is dense in $W^{2,2}(H,\nu)$.

2 SELF–ADJOINTNESS OF N_2

We shall denote by U_α, $\alpha > 0$, the *Yosida approximations* of U:

$$U_\alpha(x) = \inf\left\{U(y) + \frac{1}{2\alpha}\,|x-y|^2 : y \in H\right\}, \quad x \in H,\ \alpha > 0. \tag{2.1}$$

It is well known, see e. g. [1], that U_α is convex and of class $C^{1,1}$ and $U_\alpha(x) \le U(x)$, $x \in H$. Moreover DU_α is given by

$$DU_\alpha(x) = \frac{1}{\alpha}\,(J_\alpha(x) - x), \quad x \in H,\ \alpha > 0,$$

where

$$J_\alpha = (1 - \alpha DU)^{-1}, \quad x \in H,\ \alpha > 0.$$

Finally,

$$\lim_{\alpha \to 0} U_\alpha(x) = U(x),\ \lim_{\alpha \to 0} DU_\alpha(x) = DU(x), \quad x \in H, \tag{2.2}$$

and

$$|DU_\alpha(x)| \le |DU(x)|, \quad x \in H. \tag{2.3}$$

It is convenient to introduce a C^∞ approximation of U_α. Let $\rho \in C_0^\infty(H)$ non-negative and such that

$$\int_H \rho(x)dx = 1.$$

For any $\alpha, \beta > 0$ we set

$$U_{\alpha,\beta}(x) = \beta^{-n} \int_H U_\alpha(x-y)\rho(y/\beta)dy = \beta^{-n} \int_H U_\alpha(y)\rho((x-y)/\beta)dy,\ x \in H. \tag{2.4}$$

Then $U_{\alpha,\beta}$ is of class C^∞ and convex. Since DU_α is Lipschitz continuous, then $DU_{\alpha,\beta}$ is Lipschitz continuous as well. Therefore all derivatives of $U_{\alpha,\beta}$ of order greater than 1 are bounded.

THEOREM 2.1. *Assume that Hypothesis* 1.1 *holds. Then* N_2 *is self-adjoint on* $L^2(H,\nu)$.

Proof. Let $f \in C_b^\infty(H)$ and $\lambda, \alpha, \beta > 0$. Let us consider the following approximating equation

$$\lambda \varphi_{\alpha,\beta} - \frac{1}{2}\Delta \varphi_{\alpha,\beta} + \langle DU_{\alpha,\beta}, D\varphi_{\alpha,\beta}\rangle = f. \tag{2.5}$$

By using the Itô formula, it is easy to check that equation (2.5) has a unique solution $\varphi_{\alpha,\beta} \in C_b^\infty(H)$, given by

$$\varphi_{\alpha,\beta}(x) = \int_0^{+\infty} e^{-\lambda t}\mathbb{E}[f(X_{\alpha,\beta}(t,x))]dt, \quad x \in H, \tag{2.6}$$

where $X_{\alpha,\beta}(t,x)$ is the solution of the differential stochastic equation

$$\begin{cases} dX_{\alpha,\beta}(t,x) = -DU_{\alpha,\beta}(X_{\alpha,\beta}(t,x))dt + dW(t), \quad t \ge 0, \\ X_{\alpha,\beta}(0,x) = x, \end{cases} \tag{2.7}$$

$\mathbb{E}$ means the expectation and $W(t)$ is the standard Brownian motion in $\mathbb{R}^n$.

Moreover, setting $DX_{\alpha,\beta}(t,x) \cdot h = \eta^h_{\alpha,\beta}(t,x)$, it is easy to see that $\eta^h_{\alpha,\beta}$ is the solution of the initial value problem

$$\begin{cases} D_t\eta^h_{\alpha,\beta}(t,x) = -D^2U_{\alpha,\beta}(X_{\alpha,\beta}(t,x))\eta^h_{\alpha,\beta}(t,x), \quad t \ge 0, \\ \eta^h_{\alpha,\beta}(0,x) = h, \end{cases} \tag{2.8}$$

and that

$$\langle D\varphi_{\alpha,\beta}(x), h\rangle = \int_0^{+\infty} e^{-\lambda t}\mathbb{E}[\langle Df(X_{\alpha,\beta}(t,x)), \eta^h_{\alpha,\beta}(t,x)\rangle]dt, \quad x,h \in H,\ t \geq 0, \tag{2.9}$$

see e.g. [9].

Multiplying the first equation in (2.8) by $\eta^h_{\alpha,\beta}(t,x)$ and taking into account the convexity of U yields

$$\frac{1}{2}\, D_t|\eta^h_{\alpha,\beta}(t,x)|^2 \leq 0, \quad x,h \in H,\ t \geq 0$$

which implies

$$|\eta^h_{\alpha,\beta}(t,x)| \leq |h|, \quad x,h \in H,\ t \geq 0. \tag{2.10}$$

Consequently by (2.9), taking into account the arbitrariness of h, it follows that

$$\|D\varphi_{\alpha,\beta}\|_0 \leq \frac{1}{\lambda}\, \|f\|_1, \quad \alpha,\beta > 0. \tag{2.11}$$

We have used the notations

$$\|\varphi\|_0 = \sup_{x\in H} |\varphi(x)|, \quad \|\varphi\|_1 = \|\varphi\|_0 + \sup_{x\in H} |D\varphi(x)|.$$

Now, since $\varphi_{\alpha,\beta} \in C_b^\infty(H)$, we can write

$$\lambda\varphi_{\alpha,\beta} - N_0\varphi_{\alpha,\beta} = f + \langle DU - DU_{\alpha,\beta}, D\varphi_{\alpha,\beta}\rangle. \tag{2.12}$$

We claim that

$$\lim_{\alpha\to 0}\lim_{\beta\to 0}\langle DU - DU_{\alpha,\beta}, D\varphi_{\alpha,\beta}\rangle = 0 \quad \text{in } L^2(H,\nu). \tag{2.13}$$

This will imply that the closure of the range of $\lambda - N_0$ includes $C_b^\infty(H)$ and so that it is dense in $L^2(H,\nu)$. Therefore the conclusion will follow from the Lumer–Phillips theorem, see [10].

To prove the claim let us notice that, in view of (2.11), we have

$$\begin{aligned}\int_H |\langle DU - DU_{\alpha,\beta}, D\varphi_{\alpha,\beta}\rangle|^2 d\nu &\leq \frac{1}{\lambda^2}\, \|f\|_1^2 \int_H |DU - DU_{\alpha,\beta}|^2 d\nu \\ &\leq \frac{2}{\lambda^2}\, \|f\|_1^2 \left(\int_H |DU - DU_\alpha|^2 d\nu + \int_H |DU_\alpha - DU_{\alpha,\beta}|^2 d\nu\right).\end{aligned} \tag{2.14}$$

Since DU_α is Lipschitz continuous, there exists a constant $k_\alpha > 0$ such that

$$|DU_\alpha(x)| \leq k_\alpha(1 + |x|), \quad x \in H. \tag{2.15}$$

Consequently, there exists a constant $k'_\alpha > 0$ such that

$$|DU_{\alpha,\beta}(x)| \leq k'_\alpha(1 + |x|), \quad x \in H,\ \beta \leq 1.$$

Taking into account Hypothesis 1.1–(ii) we have, by the dominated convergence theorem

$$\lim_{\beta\to 0}\int_H |DU_\alpha - DU_{\alpha,\beta}|^2 d\nu = 0, \tag{2.16}$$

for any $\alpha > 0$. Moreover, taking into account (2.3), we have

$$\lim_{\alpha\to 0}\int_H |DU - DU_\alpha|^2 d\nu = 0, \tag{2.17}$$

again by the dominated convergence theorem. Now the claim follows from (2.16) and (2.17). The proof is complete. ∎

COROLLARY 2.2. *Assume that Hypothesis* 1.1 *holds and let* $\lambda > 0$. *Then the following statements hold:*

(i) For any $f \in C_b^\infty(H)$ *we have*

$$\lim_{\alpha\to 0}\lim_{\beta\to 0}\varphi_{\alpha,\beta} = R(\lambda, N_2)f \quad \text{in } L^2(H,\nu), \tag{2.18}$$

where $\varphi_{\alpha,\beta}$ *is the solution of (2.5).*

(ii) We have

$$R(\lambda, N_2)1 = \frac{1}{\lambda}, \qquad \lambda > 0. \tag{2.19}$$

(iii) If $f, g \in L^2(H,\nu)$ *and* $f \geq g, \nu$ *a.s., then* $R(\lambda, N_2)f \geq R(\lambda, N_2)g$, ν *a.s.*

Proof. (i). By (2.12) it follows that

$$\varphi_{\alpha,\beta} = R(\lambda, N_2)f + R(\lambda, N_2)\langle DU - DU_{\alpha,\beta}, D\varphi_{\alpha,\beta}\rangle.$$

Then the conclusion follows from (2.13).

(ii). If $f = 1$ by (2.6) we have $\varphi_{\alpha,\beta} = \frac{1}{\lambda}$. So (ii) follows from (i).

(iii). Since $C_b^\infty(H)$ is dense in $L^2(H,\nu)$, it is enough to show (iii) for $f, g \in C_b^\infty(H)$. In this case the conclusion follows again from (i). ∎

COROLLARY 2.3. *Assume that Hypothesis* 1.1 *holds. Then*

(i) $D(N_2) \subset W^{1,2}(H,\nu)$ *and for all* $\varphi \in D(N_2)$ *we have*

$$\int_H N_2\varphi\,\varphi\, d\nu = -\frac{1}{2}\int_H |D\varphi|^2 d\nu. \tag{2.20}$$

(ii) For any φ, $\psi \in D(N_2)$ *we have*

$$\int_H N_2\varphi\,\psi d\nu = -\frac{1}{2}\int_H \langle D\varphi, D\psi\rangle d\nu. \tag{2.21}$$

(iii) We have $D(\sqrt{-N_2}\,) = W^{1,2}(H,\nu)$.

Proof. (i). Let $\varphi \in D(N_2)$. Then there exists a sequence $\{\varphi_n\} \subset C_b^\infty(H)$ such that

$$\varphi_n \to \varphi, \quad N_0\varphi_n \to N_2\varphi \quad \text{in } L^2(H,\nu).$$

From (1.8) it follows for any $m, n \in \mathbb{N}$,

$$\int_H N_0(\varphi_n - \varphi_m)\,(\varphi_n - \varphi_m)d\nu = -\frac{1}{2}\int_H |D\varphi_n - D\varphi_m|^2 d\nu.$$

Consequently $\{D\varphi_n\}$ is Cauchy in $L^2(H,\nu;H)$ so that $\varphi \in W^{1,2}(H,\nu)$ as claimed.

Finally, (ii) follows easily by (1.8) and (i) and (iii) follow, by standard arguments, from the identity

$$\int_H |\sqrt{-N_2}\,\varphi|^2 d\nu = \frac{1}{2}\int_H |D\varphi|^2 d\nu, \quad \varphi \in D(N_2).$$

■

2.1 The semigroup generated by N_2

We shall denote by P_t the semigroup generated by N_2.

PROPOSITION 2.4. *The following results hold:*

(i) $P_t 1 = 1$.

(ii) If $f, g \in L^2(H,\nu)$ *and* $f \geq g$ ν *a.s., then* $P_t f \geq P_t g$ ν *a.s.*

(iii) The measure ν *is invariant for* P_t*, that is*

$$\int_H P_t\varphi d\nu = \int_H \varphi d\nu, \tag{2.22}$$

for all $\varphi : H \to H$ *uniformly continuous and bounded.*

(iv) ν *is ergodic and strongly mixing, that is*

$$\lim_{t\to+\infty} P_t\varphi(x) = \int_H \varphi(y)\nu(dy) \quad \text{in } L^2(H,\nu), \tag{2.23}$$

for all $\varphi \in L^2(H,\nu)$.

Proof. (i) and (ii) follow from Corollary 2.2. Let us prove (iii). We have, taking into account that P_t is symmetric and $P_t 1 = 1$,

$$\int_H P_t\varphi d\nu = \langle P_t\varphi, 1\rangle_{L^2(H,\nu)} = \langle \varphi, 1\rangle_{L^2(H,\nu)} = \int_H \varphi d\nu,$$

and (2.22) follows.

Now, let us show (iv). To prove ergodicity we have to show that if $N_2\varphi = 0$ then φ is constant. In fact if $N_2\varphi = 0$ then by (1.8) it follows that $\overline{D}\varphi = 0$. Therefore φ is constant by Proposition 1.6. Finally, let us recall that by (2.20) the function

$$t \to \int_H (P_t\varphi)^2 d\nu,$$

is not increasing. Consequently, there exists the limit

$$\lim_{t\to+\infty} \int_H (P_t\varphi)^2 d\nu = \lim_{t\to+\infty} \langle P_{2t}\varphi, \varphi\rangle_{L^2(H,\nu)}.$$

Therefore, by a standard argument, there exists a symmetric nonnegative operator $Q \in L(H)$ such that

$$\lim_{t\to+\infty} P_t\varphi = Q\varphi \quad \text{in } L^2(H,\nu), \quad \varphi \in L^2(H,\nu).$$

By the ergodicity of ν and the Von Neumann theorem it follows that $Q\varphi = \int_H \varphi(y)\nu(dy)$, so that ν is strongly mixing. ∎

PROPOSITION 2.5. *Let $\varphi \in L^2(H,\nu)$ and $u(t,x) = P_t\varphi(x)$. Then for any $T > 0$ we have*

$$u \in C([0,T]; L^2(H,\nu)) \cap L^2(0,T; W^{1,2}(H,\nu)).$$

Moreover the following identity holds

$$\int_H |u(T,x)|^2\nu(dx) + \frac{1}{2}\int_H |Du(t,x)|^2\nu(dx) = \int_H |\varphi(x)|^2\nu(dx). \tag{2.24}$$

Proof. Since N_2 is self–adjoint, it is well known that

$$u \in C([0,T]; L^2(H,\nu)) \cap L^2(0,T; D(\sqrt{-N_2})),$$

and

$$\int_H |u(T,x)|^2\nu(dx) + \frac{1}{2}\int_H |\sqrt{-N_2}\, u(t,x)|^2\nu(dx) = \int_H |\varphi(x)|^2\nu(dx).$$

Then the conclusion follows from Corollary 2.3. ∎

3 CHARACTERIZATION OF THE DOMAIN OF N_2

3.1 The case when $U \in C^2(H)$

PROPOSITION 3.1. *Assume that Hypothesis 1.1 holds and that $U \in C^2(H)$. Let $\varphi \in D(N_2)$. Then $\varphi \in W^{2,2}(H,\nu)$, $\langle DU, D\varphi\rangle \in L^2(H,\nu)$ and*

$$N_2\varphi = \frac{1}{2}\,\Delta\varphi - \langle DU, D\varphi\rangle. \tag{3.1}$$

Moreover the following identity holds

$$\frac{1}{2}\int_H \text{Tr}\,[(D^2\varphi)^2]d\nu + \int_H \langle D^2U\cdot D\varphi, D\varphi\rangle d\nu = 2\int_H (N_2\varphi)^2 d\nu. \tag{3.2}$$

Proof. Let $\varphi \in C_b^\infty(H)$ and set $f = N_0\varphi$. Then for any $h = 1, 2, ...N$, we have

$$N_0 D_h\varphi - \sum_{k=1}^{N} D_h D_k U \; D_k\varphi = D_h f.$$

Multiplying both sides of this identity by $D_h\varphi$, integrating with respect to ν and taking into account (2.21), we obtain

$$\frac{1}{2}\int_H |DD_h\varphi|^2 d\nu + \sum_{k=1}^{N}\int_H D_h D_k U \; D_k\varphi \; D_h\varphi d\nu = -\int_H D_h\varphi D_h f d\nu.$$

Finally, summing up with respect to h and taking into account that, thanks to (2.21) (Notice that (2.21) also holds when $\varphi \in D(N_0)$ and $\psi \in C_b^1(H)$),

$$\int_H \langle D\varphi, Df\rangle d\nu = -2\int_H f^2 d\nu,$$

we see that φ fulfills (3.2).

Now, let $\varphi \in D(N_2)$ and let $\{\varphi_n\} \subset C_b^\infty(H)$ be such that

$$\varphi_n \to \varphi, \quad N_2\varphi_n \to N_2\varphi \quad \text{in } L^2(H,\nu).$$

Let moreover $m, n \in \mathbb{N}$. Then by (3.2) it follows that

$$\frac{1}{2}\int_H \text{Tr}\,[(D^2(\varphi_m - \varphi_n))^2]d\nu + \int_H \langle D^2U \cdot D(\varphi_m - \varphi_n), D(\varphi_m - \varphi_n)\rangle d\nu$$

$$= 2\int_H (N_2(\varphi_m - \varphi_n))^2 d\nu.$$

This shows that the sequence $\{\varphi_n\}$ is Cauchy in $W^{2,2}(H,\nu)$ and (3.2) follows.

Finally, since

$$\langle DU, D\varphi_n\rangle = \frac{1}{2}\,\Delta\varphi_n - N_0\varphi_n,$$

it follows that there exists the limit

$$\lim_{n\to\infty} \langle DU, D\varphi_n\rangle = \langle DU, D\varphi\rangle \quad \text{in } L^2(H,\nu).$$

■

THEOREM 3.2. *Assume that Hypothesis* 1.1 *holds and that* $U \in C^2(H)$. *Then*

$$D(N_2) = \left\{\varphi \in W^{2,2}(H,\nu) : \; \langle DU, D\varphi\rangle \in L^2(H,\nu)\right\}.$$

Proof. By Proposition 3.1 we know that

$$D(N_2) \subset \left\{\varphi \in W^{2,2}(H,\nu) : \; \langle DU, D\varphi\rangle \in L^2(H,\nu)\right\}.$$

Assume conversely that $\varphi \in W^{2,2}(H,\nu)$ and $\langle DU, D\varphi\rangle \in L^2(H,\nu)$. We have to show that $\varphi \in D(N_2)$.

Step 1. We assume in addition that φ has compact support included in $B_R = \{x \in H : |x| < R\}$.

Let $\{\varphi_n\}$ be a sequence in $C_b^\infty(H)$ such that

$$\varphi_n \to \varphi \quad \text{in } W^{2,2}(H,\nu),$$

and let $g \in C_0^\infty(H)$ such that

$$g(x) = \begin{cases} 1 & \text{if } |x| < R, \\ 0 & \text{if } |x| \geq R+1. \end{cases}$$

Then $g\varphi_n \in C_b^\infty(H)$ for all $n \in \mathbb{N}$ and $\{g\varphi_n\} \to \varphi$ in $L^2(H,\nu)$. Moreover, since

$$N_0(g\varphi_n) = N_0(g)\varphi_n + \langle D\varphi_n, Dg\rangle + \frac{1}{2}\, g\Delta\varphi_n - g\langle DU, D\varphi_n\rangle,$$

the sequence $\{N_0(g\varphi_n)\}$ is Cauchy in $L^2(H,\nu)$. Therefore $g\varphi \in D(N_2)$ as claimed.

Step 2. Conclusion.

Let $\rho \in C_0^\infty(H)$ such that $\rho(0) = 1$ and set $r_n(x) = \rho(x/n)$, $n \in \mathbb{N}$, $x \in H$. By Step 1 we have $r_n\varphi \in D(N_2)$. Moreover

$$\lim_{n\to\infty} \varphi r_n = \varphi \quad \text{in } L^2(H,\nu),$$

and, since

$$\begin{aligned} N_2(\varphi r_n) &= N_2\varphi\, r_n + \frac{1}{2}\,\varphi\Delta r_n + \langle D\varphi, Dr_n\rangle - \langle DU, Dr_n\rangle\varphi \\ &= N_2\varphi\, \rho(x/n) + \frac{1}{2n^2}\,\varphi\Delta\rho(x/n) + \frac{1}{n}\,\langle D\varphi, D\rho(x/n)\rangle + \frac{1}{n}\,\langle DU, D\rho(x/n)\rangle\varphi, \end{aligned}$$

we have

$$\lim_{n\to\infty} N_2(\varphi r_n) = N_2\varphi \quad \text{in } L^2(H,\nu).$$

This implies that $\varphi \in D(N_2)$ as required. ∎

3.2 The case when $U \in C^1(H)$

We need here an additional assumption.

HYPOTHESIS 3.3. *U is nonnegative and*

$$\int_H (1 + |x|^2 + |DU(x)|^2)e^{-2U_1(x)}dx < +\infty. \tag{3.3}$$

where U_1 is the Yosida approximation of U, defined by (2.1), at $\alpha = 1$.

THEOREM 3.4. *Assume that Hypotheses* 1.1 *and* 3.3 *hold. Then*

$$D(N_2) = \left\{\varphi \in W^{2,2}(H,\nu) : \langle DU, D\varphi\rangle \in L^2(H,\nu)\right\}.$$

Proof.

Step 1. If $\varphi \in C_b^\infty(H)$ and $f = N_0\varphi$ we have

$$\int_H \text{Tr}\,[(D^2\varphi)^2]d\nu \le 4\int_H (N_2\varphi)^2 d\nu. \tag{3.4}$$

For any $\alpha, \beta > 0$ let us consider the approximation $\{U_{\alpha,\beta}\}$ of U defined by (2.4) and set

$$\widetilde{U}_{\alpha,\beta}(x) = U_{\alpha,\beta}(x) - \alpha|x|^2, \quad x \in H.$$

Let moreover $\varphi \in C_b^\infty(H)$ and set

$$\widetilde{f}_{\alpha,\beta} = \frac{1}{2}\,\Delta\varphi - \langle D\widetilde{U}_{\alpha,\beta}, D\varphi\rangle.$$

Applying (3.2) with ν replaced by $\nu_{\alpha,\beta}$, where

$$\nu_{\alpha,\beta}(dx) = Z_{\alpha,\beta}^{-1}e^{-2\widetilde{U}_{\alpha,\beta}(x)}dx, \quad \alpha,\beta > 0$$

and

$$Z_{\alpha,\beta} = \int_H e^{-2\widetilde{U}_{\alpha,\beta}(y)}dy, \quad \alpha,\beta > 0,$$

we have

$$\frac{1}{2}\int_H \text{Tr}\,[(D^2\varphi)^2]e^{-2U_{\alpha,\beta}(x)-2\alpha|x|^2}dx \le 2\int_H \widetilde{f}^2_{\alpha,\beta}e^{-2U_{\alpha,\beta}(x)-2\alpha|x|^2}dx. \tag{3.5}$$

Obviously we have

$$\text{Tr}\,[(D^2\varphi)^2]e^{-2U_{\alpha,\beta}(x)-2\alpha|x|^2} \le \text{Tr}\,[(D^2\varphi)^2]e^{-2\alpha|x|^2}, \tag{3.6}$$

since $U_{\alpha,\beta}$ is nonnegative. Moreover, taking into account (2.15), it follows that

$$\begin{aligned}\widetilde{f}^2_{\alpha,\beta}e^{-2U_{\alpha,\beta}(x)-2\alpha|x|^2} &= \left(\frac{1}{2}\,\Delta\varphi - \langle D\widetilde{U}_{\alpha,\beta}, D\varphi\rangle\right)^2 e^{-2U_{\alpha,\beta}(x)-2\alpha|x|^2}\\ &\le \left(\frac{1}{2}\,|\Delta\varphi| + (k'_\alpha(1+|x|) + 2\alpha|x|)\,|D\varphi|^2\right)^2 e^{-\alpha|x|^2}.\end{aligned} \tag{3.7}$$

Now, in view of (3.6) and (3.7) and the dominated convergence theorem , we find, letting $\beta \to 0$ in (3.5),

$$\frac{1}{2}\int_H \text{Tr}\,[(D^2\varphi)^2]e^{-2U_\alpha(x)-2\alpha|x|^2}dx \le 2\int_H f_\alpha^2 e^{-2U_\alpha(x)-2\alpha|x|^2}dx. \tag{3.8}$$

where

$$f_\alpha = \frac{1}{2}\,\Delta\varphi - \langle DU_\alpha - 2\alpha x, D\varphi\rangle.$$

But

$$\text{Tr } [(D^2\varphi)^2]e^{-2U_\alpha(x)-2\alpha|x|^2} \leq \text{Tr } [(D^2\varphi)^2]e^{-2U_1(x)},$$

and

$$f_\alpha^2 e^{-2U_\alpha(x)-2\alpha|x|^2} \leq \left(\frac{1}{2}\,|\Delta\varphi| + (|DU| + 2\alpha|x|)|D\varphi|\right)^2 e^{-2U_1(x)}.$$

Therefore, thanks to (3.3), we obtain (3.4) letting α tend to 0 and using once again the dominated convergence theorem.

Step 2. Conclusion.

Arguing as in the proof of Proposition 3.1 we see that if $\varphi \in D(N_2)$ then $\varphi \in W^{2,2}(H,\nu)$, $\langle DU, D\varphi\rangle \in L^2(H,\nu)$ and (3.1) holds. This implies

$$D(N_2) \subset \left\{\varphi \in W^{2,2}(H,\nu) : \langle DU, D\varphi\rangle \in L^2(H,\nu)\right\}.$$

Finally, the converse inclusion can be proved in exactly the same way as in Theorem 3.2. ■

4 U IS CONTINUOUS

In this section we consider a continuous convex function U. We denote by $DU(x) := (D_1U(x), ..., D_NU(x))$ the element of minimal norm of the subdifferential of U at x. Since U is continuous, $DU(x)$ is defined for all $x \in H$.

We are still concerned with the differential operator N_0 defined by (1.1). We shall assume, besides Hypotheses 1.1, the following

HYPOTHESIS 4.1. *(i) U is convex, continuous, nonnegative and*

$$\lim_{|x|\to+\infty} U_1(x) = +\infty, \tag{4.1}$$

and

$$\int_H (1 + |x|^2)e^{-2U_1(x)}dx < +\infty. \tag{4.2}$$

(ii) We have

$$\int_H |DU_\alpha(x)|^2 e^{-2U_\alpha(x)}dx < +\infty, \quad \alpha \in [0,1]. \tag{4.3}$$

EXAMPLE 4.2. *Let* $N = 1$ *and* $U(x) = |x|$, $x \in H$. *Then*

$$U_\alpha(x) = \begin{cases} -x - \frac{\alpha}{2} & \text{if } x \leq -\alpha, \\ \frac{\alpha^2}{2} & \text{if } |x| \leq \alpha, \\ x - \frac{\alpha}{2} & \text{if } x \geq -\alpha. \end{cases}$$

Moreover

$$DU(x) = \begin{cases} 1 & \text{if } x > 0, \\ -1 & \text{if } x \leq 0. \end{cases}$$

Clearly, Hypotheses 1.1 and 4.1 are fulfilled.

We consider as before the following probability measure on H

$$\nu(dx) = Z^{-1}e^{-2U(x)}dx := \rho(x)dx,$$

where $Z = \int_H e^{-2U(x)}dx$, together with the approximated measures

$$\nu_\alpha(dx) = Z_\alpha^{-1}e^{-2U_\alpha(x)}dx,$$

where $Z_\alpha = \int_H e^{-2U_\alpha(x)}dx,\ \alpha > 0$.

Notice that U_α fulfills Hypothesis 1.1 and moreover that

$$\lim_{\alpha\to 0} Z_\alpha = Z, \tag{4.4}$$

by the dominated convergence theorem.

PROPOSITION 4.3. *Assume that Hypothesis* 4.1 *holds. Then*

(i) The measure ν is infinitesimally invariant for the operator N_0, that is

$$\int_H N_0\varphi(x)\nu(dx) = 0, \quad \varphi \in C_b^\infty(H). \tag{4.5}$$

(ii) For any $\varphi,\ \psi \in C_b^\infty(H)$ we have

$$\int_H N_0\varphi\psi d\nu = -\frac{1}{2}\int_H \langle D\varphi, D\psi\rangle d\nu. \tag{4.6}$$

(iii) N_0 is symmetric on $L^2(H,\nu)$.

Proof. Let us prove (i). Let $\varphi \in C_b^\infty(H)$, Then for any $\alpha > 0$ we have by (1.6), applied to the measure ν_α

$$\frac{1}{2}Z_\alpha^{-1}\int_H \Delta\varphi e^{-2U_\alpha(x)}dx = Z_\alpha^{-1}\int_H \langle D\varphi, DU_\alpha\rangle e^{-2U_\alpha(x)}dx.$$

Letting $\alpha \to 0$ we find, in view of Hypotheses 1.1 and 4.1 and the dominated convergence theorem,

$$\frac{1}{2}\int_H \Delta\varphi d\nu = \int_H \langle D\varphi, DU\rangle d\nu.$$

Consequently (4.5) follows. The other statements can be proved in the same way. ∎

By Proposition 4.3 it follows that N_0 is dissipative in $L^2(H,\nu)$. We shall denote by N_2 its closure.

We will need the following *integration by parts formula*.

PROPOSITION 4.4. *Let $\varphi, \psi \in C_b^\infty(H)$ and $h = 1, ..., N$. Then we have*

$$\int_H D_h\varphi\ \psi\ d\nu = -\int_H \varphi\ D_h\psi\ d\nu + 2\int_H \varphi\ \psi\ D_hU\ d\nu. \tag{4.7}$$

Proof. By (1.9) we have

$$\int_H D_h\varphi\, \psi\, e^{-2U_\alpha(x)}dx = -\int_H \varphi\, D_h\psi\, e^{-2U_\alpha(x)}dx + 2\int_H \varphi\, \psi\, D_hU_\alpha\, e^{-2U_\alpha(x)}dx.$$

The conclusion follows, recalling Hypothesis 4.1 and letting $\alpha \to 0$. ■

Using formula (4.7) we can define the Sobolev spaces $W^{1,2}(H,\nu)$ and $W^{2,2}(H,\nu)$ as in §1.

Now, the following result can be proved as Theorem 2.1.

THEOREM 4.5. *Assume that Hypothesis* 4.1 *holds. Then* N_2 *is self-adjoint on* $L^2(H,\nu)$.

Let us prove finally a characterization result for $D(N_2)$.

THEOREM 4.6. *Assume that Hypothesis* 1.1 *holds. Then*

$$D(N_2) = \{\varphi \in W^{2,2}(H,\nu) :\ \langle DU, D\varphi\rangle \in L^2(H,\nu)\}\,.$$

Proof.

Step 1. If $\varphi \in C_b^\infty(H)$ and $f = N_0\varphi$ we have

$$\int_H \mathrm{Tr}\,[(D^2\varphi)^2]d\nu \le 4Z^{-1}\int_H (N_2\varphi)^2 d\nu. \tag{4.8}$$

Let $\varphi \in C_b^\infty(H)$ and set

$$f_\alpha = \frac{1}{2}\,\Delta\varphi - \langle DU_\alpha, D\varphi\rangle.$$

Then, by Hypotheses 1.1 and 4.1, we have

$$\lim_{\alpha\to 0} f_\alpha = f \quad \text{in } L^2(H,\nu).$$

Moreover, by (3.4) it follows that

$$\frac{1}{2}\int_H \mathrm{Tr}\,[(D^2\varphi)^2]d\nu_\alpha \le 2\int_H f_\alpha^2 d\nu_\alpha = 2Z_\alpha^{-1}\int_H f_\alpha^2(x)e^{-2U_\alpha(x)}dx.$$

and (4.8) follows letting $\alpha \to 0$.

Step 2. Conclusion.

One can argue as in the second step of the proof of Theorem 3.4 ■

BIBLIOGRAPHY

1. H. Brézis, *Opérateurs maximaux monotones*, North–Holland, Amsterdam, 1973.

2. A. Chojnowska-Michalik and B. Goldys, *Generalized Ornstein-Uhlenbeck semigroups: Littlewood–Paley–Stein inequalities and the P. A. Meyer equivalence of norms*, J. Functional Analysis **182** (2001), 243–279.

3. G. Da Prato, *Regularity results for Kolmogorov equations on $L^2(H,\mu)$ spaces and applications,* Ukrainian Mathematical Journal **49** (1997), 448–457.

4. G. Da Prato, *The Ornstein–Uhlenbeck generator perturbed by the gradient of a potential,* Bollettino U.M.I **8** (1998), 501–519.

5. G. Da Prato, *Characterization of the domain of an elliptic operator of infinitely many variables in $L^2(\mu)$ spaces.* Rend. Mat. Acc. Lincei, ser. 9, **8** (1997), 101–105.

6. G. Da Prato and B. Goldys, *Elliptic operators on R^d with unbounded coefficients,* Journal Differential Equations **172** (2001), no. 2, 333–358.

7. G. Da Prato and V. Vespri, *Maximal L^p regularity for elliptic equations with unbounded coefficients*, Nonlinear Analysis TMA, to appear.

8. A. Eberle, *Uniqueness and non–uniqueness of singular diffusion operators,* Lecture Notes in Mathematics vol. 1718, Berlin, Springer–Verlag, 1999.

9. N. V. Krylov, *Introduction to the theory of diffusion processes,* Translations of Mathematical Monographs, vol. 142, American Mathematical Society, Providence, RI, 1995.

10. G. Lumer and R. S. Phillips, *Dissipative operators in a Banach space,* Pacific J. Math **11** (1961), 679–698.

11. A. Lunardi, *On the Ornstein-Uhlenbeck operator in L^2 spaces with respect to invariant measures*, Trans. Amer. Math. Soc. **349** (1997), 155–169.

12. G. Metafune, J. Prüss, A. Rhandi, and R. Schnaubelt, *The domain of the Ornstein–Uhlenbeck operator on an L^p space with invariant measure,* Report 19, Martin–Luther–Universität Halle–Wittenberg, 2001.

Differentiability of the solution semigroup for delay differential equations

GABRIELLA DI BLASIO

Dipartimento di Matematica, Università di Roma "La Sapienza"
P.le Aldo Moro, 5 – 00185 Roma, Italy
e-mail: diblasio@mat.uniroma1.it

ABSTRACT. In this paper we study the differentiability for large t of the solutions u of a class of delay equations with unbounded operators acting on delay term. This property will imply that the solution semigroup Σ is ultimately differentiable. This in turn will enable us to study the asymptotic behaviour of u through the location of spectrum of the infinitesimal generator of Σ.

0 INTRODUCTION

We consider the following delay functional differential equation

$$u'(t) = Au(t) + \int_{-r}^{0} K(s)u_t(s)ds + f(t) \ , \ t > 0 \ . \tag{0.1}$$

Here A is the infinitesimal generator of an analytic semigroup on a Banach space X and $K(s)$ are linear operators satisfying $D(A) \subseteq D(K(s))$, $s \in]-r, 0[$. Moreover $u_t(s) := u(t+s)$. For example A is the realization, in some function space, of an elliptic operator of order $2m$ with suitable boundary conditions and $K(s)$ are differential operators of order less than or equal to $2m$.

To treat (0.1) one has to know the history of u in the interval $[-r, 0]$. Hence one has to assign an initial datum in some space of functions defined on $[-r, 0]$.

In the framework of continuous initial functions existence results for equations similar to (0.1) (also in the case $D(K(s)) = D(A)$) are given by Ardito and Ricciardi [2], Prüss [19; Section III.13.6] and Sinestrari [20]. For the (easier) case where $D(A)$ is strictly contained in $D(K(s))$ we refer to the extensive bibliography in the book of Wu [22].

In this paper we are interested in situations which require the existence of the variation of constants formula. This is useful, for instance, in applications arising from control theory (see [9]). Hence we study (0.1) in a L^p setting, $1 \le p < \infty$, and consider the following initial conditions

$$u(s) = \varphi_1(s) \text{ , a.e. } s \in]-r, 0[; \quad u(0) = \varphi_0 \ . \tag{0.2}$$

There is a number of results concerning the study of (0.1)– (0.2) in the case where $D(A)$ is strictly contained in $D(K(s))$. See e.g. Wu [22], Bátkai and Piazzera [4] and the references therein.

If $D(A) = D(K(s))$ existence and stability for solutions of (0.1)– (0.2) have been investigated, in a more general setting, mainly if X is a Hilbert space and $p = 2$. See e. g. Tanabe [21; Chapter 8], Mastinšek [16] and the references therein.

If X is a Banach space and $1 \le p < \infty$ problems similar to (0.1)–(0.2), also containing discrete delay terms, are studied in [14] where existence and regularity results are given, under various assumptions on the data.

Using these results we can associate to each pair $(\varphi_1, \varphi_0) \in Y$, where Y is a suitable product space, the solution $\tilde{u}$ of (0.1)–(0.2) with $f = 0$. This will introduce the solution semigroup $(\Sigma(t))_{t \ge 0}$ in Y.

In this paper we characterize the infinitesimal generator Λ and give conditions for ultimately differentiability of $(\Sigma(t))_{t \ge 0}$. This property will enable us to study the stability of $\Sigma(\cdot)(\varphi_1, \varphi_0)$, and hence the asymptotic behaviour of $\tilde{u}$, through the location of the spectrum of Λ. Finally the stability of the solutions u of (0.1)–(0.2) can be studied via the variation of constants formula.

$$(u_t(\cdot), u(t)) = \Sigma(t)(\varphi_1(\cdot), \varphi_0) + \int_0^t \Sigma(t - \tau)(0, f(\tau))\, d\tau \ .$$

1 PRELIMINARIES

In this section we collect some definitions and results concerning those aspects of the theory of holomorphic semigroups in Banach spaces which are used in the sequel.

We begin to introduce the notion of interpolation spaces. There is no unique method for doing this and different approaches may or may not lead to different families of spaces. Here we follow the treatment of Butzer and Berens [6] which is more convenient for our purposes.

Let A generate an analytic semigroup $(S(t))_{t \ge 0}$ on a Banach space X. For simplicity we assume that S is of negative type ω. If this is not the case we can replace A by $A - aI$ and $S(t)$ by $e^{-at}S(t)$, for suitable $a \in \mathbb{R}$. Thus $A^{-1} \in \mathcal{L}(X)$ and $D(A)$ is a Banach space under the norm $|x|_{D(A)} := \|Ax\|$, where $\|\cdot\|$ denotes the norm on X.

For $0 < \theta < 1$ and $1 \le p < \infty$ we denote by $(X, D(A))_{\theta,p}$ the real interpolation spaces of X and $D(A)$ generated by the Peetre's K-method. It can be proved (see [6; Theorem 3.5.3]) that these spaces can be endowed with the following norm

$$|x|_{(X,D(A))_{\theta,p}} := \left(\int_0^{+\infty} (t^{1-\theta} \|A\, S(t)x\|)^p t^{-1} dt \right)^{\frac{1}{p}} \ . \tag{1.1}$$

We now recall the notion of extrapolation spaces. These spaces were introduced by Da Prato and Grisvard ([8]) and by Nagel ([17]), by different methods. Here we use the approach of [17] and present a brief description of those results that are used in the sequel. For more details and proofs we refer to the above mentioned papers or to the monograph [15; Chapter II.5].

For $x \in X$ we set

$$||x||_{-1} := ||A^{-1}x||$$

and denote by X_{-1} the completion of X with respect to $||\cdot||_{-1}$. It follows from the definition that $X \hookrightarrow X_{-1}$ and X_{-1} is called the extrapolation space of X with respect to A.

It can be shown that $S(t)$ can be extended over X_{-1}

$$S_{-1}(t)x := \lim_{n\to\infty} S(t)x_n \ , \ \ x \in X_{-1}$$

where $\{x_n\} \subset X$ verify $x_n \to x$ in X_{-1}. It turns out that $(S_{-1}(t))_{t\geq 0}$ is an analytic semigroup in X_{-1} of type ω. Denoting by A_{-1} its infinitesimal generator it can be proved that $D(A_{-1}) = X$ and that A is the part of A_{-1} in X. Furthermore $\rho(A_{-1}) = \rho(A)$.

For $0 < \theta < 1$ we repeat the procedure of (1.1) and denote by $(X_{-1}, X)_{\theta,p}$ the interpolation space of X_{-1} and $D(A_{-1}) = X$ with norm

$$|x|_{(X_{-1},X)_{\theta,p}} := \left(\int_0^{+\infty} (t^{1-\theta}||A_{-1}\, S_{-1}(t)x||_{-1})^p t^{-1} dt \right)^{\frac{1}{p}} . \tag{1.2}$$

Finally, for brevity in notation we set

$$X_{\alpha,p} = \begin{cases} (X_{-1}, X)_{1+\alpha,p}, & \text{if } -1 < \alpha < 0 \\ (X, D(A))_{\alpha,p}, & \text{if } \quad 0 < \alpha < 1 \end{cases} \tag{1.3}$$

and denote by $|\cdot|_{\alpha,p}$ the norm in $X_{\alpha,p}$. Hence we have $X_{\alpha,p} \subset X$, if $\alpha > 0$, whereas $X \subset X_{\alpha,p}$, if $\alpha < 0$.

Furthermore we denote by $X_{0,p}$ (see [12]) the completion of the set

$$X_p := \{x \in X, |x|_{0,p} := \left(\int_0^{+\infty} (t\,||A\,S(t)x||)^p t^{-1} dt \right)^{\frac{1}{p}} < +\infty\}$$

with respect to $|\cdot|_{0,p}$.

The spaces $X_{\alpha,p}$, for $\alpha \geq 0$, provide a useful tool for the study of the L^p-regularity of solutions of parabolic evolution equations (see [3], [7], [11] and [12]). If $\alpha < 0$ the use of these spaces allows to study parabolic equations with non regular data (see [1], [13] and [14]).

Finally, given a Banach space E, we introduce the following notation.

- $L^p(0,T;E)$, $1 \leq p < \infty$, is the space of Bochner measurable u on $]0,T[$ such that $||u(\cdot)||_E^p$ is integrable.
- $C(0,T;E)$ is the space of continuous functions on $[0,T]$.

- $W^{\alpha,p}(0,T;E)$, $0<\alpha<1$, is the Sobolev space of $u \in L^p(0,T;E)$ verifying

$$\int_0^T dt \int_0^T \|u(t)-u(\tau)\|_E^p \, |t-\tau|^{-1-p\alpha} d\tau < +\infty .$$

- $W^{1,p}(0,T;E)$ is the Sobolev space of absolutely continuous functions u having distributional derivative $u' \in L^p(0,T;E)$.

We denote by $\|\cdot\|_{L_T^p(E)}$, $\|\cdot\|_{C_T(E)}$, and $\|\cdot\|_{W_T^{\alpha,p}(E)}$, $0<\alpha\le 1$, the usual norms of these spaces.

2 HOMOGENEOUS DELAY EQUATIONS

We now turn to the study of problem (0.1)–(0.2), where A satisfies the assumptions of the preceding section and $K \in L^1(-r,0;\mathcal{L}(D(A),X))$. We note that using the transformation of variable $u(t)=e^{-at}v(t)$, for suitable $a \in \mathbb{R}$, it is not restrictive to consider (0.1) in the case where A generates a semigroup of negative type.

To investigate the behaviour of the solutions we rewrite (0.1)–(0.2), as it is usual in delay equations, as an undelayed problem in a suitable Banach space. This will require a preliminary study, for given $\varphi_1 \in L^p(-r,0;X)$ and $\varphi_0 \in X$, of the homogeneous case $f(t)\equiv 0$.

$$\text{(D)} \quad \begin{cases} u'(t) = Au(t) + \int_{-r}^0 K(s)u(t+s)ds, & t>0 \\ u(s)=\varphi_1(s), \text{ a.e. } s \in \,]-r,0[; \;\; u(0)=\varphi_0. \end{cases}$$

A function u defined on $[0,+\infty[$ is called a *strong solution* of (D) if

$$u \in W^{1,p}(0,T;X) \cap L^p(-r,T;D(A)) \quad \text{for each } T>0,$$

and satisfies (D).

It follows from the definition that the existence of strong solutions requires suitable regularity on the data. In fact in addition to $\varphi_1 \in L^p(-r,0;D(A))$ it is necessary to assume $\varphi_0 \in X_{1-1/p,p}$, see [13; section 5].

To consider also the case where the data are less regular we generalize the notion of solution (see [14]). We assume that for some $0<\theta<1$ there exist $\gamma \in L^{p'}(]-r,0[)$ satisfying

$$|K(s)x|_{-1+\theta,p} \le \gamma(s)\, |x|_{\theta,p} \tag{2.1}$$

for each $x \in D(A)$. Hence $K(s)$ can be extended into operators $\bar{K}(s)$ from $X_{\theta,p}$ to $X_{-1+\theta,p}$

$$\bar{K}(s)x = \lim K(s)x_n \tag{2.2}$$

where $x \in X_{\theta,p}$ and $x_n \in D(A)$ satisfy $x_n \to x$, in $X_{\theta,p}$.

Examples of operators K satisfying (2.1) are given in [14; Example 3.2].

Using the notation of the preceding section we introduce the extrapolated version of (D)

$$\text{(DE)} \quad \begin{cases} u'(t) = A_{-1}u(t) + \int_{-r}^0 \bar{K}(s)u(t+s)ds, & t>0 \\ u(s)=\varphi_1(s), \text{ a.e. } s \in \,]-r,0[; \;\; u(0)=\varphi_0. \end{cases}$$

As before a function u verifying $u \in W^{1,p}(0,T;X_{-1}) \cap L^p(-r,T;X_{\theta,p})$, for each $T > 0$, and satisfying (DE) is called a strong solution. It follows from the definition that the existence of such a solution requires $\varphi_1 \in L^p(-r,0;X_{\theta,p})$ and $\varphi_0 \in X_{\theta-1/p,p}$.

It can be proved (see [14; Theorem 3.6]) that if u is a strong solution of (DE) then there exist regular sequences $\{\varphi_{1_n}\}$, $\{\varphi_{0_n}\}$ and $\{u_n\}$, such that u_n is strong solution of (D) with data $(\varphi_{1_n}, \varphi_{0_n})$ and we have $\varphi_{1_n} \to \varphi_1$, $\varphi_{0_n} \to \varphi_0$ and $u_n \to u$. Such a function u is called a *weak solution* of (D). Hence it turns out that a strong solution of (DE) is a weak solution of (D). It is readily seen that if a weak solution satisfies $u \in W^{1,p}(0,T;X) \cap L^p(-r,T;D(A))$, for each $T > 0$, then u is actually a strong solution.

The notion of weak solution, via the introduction of (DE), allows to treat problem (D) with less regular data without giving up the techniques for differentiable solutions. See [14] for a general treatment. Here we recall the results which are needed in the following.

For brevity we set

$$||\varphi_1||_{r;\theta,p} := ||\varphi_1||_{L^p(-r,0;X_{\theta,p})}.$$

Then we have the following existence and uniqueness result.

THEOREM 2.1. *Let $\varphi_1 \in L^p(-r,0;X_{\theta,p})$ and $\varphi_0 \in X_{\theta-1/p,p}$. Then there exists a unique solution u of* (DE) *and we have in addition $u \in C(0,T;X_{\theta-1/p,p})$ and $u \in W^{\theta,p}(0,T;X)$, for each $T > 0$. Moreover there exists $c = c(\theta,p,T)$, which is bounded for bounded T, satisfying*

$$||u||_{L^p_T(X_{\theta,p}) \cap C_T(X_{\theta-1/p,p}) \cap W^{\theta,p}_T(X)} \le c\,[\,||\varphi_1||_{r;\theta,p} + |\varphi_0|_{\theta-1/p,p}\,].$$

Proof. The assertions are proved in [14; Theorem 3.3]. ∎

Furthermore we have the following regularity result.

THEOREM 2.2. *Let let $\varphi_1 \in W^{1,p}(-r,0;X_{\theta,p})$, $\varphi_1(0) = \varphi_0$ and*

$$A_{-1}\varphi_1(0) + \int_{-r}^{0} \bar{K}(s)\varphi_1(s)ds \in X_{\theta-1/p,p}.$$

Then for each $T > 0$ the solution of (DE) *verifies $u' \in L^p(-r,T;X_{\theta,p})$, $u' \in C(0,T;X_{\theta-1/p,p})$ and $u' \in W^{\theta,p}(0,T;X)$. Moreover we have*

$$||u'||_{L^p_T(X_{\theta,p}) \cap C_T(X_{\theta-1/p,p}) \cap W^{\theta,p}_T(X)} \le$$
$$c\,[\,||\varphi_1'||_{r;\theta,p} + |A_{-1}\varphi_1(0) + \int_{-r}^{0} \bar{K}(s)\varphi_1(s)ds|_{\theta-1/p,p}\,]$$

where c is given by Theorem 2.1.

Proof. The result is proved in [14; Theorem 3.4]. ∎

We now consider problem (D) under the following further assumption

$$\bar{K} \in W^{1,p}(-r, 0; \mathcal{L}(X_{\theta,p}, X_{-1+\theta,p})). \tag{2.3}$$

As an example of K satisfying (2.3) we can consider

$$K(s)u := k(s)\mathcal{K}u \tag{2.4}$$

where $k \in W^{1,p}(]-r, 0[)$ and $\mathcal{K}$ satisfies (2.1).

Then we have

THEOREM 2.3. *Let let $\varphi_1 \in L^p(-r, 0; X_{\theta,p})$ and $\varphi_0 \in X_{\theta-1/p,p}$. Then for each $\epsilon > 0$ and $T > 0$ the solution of problem* (DE) *satisfies*

(i) $u' \in L^p(\epsilon, T; X_{\theta,p}) \cap C(\epsilon, T; X_{\theta-1/p,p}) \cap W^{\theta,p}(\epsilon, T; X)$.

Proof. Let u denote the solution of (DE). Then

$$u(t) = S_{-1}(t)\varphi_0 + \int_0^t S_{-1}(t-\tau)g(\tau)\,d\tau \tag{2.5}$$

where we have set

$$g(t) := \int_{-r}^{0} \bar{K}(s)u(t+s)ds. \tag{2.6}$$

From Theorem 2.1 we have $u \in L^p(-r, T; X_{\theta,p})$ so that from (2.3) and

$$g'(t) = (\int_{t-r}^{t} \bar{K}(s-t)u(s)ds)'$$

it is easy to see that $g \in W^{1,p}(0, T; X_{-1+\theta,p})$.

From (2.5) u can be rewritten as

$$\begin{aligned} u(t) &= S_{-1}(t)\varphi_0 + \int_0^t S_{-1}(\tau)g(0)\,d\tau + \int_0^t S_{-1}(t-\tau)(g(\tau) - g(0))\,d\tau \\ &=: u_1 + u_2 + u_3. \end{aligned}$$

Since S_{-1} is analytic we have that u_1 satisfies (i). Furthermore

$$u_2(t) = S_{-1}(t)A_{-1}^{-1}g(0) - A_{-1}^{-1}g(0)$$

and hence we have that also u_2 satisfies (i). Finally the function $f(t) := g(t) - g(0)$ satisfies the assumptions of [13; Lemma 6.10]. Therefore we obtain that u_3 satisfies (i) with $\epsilon = 0$. ∎

3 THE SOLUTION SEMIGROUP

We now introduce the semigroup associated to the solutions of (D) and characterize its infinitesimal generator. This will enable us to rewrite (0.1)–(0.2) as an undelayed problem in an appropriate Banach space. See [5] where this approach seems to be used for the first time.

We proceed as follows. Let θ satisfy (2.1) and denote by Y the Banach space

$$Y := L^p(-r, 0; X_{\theta,p}) \times X_{\theta-1/p,p}$$

with norm

$$|\varphi|_Y = |(\varphi_1, \varphi_0)|_Y := \|\varphi_1\|_{r;\theta,p} + |\varphi_0|_{\theta-1/p,p}.$$

Further for $\varphi \in Y$ we denote by u the weak solution of (D) given by Theorem 2.1. Thus for each $T > 0$ we have $u \in L^p(-r, T; X_{\theta,p}) \cap C(0, T; X_{\theta-1/p,p})$ and we can define a solution semigroup in the following way.

For each $t > 0$ we set, for $s \in [-r, 0]$,

$$u_t(s) := u(t + s).$$

As $u_t \in L^p(-r, 0; X_{\theta,p})$ and $u(t) \in X_{\theta-1/p,p}$ we can introduce the family $(\Sigma(t))_{t\geq 0}$ of operators in Y defined as

$$\Sigma(t)\varphi := \begin{cases} (u_t, u_t(0)), & \text{if } t > 0 \\ (\varphi_1, \varphi_0), & \text{if } t = 0. \end{cases} \tag{3.1}$$

Then we have the following result.

THEOREM 3.1. *$(\Sigma(t))_{t\geq 0}$ is a strongly continuous semigroup on Y.*

Proof. The assertion can be easily proved using Theorem 2.1. ∎

We now want to characterize the infinitesimal generator of the semigroup (3.1). For brevity in notation we set

$$W^{1,p}_{\theta,p} := W^{1,p}(-r, 0; X_{\theta,p})$$

and

$$D := \{\varphi : \varphi_1 \in W^{1,p}_{\theta,p},\ \varphi_1(0) = \varphi_0,\ A_{-1}\varphi_1(0) + \int_{-r}^{0} \bar{K}(s)\varphi_1(s)ds \in X_{\theta-1/p,p}\}.$$

Then we introduce the operator Λ on Y defined as

$$D(\Lambda) = D;\ \Lambda(\varphi_1, \varphi_1(0)) = (\varphi_1', A_{-1}\varphi_1(0) + \int_{-r}^{0} \bar{K}(s)\varphi_1(s)ds). \tag{3.2}$$

The following lemmas establish some properties of the operator (3.2).

LEMMA 3.2. *For each $t > 0$ we have $\Sigma(t)\Lambda \subset \Lambda$.*

Proof. Let $\varphi \in D(\Lambda)$ and let u be the solution of (DE). Then from Theorem 2.2 we have $u_t \in W^{1,p}(-r,0;X_{\theta,p})$ and

$$A_{-1}u_t(0) + \int_{-r}^{0} \bar{K}(s)u_t(s)ds = u'(t) \in X_{\theta-1/p,p}$$

for each $t > 0$. Therefore $(u_t, u_t(0)) \in D(\Lambda)$. ■

LEMMA 3.3. *$D(\Lambda)$ is dense in Y.*

Proof. For each $\varphi \in Y$ we have

$$\lim_{\varepsilon \to 0} \varepsilon^{-1} \int_0^{\varepsilon} \Sigma(t)\varphi\, dt = \varphi. \tag{3.3}$$

Therefore it suffices to show that the left-hand side of (3.3) belongs to $D(\Lambda)$.
Let u denote the solution of (DE). From Theorem 2.1 for each $t > 0$ we have $u_t \in L^p(-r,0;X_{\theta,p})$ so that

$$\int_0^{\varepsilon} u_t\, dt \in W^{1,p}(-r,0;X_{\theta,p}).$$

Furthermore, using the fact that $u \in C(0,\varepsilon;X_{\theta-1/p,p})$, we find

$$\int_0^{\varepsilon} [A_{-1}u_t(0) + \int_{-r}^{0} \bar{K}(s)u_t(s)ds]\, dt = \int_0^{\varepsilon} u'(t)dt = u(\varepsilon) - u(0) \in X_{\theta-1/p,p}$$

and the conclusion follows. ■

LEMMA 3.4. *Λ is a closed operator on Y.*

Proof. Let $\{\varphi_n\} \subset D(\Lambda)$ satisfy $\varphi_n \to \varphi$ and $\Lambda\varphi_n \to \phi$, in Y. Thus $\varphi_{1_n} \to \varphi_1$ in $L^p(-r,0;X_{\theta,p})$ and $\varphi_{1_n}(0) \to \varphi_0$ in $X_{\theta-1/p,p}$. Furthermore $\varphi'_{1_n} \to \phi_1$ in $L^p(-r,0;X_{\theta,p})$ and

$$A_{-1}\varphi_{1_n}(0) + \int_{-r}^{0} \bar{K}\varphi_{1_n}(s)ds \to \phi_0, \text{ in } X_{\theta-1/p,p}. \tag{3.4}$$

Combining these properties we have $\varphi_1 \in W^{1,p}(-r,0;X_{\theta,p})$, $\varphi'_1 = \phi_1$ and $\varphi_{1_n} \to \varphi_1$ in $W^{1,p}(-r,0;X_{\theta,p})$. This in turn implies $\varphi_{1_n}(0) \to \varphi_1(0)$ in $X_{\theta,p}$ and hence $\varphi_1(0) = \varphi_0$.
Using (2.1) we have

$$\bar{K}\varphi_{1_n} \to \bar{K}\varphi_1 \text{ in } L^p(-r,0;X_{-1+\theta,p})$$

so that from (3.4) and the closedness of A_{-1} we obtain $\varphi_1(0) \in D(A_{-1})$ and $A_{-1}\varphi_{1_n}(0) \to A_{-1}\varphi_1(0)$, in X_{-1}. Finally using (3.4) once again we find

$$A_{-1}\varphi_1(0) + \int_{-r}^{0} \bar{K}\varphi_1(s)ds = \phi_0$$

so that $\varphi \in D(\Lambda)$ and $\Lambda\varphi = \phi$. ∎

Finally we can prove the following result.

THEOREM 3.5. *The operator Λ is the infinitesimal generator of the semigroup $(\Sigma(t))_{t\geq 0}$.*

Proof. Let $\varphi \in D(\Lambda)$ and set $v(t) := \Sigma(t)\varphi$. Then from Theorem 2.2 we have that v is right-differentiable in Y at $t = 0$ and we have $v'_+(0) = \Lambda\varphi$. Therefore the proof is accomplished by using Lemmas 3.2–3.4. ∎

There is no reason to expect that $(\Sigma(t))_{t\geq 0}$ is analytic. However if assumption (2.3) is satisfied it can be proved that, for t sufficiently large, $\Sigma(t)$ maps Y into $D(\Lambda)$.

THEOREM 3.6. *Let assumption* (2.3) *be satisfied. Then* $(\Sigma(t))_{t\geq 0}$ *is differentiable on* $]r, +\infty[$.

Proof. Let u be the solution of (DE). From Theorem 2.3 we have that $u \in W^{1,p}(\tau - r, \tau; X_{\theta,p})$ and $u' \in C(\tau - r, \tau; X_{\theta-1/p,p})$, for each $\tau > r$. Thus from (DE) we obtain, for $t = \tau$,

$$u'(\tau) = A_{-1}u(\tau) + \int_{-r}^{0} \bar{K}(s)u(\tau + s)ds \in X_{\theta-1/p,p}.$$

Therefore from (3.2) we obtain $\Sigma(\tau)\varphi \in D(\Lambda)$ for each $\tau > r$ and the result is proved. ∎

From Theorem 3.6 we have that if (2.3) holds then Σ is eventually differentiable. This in turn implies that the spectrum $\sigma(\Lambda)$ of its generator has to be limited by a function of (at most) exponential growth. See [18; Chapter 2, Thm. 4.7]

Another important consequence in this context is given by the following result.

THEOREM 3.7. *Let assumption* (2.3) *be satisfied. Denote by ω_Σ the type of the semigroup $(\Sigma(t))_{t\geq 0}$. Then*

$$\omega_\Sigma = \sup\{\mathbb{R}e\lambda : \lambda \in \sigma(\Lambda)\}.$$

Proof. The assertion is a consequence of Theorem 3.6 and the spectral mapping theorem for semigroups. See e.g. [15; Chapter IV, Section 3]. ∎

Now take $\varphi \in Y$ and consider the following Cauchy problem

$$\text{(CP)} \qquad \begin{cases} U'(t) = \Lambda U(t), & t > 0 \\ U(0) = \varphi. \end{cases}$$

Denoting by U the mild solution of (CP) and using Theorem 3.5 we have $U(t) = (u_t, u(t))$, where u is the weak solution of (D). Using this approach we can investigate the stability for (D) through the determination of the type ω of the semigroup $(\Sigma(t))_{t\geq 0}$ and hence, if assumption (2.3) is verified, through the location of the spectrum of Λ.

Using the variation of constants formula we can also study the stability of the solutions of the inhomogeneous delay problem (0.1)–(0.2). In fact, for each $T > 0$, take $f \in L^p(0,T; X_{\theta-1/p,p})$, and set

$$F(t) = (0, f(t)).$$

Then $F \in L^p(0,T;Y)$, for each $T > 0$, and we can consider the problem

$$\text{(ICP)} \qquad \begin{cases} U'(t) = \Lambda U(t) + F(t), & t > 0 \\ U(0) = \varphi. \end{cases}$$

Denoting by U the mild solution of (ICP)

$$U(t) = \Sigma(t)\varphi + \int_0^t \Sigma(t-\tau)F(\tau)\,d\tau \tag{3.5}$$

we have (see [5]) that $U(t) = (u_t, u(t))$ where u is the weak solution of the inhomogeneous delay problem (0.1)–(0.2).

EXAMPLE 3.8. We give a simple application of the preceding results to delay partial differential equations.

Let $\Omega \subset \mathbb{R}^n$ be a bounded set with smooth boundary $\partial\Omega$ and consider the problem

$$\text{(PD)} \begin{cases} \frac{\partial}{\partial t}u(t,x) = \Delta u(t,x) + \int_{-r}^{0} a\Delta u(t+s,x)\,ds + f(t,x), \text{ in } \mathbb{R}_+ \times \Omega \\ u(t,x) = 0, \text{ in } \mathbb{R}_+ \times \partial\Omega \\ u(s,x) = \varphi_1(s,x), \text{ in }]-r,0[\times\Omega; \;\; u(0,x) = \varphi_0(x) \text{ in } \Omega. \end{cases}$$

where Δ is the Laplace operator and $a \in \mathbb{R}$.

Set $X := L^p(\Omega)$ and denote by A be the following operator

$$D(A) = \{u \in W^{2,p}(\Omega) \cap W_0^{1,p}(\Omega), \Delta u \in L^p(\Omega)\}, \quad Au = \Delta u.$$

Then it is known that A satisfies the assumptions of section 1 and we can introduce the spaces $X_{\theta,p}$, X_{-1} and the operator A_{-1}, following the notation of section 1. Furthermore for $0 < \theta < 1$ we have the characterization.

$$X_{\theta,p} = W_*^{2\theta,p}(\Omega) := \begin{cases} W^{2\theta,p}(\Omega), & \text{if } 0 < \theta < 1/2 \\ B^{1,p}(\Omega), & \text{if } \theta = 1 \\ W^{2\theta,p}(\Omega) \cap W_0^{1,p}(\Omega), & \text{if } 1/2 < \theta < 1 \end{cases}$$

where $W^{2\theta,p}$ are Sobolev spaces of fractional order and $B^{1,p}$ are Besov spaces.

It can be seen that the operator $K := a\Delta$ satisfies assumptions (2.1), (2.3) for each $0 < \theta < 1$ and $1 \leq p < \infty$. Now take for simplicity in notation $p > 1$ and $\theta > 1/p$ (the case of negative $\theta - 1/p$ would require a more refined treatment) and apply the preceding result with Y chosen as

$$Y = L^p(-r, 0; W_*^{2\theta,p}(\Omega)) \times W_*^{2\theta-2/p,p}(\Omega).$$

Then we can introduce the semigroup $(\Sigma(t))_{t\geq 0}$ and its generator Λ and we can rewrite (PD) as an abstract Cauchy problem (ICP) in Y. Due to Theorem 3.7 the asymptotic behaviour of the solutions of (PD) can be studied through the location of the spectrum of Λ. To investigate this we can use the property $\sigma(A_{-1})=\sigma(A)$ and repeat, with obvious modifications, the methods used in [10; Theorem 5.13].

BIBLIOGRAPHY

1. H. Amann, *Parabolic evolution equations in interpolation and extrapolation spaces*, J. Funct. Anal. **78** (1988), 233–270.

2. A. Ardito and P. Ricciardi, *Existence and regularity for linear delay partial differential equations*, Nonlinear Anal. **4** (1980), 411–414.

3. A. Ashyralyev and P. S. Sobolevskii, *Well Posedness of Parabolic Difference Equations*, Birkhäuser, 1994.

4. A. Bátkai and S. Piazzera, *A semigroup method for delay equations with relatively bounded operators in the delay term*, Semigroup Forum **64** (2002), 71–89.

5. J. G. Borisovic and A. S. Turbabin, *On the Cauchy problem for linear inhomogeneous differential equations with retarded argument*, Dokl. Akad. Nauk SSSR **185** (1969), 741–744.

6. P. Butzer and H. Berens, *Semigroups of Operators and Approximations*, Springer–Verlag, 1967.

7. G. Da Prato and P. Grisvard, *Sommes d' opérateurs nonlinéaires et équations différentielles opérationelles*, J. Math. Pures et Appl. **54** (1975), 305–387.

8. G. Da Prato and P. Grisvard, *Maximal regularity for evolution equations by interpolation and extrapolation*, J. Funct. Anal. **58** (1984), 107–124.

9. M. C. Delfour, *The linear quadratic optimal control problem for hereditary differential systems: theory and numerical solutions*, Appl. Mat. Optim. **3** (1976/1977), 101–162.

10. G. Di Blasio, K. Kunisch, and E. Sinestrari, *Stability for abstract functional differential equations*, Israel J. Math. **50** (1985), 231–263.

11. G. Di Blasio, *Linear parabolic evolution equations in L^p-spaces*, Ann. Mat. Pura Appl. IV (1984), 55–104.

12. G. Di Blasio, *Limiting case for interpolation spaces generated by holomorphic semigroups*, Semigroup Forum **57** (1988), 359–377.

13. G. Di Blasio, *Sobolev regularity for solutions of parabolic equations by extrapolation methods*, Adv. Diff. Eqns. **4** (2001), 481–512.

14. G. Di Blasio, *Delay differential equations with unbounded operator acting on delay terms*, Nonlinear Analysis TMA, to appear.

15. K.-J. Engel and R. Nagel, *One–Parameter Semigroups for Linear Evolution Equations*, Springer, 2000.

16. M. Mastinšek, *Stability conditions for abstract functional differential equations in Hilbert spaces*, Semigroup Forum, to appear.

17. R. Nagel, *Sobolev spaces and semigroups, Semesterbericht Funktionalanalysis Tübingen*, **4** (1983), 1–19.

18. A. Pazy, *Semigroups of Linear Operators and Applications to Partial Differential Equations*, Springer, 1983.

19. J. Prüss, *Evolutionary Integral Equations and Applications*, Birkhäuser, 1993.

20. E. Sinestrari, *On a class of retarded partial differential equations*, Math. Z. **186** (1984), 223–24.

21. H. Tanabe, *Functional Analytic Methods for Partial Differential Equations*, Marcel Dekker Inc., 1997.

22. J. Wu, *Theory and Applications of Partial Functional Differential Equations*, Springer–Verlag, 1996.

Second order differential operators on C[0, 1] with Wentzell–Robin boundary conditions

KLAUS-JOCHEN ENGEL
Università di L'Aquila, Dipartimento di Matematica, Sezione Ingegneria
Località Monteluco, I-67040 Roio Poggio (AQ), Italy
e-mail: engel@ing.univaq.it

Dedicated to Jerry Goldstein in occasion of his 60th birthday.

ABSTRACT. In this note we give a simple proof that the operator A defined by

$$Af := (af')',\ D(A) := \{f \in \mathrm{C}^2[0,1] : (a f')'(j) + \beta_j f'(j) + \gamma_j f(j) = 0;\ j = 0,1\} \quad (1)$$

generates an analytic semigroup on C[0, 1]. Here $0 < \alpha \le a(\cdot) \in \mathrm{C}^1[0,1]$ is a strictly positive differentiable function and $\beta_j, \gamma_j \in \mathbb{C}$, $j = 0, 1$, are arbitrary complex numbers. Moreover, we give conditions implying positivity and stability of the generated semigroup.

1 INTRODUCTION

Recently several authors have studied the operator A defined by (1) on spaces of continuous functions and its generalizations to higher dimensions, see [2; Sect.3], [3], [4], [5], [7], [8], [9], [11]. While most of these approaches are based on form methods and/or L^p-theory, the aim of this paper is to give a simple "direct" proof that A generates an analytic semigroup on C[0, 1]. In fact, our proof is based only on similarity transformations, perturbation arguments and the fact that the part $A_0|_{\mathrm{C}_0(0,1)}$ of the operator

$$A_0 f := (af')', \quad D(A_0) := \{f \in \mathrm{C}^2[0,1] : f(j) = 0;\ j = 0,1\}$$

generates an analytic semigroup on $\mathrm{C}_0(0,1)$. We point out that our arguments seem to work equally well for degenerate boundary conditions and in higher dimensions. This will be investigated in a forthcoming paper. For the motivation concerning differential operators with Wentzell boundary conditions and further literature on this topic we refer to the above cited papers and the references therein.

2 THE MAIN RESULT

THEOREM 2.1. *The operator A defined in (1) generates a compact analytic semigroup of angle $\frac{\pi}{2}$ on* $\mathrm{C}[0,1]$.

In order to simplify our presentation we will use in the sequel the following

Notations.

- $X := \mathrm{C}[0,1]$, $\Gamma := \partial[0,1] = \{0,1\}$, and $\partial X := \mathrm{C}(\Gamma) = \mathbb{C}^2$ called the *boundary space.*
- If $\binom{f(0)}{f(1)} = x \in \partial X$ we write briefly $f|_\Gamma = x$. If $\delta_s \in X'$ denotes the Dirac measure in $s \in [0,1]$, i.e., $\delta_s f := f(s)$, and $L := \binom{\delta_0}{\delta_1} : X \to \partial X$ is the *boundary operator*, then $f|_\Gamma = x$ can also be expressed as $Lf = x$. Moreover, we use the notation δ_s' for the (unbounded) linear form defined by $\delta_s' f := f'(s)$ on $D(\delta_s') := \mathrm{C}^1[0,1]$.
- $(A_m, D(A_m))$ denotes the "maximal" operator on X given by $A_m f := (a\, f')'$ for $f \in D(A_m) := \mathrm{C}^2[0,1]$. Moreover, we consider its restriction $A_0 := A_m|_{D(A_0)}$ in X, where $D(A_0) := \ker L \cap D(A_m) = \{f \in D(A_m) : f|_\Gamma = 0\}$.

Our proof now goes in several steps.

Step 1. The space $X := \mathrm{C}[0,1]$ is isomorphic to

$$\mathcal{X} := \left\{ \binom{f}{x} \in X \times \partial X : f|_\Gamma = x \right\},$$

via the transformations

$$S : X \to \mathcal{X}, \qquad f \mapsto \binom{f}{f|_\Gamma},$$
$$S^{-1} : \mathcal{X} \to X, \qquad \binom{f}{x} \mapsto f.$$

Step 2. The operator A on X is similar to $\mathcal{A} := SAS^{-1}$ on $\mathcal{X}$ given by

$$\mathcal{A} = \begin{pmatrix} A & 0 \\ \binom{-\beta_0\delta_0'}{-\beta_1\delta_1'} & \begin{pmatrix} -\gamma_0 & 0 \\ 0 & -\gamma_1 \end{pmatrix} \end{pmatrix} =: \begin{pmatrix} A & 0 \\ B & C \end{pmatrix}, \tag{2.1}$$

$$D(\mathcal{A}) = \left\{ \binom{f}{x} \in \mathcal{X} : f \in D(A) \right\}. \tag{2.2}$$

Proof. We first observe that for $\binom{f}{x} \in \mathcal{X}$ we have

$$\binom{f}{x} \in D(SAS^{-1}) \quad \Longleftrightarrow \quad S^{-1}\binom{f}{x} = f \in D(A),$$

proving (2.2). Now take $\binom{f}{x} \in D(\mathcal{A})$, i.e. $f|_\Gamma = x$ and $f \in D(A)$. Then, using the fact that $Af|_\Gamma = Bf + Cx$, we obtain

$$SAS^{-1}\binom{f}{x} = SAf = \binom{Af}{Af|_\Gamma} = \binom{Af}{Bf + Cx} = \begin{pmatrix} A & 0 \\ B & C \end{pmatrix}\binom{f}{x} = \mathcal{A}\binom{f}{x}$$

as claimed. ∎

Step 3. The matrix $\mathcal{A}$ can be represented as the part of $\tilde{\mathcal{A}}$ in $\mathcal{X}$, i.e., $\mathcal{A} = \tilde{\mathcal{A}}|_{\mathcal{X}}$, where $\tilde{\mathcal{A}}$ on

$$\tilde{\mathcal{X}} := \mathrm{C}[0,1] \times \mathbb{C}^2 = X \times \partial X$$

is defined by

$$\tilde{\mathcal{A}} := \begin{pmatrix} A_0 & 0 \\ B & C + BL_0 \end{pmatrix} \begin{pmatrix} I & -L_0 \\ 0 & I \end{pmatrix}, \tag{2.3}$$

$$D(\tilde{\mathcal{A}}) := \left\{ \binom{f}{x} \in \tilde{\mathcal{X}} : f - L_0 x \in D(A_0) \right\}. \tag{2.4}$$

Here $L_0 : \mathbb{C}^2 \to \ker A_m$ is the *Dirichlet operator*, i.e., $L_0 = (L|_{\ker A_m})^{-1}$. It is given by

$$L_0 \binom{x_0}{x_1} := x_0 \cdot \varepsilon_0 + x_1 \cdot \varepsilon_1, \quad \text{where}$$

$$\varepsilon_0(s) := 1 - \int_0^s \frac{c}{a(r)}\, dr,$$

$$\varepsilon_1(s) := \int_0^s \frac{c}{a(r)}\, dr \qquad \text{for}$$

$$c := \left(\int_0^1 \frac{1}{a(r)}\, dr \right)^{-1}.$$

Proof. The fact that $L_0 = (L|_{\ker A_m})^{-1}$ is given by the above formulas follows easily from the fact that ε_0, ε_1 are the unique elements in $\ker A_m$ satisfying $\varepsilon_i(j) = \delta_{ij}$ for $i, j = 0, 1$.

Moreover, we have

$$\begin{aligned}
\binom{f}{x} \in D(\tilde{\mathcal{A}}|_{\mathcal{X}}) &\iff \binom{f}{x} \in D(\tilde{\mathcal{A}}) \cap \mathcal{X} \quad \text{and} \quad \tilde{\mathcal{A}}\binom{f}{x} \in \mathcal{X} \\
&\iff f - L_0 x \in D(A_0) \quad \text{and} \\
&\qquad \tilde{\mathcal{A}}\binom{f}{x} = \begin{pmatrix} A_0 & 0 \\ B & C + BL_0 \end{pmatrix} \begin{pmatrix} f - L_0 x \\ x \end{pmatrix} = \begin{pmatrix} A_m f \\ Bf + Cx \end{pmatrix} \in \mathcal{X} \\
&\iff \binom{f}{x} \in \mathcal{X},\ f \in D(A_m) \quad \text{and} \quad (A_m f)|_\Gamma = Bf + Cf|_\Gamma \\
&\iff \binom{f}{x} \in \mathcal{X} \quad \text{and} \quad f \in D(A), \\
&\iff \binom{f}{x} \in D(\mathcal{A}),
\end{aligned}$$

where the second equivalence also shows that $\tilde{\mathcal{A}}|_{\mathcal{X}}\binom{f}{x} = \mathcal{A}\binom{f}{x}$. ∎

Step 4. Let $\tilde{\mathcal{T}} := \left(\begin{smallmatrix} I & -L_0 \\ 0 & I \end{smallmatrix}\right) \in \mathcal{L}(\tilde{\mathcal{X}})$. Then $\tilde{\mathcal{T}}$ is invertible with inverse $\tilde{\mathcal{T}}^{-1} = \left(\begin{smallmatrix} I & L_0 \\ 0 & I \end{smallmatrix}\right) \in \mathcal{L}(\tilde{\mathcal{X}})$. Moreover, $\tilde{\mathcal{T}}$ maps $\mathcal{X}$ onto $\mathcal{X}_0 := X_0 \times \partial X := \mathrm{C}_0(0,1) \times \mathbb{C}^2$ and hence its restriction $\mathcal{T} := \tilde{\mathcal{T}}|_{\mathcal{X}} : \mathcal{X} \to \mathcal{X}_0$ is invertible as well with inverse $\mathcal{T}^{-1} = \tilde{\mathcal{T}}^{-1}|_{\mathcal{X}_0} : \mathcal{X}_0 \to \mathcal{X}$. For these operators and spaces we have

$$\mathcal{T}\mathcal{A}\mathcal{T}^{-1} = (\tilde{\mathcal{T}}\tilde{\mathcal{A}}\tilde{\mathcal{T}}^{-1})\big|_{\mathcal{X}_0}. \tag{2.5}$$

Proof. Since $D(\tilde{\mathcal{A}}) \subset \mathcal{X}$ equation (2.5) follows immediately from Lemma 4.1. ■

Step 5. The operator $\hat{\mathcal{A}} := \tilde{\mathcal{T}}\tilde{\mathcal{A}}\tilde{\mathcal{T}}^{-1}$ on $\tilde{\mathcal{X}}$ is given by

$$\hat{\mathcal{A}} = \begin{pmatrix} A_0 - L_0 B & -L_0(C + BL_0) \\ B & C + BL_0 \end{pmatrix}, \tag{2.6}$$

$$D(\hat{\mathcal{A}}) = D(A_0) \times \partial X. \tag{2.7}$$

Moreover, it can be decomposed as

$$\hat{\mathcal{A}} = \begin{pmatrix} A_0 & 0 \\ 0 & 0 \end{pmatrix} + \begin{pmatrix} -L_0 B & -L_0(C + BL_0) \\ B & C + BL_0 \end{pmatrix} =: \hat{\mathcal{A}}_0 + \hat{\mathcal{B}},$$

where $\hat{\mathcal{A}}_0$ is sectorial in the sense of Definition 4.2 and $\hat{\mathcal{B}}$ is relatively $\hat{\mathcal{A}}_0$-bounded with relative $\hat{\mathcal{A}}_0$-bound zero.

Proof. The representation of $\hat{\mathcal{A}}$ in (2.6) and (2.7) follows from the definition of $\tilde{\mathcal{A}}$ in (2.3) and (2.4). In fact, we have

$$\hat{\mathcal{A}} = \tilde{\mathcal{T}}\tilde{\mathcal{A}}\tilde{\mathcal{T}}^{-1} = \begin{pmatrix} I & -L_0 \\ 0 & I \end{pmatrix} \begin{pmatrix} A_0 & 0 \\ B & C + BL_0 \end{pmatrix} = \begin{pmatrix} A_0 - L_0 B & -L_0(C + BL_0) \\ B & C + BL_0 \end{pmatrix},$$

$$D(\hat{\mathcal{A}}) = D(A_0) \times \partial X.$$

To prove that $\hat{\mathcal{B}}$ is relatively $\hat{\mathcal{A}}_0$-bounded with bound zero we first observe that the operators in the right column of the operator matrix $\hat{\mathcal{B}}$ are bounded. Moreover, we note that the first derivative on $\mathrm{C}_0(0,1)$ has relative bound zero with respect to the second derivative, cf. [6; Expl.III.2.2 and the following remark]. Since $a(s) \geq \alpha > 0$ for all $s \in [0,1]$ the assertion then follows from the fact that $A_0 f = (a\,f')' = a\,f'' + a'f'$ and the definition of B in (2.1). Finally, A_0 is sectorial of angle $\frac{\pi}{2}$ by the proof of [6; Thm.VI.4.6] and hence the diagonal matrix $\hat{\mathcal{A}}_0$ is sectorial of angle $\frac{\pi}{2}$ as well. ■

Step 6. *Proof of Theorem 2.1.* By Step 5 the matrix $\hat{\mathcal{A}}_0$ is sectorial of angle $\frac{\pi}{2}$ and $\hat{\mathcal{B}}$ is relatively $\hat{\mathcal{A}}_0$-bounded with bound zero. Hence, by Lemma 4.3, $\hat{\mathcal{A}} = \hat{\mathcal{A}}_0 + \hat{\mathcal{B}}$ is sectorial of angle $\frac{\pi}{2}$. Next observe that $\overline{D(\hat{\mathcal{A}})} = \overline{D(A_0) \times \partial X} = X_0 \times \partial X = \mathcal{X}_0$ and Lemma 4.4 implies that $\hat{\mathcal{A}}|_{\mathcal{X}_0}$ generates an analytic semigroup of angle $\frac{\pi}{2}$ on $\mathcal{X}_0$. From the similarity of $\hat{\mathcal{A}}|_{\mathcal{X}_0}$ and $\mathcal{A}$ proved in Step 4, see (2.5), we then deduce that $\mathcal{A}$ generates an analytic semigroup of angle $\frac{\pi}{2}$ on $\mathcal{X}$. Again by the similarity of A and $\mathcal{A}$ shown in Step 2 we finally obtain that A generates an analytic semigroup of angle $\frac{\pi}{2}$ on X. To show its compactness it suffices to verify that A has compact resolvent, see [6; Thm.II.4.29]. This, however, follows immediately from the compact imbedding[1] $[D(A)] \hookrightarrow \mathrm{C}^1[0,1] \overset{c}{\hookrightarrow} \mathrm{C}[0,1]$ (use the Arzela–Ascoli theorem) and [6; Prop.II.4.25]. ■

[1] Here $[D(A)] := (D(A), \|\cdot\|_A)$ and "$\hookrightarrow$", "$\overset{c}{\hookrightarrow}$" denote continuous and compact injections, respectively.

3 ASYMPTOTIC BEHAVIOR

THEOREM 3.1. *The semigroup generated by A is positive if $\gamma_0, \gamma_1 \in \mathbb{R}$ and $\beta_0 \le 0$, $\beta_1 \ge 0$.*

Proof. By [10; B-II, Thm.1.6] the semigroup generated by A is positive if (and only if) A satisfies the *positive minimum principle*, that is

$$\begin{cases} \text{for every } 0 \le f \in D(A) \text{ and } s \in [0,1], \\ f(s) = 0 \text{ implies } (af')'(s) \ge 0. \end{cases} \tag{P}$$

We first note that (P) is verified for every $s \in (0,1)$. In fact, for f as in (P) we have $f'(s) = 0$ and $f''(s) \ge 0$, otherwise f cannot be non-negative. Hence we obtain $(af')'(s) = a(s)f''(s) + a'(s)f'(s) = a(s)f''(s) \ge 0$ since a is positive.

If $s = 0$ then $f'(0) \ge 0$ and the boundary conditions for functions in $D(A)$ imply

$$(af')'(0) = -\beta_0 f'(0) \ge 0.$$

For $s = 1$ the condition (P) can be verified similarly, hence the proof is complete. ∎

COROLLARY 3.2. *If $\gamma_0, \gamma_1 \in \mathbb{R}$, $\beta_0 \le 0$ and $\beta_1 \ge 0$, the semigroup generated by A is uniformly exponentially stable if*[2]

$$\mathrm{s}(C + BL_0) = \mathrm{s}\begin{pmatrix} \frac{\beta_0 c}{a(0)} - \gamma_0 & -\frac{\beta_0 c}{a(0)} \\ \frac{\beta_1 c}{a(1)} & -\frac{\beta_1 c}{a(1)} - \gamma_1 \end{pmatrix} < 0. \tag{3.1}$$

Proof. From the previous theorem we conclude that A generates a positive semigroup. By [10; B-III, Cor.1.3 & B-IV, Thm.1.1] this semigroup is uniformly exponentially stable if (and only if) A^{-1} exists and is negative, which clearly follows if $\tilde{A}^{-1} \le 0$. Now A_0 is injective, hence invertible and by (2.3) we obtain that $\tilde{A}$ is invertible with inverse

$$\tilde{A}^{-1} = \begin{pmatrix} I & L_0 \\ 0 & I \end{pmatrix} \begin{pmatrix} A_0^{-1} & 0 \\ -(C + BL_0)^{-1} BA_0^{-1} & (C + BL_0)^{-1} \end{pmatrix} \tag{3.2}$$

Here, by definition (cf. Step 3), $L_0 \ge 0$, while $A_0^{-1} \le 0$ by the weak maximum principle. Moreover, for $0 \le g \in D(A_0)$ it is clear that $g'(0) \ge 0$ and $g'(1) \le 0$, and hence $BA_0^{-1} \le 0$. Finally, $C + BL_0$ generates a positive semigroup on $\mathbb{C}^2$ and therefore $\mathrm{s}(C + BL_0) < 0$ implies that $C + BL_0 \le 0$. This shows that the left factor in the representation of $\tilde{A}^{-1}$ in (3.2) is positive while the right one is negative, implying that $\tilde{A}^{-1} \le 0$. ∎

4 SOME AUXILIARY RESULTS

LEMMA 4.1. *Let $\tilde{X}$ be a Banach space and X, X_0 two closed subspaces of $\tilde{X}$. Moreover, let $\tilde{T} \in \mathcal{L}(\tilde{X})$ be an invertible operator with $\tilde{T}X = X_0$. Then $T := \tilde{T}|_X \in \mathcal{L}(X, X_0)$ is invertible and for every operator $\tilde{A}$ on $\tilde{X}$ satisfying $D(\tilde{A}) \subseteq X$ one has*

$$TAT^{-1} = (\tilde{T}\tilde{A}\tilde{T}^{-1})|_{X_0}, \tag{4.1}$$

where $A := \tilde{A}|_X$.

[2]Note that (3.1) is satisfied if and only if $\frac{\beta_0 c}{a(0)} - \gamma_0 < 0$ and $\det(C + BL_0) > 0$.

Proof. Let A_l and A_r denote the operator on the left and on the right side, respectively, of equation (4.1). Then

$$\begin{aligned} f \in D(A_r) &\iff f \in X_0,\ \tilde{T}^{-1}f \in D(\tilde{A}) \quad \text{and} \quad \tilde{T}\tilde{A}\tilde{T}^{-1}f \in X_0 \\ &\iff f \in X_0,\ \tilde{T}^{-1}f \in D(\tilde{A}) = D(\tilde{A}) \cap X \quad \text{and} \quad \tilde{A}\tilde{T}^{-1}f \in X \\ &\iff f \in X_0 \quad \text{and} \quad T^{-1}f \in D(\tilde{A}|_X) = D(A) \\ &\iff f \in D(A_l). \end{aligned}$$

Now it is easy to verify that $A_l f = A_r f$ for all $f \in D(A_l) = D(A_r)$ and (4.1) follows ∎

DEFINITION 4.2. A closed linear operator A on a Banach space X is called *sectorial*[3] (of angle δ) if there exist $r \geq 0$ and $0 < \delta \leq \frac{\pi}{2}$ such that

$$\tilde{\Sigma}_{\frac{\pi}{2}+\delta,r} := \left\{\lambda \in \mathbb{C} : |\arg\lambda| < \frac{\pi}{2} + \delta\right\} \cap \{z \in \mathbb{C} : |z| > r\}$$

is contained in the resolvent set $\rho(A)$, and if for each $\varepsilon \in (0,\delta)$ there exists $M_\varepsilon \geq 1$ such that

$$||R(\lambda, A)|| \leq \frac{M_\varepsilon}{|\lambda|} \quad \text{for all } 0 \neq \lambda \in \left(\tilde{\Sigma}_{\frac{\pi}{2}+\delta-\varepsilon,r}\right)^{-}.$$

LEMMA 4.3. *If A_0 is sectorial of angle δ and B is relatively A_0-bounded with relative A_0-bound zero, then $A := A_0 + B$ is sectorial of angle δ as well.*

Proof. This follows from [6; Lem.III.2.6]. ∎

LEMMA 4.4. *The part of a sectorial operator A of angle δ in $X_0 := \overline{D(A)}$ generates an analytic semigroup of angle δ on X_0.*

Proof. This follows from [1; Rem.3.7.13]. ∎

BIBLIOGRAPHY

1. W. Arendt, C. J. K. Batty, M. Hieber, and F. Neubrander, *Vector-valued Laplace Transforms and Cauchy Problems*, Monographs Math., vol. 96, Birkhäuser Verlag, 2001.

2. H. Amann and J. Escher, *Strongly continuous dual semigroups*, Ann. Mat. Pura Appl. **171** (1996), 41–62.

3. W. Arendt, G. Metafune, D. Pallara, and S. Romanelli, *The Laplacian with Wentzell–Robin boundary conditions on spaces of continuous functions*, preprint, 2002.

4. M. Campiti and G. Metafune, *Ventcel's boundary conditions and analytic semigroups*, Arch. Math. **70** (1998), 377–390.

[3] In contrast to [6; Def.II.4.1] here we do not require that A is densely defined.

5. M. Campiti, G. Metafune, D. Pallara, and S. Romanelli, *Semigroups for ordinary differential operators*, In: One-Parameter Semigroups for Linear Evolution Equations (K.-J. Engel and R. Nagel, eds.), Graduate Texts in Math., vol. 194, Springer-Verlag, 2000, pp. 383–404.

6. K.-J. Engel and R. Nagel, *One-Parameter Semigroups for Linear Evolution Equations*, Graduate Texts in Math., vol. 194, Springer-Verlag, 2000.

7. A. Favini, G. Ruiz Goldstein, J. A. Goldstein, and S. Romanelli, *C_0-semigroups generated by second order differential operators with generalized Wentzell boundary conditions*, Proc. Amer. Math. Soc. **128** (2000), 1981–1989.

8. A. Favini, G. Ruiz Goldstein, J. A. Goldstein, and S. Romanelli, *The heat equation with generalized Wentzell boundary condition*, J. Evol. Equ. **2** (2002), 1–19.

9. A. Favini and A. Yagi, *Degenerate Differential Equations in Banach Spaces*, Pure and Applied Mathematics, vol. 215, Marcel Dekker, 1999.

10. R. Nagel (ed.), *One-parameter Semigroups of Positive Operators*, Lect. Notes in Math., vol. 1184, Springer-Verlag, 1986.

11. M. Warma, *Wentzell–Robin boundary conditions on* C[0, 1], Semigroup Forum (2002), to appear.

A new approach to the regularity of solutions for parabolic equations

JOACHIM ESCHER

Institute for Applied Mathematics
University of Hannover
D-30167 Hannover, Germany
e-mail: escher@ifam.uni-hannover.de

JAN PRÜSS

Fachbereich Mathematik und Informatik
Martin-Luther-Universität Halle-Wittenberg
D-60120 Halle, Germany
e-mail: anokd@volterra.mathematik.uni-halle.de

GIERI SIMONETT

Department of Mathematics
Vanderbilt University
Nashville, TN 37240, USA
e-mail: simonett@math.vanderbilt.edu

Dedicated to Jerry Goldstein on the occasion of his 60th birthday

ABSTRACT. In this note we describe a new approach to establish regularity properties for solutions of parabolic equations. It is based on maximal regularity and the implicit function theorem.

1 INTRODUCTION

In this note we describe a new approach to establish regularity properties for a wide array of parabolic evolution equations. It is based on the theory of maximal regularity and the implicit function theorem. The thrust of this approach is manifold.

- It allows to solve a given partial differential equation without loss of derivatives, thus permitting to handle fully nonlinear equations.
- It allows to resort to the implicit function theorem to study further properties of solutions, such as smooth dependence on given data.
- It allows to study the regularity of solutions by merely applying scaling arguments in conjunction with the implicit function theorem.

In order to explain the main idea of our approach, let us consider the model problem of a family of graphs $\{\Gamma(t) = \text{graph}(u(\cdot,t))\,;\, 0 \le t \le T\}$ over $\mathbb{R}^n$, evolving according to the mean curvature flow

$$\partial_t u - \left(\delta_{ij} - \frac{\partial_i u \partial_j u}{1+|\nabla u|^2}\right)\partial_i\partial_j u = 0, \qquad u(0) = u_0, \tag{1.1}$$

where $1 \le i, j \le n$, and where δ_{ij} denotes the Kronecker delta. Equation (1.1) is a quasilinear parabolic evolution equation of second order. To economize notation we set

$$F(u) := -\left(\delta_{ij} - \frac{\partial_i u \partial_j u}{1+|\nabla u|^2}\right)\partial_i\partial_j u$$

and restate equation (1.1) as

$$\partial_t u + F(u) = 0, \quad u(0) = u_0. \tag{1.2}$$

Let $E_j := buc^{2j+s}(\mathbb{R}^n)$, $j = 0, 1$, be the little Hölder spaces defined in (2.8). The mapping F is real analytic, that is,

$$F \in C^\omega(E_1, E_0). \tag{1.3}$$

Given that F is differentiable, one can consider the linearized problem

$$\partial_t v + F'(u)v = f, \quad v(0) = v_0, \tag{1.4}$$

where $F'(u)$ is the Fréchet derivative of F at $u \in E_1$. Next we introduce the anisotropic spaces

$$\mathbb{E}_0(I) := C(I, E_0), \quad \mathbb{E}_1(I) := C^1(I, E_0) \cap C(I, E_1),$$

where $I = [0, T]$ is a fixed interval. Clearly, the trace operator $\gamma_0 : \mathbb{E}_1(I) \to E_1$, $v \mapsto v(0)$ is linear and continuous. It can be shown, and this is the essential part of the analysis, that the linear problem (1.4) enjoys the property of *maximal regularity*. By definition, this means that

$$(\partial_t + F'(u), \gamma_0) \in \text{Isom}(\mathbb{E}_1(I), \mathbb{E}_0(I) \times E_1) \tag{1.5}$$

for any function $u \in E_1$. That is, the linear mapping $(\partial_t + F'(u), \gamma_0)$ is a topological isomorphism between the indicated spaces. It is here where maximal regularity begins to unfold. It implies that the linear problem (1.4) has a unique solution $v \in \mathbb{E}_1(I)$ for any given right hand side $(f, v_0) \in \mathbb{E}_0(I) \times E_1$. The solution v has optimal regularity, and therefore, no loss of regularity can occur for the linearized problem. Existence of a unique solution in $\mathbb{E}_1(I)$ to the nonlinear problem (1.2) can now be obtained by a reiteration argument and the contraction principle. As an immediate outcome, one sees that there is also no 'loss of derivatives' for the nonlinear problem. (This is also true if F is fully nonlinear). It should be noted that iteration techniques based on the Nash-Moser implicit function theorem usually result in a loss of derivatives.

We give a brief account on how the property of maximal regularity in conjunction with a scaling argument (or a parameter trick) will show that the solution $u \in \mathbb{E}_1(I)$ of (1.2) is real analytic in space and time for any positive time.

Let u be the unique solution of (1.2) defined on a maximal interval of existence $[0, t^+(u_0))$. Let $T \in (0, t^+(u_0))$ be a fixed number and set $I := [0, T]$. For any given parameters $(\lambda, \mu) \in \mathbb{R} \times \mathbb{R}^n$ with $\lambda \in (-\varepsilon_0, \varepsilon_0)$ one can set

$$u_{\lambda,\mu}(t, x) := u(t + t\lambda, x + t\mu), \quad (t, x) \in I \times \mathbb{R}^n. \tag{1.6}$$

It is easy to see that $u_{\lambda,\mu} \in \mathbb{E}_1(I)$ for all (λ, μ), provided ε_0 is sufficiently small. Since the mapping F commutes with translations, that is,

$$\tau_a F(u) = F(\tau_a u), \quad u \in E_1, \quad a \in \mathbb{R}^n, \tag{1.7}$$

one finds that $v := u_{\lambda,\mu} \in \mathbb{E}_1(I)$ satisfies the parameter dependent equation

$$\partial_t v + (1 + \lambda)F(v) - (\mu|\nabla v) = 0, \quad v(0) = u_0,$$

or equivalently, that $v := u_{\lambda,\mu}$ solves

$$\Phi(v, (\lambda, \mu)) = 0 \tag{1.8}$$

where $\Phi(v, (\lambda, \mu)) := (\partial_t v + (1 + \lambda)F(v) - (\mu|\nabla v), \gamma_0 v - u_0)$. It follows from (1.3) that the mapping

$$\Phi : \mathbb{E}_1(I) \times ((-\varepsilon_0, \varepsilon_0) \times \mathbb{R}^n) \to \mathbb{E}_0(I) \times E_1$$

is real analytic. Moreover, $\Phi(\bar{u}, (0, 0)) = (0, 0)$, where $\bar{u} := u|_I$. It is a consequence of the maximal regularity property (1.5) that the Fréchet derivative $D_1\Phi(\bar{u}, (0, 0))$ of Φ with respect to v satisfies

$$D_1\Phi(\bar{u}, (0, 0)) = (\partial_t + F'(\bar{u}), \gamma_0) \in \mathrm{Isom}(\mathbb{E}_1(I), \mathbb{E}_0(I) \times E_1). \tag{1.9}$$

The implicit function theorem now allows to solve equation (1.8) for v in terms of (λ, μ) in an open neighborhood U of $(0, 0) \in \mathbb{R} \times \mathbb{R}^n$. One concludes that

$$[(\lambda, \mu) \mapsto u_{\lambda,\mu}] \in C^\omega(U, \mathbb{E}_1(I)). \tag{1.10}$$

Consequently, the mapping

$$[(\lambda, \mu) \mapsto u_{\lambda,\mu}(t_0, x_0) = u(t_0 + t_0\lambda, x_0 + t_0\mu)] : U \to \mathbb{R} \tag{1.11}$$

is real analytic for any fixed $(t_0, x_0) \in I \times \mathbb{R}^n$ with $t_0 > 0$. Hence, the solution u of the mean curvature flow (1.1) is analytic in space and time for any positive time $t \in (0, t^+(u_0))$.

It is now clear that the only properties needed to carry through the arguments are (1.3), (1.7), and the crucial maximal regularity property (1.5). The nature of the mapping F is completely immaterial: it can be fully nonlinear, can act as a nonlocal mapping, and it can be of any order.

The idea of using parameters to prove regularity properties of solutions goes back to Angenent [3, 4]. The strategy of using translations to show analyticity in space was first employed in [8] for a free boundary problem for the flow of an incompressible fluid in a porous medium of infinite extent. In that context the mapping F happens to be fully nonlinear, nonlocal, and of first order. Translations were also used in [7] for the Stefan problem with surface tension in the case where the free interface is represented as the graph of a function over $\mathbb{R}^n$.

The advantage of applying maximal regularity lies in the fact that one can resort to the implicit function theorem. The difficulty, of course, lies in establishing maximal regularity for a given partial differential equation.

Our approach described so far relies on the fact that we can use translations on $\mathbb{R}^n$, and that the mapping F is equivariant with respect to translations. The approach can be generalized in two directions. First, it can be generalized to parabolic equations on a symmetric Riemannian manifold M, where one assumes that the nonlinear mapping F is equivariant with respect to the Lie group which acts as a transformation group on M. This has been done in [9]. In this note we show how the translation-parameter trick can be localized. In order to do so, we pick $(t_0, x_0) \in J \times \mathbb{R}^n$ and choose smooth cut-off functions $\chi \in \mathcal{D}(\mathbb{R}^n)$ and $\zeta \in \mathcal{D}(J)$ with

$$\operatorname{supp}(\chi) \subset \mathbb{B}(x_0, \varepsilon_0), \qquad \operatorname{supp}(\zeta) \subset (t_0 - \varepsilon_0, t_0 + \varepsilon_0), \tag{1.12}$$

where ε_0 can be chosen as small as we wish for. Instead of (1.6) we can now consider the parameter-dependent function

$$u_{\lambda,\mu}(t,x) := u(t + \zeta(t)\lambda, x + \zeta(t)\chi(x)\mu), \qquad (t,x) \in J \times \mathbb{R}^n. \tag{1.13}$$

The function $v := u_{\lambda,\mu}$ also satisfies a parameter-dependent equation

$$\partial_t v + F_{\lambda,\mu}(v) = 0, \qquad v(0) = u_0. \tag{1.14}$$

The new difficulty now lies in showing that the mapping $[(v,(\lambda,\mu)) \mapsto F_{\lambda,\mu}(v)]$ is analytic. This will be done in the following sections. The current note will serve as the basis to establish regularity results for free boundary problems, such as the Stefan problem with surface tension, and the Navier-Stokes equations with surface tension.

2 PARAMETER-DEPENDENT DIFFEOMORPHISMS

In the following, we assume that X is an open set in $\mathbb{R}^n$. Moreover, we assume that $x_0 \in X$ is fixed. Let $\varepsilon_0 > 0$ be chosen such that $\overline{\mathbb{B}}(x_0, 3\varepsilon_0) \subset X$ and let $\chi \in \mathcal{D}(\mathbb{B}(x_0, 2\varepsilon_0), \mathbb{R})$ be a smooth cut-off function with $\chi \equiv 1$ on $\overline{\mathbb{B}}(x_0, \varepsilon_0)$ and with $0 \le \chi \le 1$. We define the parameter dependent mapping

$$\Theta_\mu(x) := x + \chi(x)\mu, \quad x \in \mathbb{R}^n, \quad \mu \in \mathbb{C}^n. \tag{2.1}$$

Here and in the following, $\mathbb{B}(x_0, r)$ denotes the ball of radius r and center x_0 with respect to the Euclidean norm in $\mathbb{R}^n$, and $\mathbb{B}_{\mathbb{C}^n}(x_0, r)$ stands for the corresponding ball in $\mathbb{C}^n$.

LEMMA 2.1. *There exists a positive number r_0 such that*

(a) $\Theta_\mu(\mathbb{B}(x_0, 3\varepsilon_0)) \subset \mathbb{B}_{\mathbb{C}^n}(x_0, 3\varepsilon_0)$ *for any* $\mu \in \mathbb{B}_{\mathbb{C}^n}(0, r_0)$.

(b) $\Theta_\mu(\overline{\mathbb{B}}(x_0, 3\varepsilon_0)) \subset \overline{\mathbb{B}}_{\mathbb{C}^n}(x_0, 3\varepsilon_0)$ *for any* $\mu \in \mathbb{B}_{\mathbb{C}^n}(0, r_0)$.

(c) $|\Theta_\mu(x) - \Theta_{\mu_0}(y)| \le 3/2|x-y| + |\mu - \mu_0|, \ \forall x, y \in X, \ \forall \mu, \mu_0 \in \mathbb{B}_{\mathbb{C}^n}(0, r_0)$.

Proof. (a) Choose $r_0 < \varepsilon_0$ and let $x \in \mathbb{B}(x_0, 2\varepsilon_0)$ and $\mu \in \mathbb{B}_{\mathbb{C}^n}(0, r_0)$ be given. Then we have

$$|\Theta_\mu(x) - x_0| \le |x - x_0| + \chi(x)\,|\mu| < 3\varepsilon_0,$$

showing that $\Theta_\mu(\mathbb{B}(x_0, 2\varepsilon_0)) \subset \mathbb{B}_{\mathbb{C}^n}(x_0, 3\varepsilon_0)$. Since

$$\Theta_\mu(x) = x \qquad \text{for } x \in \mathbb{B}(x_0, 3\varepsilon_0) \setminus \mathbb{B}(x_0, 2\varepsilon_0)$$

we obtain the assertion in (a).

(b) is a consequence of (a).

(c) It follows from the mean value theorem that $|\chi(x) - \chi(y)| \le \|\nabla\chi\|_\infty |x-y|$ for $x, y \in X$. A simple computation then yields

$$\begin{aligned} |\Theta_\mu(x) - \Theta_{\mu_0}(y)| &\le |x-y| + |\chi(x) - \chi(y)|\,|\mu| + |\chi(y)|\,|\mu - \mu_0| \\ &\le (1 + \|\nabla\chi\|_\infty\, r_0)|x-y| + |\mu - \mu_0|. \end{aligned}$$

We can assume that r_0 is already chosen small enough such that $\|\nabla\chi\|_\infty\, r_0 \le 1/2$ and this implies (c). ∎

PROPOSITION 2.2. *There exists a positive number r_0 such that*

$$\Theta_\mu \in \mathit{Diff}^\infty(X), \qquad \mu \in \mathbb{B}(0, r_0).$$

Proof. Let $\mu \in \mathbb{R}^n$ be given. Clearly, the mapping Θ_μ is smooth in x. Its derivative is given by

$$D\Theta_\mu = I + R_\mu \quad \text{with} \quad R_\mu(x) = [\nabla\chi(x) \otimes \mu]. \tag{2.2}$$

Let r_0 be the number of Lemma 2.1. We can assume that

$$\sup_{x \in X} \|R_\mu(x)\| \le 1/2, \qquad \mu \in \mathbb{B}(0, r_0). \tag{2.3}$$

Note that equations (2.2)–(2.3) imply that the derivative $D\Theta_\mu(x)$ is invertible for $x \in X$. Lemma 2.1(a) shows that $W_\mu := \Theta_\mu(\mathbb{B}(x_0, 3\varepsilon_0)) \subset \mathbb{B}(x_0, 3\varepsilon_0)$ for any $\mu \in \mathbb{B}(0, r_0)$. We can then infer from the inverse function theorem, applied to the mapping $\Theta_\mu : \mathbb{B}(x_0, 3\varepsilon_0) \to \mathbb{B}(x_0, 3\varepsilon_0)$, that

$$W_\mu \subset \mathbb{B}(x_0, 3\varepsilon_0) \text{ is open}, \quad \mu \in \mathbb{B}(0, r_0). \tag{2.4}$$

We claim that $W_\mu = \mathbb{B}(x_0, 3\varepsilon_0)$ and that Θ_μ is injective. Since $\mathbb{B}(x_0, 3\varepsilon_0)$ is convex, we may apply the mean value theorem, yielding

$$x - y = \Theta_\mu(x) - \Theta_\mu(y) - \int_0^1 R_\mu(y + \tau(x-y))\,d\tau\,(x-y) \tag{2.5}$$

for $x, y \in \mathbb{B}(x_0, 3\varepsilon_0)$. It follows from (2.3) that

$$|x - y| \le 2\,|\Theta_\mu(x) - \Theta_\mu(y)| \tag{2.6}$$

for every $x, y \in \mathbb{B}(x_0, 3\varepsilon_0)$ and every $\mu \in \mathbb{B}(0, r_0)$. We conclude that Θ_μ is injective and that W_μ is closed in $\mathbb{B}(x_0, 3\varepsilon_0)$. Since $\mathbb{B}(x_0, 3\varepsilon_0)$ is connected, (2.4) implies that W_μ coincides with $\mathbb{B}(x_0, 3\varepsilon_0)$. It follows from the inverse function theorem that

$$\Theta_\mu \in \mathrm{Diff}^\infty(\mathbb{B}(x_0, 3\varepsilon_0)), \qquad \mu \in \mathbb{B}(x_0, r_0).$$

Since $\Theta_\mu(x) = x$ for $x \in X \setminus \mathbb{B}(x_0, 2\varepsilon_0)$ and $\Theta_\mu \in C^\infty(X)$ we obtain $\Theta_\mu \in \mathrm{Diff}^\infty(X)$, and the proof is now complete. ∎

REMARKS 2.3. (a) It follows from Proposition 2.2 and the definition of Θ_μ that

$$\Theta_\mu \in \mathrm{Diff}^\infty(\mathbb{B}(x_0, 2\varepsilon_0)), \qquad \mu \in \mathbb{B}(0, r_0).$$

(b) It is clear that $\Theta_\mu \in \mathrm{Diff}^\infty(U)$ for any open set U with $\overline{\mathbb{B}}(x_0, 3\varepsilon_0) \subset U$.

In the following we assume that U is an open set in $\mathbb{R}^n$ such that

- $\overline{\mathbb{B}}(x_0, 3\varepsilon_0) \subset U$,
- U is either bounded and has a smooth boundary, or $U = \mathbb{R}^n$. (2.7)

Let $s \ge 0$. The *little Hölder spaces* are defined by

$$buc^s(U) := \begin{cases} BUC^s(U), & \text{if } s \in \mathbb{N} \\ \text{the closure of } BUC^{[s]+1}(U) \text{ in } BUC^s(U), & \text{if } s \notin \mathbb{N} \end{cases} \tag{2.8}$$

where $[s]$ denotes the integer part of s, and where $BUC^s(U)$ are the classical Hölder spaces. Moreover, for $1 < p < \infty$ let $W_p^s(U)$ denote the *Sobolev-Slobodecki spaces*, and let $H_p^s(U)$ be the *Bessel-potential spaces*.

Let $m \in \mathbb{N}$ be given and let $s \in (0, m)$. The following interpolation results are well-known, see [11, 13, 14], and also [1, Section I.2] for a short account of interpolation theory,

$$\begin{aligned} (BUC(U), BUC^m(U))_{s/m,\infty} &= BUC^s(U), \quad s \notin \mathbb{N}, \\ (BUC(U), BUC^m(U))^0_{s/m,\infty} &= buc^s(U), \quad s \notin \mathbb{N}, \\ (L_p(U), W_p^m(U))_{s/m,p} &= W_p^s(U), \quad s \notin \mathbb{N}, \\ [L_p(U), W_p^m(U)]_{s/m} &= H_p^s(U). \end{aligned} \tag{2.9}$$

Moreover, we have the interpolation inequalities

$$\|u\|_{\mathfrak{F}^s} \le c(s)\|u\|_{\mathfrak{F}^0}^{1-s/m}\|u\|_{\mathfrak{F}^m}^{s/m}, \qquad s \in (0, m), \quad u \in \mathfrak{F}^m, \tag{2.10}$$

where $\mathfrak{F} \in \{buc(U), BUC(U), W_p(U), H_p(U)\,;\, 1 < p < \infty\}$. Our notation indicates that we choose one of the symbols in $\{buc(U), BUC(U), W_p(U), H_p(U)\}$, and then use this symbol exclusively throughout formula (2.10). We recall that

$$buc^0(U) := buc(U) = BUC(U) =: BUC^0(U), \quad W_p^0(U) = H_p^0(U) = L_p(U).$$

We also recall that

$$\mathfrak{F}^m \subset \mathfrak{F}^s \quad \text{is dense for} \quad \mathfrak{F} \in \{buc(U), W_p(U), H_p(U)\,;\, 1 < p < \infty\}. \tag{2.11}$$

It is well-known that

$$\partial_j \in \mathcal{L}(\mathfrak{F}^{s+1}, \mathfrak{F}^s), \quad \mathfrak{F} \in \{buc(U), BUC(U), W_p(U), H_p(U)\}, \quad s \geq 0. \tag{2.12}$$

Moreover, point-wise multiplication $[(a,u) \mapsto au]$ is bilinear and continuous for the spaces

$$\begin{aligned} &BUC^\rho(U) \times \mathfrak{F}^s(U) \to \mathfrak{F}^s(U), \quad \mathfrak{F} \in \{buc, W_p, H_p\}, \quad 0 \leq s < \rho, \\ &BUC^m(U) \times \mathfrak{F}^m(U) \to \mathfrak{F}^m(U), \quad \mathfrak{F} \in \{BUC, W_p\}, \quad m \in \mathbb{N}, \\ &BUC^s(U) \times BUC^s(U) \to BUC^s(U), \quad s \geq 0, \\ &buc^s(U) \times buc^s(U) \to buc^s(U), \quad s \geq 0. \end{aligned} \tag{2.13}$$

Given a function $u \in L_{1,\mathrm{loc}}(U)$ we define the pull-back and the push-forward operator, respectively, induced by the diffeomorphism Θ_μ:

$$\begin{aligned} &\Theta_\mu^* u := u \circ \Theta_\mu, \\ &\Theta_*^\mu u := u \circ (\Theta_\mu)^{-1}, \qquad \mu \in \mathbb{B}(0, r_0). \end{aligned} \tag{2.14}$$

In the following Proposition we collect some useful properties for the operators Θ_μ^*. We show that Θ_μ^* induces an isomorphism on all the function spaces introduced above, and we study the dependence on the parameter μ.

For future reference, the results are stated in a more general form than actually needed in the present note.

PROPOSITION 2.4. *Let $m \in \mathbb{N}$ and $s \in [0, m]$.*

(a) Suppose $\mathfrak{F} \in \{buc(U), BUC(U), W_p(U), H_p(U)\}$. Then

$$\Theta_\mu^* \in \mathrm{Isom}(\mathfrak{F}^s), \qquad [\Theta_\mu^*]^{-1} = \Theta_*^\mu, \qquad \mu \in \mathbb{B}(0, r_0).$$

Moreover, there exists a positive constant $M = M(m)$ such that

$$\|\Theta_\mu^*\|_{\mathcal{L}(\mathfrak{F}^s)} \leq M, \qquad \mu \in \mathbb{B}(0, r_0). \tag{2.15}$$

(b) Suppose $\mathfrak{F} \in \{buc(U), W_p(U), H_p(U)\}$. Then

$$[\mu \mapsto \Theta_\mu^* u] \in C(\mathbb{B}(0, r_0), \mathfrak{F}^s) \quad \textit{for any} \quad u \in \mathfrak{F}^s. \tag{2.16}$$

(c) Suppose $\mathfrak{F} \in \{buc(U), W_p(U), H_p(U)\}$. Then

$$[\mu \mapsto \Theta_\mu^* u] \in C^1(\mathbb{B}(0, r_0), \mathfrak{F}^s) \quad \textit{for any} \quad u \in \mathfrak{F}^{s+1}. \tag{2.17}$$

The partial derivatives are given by

$$\partial_{\mu_j}[\Theta_\mu^* u] = \chi[\Theta_\mu^* \partial_j u], \qquad u \in \mathfrak{F}^{s+1}, \quad j \in \{1, \cdots, n\}. \tag{2.18}$$

Proof. (a) (i) Pick $\mu \in \mathbb{B}(0, r_0)$ and $u \in BUC(U)$. We conclude from Lemma 2.1(c) and from Proposition 2.2 that

$$\Theta_\mu^* u \in BUC(U), \qquad \|\Theta_\mu^* u\|_{BUC(U)} \leq \|u\|_{BUC(U)}, \quad \mu \in \mathbb{B}(0, r_0). \tag{2.19}$$

Next, let $u \in BUC^m(U)$. It is evident that $\Theta_\mu^* u \in C^m(U)$, and a straightforward computation shows that

$$\partial^\beta[\Theta_\mu^* u] = \sum_{|\gamma|\le|\beta|} b_{\beta,\gamma}(\mu,\cdot)\,[\Theta_\mu^*\partial^\gamma u], \qquad |\beta| \le m, \tag{2.20}$$

where $b_{\beta,\gamma} \in BUC(\mathbb{B}(0,r_0)\times U)$. (We have, in fact, $b_{\beta,\gamma} \in BUC^\infty(\mathbb{B}(0,r_0)\times U)$). We conclude from (2.19) and (2.20) that

$$\Theta_\mu^* u \in BUC^m(U), \qquad \|\Theta_\mu^* u\|_{BUC^m(U)} \le M\|u\|_{BUC^m(U)}, \quad \mu \in \mathbb{B}(0,r_0),$$

for an appropriate constant M. Clearly, Θ_μ^* is linear for every fixed $\mu \in \mathbb{B}(0,r_0)$, and it follows from (2.19)-(2.20) that

$$\Theta_\mu^* \in \mathcal{L}(BUC^l(U)), \qquad \|\Theta_\mu^*\|_{\mathcal{L}(BUC^l(U))} \le M, \quad l \in [0,m]\cap\mathbb{N}. \tag{2.21}$$

It is clear that $[\Theta_\mu^*]^{-1} = \Theta_*^\mu$, and the open mapping theorem yields $\Theta_\mu^* \in BUC^l(U)$ for $l \in [0,m]\cap\mathbb{N}$. The case $\mathfrak{F}^s \in \{buc^s(U), BUC^s(U)\}$ for $s \in (0,m)\setminus\mathbb{N}$ follows from (2.9) and (2.21) by interpolation.

(ii) It is a consequence of the transformation rule, Remark 2.3, and equations (2.2)–(2.3) that

$$\Theta_\mu^* \in \mathcal{L}(L_p(U)), \quad \|\Theta_\mu^*\|_{\mathcal{L}(L_p(U))} \le M_1, \quad \mu \in \mathbb{B}(0,r_0). \tag{2.22}$$

It is not difficult to show (by approximating) that formula (2.20) remains valid for $u \in W_p^m(U)$. One can then conclude that

$$\Theta_\mu^* \in \mathcal{L}(W_p^m(U)), \quad \|\Theta_\mu^*\|_{\mathcal{L}(W_p^m(U))} \le M_2, \quad \mu \in \mathbb{B}(0,r_0). \tag{2.23}$$

As in (i) we obtain the assertion for $\mathfrak{F}^s \in \{W_p^s(U), H_p^s(U)\}$ by interpolation.

(b) (i) We first consider $u \in BUC(U)$. Since u is uniformly continuous we find for every $\varepsilon > 0$ a number $\delta > 0$ such that $|u(y) - u(z)| < \varepsilon$ whenever $y, z \in U$ and $|y - z| < \delta$. Lemma 2.1(c) then shows that

$$|(\Theta_\mu^* u)(x) - (\Theta_{\mu_0}^* u)(x)| < \varepsilon,$$

whenever $x \in U$, $\mu, \mu_0 \in \mathbb{B}(0,r_0)$ and $|\mu - \mu_0| < \delta$. We have, thus, proved that

$$[\mu \mapsto \Theta_\mu^* u] \in C(\mathbb{B}(0,r_0), BUC(U)), \qquad u \in BUC(U). \tag{2.24}$$

The assertion in (b) follows now from (2.20) and (2.24) for $\mathfrak{F}^l = BUC^l$, $l \in \{0,m\}$. Suppose that $s \in (0,m)$ and let $u \in buc^s(U)$. Let $\varepsilon > 0$ be given. According to (2.11) we find a function v such that

$$v \in BUC^m(U), \quad \|u - v\|_s < \varepsilon/3M, \tag{2.25}$$

where M is the constant of equation (2.15). Equations (2.15) and (2.10) yield

$$\begin{aligned}
\|\Theta_\mu^* u - \Theta_{\mu_0}^* u\|_s &\le \|\Theta_\mu^*(u-v)\|_s + \|\Theta_\mu^* v - \Theta_{\mu_0}^* v\|_s + \|\Theta_{\mu_0}^*(u-v)\|_s \\
&\le 2M\|u-v\|_s + c\|\Theta_\mu^* v - \Theta_{\mu_0}^* v\|_m^{s/m}\|\Theta_\mu^* v - \Theta_{\mu_0}^* v\|_0^{1-s/m} \\
&\le 2M\|u-v\|_s + c(2M\|v\|_m)^{s/m}\|\Theta_\mu^* v - \Theta_{\mu_0}^* v\|_0^{1-s/m}
\end{aligned}$$

for any $\mu, \mu_0 \in \mathbb{B}(0, r_0)$, where we use $\|\cdot\|_s := \|\cdot\|_{BUC^s(U)}$. The case $\mathfrak{F}^s = buc^s(U)$ is now a consequence of (2.24) and (2.25).

(ii) Let $u \in L_p(U)$ and let $\varepsilon > 0$ be given. There exists a function v with

$$v \in C_c(U), \qquad \|u - v\|_p \leq \varepsilon/3M. \tag{2.26}$$

Using Lemma 2.1 and Proposition 2.2 it is easy to see that there exists a compact set K contained in U such that supp $(\Theta_\mu^* v) \subset K$ for any $\mu \in \mathbb{B}(0, r_0)$. We conclude that

$$\|\Theta_\mu^* v - \Theta_{\mu_0}^* v\|_p \leq (\lambda_n(K))^{1/p} \|\Theta_\mu^* v - \Theta_{\mu_0}^* v\|_{BUC(U)} \tag{2.27}$$

where $\lambda_n(K)$ denotes the Lebesgue measure of K. It follows from (2.15) that

$$\begin{aligned} \|\Theta_\mu^* u - \Theta_{\mu_0}^* u\|_p &\leq \|\Theta_\mu^*(u - v)\|_p + \|\Theta_\mu^* v - \Theta_{\mu_0}^* v\|_p + \|\Theta_{\mu_0}^*(u - v)\|_p \\ &\leq 2M\|u - v\|_p + \|\Theta_\mu^* v - \Theta_{\mu_0}^* v\|_p \end{aligned}$$

and we infer from (2.24) and (2.26)-(2.27) that

$$[\mu \mapsto \Theta_\mu^* u] \in C(\mathbb{B}(0, r_0), L_p(U)), \qquad u \in L_p(U). \tag{2.28}$$

The case $\mathfrak{F}^s \in \{W_p^s(U), H_p^s(U)\}$ follows in the same way as in step (b)(i).

(c) Pick $u \in \mathfrak{F}^{s+1}$. We infer from (2.12)–(2.13) and from part (b) that

$$[\mu \mapsto \chi\Theta_\mu^* \partial_j u] \in C(\mathbb{B}(0, r_0), \mathfrak{F}^s), \qquad j \in \{1, \cdots, n\}. \tag{2.29}$$

Let $\mu \in \mathbb{B}(0, r_0)$ be fixed, and choose $\varepsilon > 0$ small enough such that $\mu + he_j \in \mathbb{B}(0, r_0)$ for $h \in (-\varepsilon, \varepsilon)$.

(i) Let us temporarily assume that $u \in C^\infty(U) \cap \mathfrak{F}^{s+1}$. It follows from (3.7) that

$$\frac{1}{h}[\Theta_{\mu+he_j}^* u - \Theta_\mu^* u] - \chi\Theta_\mu^* \partial_j u = \int_0^1 (\chi\Theta_{\mu+\tau he_j}^* \partial_j u - \chi\Theta_\mu^* \partial_j u)\, d\tau \quad \text{in } \mathfrak{F}^s. \tag{2.30}$$

(ii) An approximation argument shows that the statement in (2.30) is also valid for $u \in \mathfrak{F}^{s+1}$. The assertion in (c) can now be obtained from (2.29) by analogous arguments as in step (i) of the proof of Proposition 3.2. ∎

REMARKS 2.5.

(a) The assumption that U be a 'smooth' open set is not indispensable. It is only required for the interpolation and and multiplier results (2.9)–(2.13) which are used in the proof of Proposition 2.4.

(b) The proof of Proposition 2.4 shows that the assertions are valid in the case

$$\mathfrak{F}^l \in \{BUC^l(U), W_p^l(U)\}, \quad l \in \mathbb{N},$$

for any open set U with $\overline{\mathbb{B}}(x_0, 3r_0) \subset U$.

(c) The assertions of Proposition 2.4 also remain valid if

$$\mathfrak{F}^s \in \{buc^s(U), BUC^s(U), W_p^s(U)\}$$

for any open set U with $\overline{\mathbb{B}}(x_0, 3r_0) \subset U$.

3 HIGHER REGULARITY

In this section we show that the mapping $[\mu \mapsto \Theta_\mu^* u]$ enjoys more regularity than stated in Proposition 2.4, provided the function u has better regularity properties. In the following, U and X are open sets as considered in section 2. We begin with a technical Lemma.

LEMMA 3.1. *Let $m \in \mathbb{N}$ and $k \in \mathbb{N}^* \cup \{\infty\}$. Suppose that $a \in C^{m+k}(X)$. Then*

$$[\mu \mapsto \chi^{|\alpha|}\Theta_\mu^* \partial^\alpha a] \in C(\mathbb{B}(0, r_0), BUC^m(U)), \qquad 0 < |\alpha| \leq k.$$

Proof. Since $\operatorname{supp}(\chi) \subset \mathbb{B}(x_0, 3\varepsilon_0) \subset U \cap X$, the assertion of the Lemma is meaningful. Let $v \in C(X)$ be given, and let $\zeta \in \mathcal{D}((\mathbb{B}(x_0, 2\varepsilon_0), \mathbb{R})$. Since v is uniformly continuous on $\overline{\mathbb{B}}(x_0, 3\varepsilon_0)$ we find for every given $\varepsilon > 0$ a number $\delta > 0$ such that

$$|v(y) - v(z)| < \varepsilon, \quad y, z \in \overline{\mathbb{B}}(x_0, 3\varepsilon_0), \quad |y - z| < \delta.$$

We can now deduce from Lemma 2.1(b)–(c) that

$$|(\zeta\Theta_\mu^* v)(x) - (\zeta\Theta_{\mu_0}^* v)(x)| = |\zeta(x)|\,|v(\Theta_\mu(x)) - v(\Theta_{\mu_0}(x))| < \varepsilon,$$

whenever $x \in U$, $\mu, \mu_0 \in \mathbb{B}(0, r_0)$ and $|\mu - \mu_0| < \delta$. We have shown that

$$[\mu \mapsto \zeta\Theta_\mu^* v] \in C(\mathbb{B}(0, r_0), BUC(U)). \tag{3.1}$$

Now let $a \in C^{m+k}(X)$. Let $\eta \in \mathbb{N}^n$ be a fixed multi-index with $|\eta| \leq m$. If follows from Leibniz' rule and from (2.20) that

$$\partial^\eta(\chi^{|\alpha|}\,[\Theta_\mu^* \partial^\alpha a]) = \sum_{\beta \leq \eta} \sum_{|\gamma| \leq |\beta|} \binom{\eta}{\beta} b_{\beta,\gamma}(\mu, \cdot)(\partial^{\eta-\beta}\chi^{|\alpha|})[\Theta_\mu^* \partial^{\alpha+\gamma} a]. \tag{3.2}$$

Note that $v := \partial^{\alpha+\gamma} a \in C(U)$ and $\zeta := \partial^{\eta-\gamma}\chi^{|\alpha|} \in \mathcal{D}((\mathbb{B}(x_0, 2\varepsilon_0), \mathbb{R})$. The claim in (c) follows from (3.1) and (3.2). ∎

PROPOSITION 3.2. *Let $m \in \mathbb{N}$ and $k \in \mathbb{N} \cup \{\infty, \omega\}$. Suppose that*

$$a \in C^{m+k}(X) \cap BUC^m(U).$$

Then we have

$$[\mu \mapsto \Theta_\mu^* a] \in C^k(\mathbb{B}(0, r_0), BUC^m(U)) \tag{3.3}$$

and

$$\partial_\mu^\alpha[\Theta_\mu^* a] = \chi^{|\alpha|}[\Theta_\mu^* \partial^\alpha a], \qquad |\alpha| \leq k.$$

Proof. Lemma 2.4 shows that the mapping

$$g := [\mu \mapsto \Theta_\mu^* a] \in C(\mathbb{B}(0, r_0), BUC^m(U))$$

is well-defined. Moreover, Lemma 3.1 shows that

$$[\mu \mapsto \chi^{|\alpha|}\Theta_\mu^* \partial^\alpha a] \in C(\mathbb{B}(0, r_0, BUC^m(U)), \qquad 0 < |\alpha| \leq k. \tag{3.4}$$

(i) Let $\mu \in \mathbb{B}(0, r_0)$ be fixed, and choose $\varepsilon > 0$ small enough such that $\mu + he_j \in \mathbb{B}(0, r_0)$ for $h \in (-\varepsilon, \varepsilon)$. Let $x \in X$ be given. Then the mean value theorem yields

$$\frac{1}{h}[(\Theta^*_{\mu+he_j} a)(x) - (\Theta^*_\mu a)(x)] = \int_0^1 (\chi\Theta^*_{\mu+\tau he_j}\partial_j a)(x)\, d\tau. \tag{3.5}$$

Note that both sides of equation (3.5) vanish if $x \notin \operatorname{supp}(\chi)$. Consequently, formula (3.5) is also valid for any $x \in U$. It is not difficult to verify (by resorting to Riemann sums, for instance) that

$$\int_0^1 (\chi\Theta^*_{\mu+\tau he_j}\partial_j a)(x)\, d\tau = \Big(\int_0^1 \chi\Theta^*_{\mu+\tau he_j}\partial_j a\, d\tau\Big)(x), \quad x \in U, \tag{3.6}$$

where the integral $\int_0^1 (\chi\Theta^*_{\mu+\tau he_j}\partial_j a)\, d\tau$ exists in $BUC^m(U)$. We conclude that

$$\frac{1}{h}[g(\mu + he_j) - g(\mu)] - [\chi\Theta^*_\mu\partial_j a] = \int_0^1 (\chi\Theta^*_{\mu+\tau he_j}\partial_j a - \chi\Theta^*_\mu\partial_j a)\, d\tau \tag{3.7}$$

in $BUC^m(U)$. Lemma 3.1 implies that

$$(\chi\Theta^*_{\mu+\tau he_j}\partial_j a - \chi\Theta^*_\mu\partial_j a) \to 0 \quad \text{in } BUC^m(U) \text{ as } \varepsilon \to 0,$$

uniformly in $\tau \in [0, 1]$. It follows that

$$\int_0^1 (\chi\Theta^*_{\mu+\tau he_j}\partial_j a - \chi\Theta^*_\mu\partial_j a)\, d\tau \to 0 \quad \text{in } BUC^m(U) \text{ as } \varepsilon \to 0. \tag{3.8}$$

Consequently, also the left side of equation (3.7) converges to 0 in $BUC^m(U)$ as $\varepsilon \to 0$. We have, thus, proved that the partial derivative $\partial_{\mu_j} g(\mu)$ exists and is given by $\partial_{\mu_j} g(\mu) = \chi\Theta^*_\mu\partial_j a$. In addition, Lemma 3.1 shows that

$$g \in C^1(\mathbb{B}(0, r_0), BUC^m(U)).$$

(ii) We can now repeat the steps above with a replaced by $\chi\Theta^*_\mu\partial_j a$ to obtain

$$\partial_{\mu_i}\partial_{\mu_j}[\Theta^*_\mu a] = \chi^2[\Theta^*_\mu\partial_i\partial_j a].$$

An induction argument yields

$$g \in C^k(\mathbb{B}(0, r_0), BUC^m(U)) \quad \text{and} \quad \partial^\alpha_\mu g(\mu) = \chi^{|\alpha|}\Theta^*_\mu\partial^\alpha a$$

for $|\alpha| \le k$, where $k \in \mathbb{N} \cup \{\infty\}$.

(iii) Suppose now that $k = \omega$. Since a is (real) analytic, there exists an open neighborhood $U_\mathbb{C}$ of U in $\mathbb{C}^n$ and a (unique) mapping

$$a_\mathbb{C} \in C^\omega(U_\mathbb{C}, \mathbb{C}) \quad \text{such that} \quad U_\mathbb{C} \cap \mathbb{R}^n = U \quad \text{and} \quad a_\mathbb{C}|U = a. \tag{3.9}$$

We may assume that ε_0 is small enough such that $\mathbb{B}_{\mathbb{C}^n}(x_0, 3\varepsilon_0) \subset U_{\mathbb{C}}$. It follows from Lemma 2.1(a) and the definition of Θ_μ that $\Theta_\mu(U) \subset U_{\mathbb{C}}$ for any $\mu \in \mathbb{B}_{\mathbb{C}^n}(0, r_0)$, and consequently, the mapping

$$g_{\mathbb{C}}(\mu)(x) := (\Theta_\mu^* a_{\mathbb{C}})(x) := a_{\mathbb{C}}(\Theta_\mu(x)), \tag{3.10}$$

is well-defined for $x \in U$ and $\mu \in \mathbb{B}_{\mathbb{C}^n}(0, r_0)$.
(iv) Since $a_{\mathbb{C}} \in C^\omega(U_{\mathbb{C}}, \mathbb{C})$, it is clear that $\Theta_\mu^* a_{\mathbb{C}} \in C^m(U, \mathbb{C})$. We claim that

$$[\mu \mapsto \Theta_\mu^* a_{\mathbb{C}}] \in C(\mathbb{B}_{\mathbb{C}^n}(0, r_0), BUC^m(U, \mathbb{C})). \tag{3.11}$$

Let $\zeta \in \mathcal{D}(\mathbb{B}(x_0, 3\varepsilon_0))$ be a smooth cut-off function with $\zeta \equiv 1$ on $\operatorname{supp}(\chi)$. As in the proof of Lemma 3.1 one shows that

$$[\mu \mapsto \zeta\Theta_\mu^* a_{\mathbb{C}}] \in C(\mathbb{B}_{\mathbb{C}^n}(0, r_0), BUC^m(U, \mathbb{C})). \tag{3.12}$$

In more detail, we have

$$\partial_x^\beta[\Theta_\mu^* a_{\mathbb{C}}](x) = \sum_{|\gamma| \le |\beta|} b_{\beta,\gamma}(\mu, x)\,[\Theta_\mu^* \partial_x^\gamma a_{\mathbb{C}}](x), \qquad x \in U,\ |\beta| \le m, \tag{3.13}$$

with appropriate functions $b_{\mu,\gamma} \in BUC(\mathbb{B}_{\mathbb{C}^n}(0, r_0) \times U, \mathbb{C})$. Let $\gamma \in \mathbb{N}^n$ be a fixed multi-index with $|\gamma| \le m$. We know that the real partial derivatives $\partial_x^\gamma a_{\mathbb{C}}$ are continuous on $U_{\mathbb{C}}$, and therefore are uniformly continuous on the compact set $\overline{\mathbb{B}}_{\mathbb{C}^n}(x_0, 3\varepsilon_0)$. That is to say that for any $\varepsilon > 0$ there is a number $\delta > 0$ such that

$$|\partial_x^\gamma a_{\mathbb{C}}(z_1) - \partial_x^\gamma a_{\mathbb{C}}(z_2)| < \varepsilon, \quad z_1, z_2 \in \overline{\mathbb{B}}_{\mathbb{C}^n}(x_0, 3\varepsilon_0), \quad |z_1 - z_2| < \delta.$$

Lemma 2.1(b)-(c) then implies that

$$|\Theta_\mu^* \partial_x^\gamma a_{\mathbb{C}}(x) - \Theta_\mu^* \partial_x^\gamma a_{\mathbb{C}}(y)| < \varepsilon, \quad x, y \in \overline{\mathbb{B}}(x_0, 3\varepsilon_0),\ |x - y| < (2/3)\delta, \tag{3.14}$$

and $\mu \in \mathbb{B}_{\mathbb{C}^n}(0, r_0)$, as well as

$$|\Theta_\mu^* \partial_x^\gamma a_{\mathbb{C}}(x) - \Theta_{\mu_0}^* \partial_x^\gamma a_{\mathbb{C}}(x)| < \varepsilon, \quad \mu, \mu_0 \in \mathbb{B}_{\mathbb{C}^n}(x_0, r_0),\ |\mu - \mu_0| < \delta, \tag{3.15}$$

uniformly in $x \in \overline{\mathbb{B}}(x_0, 3\varepsilon_0)$. Equation (3.12) follows now from Leibniz' rule and from (3.13)–(3.15). Recall that $\Theta_\mu(x) = x$ for $x \notin \operatorname{supp}(\chi)$. If follows from the fact that $\zeta \equiv 1$ on $\operatorname{supp}(\chi)$ and from (3.9) that

$$(1 - \zeta)\Theta_\mu^* a_{\mathbb{C}} = (1 - \zeta)a, \qquad \mu \in \mathbb{B}_{\mathbb{C}^n}(0, r_0). \tag{3.16}$$

Since $a \in BUC^m(U)$ by assumption, we evidently have

$$[\mu \mapsto (1 - \zeta)\Theta_\mu^* a_{\mathbb{C}}] \in C(\mathbb{B}_{\mathbb{C}^n}(0, r_0), BUC^m(U)) \tag{3.17}$$

and the assertion in (3.11) follows from (3.12) and (3.17).
(v) Let $\partial_{z_j} a_{\mathbb{C}}$ denote a complex partial derivative of $a_{\mathbb{C}}$, where we use the notation $z = (z_1, \cdots, z_j, \cdots, z_n) \in \mathbb{C}^n$. Since $\partial_{z_j} a_{\mathbb{C}} \in C^m(U_{\mathbb{C}}, \mathbb{C})$ we can conclude as in (i) that

$$[\mu \mapsto \chi\Theta_\mu^* \partial_{z_j} a_{\mathbb{C}}] \in C(\mathbb{B}_{\mathbb{C}^n}(0, r_0), BUC^m(U, \mathbb{C})). \tag{3.18}$$

(vi) Next we show that

$$g_{\mathbb{C}} \in C^1(\mathbb{B}_{\mathbb{C}^n}(0,r_0), BUC^m(U,\mathbb{C})) \quad \text{and} \quad \partial_{\mu_j} g_{\mathbb{C}} = \chi\Theta_\mu^* \partial_{z_j} a_{\mathbb{C}}. \tag{3.19}$$

In fact, the assertion follows by the same arguments as in step (i) of the proof. We now have $\mu \in \mathbb{B}_{\mathbb{C}^n}(0,r_0)$, $h \in \mathbb{B}_{\mathbb{C}}(0,\varepsilon)$, and we replace g and a by $g_{\mathbb{C}}$ and $a_{\mathbb{C}}$, respectively.

(vii) We infer from step (iii) – and the well-known fact that a holomorphic function is complex analytic – that $g_{\mathbb{C}} \in C^\omega(\mathbb{B}_{\mathbb{C}^n}(0,r_0), BUC^m(U,\mathbb{C}))$. Consequently,

$$g = g_{\mathbb{C}}|\mathbb{B}(0,r_0) \in C^\omega(\mathbb{B}(0,r_0), BUC^m(U))$$

and the proof is now complete. ∎

In order to be able to treat differential operators in various function spaces, we present the following result, which generalizes Proposition 3.2.

THEOREM 3.3. *Let $m \in \mathbb{N}$ and $k \in \mathbb{N} \cup \{\infty, \omega\}$. Suppose that*

$$a \in C^{m+k}(X) \cap \mathfrak{F}^s$$

where $\mathfrak{F} \in \{buc(U), W_p(U), H_p(U)\,;\, 1 < p < \infty\}$, $s \in [0,m]$. Then we have

$$[\mu \mapsto \Theta_\mu^* a] \in C^k(\mathbb{B}(0,r_0), \mathfrak{F}^s).$$

Proof. (i) Let $k \in \mathbb{N} \cup \{\infty\}$. Lemma 2.4 asserts that the mapping

$$g := [\mu \mapsto \Theta_\mu^* a] \in C(\mathbb{B}(0,r_0), \mathfrak{F}^s)$$

is well-defined. Next, observe that equation (3.8) remains valid in the present context, since the proof only relies on the property that $a \in C^{m+k}(X)$. We conclude from

$$\operatorname{supp}\Bigl(\int_0^1 (\chi\Theta_{\mu+\tau h e_j}^* \partial_j a - \chi\Theta_\mu^* \partial_j a)\, d\tau\Bigr) \subset \operatorname{supp}(\chi), \qquad \mu \in \mathbb{B}(0,r_0),$$

and from (3.8) that

$$\int_0^1 (\chi\Theta_{\mu+\tau h e_j}^* \partial_j a - \chi\Theta_\mu^* \partial_j a)\, d\tau \to 0 \quad \text{in } \mathfrak{F}^s \text{ as } \varepsilon \to 0 \tag{3.20}$$

and the assertions follow from (3.7) and (3.20).

(ii) Suppose $k = \omega$. An inspection of step (iii) in the proof of Proposition 3.2 shows that equations (3.12) and (3.17) are also satisfied for the spaces $\mathfrak{F}^s(U,\mathbb{C})$. The proof proceeds now along the lines of steps (v)–(vii) of the proof of Proposition 3.2. ∎

The following result shows that our method can be used to characterize smoothness.

THEOREM 3.4. *Let $X \subset \mathbb{R}^n$ be an open set and let $k \in \mathbb{N} \cup \{\infty, \omega\}$. Suppose that $u \in BUC(X)$. Then $u \in C^k(X)$ iff for any $x_0 \in X$ there exists $r_0 = r_0(x_0) > 0$ such that*

$$[\mu \mapsto \Theta_\mu^* u] \in C^k(\mathbb{B}(0,r_0), BUC(X)).$$

Proof. (i) Assume that $u \in C^k(X)$. Let x_0 be fixed and choose $\varepsilon_0 > 0$ such that $\mathbb{B}(x_0, 3\varepsilon_0) \subset X$. The assertion follows now from Remark 2.5 and Theorem 3.3.
(ii) Let $x_0 \in X$ be fixed and suppose that

$$[\mu \mapsto \Theta_\mu^* u] \in C^k(\mathbb{B}(0, r_0), BUC(X)), \tag{3.21}$$

for some number $r_0 > 0$, where Θ_μ is defined in (2.1). Let $\mathcal{E} : BUC(X) \to \mathbb{R}$, $\mathcal{E}v := v(x_0)$, and observe that $\mathcal{E} \in \mathcal{L}(BUC(X), \mathbb{R})$. Hence $\mathcal{E} \in C^\omega(BUC(X), \mathbb{R})$ and we conclude from (3.21) that

$$[\mu \mapsto (\Theta_\mu^* u)(x_0) = u(x_0 + \mu)] \in C^k(\mathbb{B}(0, r_0), \mathbb{R}). \tag{3.22}$$

Equation (3.22) means that $u \in C^k(\mathbb{B}(x_0, r_0), \mathbb{R})$. Since this is true for any point $x_0 \in X$ we have proved that $u \in C^k(X)$, and the proof is now complete. ∎

4 DIFFERENTIAL OPERATORS

For later use we study how differential operators transform under a change of coordinates induced by Θ_μ.
We will first consider differential operators with constant coefficients and we set

$$A^\alpha(\mu) := \Theta_\mu^* (\partial^\alpha (\Theta_*^\mu \cdot)), \quad \alpha \in \mathbb{N}^n, \quad \mu \in \mathbb{B}(0, r_0). \tag{4.1}$$

PROPOSITION 4.1. *Suppose that* $\mathfrak{F} \in \{buc(U), BUC(U), W_p(U), H_p(U)\}$. *Let* $l \in \mathbb{N}$. *Then*

$$[\mu \mapsto A^\alpha(\mu)] \in C^\omega(\mathbb{B}(0, r_0), \mathcal{L}(\mathfrak{F}^{s+l}, \mathfrak{F}^s)), \qquad |\alpha| \le l.$$

Proof. (i) Let $A_j(\mu) := \Theta_\mu^* (\partial_j (\Theta_*^\mu \cdot))$. An easy computation shows that

$$A_j(\mu) u = ((D\Theta_\mu)^{-1} e_j | \nabla u), \quad u \in C^1(U), \quad \mu \in \mathbb{B}(0, r_0), \tag{4.2}$$

where e_j is the j-th canonical basis vector in $\mathbb{R}^n$, and where $(\cdot | \cdot)$ denotes the inner product in $\mathbb{R}^n$. It is not difficult to see that formula (4.2) holds true for any $u \in \mathfrak{F}^t$ with $t \ge 1$. In fact, this is evident for $\mathfrak{F} \in \{buc, BUC\}$, and follows by approximation in the other cases. It follows from (2.2)–(2.3) and from Cramer's rule (for instance) that

$$[\mu \mapsto (D\Theta_\mu)^{-1}] \in C^\omega(\mathbb{B}(0, r_0), BUC^m(U, \mathcal{L}(\mathbb{R}^n))) \tag{4.3}$$

where we take $m = [s] + l$. We conclude from (2.12) and (2.13) that

$$[B \mapsto (Be_j | \nabla \cdot)] \in C^\omega(BUC^m(U, \mathcal{L}(\mathbb{R}^n)), \mathcal{L}(\mathfrak{F}^t, \mathfrak{F}^{t-1})), \quad 1 \le t \le s + l, \tag{4.4}$$

since all the operations involved are linear [or bilinear] and continuous. It is now a straightforward consequence of (4.2)–(4.4) that

$$[\mu \mapsto A_j(\mu)] \in C^\omega(\mathbb{B}(0, r_0), \mathcal{L}(\mathfrak{F}^t, \mathfrak{F}^{t-1})), \quad 1 \le t \le s + l. \tag{4.5}$$

(ii) Suppose $\alpha = e_j + e_k$. Then we have

$$A^\alpha(\mu) u = \Theta_\mu^* (\partial_k \partial_j (\Theta_*^\mu u)) = \Theta_\mu^* (\partial_k \Theta_*^\mu \Theta_\mu^* \partial_j (\Theta_*^\mu u)) = (A_k(\mu) A_j(\mu)) u. \tag{4.6}$$

The mapping

$$\mathcal{L}(\mathfrak{F}^t, \mathfrak{F}^{t-1}) \times \mathcal{L}(\mathfrak{F}^{t-1}, \mathfrak{F}^{t-2}) \to \mathcal{L}(\mathfrak{F}^t, \mathfrak{F}^{t-2}), \qquad (A, B) \mapsto BA \tag{4.7}$$

is bilinear and continuous, and hence it is analytic. We infer from (4.5)–(4.7) that

$$[\mu \mapsto A^\alpha(\mu)] \in C^\omega(\mathbb{B}(0, r_0), \mathcal{L}(\mathfrak{F}^t, \mathfrak{F}^{t-2})), \quad 2 \leq t \leq s + l.$$

(iii) Let $\alpha = (\alpha_1, \cdots, \alpha_n)$. Then

$$A^\alpha(\mu) = (A_1(\mu))^{\alpha_1} \cdots (A_n(\mu))^{\alpha_n}$$

and the claim follows from (4.5) and (4.7) by induction. ∎

We will now consider differential operators with variable coefficients. That is, we consider the differential operator

$$A := \sum_{|\alpha| \leq l} a_\alpha \, \partial^\alpha, \qquad a_\alpha \in C(U, \mathbb{R}), \qquad |\alpha| \leq l, \tag{4.8}$$

where l is a positive integer. The parameter-dependent family of diffeomorphisms $\{\Theta_\mu \,;\, \mu \in \mathbb{B}(0, r_0)\}$ generate a parameter-dependent family $\{A_\mu \,;\, \mu \in \mathbb{B}(0, r_0)\}$ of differential operators (the transformed differential operators), given by

$$A_\mu := \sum_{|\alpha| \leq l} (\Theta_\mu^* a_\alpha) \Theta_\mu^* (\partial^\alpha (\Theta_*^\mu \cdot)), \qquad \mu \in \mathbb{B}(0, r_0). \tag{4.9}$$

We shall show that regularity properties of the coefficients a_α translate into regularity properties for the map $[\mu \mapsto A_\mu]$.

THEOREM 4.2. *Let $m \in \mathbb{N}$ and $k \in \mathbb{N} \cup \{\infty, \omega\}$.*

(a) Suppose that $a_\alpha \in C^{m+k}(X) \cap BUC^m(U)$ for every $|\alpha| \leq l$. Then

$$[\mu \mapsto A_\mu] \in C^k(\mathbb{B}(0, r_0), \mathcal{L}(\mathfrak{F}^{s+l}(U), \mathfrak{F}^s(U))),$$

where $\mathfrak{F} \in \{buc, BUC, W_p, H_p\}$ and $s \in [0, m]$.

(b) Suppose that $a_\alpha \in C^{m+k}(X) \cap buc^s(U)$ for every $|\alpha| \leq l$, where $s \in [0, m]$ is fixed. Then

$$[\mu \mapsto A_\mu] \in C^k(\mathbb{B}(0, r_0), \mathcal{L}(\mathfrak{F}^{s+l}(U), \mathfrak{F}^s(U))),$$

where $\mathfrak{F}^s(U) = buc^s(U)$.

Proof. (a) It follows from (2.13) that the mapping

$$\begin{aligned} &f : BUC^m(U) \times \mathcal{L}(\mathfrak{F}^{s+l}(U), \mathfrak{F}^s(U)) \to \mathcal{L}(\mathfrak{F}^{s+l}(U), \mathfrak{F}^s(U)), \\ &f(a, T)(u) := a(Tu) \end{aligned} \tag{4.10}$$

is bilinear and continuous, and hence analytic. We can now conclude from (3.3) and Proposition 4.1 that

$$[\mu \mapsto \sum_{|\alpha| \leq l} f(\Theta_\mu^* a_\alpha, A^\alpha(\mu))] \in C^k(\mathbb{B}(0, r_0), \mathcal{L}(\mathfrak{F}^{s+l}(U), \mathfrak{F}^s(U))), \tag{4.11}$$

and this is exactly the assertion in (a).

The proof of (b) follows from (2.13) and Theorem 3.3 by analogous arguments. ∎

5 TIME DEPENDENCE

We will now consider the situation where time is an additional variable.

In the following we use the notation $I := [0, T]$, where T is a fixed positive number. Let J be an open interval in $(0, T)$. Let $t_0 \in J$ be fixed and choose ε_0 such that $[t_0 - 3\varepsilon_0, t_0 + 3\varepsilon_0] \subset J$. Moreover, let $\zeta \in \mathcal{D}(t_0 - 2\varepsilon_0, t_0 + 2\varepsilon_0)$ be a smooth cut-off function with $\zeta \equiv 1$ on $[t_0 - \varepsilon_0, t_0 + \varepsilon_0]$ and with $0 \le \zeta \le 1$. Of course, we can – and we will – assume that the number ε_0 also satisfies the assumptions stated at the beginning of section 2.

It turns out to be convenient to introduce the mapping

$$\theta_\lambda(t) := t + \zeta(t)\lambda, \quad t \in I, \quad \lambda \in \mathbb{R}. \tag{5.1}$$

Proposition 2.2 shows that there is a positive number r_0 such that

$$\theta_\lambda \in \mathrm{Diff}^{\,\infty}(J), \qquad \lambda \in (-r_0, r_0). \tag{5.2}$$

In order to obtain regularity results in time and space for parabolic equations, we define the parameter-dependent mapping

$$\Phi_{\lambda,\mu}(t, x) := (t + \zeta(t)\lambda, x + \zeta(t)\chi(x)\mu), \quad (t, x) \in J \times U, \quad (\lambda, \mu) \in \mathbb{R}^{n+1}. \tag{5.3}$$

A straightforward modification of the proof of Proposition 2.2 shows that there exists a number $r_0 > 0$ such that

$$\Phi_{\lambda,\mu} \in \mathrm{Diff}^{\,\infty}(J \times U), \qquad (\lambda, \mu) \in \mathbb{B}^{n+1}(0, r_0). \tag{5.4}$$

We can assume that all the results of sections 2 and 3 remain valid for the same number r_0. Given a function $u : I \times U \to \mathbb{R}$ we set

$$u_{\lambda,\mu} := \Phi^*_{\lambda,\mu} u, \qquad (\lambda, \mu) \in \mathbb{B}^{n+1}(0, r_0). \tag{5.5}$$

The parameter-dependent function $u_{\lambda,\mu}$ can also be written as

$$u_{\lambda,\mu}(t) = T_\mu(t)\theta^*_\lambda u(t, \cdot) \quad \text{where} \quad T_\mu(t) := \Theta^*_{\zeta(t)\mu}, \quad t \in I. \tag{5.6}$$

It is important to note that

$$u_{\lambda,\mu}(0, \cdot) = u(0, \cdot) \quad \text{for any function } u \text{ and any } (\lambda, \mu). \tag{5.7}$$

We will first prove the following useful extension result.

LEMMA 5.1. *Let E be a Banach space. Suppose that*

$$[\mu \mapsto f(\mu)] \in C^k(\mathbb{B}(0, r_0), E), \qquad k \in \mathbb{N}^* \cup \{\infty, \omega\}.$$

Let $F(\mu)(t) := f(\zeta(t)\mu)$ for $\mu \in \mathbb{B}(0, r_0)$ and $t \in I$. Then we have

$$[\mu \mapsto F(\mu)] \in C^k(\mathbb{B}(0, r_0), C(I, E)).$$

Proof. To shorten the notation we set $W := \mathbb{B}(0, r_0)$. Since $0 \le \zeta \le 1$ we see that $F(\mu) \in C(I, E)$ for each $\mu \in W$. We will focus on the case $k = \omega$. It will be clear from the proof how to proceed for $k \in \mathbb{N}^* \cup \{\infty\}$.

Let us assume that $f \in C^\omega(W, E)$. Then there exists an open neighborhood $W_{\mathbb{C}}$ of W in $\mathbb{C}^n$ and a unique mapping

$$f_{\mathbb{C}} \in C^\omega(W_{\mathbb{C}}, E_{\mathbb{C}}) \quad \text{such that} \quad W_{\mathbb{C}} \cap \mathbb{R}^n = W \quad \text{and} \quad f_{\mathbb{C}}|W = f,$$

where $E_{\mathbb{C}}$ is the complexification of E. We can assume without loss of generality that $W_{\mathbb{C}} = \mathbb{B}_{\mathbb{C}^n}(0, r_0)$. This implies that $\zeta(t)\mu \in W_{\mathbb{C}}$ whenever $\mu \in W_{\mathbb{C}}$ and $t \in I$. It is then clear that

$$F_{\mathbb{C}}(\mu) := f_{\mathbb{C}}(\zeta(\cdot)\mu) \in C(I, E_{\mathbb{C}}), \qquad \mu \in W_{\mathbb{C}}.$$

Let $\mu \in W_{\mathbb{C}}$ be fixed and choose $\varepsilon > 0$ such that $\mu + he_j \in W_{\mathbb{C}}$ for all $h \in \mathbb{B}_{\mathbb{C}}(0, \varepsilon)$. It follows from the mean value theorem that

$$\frac{1}{h}[F_{\mathbb{C}}(\mu + he_j) - F_{\mathbb{C}}(\mu)](t) = \int_0^1 \zeta(t)\partial_{z_j} f_{\mathbb{C}}(\zeta(t)(\mu + \tau he_j))\, d\tau.$$

It is easy to see that the quotient on the left side converges to $\zeta(t)\partial_{z_j} f_{\mathbb{C}}(\zeta(t)\mu)$ uniformly in $t \in I$ as $h \to 0$, and we conclude that the partial derivatives $\partial_{\mu_j} F_{\mathbb{C}}$ exist in $C(I, E_{\mathbb{C}})$ and are given by

$$\partial_{\mu_j} F_{\mathbb{C}} = \zeta(\cdot)\partial_{z_j} f_{\mathbb{C}}(\zeta(\cdot)\mu) \quad \mu \in W_{\mathbb{C}}, \quad j \in \{1, \ldots, n\}.$$

A moment of reflection shows that

$$[\mu \mapsto \zeta(\cdot)\partial_{z_j} f_{\mathbb{C}}(\zeta(\cdot)\mu)] \in C(W, C(I, E_{\mathbb{C}})).$$

Therefore, the mapping $F_{\mathbb{C}}$ is (continuously) complex differentiable and we obtain

$$[\mu \mapsto F_{\mathbb{C}}(\mu)] \in C^\omega(W_{\mathbb{C}}, C(I, E_{\mathbb{C}})).$$

We conclude that $F = F_{\mathbb{C}}|W \in C^\omega(W, C(I, E))$, and this completes the proof. ■

PROPOSITION 5.2.

(a) *Let $m \in \mathbb{N}$ and $k \in \mathbb{N} \cup \{\infty, \omega\}$. Suppose that*

$$a \in C^{m+k}(X) \cap \mathfrak{F}^s \tag{5.8}$$

where $\mathfrak{F} \in \{buc(U), W_p(U), H_p(U)\,;\, 1 < p < \infty\}$, $s \in [0, m]$. Then we have

$$[\mu \mapsto T_\mu a] \in C^k(\mathbb{B}(0, r_0), C(I, \mathfrak{F}^s)).$$

Moreover, $\partial_\mu^\alpha [T_\mu a] = (\zeta\chi)^{|\alpha|}[T_\mu \partial^\alpha a]$ for every $|\alpha| \le k$.

(b) *Suppose $\mathfrak{F} \in \{buc(U), BUC(U), W_p(U), H_p(U)\}$. Then*

$$[\mu \mapsto T_\mu \partial^\alpha T_\mu^{-1}] \in C^\omega\big(\mathbb{B}(0, r_0), C(I, \mathcal{L}(\mathfrak{F}^{s+l}, \mathfrak{F}^s))\big), \quad |\alpha| \le l.$$

Proof. (a) Let $E = \mathfrak{F}^s$ and define $f(\mu) := \Theta_\mu^* a$. The assertions follow from Theorem 3.3 and Lemma 5.1.

(b) Let $E := \mathcal{L}(\mathfrak{F}^{s+l}, \mathfrak{F}^s)$ and let $f(\mu) := A^\alpha(\mu)$, where $A^\alpha(\mu)$ is defined in (4.1). We can now apply Proposition 4.1 and Lemma 5.1. ■

PROPOSITION 5.3. *Let $l \in \mathbb{N}^*$ be fixed and let $\mathfrak{F} \in \{buc(U), W_p(U), H_p(U)\}$.*

(a) Suppose $u \in C^1(I, \mathfrak{F}^s) \cap C(I, \mathfrak{F}^{s+l})$. Then $u_{\lambda,\mu} \in C^1(I, \mathfrak{F}^s) \cap C(I, \mathfrak{F}^{s+l})$.

(b) Suppose $u \in W_p^1(I, \mathfrak{F}^s) \cap L_p(I, \mathfrak{F}^{s+l})$. Then $u_{\lambda,\mu} \in W_p^1(I, \mathfrak{F}^s) \cap L_p(I, \mathfrak{F}^{s+l})$.

In both cases, the time derivative is given by

$$\partial_t u_{\lambda,\mu} = (1 + \zeta'\lambda) T_\mu \theta_\lambda^* \partial_t u + B_\mu u_{\lambda,\mu}\,, \tag{5.9}$$

where

$$[\mu \mapsto B_\mu] \in C^\omega(\mathbb{B}(0, r_0), C(I, \mathcal{L}(\mathfrak{F}^{s+l}, \mathfrak{F}^s))). \tag{5.10}$$

Proof. (i) We first observe that the parameter-dependent mapping θ_λ has the same properties as the mapping Θ_μ of section 2, where the set U is now replaced by the interval I. The fact that I is closed does not create any additional difficulties. It is clear that the proof of Proposition 2.4(a) also works for the vector-valued spaces $BUC^m(I, E)$ and $W_p^m(I, E)$, where E is some Banach space. Note that we have $C^m(I, E) = BUC^m(I, E)$ due to the fact that I is compact. We conclude that

$$\theta_\lambda^* u \in \mathbb{E}_1(I) \quad \text{and} \quad \partial_t \theta_\lambda^* u = (1 + \zeta'\lambda)\theta_\lambda^* \partial_t u \quad \text{for any} \quad u \in \mathbb{E}_1(I), \tag{5.11}$$

where $\mathbb{E}_1(I) \in \{C^1(I, \mathfrak{F}^s) \cap C(I, \mathfrak{F}^{s+l}), W_p^1(I, \mathfrak{F}^s) \cap L_p(I, \mathfrak{F}^{s+l})\}$.

(ii) Let $v \in C^1(I, \mathfrak{F}^s) \cap C(I, \mathfrak{F}^{s+l})$.
We obtain from (2.15)–(2.18) that $T_\mu v \in C^1(I, \mathfrak{F}^s) \cap C(I, \mathfrak{F}^{s+l})$, and also that

$$\partial_t T_\mu v = T_\mu \partial_t v + \sum_j \zeta' \chi \mu_j T_\mu \partial_j v. \tag{5.12}$$

(iii) Let $v \in L_p(I, \mathfrak{F}^{s+l})$. It follows from (2.16) that $T_\mu : I \to \mathcal{L}(\mathfrak{F}^{s+l})$ is strongly continuous. A well-known property then asserts that $T_\mu v : I \to \mathfrak{F}^{s+l}$ is measurable and (2.15) implies that $T_\mu v \in L_p(I, \mathfrak{F}^{s+l})$.

(iv) Suppose that $v \in W_p^1(I, \mathfrak{F}^s) \cap L_p(I, \mathfrak{F}^{s+l})$. Based on the property that v is absolutely continuous and has a derivative in $L_p(I, \mathfrak{F}^s)$ almost everywhere, we obtain by similar arguments as in (ii) that

$$T_\mu v \in W_p^1(I, \mathfrak{F}^s), \qquad \partial_t T_\mu v = T_\mu \partial_t v + \sum_j \zeta' \chi \mu_j T_\mu \partial_j v. \tag{5.13}$$

(v) Let

$$B_\mu v := \sum_{j=1}^n \zeta' \chi \mu_j [T_\mu \partial_j T_\mu^{-1}] v, \quad \mu \in \mathbb{B}(0, r_0), \quad v \in \mathbb{E}_1(I). \tag{5.14}$$

Proposition 5.3 follows now from (5.11)–(5.14) and Proposition 5.2. ∎

PROPOSITION 5.4. *Let $k \in \mathbb{N}^* \cup \{\infty, \omega\}$.*
Suppose that $a \in C^{m+k}(J \times X) \cap BUC^m(I \times U)$. Then

$$[(\lambda, \mu) \mapsto a_{\lambda,\mu}] \in C^k(\mathbb{B}^{n+1}(0, r_0), BUC^m(I \times U)). \tag{5.15}$$

Proof. The assertion follows by the same arguments as in the proof of Proposition 3.2. ∎

6 EXAMPLES

In this section we collect four simple examples which show the flexibility and power of our approach.

In our first example we show how the results in sections 2 and 3 can be used to prove regularity properties for elliptic equations.

Let us consider the second order elliptic equation

$$a\,\Delta u = f \quad \text{in} \quad X, \tag{6.1}$$

where X is an open set in $\mathbb{R}^n$. We assume that the differential operator $A := a\,\Delta$ is uniformly strongly elliptic, that is, we assume that there exists a positive number δ such that $a(x) \geq \delta$ for every $x \in X$.

EXAMPLE 6.1. Suppose that $(a, f) \in C^\omega(X)$ and that $u \in C^2(X)$ is a solution of (6.1). Then $u \in C^\omega(X)$.

Proof. Pick $x_0 \in X$. Choose $\varepsilon_0 > 0$ with $\overline{\mathbb{B}}(x_0, 3\varepsilon_0) \subset X$. Let $\{\Theta_\mu\,;\, \mu \in \mathbb{B}(0, r_0)\}$ be the family of diffeomorphisms introduced in section 2. Let $U := \mathbb{B}(x_0, 3\varepsilon_0)$ and observe that

$$u \in W_p^2(U), \quad a \in C^\omega(U) \cap BUC(U), \quad f \in C^\omega(U) \cap L_p(U). \tag{6.2}$$

Next set $g := \gamma u$, where $\gamma \in \mathcal{L}(W_p^2(U), W_p^{2-1/p}(\Gamma))$ denotes the trace operator for $\Gamma := \partial U$, see [13, 14]. Clearly, u solves the elliptic boundary value problem

$$Au = f \quad \text{in} \quad U, \qquad \gamma u = g \quad \text{on} \quad \Gamma.$$

For later use we note that

$$(A, \gamma) \in \operatorname{Isom}(W_p^2(U), L_p(U) \times W_p^{2-1/p}(\Gamma)) \tag{6.3}$$

see [14, Theorem 4.3.3.(ii)], for instance. We introduce the transformed quantities

$$u_\mu := \Theta_\mu^* u, \quad A_\mu := \Theta_\mu^*(A\,\Theta_*^\mu\,\cdot), \quad f_\mu := \Theta_\mu^* f, \qquad \mu \in \mathbb{B}(0, r_0). \tag{6.4}$$

It follows from (6.2), from Theorem 3.3 and from Theorem 4.2(a) that

$$[\mu \mapsto (A_\mu, f_\mu)] \in C^\omega(\mathbb{B}(0, r_0), \mathcal{L}(W_p^2(U), L_p(U)) \times L_p(U)). \tag{6.5}$$

Moreover, we know from Proposition 2.4 that $u_\mu \in W_p^2(U)$ for $\mu \in \mathbb{B}(0, r_0)$. We note that equation (6.4) yields $A_\mu u_\mu = \Theta_\mu^*(Au) = \Theta_\mu^* f$, and that $\gamma u_\mu = \gamma u = g$. We conclude that u_μ solves the transformed elliptic problem

$$(A_\mu v, \gamma v) = (f_\mu, g), \qquad \mu \in \mathbb{B}(0, r_0). \tag{6.6}$$

We finally introduce the function

$$\begin{aligned}
&\Phi : W_p^2(U) \times \mathbb{B}(0, r_0) \to L_p(U) \times W_p^{2-1/p}(\Gamma), \\
&\Phi(v, \mu) := (A_\mu v - f_\mu, \gamma v - g).
\end{aligned}$$

It is a consequence of (6.5) that

$$\Phi \in C^\omega(W_p^2(U) \times \mathbb{B}(0, r_0), L_p(U) \times W_p^{2-1/p}(\Gamma)) \tag{6.7}$$

and it follows from equations (6.3) and (6.6) that

$$\begin{aligned} \Phi(u_\mu, \mu) &= (0,0), \quad \mu \in \mathbb{B}(0, r_0), \\ D_1\Phi(u_0, 0) &= (A, \gamma) \in \mathrm{Isom}(W_p^2(U), L_p(U) \times W_p^{2-1/p}(\Gamma) \end{aligned} \tag{6.8}$$

where, of course, $u_0 = u$. We conclude from (6.7)–(6.8) and the implicit function theorem that there exists a number $r = r(x_0) \in (0, r_0)$ such that

$$[\mu \mapsto u_\mu] \in C^\omega(\mathbb{B}(0,r), W_p^2(U)).$$

Let us assume that p is chosen large enough such that $W_p^2(U) \hookrightarrow BUC(U)$. Since x_0 can be taken arbitrary we obtain that $u \in C^\omega(X)$ from Theorem 3.4. ∎

In our second example we consider the linear parabolic equation

$$\partial_t u - a\,\Delta u = 0 \quad \text{in} \quad \mathbb{R}^n, \qquad u(0) = u_0. \tag{6.9}$$

We assume that X is an open subset of $\mathbb{R}^n$, that

$$a \in C^\omega(X) \cap BUC(\mathbb{R}^n), \tag{6.10}$$

and that the differential operator $A := a\,\Delta$ is uniformly strongly elliptic.
Let $T > 0$ be fixed and set $I := [0, T]$ and $J := (0, T)$. Finally, let $p \in (1 + n/2, \infty)$.

EXAMPLE 6.2. *Let $u_0 \in W_p^{2-2/p}(\mathbb{R}^n)$ be given. Then equation (6.9) has a unique solution $u \in W_p^1(I, L_p(\mathbb{R}^n)) \cap L_p(I, W_p^2(\mathbb{R}^n))$ with $u \in C^\omega(J \times X)$.*

Proof. The proof is based on the maximal regularity result

$$(\partial_t + (\nu - A), \gamma_0) \in \mathrm{Isom}\,(\mathbb{E}_1(I), \mathbb{E}_0(I) \times W_p^{2-2/p}(\mathbb{R}^n)) \tag{6.11}$$

where $\nu > 0$ is an appropriate constant, where $\gamma_0 v := v(0)$, and where

$$\mathbb{E}_1(I) := W_p^1(I, L_p(\mathbb{R}^n)) \cap L_p(I, W_p^2(\mathbb{R}^n)), \quad \mathbb{E}_0(I) := L_p(I, L_p(\mathbb{R}^n)),$$

see [6, Corollary 6.2] and [1, Theorem III.4.10.7].
Let $v \in \mathbb{E}_1(I)$ be the (unique) solution of

$$(\partial_t v + (\nu - A)v, \gamma_0 v) = (0, u_0). \tag{6.12}$$

Pick $(t_0, x_0) \in J \times X$ and let

$$v_{\lambda,\mu}(t,x) := v(t + \zeta(t)\lambda, x + \zeta(t)\chi(x)\mu), \qquad (t,x) \in I \times \mathbb{R}^n.$$

It follows from Proposition 5.3 that $v_{\lambda,\mu} \in \mathbb{E}_1(I)$. We conclude from (5.9) and (6.12) that

$$\begin{aligned} \partial_t v_{\lambda,\mu} &= -(1 + \zeta'\lambda)\,[\nu - (T_\mu a)T_\mu \Delta T_\mu^{-1}]T_\mu \theta_\lambda^* v + B_\mu v_{\lambda,\mu} \\ &= -(1 + \zeta'\lambda)\,[\nu - (T_\mu a)T_\mu \Delta T_\mu^{-1}]v_{\lambda,\mu} + B_\mu v_{\lambda,\mu}. \end{aligned} \tag{6.13}$$

Consequently, $v_{\lambda,\mu}$ is a solution of the parameter-dependent equation

$$(\partial_t w + A_{\lambda,\mu} w, \gamma_0 w) = (0, u_0),$$

where

$$A_{\lambda,\mu} w := (1 + \zeta'\lambda)\,[\nu - (T_\mu a) T_\mu \Delta T_\mu^{-1}] w - B_\mu w.$$

We infer from (5.10) and Proposition 5.2 that

$$[(\lambda,\mu) \mapsto A_{\lambda,\mu}] \in C^\omega(\mathbb{B}^{n+1}(0,r_0), \mathcal{L}(\mathbb{E}_1(I), \mathbb{E}_0(I))).$$

The implicit function theorem shows that

$$[(\lambda,\mu) \mapsto v_{\lambda,\mu}] \in C^\omega(\mathbb{B}^{n+1}(0,r_0), \mathbb{E}_1(I)).$$

Since $\mathbb{E}_1(I) \hookrightarrow BUC(I \times \mathbb{R}^n)$ we conclude that

$$[(\lambda,\mu) \mapsto v_{\lambda,\mu}(t_0,x_0) = v(t_0+\lambda, x_0+\mu)] \in C^\omega(\mathbb{B}^{n+1}(0,r_0), \mathbb{R}),$$

showing that v is in fact analytic on a neighborhood of (t_0,x_0). Since (t_0,x_0) can be chosen anywhere in $J \times X$ we have shown that $v \in C^\omega(J \times X)$. It remains to observe that $u(t) := e^{\nu t} v(t)$ solves the parabolic equation (6.9) and that u has the same regularity properties as v. ∎

REMARKS 6.3. (a) By relying on maximal regularity results in little Hölder spaces of negative order [2], the regularity assumptions on the initial value u_0 can be considerably relaxed (at the expense of imposing slightly more regularity on the coefficient a).

(b) It is clear that we can also treat much more general parabolic systems which satisfy the condition of normal ellipticity, see [2, 6]. In addition, we can also admit time dependent coefficients and time dependent source terms.

In our next example we presuppose existence of a classical solution for the non-autonomous parabolic equation

$$\partial_t u - a\,\Delta u = f \quad \text{in} \quad J \times X, \tag{6.14}$$

where X is an arbitrary open subset of $\mathbb{R}^n$ and $J = (0,T)$ for some $T > 0$, and we will be concerned with the regularity properties of u. We assume that

$$(a, f) \in C^\omega(J \times X), \tag{6.15}$$

and that the differential operator $A := a\,\Delta$ is uniformly strongly elliptic.

EXAMPLE 6.4. *Suppose that u is a classical solution of the parabolic equation (6.14). Then $u \in C^\omega(J \times X)$.*

Proof. Let $(t_0,x_0) \in J \times X$ be fixed and choose $\varepsilon_0 > 0$ such that $\overline{\mathbb{B}}(x_0, 3\varepsilon_0) \subset U$ as well as $[t_0 - 3\varepsilon_0, t_0 + 3\varepsilon_0] \subset J$. In addition, choose τ small enough such that $\tau \notin \operatorname{supp}(\chi)$ and define $I := [0, T_0 - \tau]$ where T_0 is slightly smaller than T. Let $U := \mathbb{B}(x_0, 3\varepsilon_0)$ and let $\Gamma := \partial U$. Moreover, let $p \in (1 + n/2, \infty)$. The basic maximal regularity result for the present situation is

$$(\partial_t - b\Delta, \gamma_\Gamma, \gamma_0) \in \operatorname{Isom}(\mathbb{E}_1(I), \mathbb{E}_0(I)), \tag{6.16}$$

where we set $b(t,x) = a(\tau + t, x)$ for $t \in I$, and where

$$\mathbb{E}_1(I) := W_p^1(I, L_p(U)) \cap L_p(I, W_p^2(U)),$$
$$\mathbb{E}_0(I) := \{(g_1, g_2, w_0) \in \mathbb{F}_0(I) \cap W_p^{2-2/p}(U)\,;\, g_2(0) = w_0|_\Gamma\},$$
$$\mathbb{F}_0(I) := L_p(I, L_p(U)) \times \left(W_p^{1-1/2p}(I, L_p(\Gamma))\right) \cap L_p(I, W_p^{2-1/p}(\Gamma))$$

see [10, Section IV.9], and also [5], [12]. Next we set

$$v_0 := u(\tau), \quad f_1(t) := f(t+\tau)|_U, \quad f_2(t) := u(t+\tau)|_\Gamma, \quad t \in I.$$

Since u is classical solution of equation (6.14), we obtain $(f_1, f_2, v_0) \in \mathbb{E}_0(I)$. Let $v(t) := u(t+\tau)$. We conclude that $v \in \mathbb{E}_1(I)$, and that v is the (unique) solution of $(\partial_t - b\Delta, \gamma_\Gamma, \gamma_\tau)v = (f_1, f_2, v_0)$. Let

$$v_{\lambda,\mu}(t,x) := v(t + \zeta(t)\chi(x)\lambda, x + \zeta(t)\chi(x)\mu), \qquad (t,x) \in I \times U.$$

It can be shown that the pertinent results of section 5 do also hold for the transformation $\hat{\Phi}(t,x) := (t + \zeta(t)\chi(x)\lambda, x + \zeta(t)\chi(x)\mu)$ and we can, once again, conclude that

$$[(\lambda,\mu) \mapsto v_{\lambda,\mu}] \in C^\omega(\mathbb{B}^{n+1}(0, r_0), \mathbb{E}_1(I)).$$

The assertion follows as in the previous example. ∎

In our last example we consider the nonlinear parabolic equation

$$\partial_t u - a\left(\delta_{ij} - \frac{\partial_i u \partial_j u}{1 + |\nabla u|^2}\right)\partial_i\partial_j u = 0 \quad \text{on} \quad \mathbb{R}^n, \qquad u(0) = u_0. \tag{6.17}$$

We assume that the coefficient a satisfies the assumptions

$$a \in buc^s(\mathbb{R}^n) \cap C^\omega(X), \qquad a(x) \geq \delta, \quad x \in X, \tag{6.18}$$

where X is an open subset of $\mathbb{R}^n$, and where $\delta > 0$. Equation (6.17) coincides with the mean curvature flow described in the introduction in case that $a \equiv 1$.

EXAMPLE 6.5. *Let $u_0 \in buc^{2+s}(\mathbb{R}^n)$ be given. Then there exists a number $T > 0$ such that equation (6.17) has a unique solution*

$$u \in C^1([0,T], buc^s(\mathbb{R}^n)) \cap C([0,T], buc^{2+s}(\mathbb{R}^n)). \tag{6.19}$$

The solution has the additional regularity property $u \in C^\omega((0,T) \times X)$.

Proof. Let

$$F(v) := -a\left(\delta_{ij} - \frac{\partial_i v \partial_j v}{1 + |\nabla u|^2}\right)\partial_i\partial_j v.$$

One can show that (1.5) remains valid for our new function F. Based on equation (1.5) we obtain a unique solution u in the class (6.19) for the parabolic equation (6.17). Let $(t_0, x_0) \in (0,T) \times X$ be fixed and set

$$u_{\lambda,\mu}(t,x) := u(t + \zeta(t)\lambda, x + \zeta(t)\chi(x)\mu), \qquad (t,x) \in I \times \mathbb{R}^n,$$

where $I := [0, T]$. It follows from Proposition 5.3 that $v = u_{\lambda,\mu}$ satisfies the parameter dependent equation

$$\partial_t v + F_{\lambda,\mu}(v) = 0, \quad v(0) = u_0,$$

where

$$F_{\lambda,\mu}(v) := (1 + \zeta'\lambda) T_\mu F(T_\mu^{-1} v) - B_\mu v.$$

Based on Proposition 5.2 we conclude that

$$[(v, (\lambda, \mu)) \mapsto F_{\lambda,\mu}(v)] \in C^\omega(\mathbb{E}_1(I) \times \mathbb{B}^{n+1}(0, r_0), \mathbb{E}_0(I))),$$

where the spaces $\mathbb{E}_1(I)$ and $\mathbb{E}_0(I)$ have the same meaning as in the introduction. The implicit function theorem lets us once more conclude that

$$[(\lambda, \mu) \mapsto u_{\lambda,\mu}] \in C^\omega(\mathbb{B}^{n+1}(0, r_0), \mathbb{E}_1(I))$$

and we obtain as in the previous examples that $u \in C^\omega(J \times X)$. ∎

BIBLIOGRAPHY

1. H. Amann, *Linear and Quasilinear Parabolic Problems. Volume I: Abstract Linear Theory*, Birhäuser, Basel, 1995.

2. H. Amann, *Elliptic operators with infinite-dimensional state spaces*, J. Evolution Equations **1** (2001), 143–188.

3. S. B. Angenent, *Nonlinear analytic semiflows*, Proc. Roy. Soc. Edinburgh **115A** (1990), 91–107.

4. S. B. Angenent, *Parabolic equations for curves on surfaces, Part I. Curves with p−integrable curvature*, Ann. of Math. **132** (1990), 451–483.

5. R. Denk, M. Hieber, J. Prüss, *R-Boundedness, Fourier Multipliers, and Problems of Elliptic and Parabolic Type*, Mem. Amer. Math. Soc., to appear.

6. X. T. Duong, G. Simonett, *H_∞-calculus for elliptic operators with non-smooth coefficients*, Differential Integral Equations **10** (1997), 201–217.

7. J. Escher, J. Prüss, G. Simonett, *Analytic solutions for a Stefan problem with Gibbs-Thomson correction*, (submitted).

8. J. Escher, G. Simonett, *Analyticity of the interface in a free boundary problem*, Math. Ann. **305** (1996), 439–459.

9. J. Escher, G. Simonett, *Analyticity of solutions to fully nonlinear parabolic evolution equations on symmetric spaces*, (submitted).

10. O. A. Ladyženskaya, V. A. Solonnikov, and N. N. Ural'ceva, *Linear and Quasilinear Equations of Parabolic Type*, Transl. Math. Monographs vol. 23. Amer. Math. Soc., 1968.

11. A. Lunardi, *Analytic Semigroups and Optimal Regularity in Parabolic Equations*, Birkhäuser, Basel, 1995.

12. J. Prüss, *Maximal regularity for abstract parabolic problems with inhomogeneous boundary data in L_p-spaces*, Proceedings Equadiff 10, 2001, Math. Bohemia, to appear.

13. H. Triebel, *Interpolation Theory, Function Spaces, Differential Operators*, North-Holland, Amsterdam, 1978.

14. H. Triebel, *Theory of Function Spaces*, Birkhäuser, Basel, 1983.

The regulator problem for a singular control system

ANGELO FAVINI*
Universitá degli Studi di Bologna, Dipartimento di Matematica
I-40126 Bologna, Italia
e-mail: favini@dm.unibo.it

Dedicated to my dearest great friend Jerry Goldstein for his 60th birthday.

ABSTRACT. The regulator problem for a degenerate differential equation in a Hilbert space is investigated. It is shown that a suitable change of variable argument reduces the minimum problem to the one for a regular system. Examples of application to ordinary differential equations and partial differential equations are given.

0 INTRODUCTION

Consider the singular system

$$\frac{d}{dt}(My)(t) + Ly(t) = Bu(t), \quad 0 < t < \tau, \tag{0.1}$$

$$My(0) = My_0, \tag{0.2}$$

where L, M are closed linear operators in the real Hilbert space H, B is a continuous linear operator from the Hilbert space U into H, the domain $\mathcal{D}(L)$ of L is contained in the domain $\mathcal{D}(M)$ of M, L has a bounded inverse, $u(\cdot)$ is the control and $y(\cdot)$ is the solution. Of course, y_0 is a given element in $\mathcal{D}(L)$.
Moreover, we assume that $z = 0$ is a simple pole for the resolvent $(z+T)^{-1}$, where $T = ML^{-1}$, or, equivalently,

$$\|M(zL+M)^{-1}\|_{L(H)} \leq C|z|^{-1} \text{ for } 0 < |z| \leq \epsilon_0$$

*Work supported by M.I.U.R. 60% and by G.N.A.M.P.A.

for some suitable $\epsilon_0 > 0$, where $L(H)$ is the space of bounded linear operators from H into itself, endowed with the uniform norm. Analogously, $L(U, H)$ denotes the space of all bounded linear operators from U to H.

We associate to (0.1), (0.2) the quadratic cost functional

$$J(u) = \int_0^{\tau} \{\langle Ky(t), y(t)\rangle_H + \langle Nu(t), u(t)\rangle_U\} dt. \tag{0.3}$$

Here and in the following, $\langle , \rangle_Y$ shall denote the inner product in the Hilbert space Y. In (0.3) the operators $K = K^* \geq 0$, $N = N^* > 0$ are self-adjoint linear operators in H and in U, respectively.

Our aim here is to show that $J(u)$ is minimized over $L^2(0, \tau; U)$, the space of all [classes of] U-valued strongly measurable functions u over $(0, \tau)$ such that

$$\int_{(0,\tau)} \|u(t)\|_U^2 dt < \infty$$

endowed with the norm

$$\|u\|_{L^2(0,\tau;U)} = \left(\int_{(0,\tau)} \|u(t)\|_U^2 dt\right)^{1/2},$$

i.e., there exists a unique control $u^* \in L^2(0, \tau; U)$ such that

$$J(u^*) = \inf_u J(u). \tag{0.4}$$

Moreover, u^* is determined by means of a feedback law related to a suitable Riccati equation.

In fact, in this paper we will solve the problem under an additional assumption on the operators L, M, see below, but it is proven in a forthcoming paper [2] that a deeper analysis allows to avoid this. However, the technique is basically the same once the well-known results by J.L. Lions [10] are used.

The minimum problem (0.1)~(0.4) is widely considered in literature. In the last decades, finite dimensional cases were first considered, having in mind above all descriptor systems in the theory of automatic control. We confine ourselves to quoting Bender and Laub [3], the survey paper [9] by Lewis, Cobb [5], Pandolfi [11] and the monographs by Campbell [4] and Dai [6]. More recently, based on the pioneering work by J.L. Lions [10], some authors have studied problem (0.1)~(0.4) or related ones in Hilbert space, see Sviridyuk and Efremov [12], Barbu and Favini [1]. Here, we give a completely new approach.

The plan of the paper is as follows. In Section 1 we indicate how to reduce (0.1)~(0.4) to an equivalent and more convenient problem. A change of variable argument in the control parameter permits to transform the new cost functional to a quadratic one, whose treatment is well-known in literature. To this end, to simplify the exposition, it is assumed that the projection operator P of H onto the null space $N(T)$ of $T = ML^{-1}$ is self-adjoint. The optimal control is determined by a feedback law, too. The argument could work for a general cost functional

$$\int_0^{\tau} \{\langle Ky(t), y(t)\rangle_H + 2\langle y(t), Ru(t)\rangle_H + \langle Nu(t), u(t)\rangle_U\} dt,$$

where $R \in L(U;H)$, provided that the operator matrix

$$\begin{bmatrix} K & R \\ R^* & N \end{bmatrix} \in L(H \times U)$$

is non-negative in the product space $H \times U$ endowed with the usual inner product. This case will be handled elsewhere.
In Section 2 we present two applications and examples from ordinary and partial differential equations to point out the scope and generality of the abstract approach.

1 MAIN RESULTS

To begin with, we recall that the assumption on the resolvent $(z+T)^{-1}$ implies the direct sum representation (see [7], p. 157)

$$H = N(T) \oplus R(T), \tag{1.1}$$

where $N(T)$, $R(T)$ denote the kernel (null space) of T and the range of T, respectively. Here $N(T)$ and $R(T)$ are endowed with the induced norm by H. Notice that in this case $R(T)$ is a closed subspace of H.
We will denote by P the projection onto $N(T)$ along $R(T)$ and we will assume that

$$P \text{ is self-adjoint.} \tag{1.2}$$

The change of variable $Ly = x$ transforms problem (0.1),(0.2) into

$$\frac{d}{dt}(Tx)(t) + x(t) = Bu(t), \quad 0 < t < \tau, \tag{1.3}$$

$$(Tx)(0) = Tx_0, \tag{1.4}$$

where

$$x_0 = Ly_0. \tag{1.5}$$

Observe that $J(u)$ reads

$$J(u) = \int_0^\tau \{\langle L^{*^{-1}} KL^{-1}x(t), x(t)\rangle_H + \langle Nu(t), u(t)\rangle_U\}dt. \tag{1.6}$$

Moreover, it is readily seen that the restriction $\tilde{T}$ of T to $R(T)$ has a bounded inverse $\tilde{T}^{-1} \in \mathcal{L}(R(T))$.
We also notice that in view of assumption (1.2), if $Q = L^{*^{-1}}KL^{-1}$, then

$$\begin{aligned} \langle L^{*^{-1}} KL^{-1}x, x\rangle_H = {} & \langle (I-P)Q(I-P)x, (I-P)x\rangle_H + \langle PQ(I-P)x, Px\rangle_H \\ & + \langle (I-P)QPx, (I-P)x\rangle_H + \langle PQPx, Px\rangle_H. \end{aligned} \tag{1.7}$$

On the other hand, system (1.3), (1.4) is equivalent to the algebraic-differential problem (observe that $\tilde{T}$ has a bounded inverse in $R(T)$)

$$\frac{d}{dt}((I-P)x(t)) + \tilde{T}^{-1}(I-P)x(t) = \tilde{T}^{-1}(I-P)Bu(t), \; 0 < t < \tau, \tag{1.8}$$

$$(I-P)x(0) = (I-P)x_0, \tag{1.9}$$

$$Px(t) = PBu(t), \quad 0 < t < \tau. \tag{1.10}$$

If expression (1.10) for $Px(\cdot)$ is substituted into (1.7), we get

$$\begin{aligned}\langle L^{*^{-1}}KL^{-1}x(t),x(t)\rangle_H &= \langle (I-P)Q(I-P)x(t),(I-P)x(t)\rangle_H \\ &\quad + \langle PQ(I-P)x(t),PBu(t)\rangle_H + \langle QPBu(t),(I-P)x(t)\rangle_H \\ &\quad + \langle QPBu(t),PBu(t)\rangle_H \\ &= \langle (I-P)Q(I-P)x(t),(I-P)x(t)\rangle_H + \langle (I-P)QPBu(t),(I-P)x(t)\rangle_H \\ &\quad + \langle B^*PQ(I-P)x(t),u(t)\rangle_U + \langle B^*PQBu(t),u(t)\rangle_U. \end{aligned} \tag{1.11}$$

Let us introduce the inner product

$$\langle ((I-P)x,u),((I-P)y,v)\rangle = \langle (I-P)x,(I-P)y\rangle_H + \langle u,v\rangle_U$$

in the space $R(T)\times U$, $(x,y)\in H\times H$, $(u,v)\in U\times U$. Then

$$J(u) = \int_0^\tau \langle \begin{bmatrix} \bar F & \mathcal{H} \\ \mathcal{H}^* & \bar G \end{bmatrix} \begin{bmatrix} (I-P)x(t) \\ u(t) \end{bmatrix}, \begin{bmatrix} (I-P)x(t) \\ u(t) \end{bmatrix} \rangle dt, \tag{1.12}$$

where

$$\bar F = (I-P)Q(I-P), \quad Q = L^{*^{-1}}KL^{-1}, \tag{1.13}$$

$$\mathcal{H} = (I-P)QPB, \tag{1.14}$$

$$\bar G = N + B^*PQPB. \tag{1.15}$$

We continue our analysis observing that $\bar G$ in (1.15) has a bounded inverse, so that $J(u)$ coincides with

$$J(u) = \int_0^\tau \langle \begin{bmatrix} \tilde F & 0 \\ 0 & \bar G \end{bmatrix} \begin{bmatrix} (I-P)x(t) \\ w(t) \end{bmatrix}, \begin{bmatrix} (I-P)x(t) \\ w(t) \end{bmatrix} \rangle dt, \tag{1.16}$$

where

$$\tilde F = \bar F - \mathcal{H}\bar G^{-1}\mathcal{H}^*, \tag{1.17}$$

$$w = u + \bar G^{-1}\mathcal{H}^*(I-P)x. \tag{1.18}$$

On the other hand,

$$\begin{bmatrix} \bar F & \mathcal{H} \\ \mathcal{H}^* & \bar G \end{bmatrix} = \begin{bmatrix} I & \mathcal{H}\bar G^{-1} \\ 0 & I \end{bmatrix} \begin{bmatrix} \tilde F & 0 \\ 0 & \bar G \end{bmatrix} \begin{bmatrix} I & 0 \\ \bar G^{-1}\mathcal{H}^* & I \end{bmatrix}, \tag{1.19}$$

so that

$$\begin{bmatrix} \tilde F & 0 \\ 0 & \bar G \end{bmatrix} = \begin{bmatrix} I & -\mathcal{H}\bar G^{-1} \\ 0 & I \end{bmatrix} \begin{bmatrix} \bar F & \mathcal{H} \\ \mathcal{H}^* & \bar G \end{bmatrix} \begin{bmatrix} I & 0 \\ -\bar G^{-1}\mathcal{H}^* & I \end{bmatrix}. \tag{1.20}$$

This yields

$$\langle \begin{bmatrix} \tilde{F} & 0 \\ 0 & \bar{G} \end{bmatrix} \begin{bmatrix} (I-P)x(t) \\ w(t) \end{bmatrix}, \begin{bmatrix} (I-P)x(t) \\ w(t) \end{bmatrix} \rangle$$

$$= \langle \begin{bmatrix} I & 0 \\ -\bar{G}^{-1}\mathcal{H}^* & I \end{bmatrix}^* \begin{bmatrix} \bar{F} & \mathcal{H} \\ \mathcal{H}^* & \bar{G} \end{bmatrix} \begin{bmatrix} I & 0 \\ -\bar{G}^{-1}\mathcal{H}^* & I \end{bmatrix} \begin{bmatrix} (I-P)x(t) \\ w(t) \end{bmatrix}, \begin{bmatrix} (I-P)x(t) \\ w(t) \end{bmatrix} \rangle$$

$$= \langle \begin{bmatrix} \bar{F} & \mathcal{H} \\ \mathcal{H}^* & \bar{G} \end{bmatrix} \begin{bmatrix} I & 0 \\ -\bar{G}^{-1}\mathcal{H}^* & I \end{bmatrix} \begin{bmatrix} (I-P)x(t) \\ w(t) \end{bmatrix}, \begin{bmatrix} I & 0 \\ -\bar{G}^{-1}\mathcal{H}^* & I \end{bmatrix} \begin{bmatrix} (I-P)x(t) \\ w(t) \end{bmatrix} \rangle. \tag{1.21}$$

Moreover,

$$\begin{bmatrix} \bar{F} & \mathcal{H} \\ \mathcal{H}^* & \bar{G} \end{bmatrix} = \begin{bmatrix} I-P & 0 \\ B^*P & I \end{bmatrix} \begin{bmatrix} Q & 0 \\ 0 & N \end{bmatrix} \begin{bmatrix} I-P & PB \\ 0 & I \end{bmatrix}. \tag{1.22}$$

Therefore

$$\langle \begin{bmatrix} \tilde{F} & 0 \\ 0 & \bar{G} \end{bmatrix} \begin{bmatrix} (I-P)x(t) \\ w(t) \end{bmatrix}, \begin{bmatrix} (I-P)x(t) \\ w(t) \end{bmatrix} \rangle$$

$$= \langle \begin{bmatrix} \bar{F} & \mathcal{H} \\ \mathcal{H}^* & \bar{G} \end{bmatrix} \begin{bmatrix} I & 0 \\ -\bar{G}^{-1}\mathcal{H}^* & I \end{bmatrix} \begin{bmatrix} (I-P)x(t) \\ w(t) \end{bmatrix},$$

$$\begin{bmatrix} I & 0 \\ -\bar{G}^{-1}\mathcal{H}^* & I \end{bmatrix} \begin{bmatrix} (I-P)x(t) \\ w(t) \end{bmatrix} \rangle \tag{1.23}$$

$$= \langle \begin{bmatrix} I-P & 0 \\ B^*P & I \end{bmatrix} \begin{bmatrix} Q & 0 \\ 0 & N \end{bmatrix} \begin{bmatrix} I-P & PB \\ 0 & I \end{bmatrix} \begin{bmatrix} I & 0 \\ -\bar{G}^{-1}\mathcal{H}^* & I \end{bmatrix} \begin{bmatrix} (I-P)x(t) \\ w(t) \end{bmatrix},$$

$$\begin{bmatrix} I & 0 \\ -\bar{G}^{-1}\mathcal{H}^* & I \end{bmatrix} \begin{bmatrix} (I-P)x(t) \\ w(t) \end{bmatrix} \rangle$$

$$= \langle \begin{bmatrix} Q & 0 \\ 0 & N \end{bmatrix} \begin{bmatrix} I-P & PB \\ 0 & I \end{bmatrix} \begin{bmatrix} I & 0 \\ -\bar{G}^{-1}\mathcal{H}^* & I \end{bmatrix} \begin{bmatrix} (I-P)x(t) \\ w(t) \end{bmatrix},$$

$$\begin{bmatrix} I-P & PB \\ 0 & I \end{bmatrix} \begin{bmatrix} I & 0 \\ -\bar{G}^{-1}\mathcal{H}^* & I \end{bmatrix} \begin{bmatrix} (I-P)x(t) \\ w(t) \end{bmatrix} \rangle \geq 0. \tag{1.24}$$

Hence

$$J(u) = J(w) = \int_0^\tau \{\langle \tilde{F}(I-P)x(t), (I-P)x(t)\rangle_{R(T)} + \langle \bar{G}w(t), w(t)\rangle_U\} dt \tag{1.25}$$

where $\bar{G} > 0$, $\tilde{F} \geq 0$. Since, by (1.18), $w = u + \bar{G}^{-1}\mathcal{H}^*(I-P)x$, we have $u = w - \bar{G}^{-1}\mathcal{H}^*(I-P)x$, so that equation (1.8) reads

$$\frac{d}{dt}(I-P)x(t) = -\tilde{T}^{-1}(I+(I-P)B\bar{G}^{-1}\mathcal{H}^*)(I-P)x(t) + \tilde{T}^{-1}(I-P)Bw(t), 0 < t < \tau. \tag{1.26}$$

It follows that the original problem (0.1)~(0.4) is equivalent to minimizing $J(w)$ in (1.24), where $(I-P)x(\cdot)$ and $w(\cdot)$ are related by equation (1.25) and initial condition (1.9). But this last problem is regular and can be solved by standard methods. We then establish our main result as follows.

THEOREM 1.1. *Let us assume $z=0$ to be a simple pole for $M(zL+M)^{-1}$ and suppose that (1.2) holds.*
Then there exists a unique optimal pair (y^*,u^*) for problem (0.1)~(0.4) *such that $y^*\in L^2(0,\tau;H)$, $My^*\in H^1(0,\tau;H)$, $u^*\in L^2(0,\tau;U)$.*

If the operators L and M are bounded and self-adjoint, $L>0$, one could consider problem (0.1)~(0.4) in an equivalent topology on H such that the projection P is self-adjoint, i.e., (1.2) holds. For a similar idea, see Kato [8], p. 419.
Indeed, let us introduce the inner product in H

$$(x,y)=\langle Lx,y\rangle_H,\quad x,y\in H. \tag{1.27}$$

Then equation (0.1) is equivalent to

$$\frac{d}{dt}(L^{-1}My(t))+y(t)=L^{-1}Bu(t),\quad 0<t<\tau, \tag{1.28}$$

where $L^{-1}M\in\mathcal{L}(H)$ is self-adjoint with respect to the new inner product (1.26).
Moreover,

$$\int_0^\tau \langle Ky(t),y(t)\rangle_H dt=\int_0^\tau (L^{-1}Ky(t),y(t))dt$$

and $L^{-1}K$ is self-adjoint ≥ 0.
Therefore (0.1)~(0.4) reduces to minimizing $J(u)$, where

$$J(u)=\int_0^\tau \{(L^{-1}Ky(t),y(t))+\langle Nu(t),u(t)\rangle_U\}dt$$

and the state $y(\cdot)$ and the control $u(\cdot)$ are related with (1.27) together with the initial condition

$$(L^{-1}My)(0)=L^{-1}My_0.$$

Since

$$(z+L^{-1}M)^{-1}=(zL+M)^{-1}L,$$

$z=0$ is a simple pole for the resolvent of $L^{-1}M$ and the corresponding projector P is self-adjoint.

REMARK 1.2. If L, M are self-adjoint and, moreover, have the same domain and commute in the resolvent sense, then T is self-adjoint too, so that P is self-adjoint.

Reduction of problem (0.1)~(0.4) to the minimum problem for $J(w)$ in (1.24), where $(I-P)x(t)$ and $w(t)$ are related by (1.25) and (1.9), allows to write a Riccati equation (in $L(R(T))$) for our optimal pair. To this end, we introduce the operators

$$A=-\tilde{T}^{-1}(I+(I-P)B\bar{G}^{-1}H^*)(I-P)\in L(R(T)), \tag{1.29}$$

$$C=-\tilde{T}^{-1}(I-P)B\in L(U;R(T)). \tag{1.30}$$

In what follows we will identify the adjoint $R(T)^*$ of $R(T)$ and we will denote by A^* and C^* the adjoint operators of A and C, respectively, after this identification. We follow Zabczyk [13], pp. 133-134 for the links between optimal control and Riccati equation.
Let $P(t) = P(t)^* \geq 0$ be the unique strongly differentiable solution ($\in \mathcal{L}(R(T))$) to the Riccati equation

$$\frac{d}{dt}P(t)\tilde{F} + P(t)A + A^*P(t) - P(t)C\bar{G}^{-1}C^*P(t), \quad 0 \leq t \leq \tau, \tag{1.31}$$

$$P(0) = 0. \tag{1.32}$$

We then have the result stated below.

THEOREM 1.3. *If $z = 0$ is a simple pole for $(z+T)^{-1}$, $T = ML^{-1}$, and* (1.2) *holds, then the elements of the optimal pair (y^*, u^*) from Theorem 1.1 are related by*

$$u^*(t) = -\bar{G}^{-1}\{C^*P(\tau - t) + \mathcal{H}^*\}(I - P)Ly^*(t), \tag{1.33}$$

where $P(t)$ is the solution to (1.30),(1.31).

REMARK 1.4. The appearance of the operator L in front of y^* in (1.32) might seem not natural when compared with the regular case $M = I$. However, this fact is strictly connected with the degeneration of the involved equation. Notice that if $\mathcal{H} = 0$, $P = 0$, $M = I$, then the operator L is necessarily bounded, $A = -L$, $C = LB$, so that (1.30) reduces to

$$\frac{d}{dt}P(t) = {L^*}^{-1}KL^{-1} - P(t)L - L^*P(t) - P(t)LBN^{-1}B^*L^*P(t)$$

and if we introduce $P_1(t) := L^*P(t)L$, $t \geq 0$, then $P_1(t)$ satisfies the usual Riccati equation

$$\frac{d}{dt}P_1(t) = K - P_1(t)L - L^*P_1(t) - P_1(t)BN^{-1}B^*P_1(t), \quad t \geq 0, \tag{1.34}$$

together with

$$P_1(0) = 0. \tag{1.35}$$

Then the optimal control u^* is expressed in feedback form

$$\begin{aligned} u^*(t) &= -N^{-1}B^*L^*P(\tau - t)Ly^*(t) \\ &= -N^{-1}B^*L^*{L^*}^{-1}P_1(\tau - t)L^{-1}Ly^*(t) = -N^{-1}B^*P_1(\tau - t)y^*(t), \end{aligned} \tag{1.36}$$

and in (1.35) operator L in fact disappears.

Proof of Theorem 1.3. We have reduced (0.1)~(0.4) to find $\min J(w)$ in (1.24) together with (1.9), (1.25). Then it is well known that the corresponding optimal control w^* is given by

$$w^*(t) = -\bar{G}^{-1}C^*P(\tau - t)(I - P)x^*(t), \quad 0 \leq t \leq \tau, \tag{1.37}$$

where $(I-P)x^*(t)$ satisfies the Cauchy problem

$$\frac{d}{dt}(I-P)x(t) = (A - C\bar{G}^{-1}C^*P(\tau - t))(I-P)x(t), \quad 0 \leq t \leq \tau, \tag{1.38}$$

$$(I-P)x(0) = (I-P)x_0. \tag{1.39}$$

Taking into account (1.18), the just optimal control $u^*(\cdot)$ is given by

$$u^*(t) = -\bar{G}^{-1}\{C^*P(\tau - t) + \mathcal{H}^*\}(I-P)x^*(t). \tag{1.40}$$

Since $Ly(t) = x(t)$, we find (1.32). ∎

REMARK 1.5. The explicit formula (1.39) guarantees that the optimal control u^* is continuous on $[0, \tau]$.

2 APPLICATIONS AND EXAMPLES

Many problems in applied mathematics can be handled on the ground of the preceding results. We refer to the quoted references [4], [6], [9], [12]. Here, we describe two possible ones.
The first example in its simplicity may be useful to clarify how Theorem 1.1 provides a precise method to solve a minimum problem for degenerate equations.

EXAMPLE 2.1. Let us consider the system

$$\frac{d}{dt}(x+y)(t) + x(t) = 2u(t), \quad 0 < t < 1, \tag{2.1}$$

$$\frac{d}{dt}(x+y)(t) + y(t) = u(t), \quad 0 < t < 1, \tag{2.2}$$

$$(x+y)(0) = \xi_0. \tag{2.3}$$

Here ξ_0 is a given real number. Theorems 1.1 and 1.3 apply immediately. What follows shows that also a direct approach would involve a trick like the one used in the proof to Theorem 1.1.
Subtracting and then summing (2.2) to (2.1), we get

$$x(t) - y(t) = u(t), \tag{2.4}$$

$$2\frac{d}{dt}(x(t) + y(t)) + (x(t) + y(t)) = 3u(t). \tag{2.5}$$

Let $x + y = \xi$. Then, by (2.4),

$$x = \frac{1}{2}(\xi + u), \quad y = \frac{1}{2}(\xi - u), \tag{2.6}$$

where $\xi(\cdot)$ satisfies the Cauchy problem

$$\frac{d\xi}{dt}(t) = -\xi(t)/2 + \frac{3}{2}u(t), \quad 0 < t < 1, \tag{2.7}$$

$$\xi(0) = \xi_0 \tag{2.8}$$

We want to minimize $J(u)$ where

$$J(u) = \int_0^1 (x^2 + 2y^2 + u^2)dt. \tag{2.9}$$

One observes that

$$\begin{aligned} J(u) &= \int_0^1 \{\frac{1}{4}(\xi + u)^2 + \frac{1}{2}(\xi - u)^2 + u^2\}dt \\ &= \int_0^1 \langle \begin{bmatrix} 3/4 & -1/4 \\ -1/4 & 7/4 \end{bmatrix} \begin{bmatrix} \xi \\ u \end{bmatrix}, \begin{bmatrix} \xi \\ u \end{bmatrix} \rangle dt. \end{aligned} \tag{2.10}$$

Let (cf. the proof to Theorem 1.1)

$$w = u - \xi/7. \tag{2.11}$$

Then

$$J(u) = J(w) = \int_0^1 (\frac{5}{7}\xi^2 + \frac{7}{4}w^2)dt, \tag{2.12}$$

while problem (2.1)∼(2.3) transforms into

$$\xi'(t) = \frac{d\xi}{dt}(t) = -\frac{2}{7}\xi(t) + \frac{3}{2}w(t), \quad 0 < t < 1, \tag{2.13}$$

$$\xi(0) = \xi_0 \tag{2.14}$$

$$\inf_w J(w) = \inf_w (2.10). \tag{2.15}$$

(2.13)∼(2.15) is a standard regulator problem and it is well known that it has a unique solution (ξ^*, w^*) characterized by the feedback law

$$w(t) = -\frac{6}{7}P(1-t)\xi(t), \tag{2.16}$$

where $P(t)$ satisfies the Riccati equation

$$\frac{d}{dt}P(t) = \frac{5}{7} - \frac{4}{7}P(t) - \frac{9}{7}P(t)^2, \quad 0 \le t \le 1, \tag{2.17}$$

together with

$$P(t) \ge 0, \quad P(0) = 0. \tag{2.18}$$

Moreover, $\xi^*(t)$ satisfies

$$\frac{d\xi}{dt}(t) = -(\frac{2}{7} + \frac{9}{7}P(1-t))\xi(t), \quad 0 < t < 1, \quad \xi(0) = \xi_0. \tag{2.19}$$

In our case, it is readily seen that

$$P(t) = 5(e^{2t} - 1)(9e^{2t} + 5)^{-1}, \quad 0 \le t \le 1. \tag{2.20}$$

From (2.11), (2.16) it follows that

$$u^*(t) = \frac{1}{7}(1 - 6P(1-t))\xi^*(t), \quad 0 \le t \le 1. \tag{2.21}$$

where $P(t)$ is given in (2.20) and $\xi^*(t)$ is the unique solution to (2.19). Correspondingly, the optimal state $(x^*(\cdot), y^*(\cdot))$ of the system is

$$x^*(t) = \frac{1}{2}(u^*(t) + \xi^*(t)), \quad y^*(t) = \frac{1}{2}(\xi^*(t) - u^*(t)). \tag{2.22}$$

This provides the synthesis of the problem. On the other hand, this is precisely the procedure followed to obtain Theorems 1.1 and 1.3.

EXAMPLE 2.2. Let us consider the abstract equation in the Hilbert space H

$$\frac{d}{dt}((A - z_0)y)(t) = Ay(t) + Bu(t), \quad 0 < t < \tau, \tag{2.23}$$

where $B \in L(U, H)$, U another Hilbert space, A is a closed linear, densely defined operator in H, with a bounded inverse and $z = z_0$ is a simple pole for the resolvent $(z - A)^{-1}$. Take $L = -A$, $M = A - z_0$. Then $T = ML^{-1} = z_0A^{-1} - I$ and

$$(z - T)^{-1} = (z+1)^{-1}A(A - z_0 + z_0 z(z+1)^{-1})^{-1} \tag{2.24}$$

for $0 < |z| \le \epsilon_0$, with ϵ_0 small. Since

$$A(A - z_0 + z_0 z(1+z)^{-1})^{-1} = I + z_0(z+1)^{-1}(A - z_0 + z_0 z(z+1)^{-1})^{-1}$$

and

$$\|A(A - z_0 + z_0 z(1+z)^{-1})^{-1}\|_{L(H)} \le 1 + C|z|^{-1} \le C'|z|^{-1}, \quad 0 < |z| \le \epsilon_0$$

we conclude from (2.24) that for some $C_1 > 0$

$$\|(z - T)^{-1}\|_{L(H)} \le C_1|z|^{-1}, \quad 0 < |z| \le \epsilon_0,$$

so that all the results in Section 1 apply provided that A is self-adjoint (this last assumption might be avoided, too).

Equation (2.23) is the model of various partial differential equations if one takes $H = L^2(\Omega)$, where Ω is a bounded domain in R^n with a smooth boundary, A is an elliptic differential operator on Ω of order $2m$, $m = 1, 2, \ldots$, with boundary conditions of Dirichlet, Neumann or mixed type, and $z = z_0$ is an isolated simple eigenvalue of A (with the given boundary conditions).

BIBLIOGRAPHY

1. V. Barbu and A. Favini, *Control of degenerate differential systems*, Control and Cybernetics **28** (1999), no. 3, 397–420.

2. V. Barbu, A. Favini, and L. Pandolfi, *Optimal regulation with quadratic cost functional of a degenerate system*, to appear.

3. D.úJ. Bender and A.J. Laub, *The linear-quadratic optimal regulator for descriptor systems*, IEEE Trans. Automatic Control Vol. AC-**32**, no. 8 (1987), 672–688.

4. S. L. Campbell, *Singular systems of differential equations*, Pitman, San Francisco, 1980.

5. D. Cobb, *Descriptor variable systems and optimal state regulation*, IEEE Trans. Automatic Control Vol. AC-**28**, no. 5 (1983), 601–611.

6. L. Dai, *Singular control systems*, LN Control Information Sciences, vol. 118, Springer Verlag, Berlin-Heidelberg-New York, 1989.

7. A. Favini and A. Yagi, *Degenerate differential equations in Banach spaces*, Pure Appl. Math., vol. 215, M. Dekker, New York-Basel-Hong Kong, 1999.

8. T. Kato, *Perturbation theory for linear operators*, Springer Verlag, Berlin-Heidelberg-New York, 1966.

9. F. L. Lewis, *A survey of linear singular systems*, Circuits Syst. & Signal Process. **5**, no. 1 (1986), 3–36.

10. J. L. Lions, *Contrôle optimal des systèmes gouvernés par des équations aux dérivées partielles*, Dunod, Paris, 1968.

11. L. Pandolfi, *On the regulator problem for linear degenerate control systems*, J. Opt. Theory Appl. **33** (1981), 241–254.

12. G. A. Sviridyuk and A. A. Efremov, *Optimal control of Sobolev type linear equations with relatively p-sectorial operators*, Diff. Uravnenia **31** (1995), 1912–1916, (Russian). English transl.: Diff. Eqs. **31** (1995), 1882–1890.

13. J. Zabczyk, *Mathematical control theory: an introduction*, Birkhäuser Verlag, Boston-Basel-Berlin, 1992.

Criteria for R-boundedness of operator families

MARIA GIRARDI

Department of Mathematics, University of South Carolina
Columbia, SC 29208, U.S.A.
e-mail: girardi@math.sc.edu

LUTZ WEIS

Mathematisches Institut I, Universität Karlsruhe
Englerstraße 2, 76128 Karlsruhe, Germany
e-mail: Lutz.Weis@math.uni-karlsruhe.de

Dedicated to Jerry Goldstein on the occasion of his 60th birthday

ABSTRACT. The main results state that large classes of convolution and Fourier multiplier operators are R-bounded. R-boundedness is of importance in connection with maximal regularity and functional calculus. Also shown is that smooth operator-valued functions have a R-bounded range, where the degree of smoothness depends on the geometry of the Banach space.

1 INTRODUCTION

Recently the notion of R-boundedness has played an important role in the functional analytic approach to partial differential equations. It is shown in [36] (see also [2, 3, 7, 8, 14, 15, 23, 26, 35]) that, for a sectorial operator A, R-boundedness of the set $\{\lambda(\lambda - A)^{-1} : \lambda \in \mathbb{C}_+\}$, i. e. R-sectoriality, characterizes maximal regularity. R-boundedness is also a very useful notion in connection with the H^∞-calculus (see [23, 24]). Furthermore it was shown in [36] (see also [8, 19, 21, 33]) that R-boundedness provides a proper setting for boundedness theorems for operator-valued Fourier multipliers. Hence, workable criteria for R-boundedness are needed. It is already

Girardi is supported in part by the Alexander von Humboldt Foundation. Weis is supported in part by *Landesforschungsschwerpunkt Evolutionsgleichungen des Landes Baden-Württenberg.*

known that sectorial operators with Gaussian estimates ([14,35]), as well as large classes of partial differential operators ([7,15,25,26]), are R-sectorial.

This leads to the natural question of whether each sectorial operator from a Banach space into itself is R-sectorial. Kalton and Lancien [22] answered this question in the negative; however, their counterexample is rather *abstract* and no *reasonable* sectorial differential operator which is not R-sectorial is known. This note gives general criteria for R-boundedness, which may explain why the common operators of analysis are R-bounded and why it is so difficult to find counterexamples. This note's R-boundedness criteria are based on the observation that the very boundedness theorems for Fourier multipliers, which have R-boundedness in their assumption, improve themselves by a trick to give R-boundedness of classes of operators that satisfy these assumptions in a uniform way. This is possible provided X has the UMD-property and Pisier's property (α). As a rule of thumb, the reflexive Banach spaces commonly appearing in analysis (such as subspaces of Sobolev spaces or Hardy spaces) have these properties. This idea is related to the fact (see [23]) that for a sectorial operator A (on such spaces X) with an H^∞-calculus, the set $\{f(A) : \|f\|_{H^\infty} \leq 1\}$ is R-bounded. We think that our approach gives an unified way to prove R-boundedness for many families of operators relevant in applications and also can be applied to situations not considered here.

The main result of Section 3, Theorem 3.2, gives that the set of all Fourier multiplier operators that satisfy the Mihlin conditions in an uniform way form an R-bounded set. This is of particular interest for scalar-valued multiplier functions. Furthermore, Theorem 3.2 leads to (see Section 4) boundedness criteria for sets of convolution operators, which include the Gauß and Poisson semigroups. The results in Section 5 shows that the range of a sufficiently smooth function from $\mathbb{R}^N$ into $\mathcal{B}(X)$ is R-bounded, where the Fourier type of X is taken into account in order to reduce the required degree of smoothness. Then further applications to semigroups and resolvents are given. Section 2 collects needed definitions and facts.

Theorem 3.2 was presented by the authors at the TULKA seminar in February 2001. Independently and with different methods, A. Venni [34] has shown a special case of Theorem 3.2.

2 DEFINITIONS AND NOTATION

Throughout this paper X, Y, and Z are complex Banach spaces. $\mathcal{B}(X,Y)$ is the space of bounded linear operators from X into Y; often $\mathcal{B}(X,X)$ is denoted by just $\mathcal{B}(X)$. The triple (Ω, Σ, μ) is a σ-finite complete measure space; corresponding to it is the usual Bochner-Lebesgue space $L_p(\Omega; X)$ of measurable functions from Ω into X with finite $L_p(\Omega; X)$-norm where $1 \leq p \leq \infty$; often $L_p(\Omega; X)$ is denoted by just $L_p(X)$ or L_p if confusion seems unlikely. If $\Omega = [0,1]$ or $\Omega = \mathbb{R}^N$ then (Ω, Σ, μ) is understood to be the usual Lebesgue measure space. The Schwartz class $\mathcal{S}(\mathbb{R}^N; X)$ is the space of rapidly decreasing smooth function from $\mathbb{R}^N$ to X. The sequence $\{r_j\}_{j=1}^\infty$ denotes a sequence of independent, symmetric, $\{1,-1\}$-valued random variables on $[0,1]$, e.g. the Rademacher functions. $\mathbb{N}$ is the set of natural numbers while $\mathbb{N}_0 = \mathbb{N} \cup \{0\}$.

R-boundedness is the central notion of this paper.

DEFINITION 2.1. Let τ be a subset of $\mathcal{B}(X,Y)$ and $p \in [1,\infty)$. The number $R_p(\tau)$ is the smallest of the constants $R \in [0,\infty]$ with the property that for each $n \in \mathbb{N}$ and subset $\{T_j\}_{j=1}^n$ of τ and subset $\{x_j\}_{j=1}^n$ of X,

$$\left\| \sum_{j=1}^{n} r_j(\cdot) T_j(x_j) \right\|_{L_p([0,1];Y)} \leq R \left\| \sum_{j=1}^{n} r_j(\cdot) x_j \right\|_{L_p([0,1];X)} .$$

The set τ is *R-bounded* provided $R_p(\tau)$ is finite for some (and thus then, by Kahane's inequality, for each) $p \in [1,\infty)$.

Thus a set τ is R-bounded provided Kahane's contraction principle holds for *operator coefficients* from τ. Note that if X and Y are q-concave Banach lattices for some finite q (e.g. $X = Y = L_q(\Omega;\mathbb{C})$ where $1 \leq q < \infty$) then R-boundedness is equivalent to the square function estimate

$$\left\| \left(\sum_{j=1}^{m} |T_j x_j|^2 \right)^{1/2} \right\|_Y \leq R \left\| \left(\sum_{j=1}^{n} |x_j|^2 \right)^{1/2} \right\|_X \tag{2.1}$$

known from harmonic analysis (cf. [28; Thm. 1.d.6]). For basic properties of R-bounded sets and further references, see [13, 36].

The space Rad(X) provides a convenient way to view R-boundedness.

DEFINITION 2.2. For a Banach space X, define

$$\mathrm{Rad}(X) := \left\{ \widetilde{x} := \{x_j\}_{j\in\mathbb{N}} \in X^{\mathbb{N}} : \sum\nolimits_{j=1}^{n} r_j(\cdot) x_j \text{ converges in } L_2([0,1];X) \right\} .$$

Often Rad(X) is denoted by just $\widetilde{X}$.

When equipped with one of the following equivalent norms, where $1 \leq p < \infty$:

$$\|\{x_j\}_{j\in\mathbb{N}}\|_{\mathrm{Rad}_p(X)} := \left\| \sum\nolimits_{j\in\mathbb{N}} r_j(\cdot)\, x_j \right\|_{L_p([0,1];X)} ,$$

$\mathrm{Rad}_p(X)$ is a Banach space. When confusion seems unlikely, $\mathrm{Rad}_p(X)$ is denoted by just Rad(X). Much can be found about Rad(X) in the literature (see, e.g. [16]).

FACT 2.3. A sequence $\{T_j\}_{j\in\mathbb{N}}$ from $\mathcal{B}(X,Y)$ is R-bounded if and only if the corresponding map

$$\widetilde{X} \ni \{x_j\}_{j\in\mathbb{N}} \xrightarrow{\ \widetilde{T}\ } \{T_j x_j\}_{j\in\mathbb{N}} \in \widetilde{Y}$$

defines an element in $\mathcal{B}\left(\widetilde{X}, \widetilde{Y}\right)$; in which case, $R_p(\{T_j\}_{j\in\mathbb{N}}) = \|\widetilde{T}\|_{\mathcal{B}(\mathrm{Rad}_p(X),\mathrm{Rad}_p(Y))}$ for each $p \in [1,\infty)$.

Fubini's theorem yields the following useful fact.

FACT 2.4. Let $p \in [1,\infty)$. The mapping

$$I_X : L_p(\Omega; \mathrm{Rad}_p(X)) \to \mathrm{Rad}_p(L_p(\Omega;X))$$

given by

$$I_X f := \{f_j\}_{j\in\mathbb{N}} \quad \text{for} \quad f(\cdot) = \{f_j(\cdot)\}_{j\in\mathbb{N}} \in L_p(\Omega; \mathrm{Rad}_p(X))$$

is an isometry. Furthermore

$$\widehat{f} = \{\widehat{f_j}\}_{j\in\mathbb{N}} \qquad \text{and} \qquad \check{f} = \{\check{f_j}\}_{j\in\mathbb{N}} \tag{2.2}$$

for each $f(\cdot) = \{f_j(\cdot)\}_{j\in\mathbb{N}} \in L_1\left(\mathbb{R}^N; \widetilde{X}\right)$.

Considering R-boundedness in $\mathcal{B}\left(\widetilde{X}, \widetilde{Y}\right)$ leads to estimating double random series; thus, an additional property of the Banach spaces X and Y is often needed.

DEFINITION 2.5. ([32; DEF. 2.1]) Let $\{\varepsilon_k\}_{k=1}^{\infty}$ (resp. $\{\varepsilon'_k\}_{k=1}^{\infty}$) be a sequence of independent, symmetric, $\{1,-1\}$-valued random variables on some probability space (Ω, Σ, μ) (resp. (Ω', Σ', μ')) with $\{\varepsilon_k\}_{k=1}^{\infty}$ and $\{\varepsilon'_k\}_{k=1}^{\infty}$ independent. For a Banach space X and $1 \le p < \infty$, the number $\alpha_p(X)$ is the smallest of the constants $\alpha \in [0,\infty]$ with the property that

$$\left\| \sum_{j,k=1}^{n} \alpha_{jk}\varepsilon_j\varepsilon'_k x_{jk} \right\|_{L_p(\Omega\times\Omega';X)} \le \alpha \left\| \sum_{j,k=1}^{n} \varepsilon_j\varepsilon'_k x_{jk} \right\|_{L_p(\Omega\times\Omega';X)} \tag{2.3}$$

for each $n \in \mathbb{N}$ and subset $\{x_{jk}\}_{j,k=1}^{n}$ of X and choices $\{\alpha_{jk}\}_{j,k=1}^{n}$ of signs $\{1,-1\}$. The space X has *property* (α) provided $\alpha_p(X)$ is finite for some (and thus then, by Kahane's inequality, for each) $p \in [1,\infty)$.

Each subspace [32] of a Banach function space of finite cotype (e.g. a subspace of $L_p(\Omega;\mathbb{C})$ with $1 \le p < \infty$) has property (α). If X has property (α), then so does $L_q(\Omega;X)$ for $1 \le q < \infty$. The estimate in (2.3) extends to R-bounded sets.

FACT 2.6. (CF. [13; LEMMA 2.3.3]) Let X and Y be Banach spaces enjoying property (α) and τ be an R-bounded subset of $\mathcal{B}(X,Y)$. Then for $p \in [1,\infty)$

$$\left\| \sum_{j,k=1}^{n} \varepsilon_j\varepsilon'_k T_{jk} x_{jk} \right\|_{L_p(\Omega\times\Omega';Y)} \le \alpha_p(X)\,\alpha_p(Y)\,R_p(\tau) \left\| \sum_{j,k=1}^{n} \varepsilon_j\varepsilon'_k x_{jk} \right\|_{L_p(\Omega\times\Omega';X)}$$

for each $n \in \mathbb{N}$ and subset $\{T_{jk}\}_{j,k=1}^{n}$ of τ and subset $\{x_{jk}\}_{j,k=1}^{n}$ of X.

The UMD property is often needed for vector-valued multiplier theorems.

DEFINITION 2.7. A Banach space X is a *UMD space* provided the Hilbert transform

$$Hf(t) = PV - \int \frac{f(s)}{t-s}\,ds \qquad \text{for} \quad f \in \mathcal{S}(X)$$

extends to a bounded linear operator on $L_p(\mathbb{R};X)$ for some (and thus then for each) $p \in (1,\infty)$.

Thus X is a UMD space if and only if $m\colon \mathbb{R}\setminus\{0\} \to \mathcal{B}(X)$ given by

$$m(t) = \operatorname{sign}(t)\, I_X$$

is a Fourier multiplier on $L_p(\mathbb{R};X)$ for some (and thus then for each) $p \in (1,\infty)$. Closed subspaces of, the dual of, and quotients spaces of a UMD space are UMD spaces. Each Hilbert space is a UMD space. If X is a UMD space and $q \in (1,\infty)$, then $L_q(\Omega;X)$ is also a UMD space.

For $f \in L_1(\mathbb{R}^N; X)$ we denote by $\widehat{f}$ or $\mathcal{F}f$ the Fourier transform of f, i.e.,

$$\widehat{f}(t) \equiv \mathcal{F}(f)(t) := \int_{\mathbb{R}^N} e^{-it\cdot s} f(s)\, ds \,.$$

In the vector-valued setting the Hausdorff-Young inequality does not hold in general. Thus one considers the following class of Banach spaces.

DEFINITION 2.8. ([30]) A Banach space X has *Fourier type* p, where $p \in [1,2]$, if the Fourier transform $\mathcal{F}$ defines a bounded linear operator from $L_p(\mathbb{R}^N; X)$ to $L_{p'}(\mathbb{R}^N; X)$ for some (and thus then for each) $N \in \mathbb{N}$. The *Fourier type constant* $\mathcal{F}_{p,N}(X)$ of X is then the norm of $\mathcal{F} \in \mathcal{B}\left(L_p(\mathbb{R}^N; X), L_{p'}(\mathbb{R}^N; X)\right)$.

The simple estimate $\|\mathcal{F}f(t)\|_X \leq \|f\|_{L_1(X)}$ shows that each Banach space X has Fourier type 1 with $\mathcal{F}_{1,N}(X) = 1$. The notion becomes more restrictive as p increases to 2. A Banach space has Fourier type 2 if and only if X is isomorphic to a Hilbert space [27]. A space $L_q(\Omega; \mathbb{R})$ has Fourier type $p = \min(q, q')$ [30]. If X have Fourier type $p \in [1,2]$ and $p \leq q \leq p'$, then $\mathcal{F}_{p,N}(X) = \mathcal{F}_{p,N}(X^*) = \mathcal{F}_{p,N}\left(L_q(\mathbb{R}^N; X)\right)$ for each $N \in \mathbb{N}$ (cf. [18]). Each closed subspace (by definition) and quotient space (by duality) of a Banach space X has the same Fourier type as X.

REMARK 2.9. Recall the following relationships between property (α), UMD spaces, Fourier type, and Rad(X).
a) Property (α) and the UMD property are independent. An $L_1(\Omega; \mathbb{C})$ space has property (α). However, infinite dimensional $L_1(\Omega; \mathbb{C})$ spaces (as well as $C(K)$-spaces) are not UMD spaces. The Schatten classes $\mathcal{S}_p$ have UMD for $p \in (1, \infty)$. However, an infinite dimensional $\mathcal{S}_p$ does not have property (α) when $1 \leq p \leq \infty$ and $p \neq 2$.
b) A UMD space has some non-trivial Fourier type $p > 1$. Indeed, a UMD space has a uniformly convex renorming [10]. A space with a uniformly convex renorming is reflexive and B-convex. A B-convex Banach space has some non-trivial Fourier type $p > 1$ [9,12].
c) If X has property (α) (resp. the UMD property, Fourier type p), then so does Rad(X). This follows from viewing Rad(X) as a subspace of $L_2([0,1]; X)$.

3 FOURIER MULTIPLIER OPERATORS

Let $q \in (1, \infty)$. A bounded measurable function $M\colon \mathbb{R}^N \setminus \{0\} \to \mathcal{B}(X, Y)$ is a *Fourier multiplier* on $L_q(\mathbb{R}^N; X)$ provided the corresponding *Fourier multiplier operator* T_M, defined by

$$T_M f := \left[M(\cdot)\, \widehat{f}(\cdot)\right]^{\vee} \qquad \text{for} \qquad f \in \mathcal{S}(\mathbb{R}^N; X)\,,$$

extends to a bounded operator $T_M \in \mathcal{B}\left(L_q(\mathbb{R}^N; X), L_q(\mathbb{R}^N; Y)\right)$.

FACT 3.1. Let τ be a R-bounded subset of $\mathcal{B}(X, Y)$ and $q \in (1, \infty)$. It is known that M is a Fourier multiplier on $L_q(\mathbb{R}^N; X)$ if X and Y are UMD spaces with

Fourier type $p \in [1,2]$ and M belongs to one of the following symbol classes:

$$\mathcal{M}^l(\tau) := \left\{M\colon \mathbb{R}^N \setminus \{0\} \to \mathcal{B}(X,Y) : |t|^{|\alpha|} M^{(\alpha)}(t) \in \tau \text{ for } t \neq 0, \alpha \in \mathbb{N}_0^N, |\alpha| \leq l\right\}$$

for some $l \in \mathbb{N}$ where $l > N/p$ (see [19,21,33,36]);

$$\mathcal{H}^N(\tau) := \left\{M\colon \mathbb{R}^N \setminus \{0\} \to \mathcal{B}(X,Y) : t^\alpha M^{(\alpha)}(t) \in \tau \text{ for } t \neq 0, \alpha \in \mathbb{N}_0^N, \alpha \leq (1,1,\ldots,1)\right\}$$

provided X and Y have property (α) (see [33]);

$$\mathcal{N}^r(\tau) := \left\{M\colon \mathbb{R} \setminus \{0\} \to \mathcal{B}(X,Y) : M(t) \in \tau, |t|^r \frac{M(t+s) - M(t)}{|s|^r} \in \tau \text{ for } s,t \neq 0\right\}$$

for some $r \in (1/p, 1)$ (see [19]). If $X = Y = \mathbb{C}$ (i.e., scalar-valued multiplier functions) and τ is the unit ball of $\mathbb{C}$, then the above classes are denoted by just $\mathcal{M}^l$, $\mathcal{H}^N$, and $\mathcal{N}^r$.

Fact 2.6 leads to our main criteria for R-boundedness of sets of Fourier multiplier operators.

THEOREM 3.2. *Let X and Y be UMD Banach spaces having property (α) and Fourier type p. Let $l = [N/p] + 1$ and $r \in (1/p, 1)$. Let τ be a R-bounded subset of $\mathcal{B}(X,Y)$. Then the sets of Fourier multiplier operators*

$$\begin{aligned} \mathcal{M} &:= \{T_M \in \mathcal{B}(L_q(\mathbb{R}^N; X), L_q(\mathbb{R}^N; Y)) : M \in \mathcal{M}^l(\tau)\} \\ \mathcal{H} &:= \{T_M \in \mathcal{B}(L_q(\mathbb{R}^N; X), L_q(\mathbb{R}^N; Y)) : M \in \mathcal{H}^N(\tau)\} \\ \mathcal{N} &:= \{T_M \in \mathcal{B}(L_q(\mathbb{R}^1; X), L_q(\mathbb{R}^1; Y)) : M \in \mathcal{N}^r(\tau)\} \end{aligned}$$

are R-bounded for each $q \in (1,\infty)$.

Proof. Fix $q \in (1,\infty)$ To simplify notation, denote $L_q(\mathbb{R}^N; Z)$ by just $L_q(Z)$ and $\mathrm{Rad}_q(Z)$ by just $\widetilde{Z}$ for the assorted Banach spaces Z appearing.

Let $\{M_j\}_{j=1}^n$ be a subset of $\mathcal{M}^l(\tau)$ (resp. $\mathcal{H}^N(\tau)$, $\mathcal{N}^r(\tau)$) and set $M_j := 0$ for $j > n$. Define

$$\widetilde{T} \in \mathcal{B}\left(\widetilde{L_q(X)}, \widetilde{L_q(Y)}\right)$$

by

$$\widetilde{T}\widetilde{g} = \{T_{M_j} g_j\}_{j=1}^\infty \quad \text{for} \quad \widetilde{g} = \{g_j\}_{j=1}^\infty \in \widetilde{L_q(X)}\,.$$

Note that $\|\widetilde{T}\| = R_q\left(\{T_{M_j}\}_{j=1}^n\right) < \infty$. By Fact 2.3, it suffices to show that there exists a constant C_* that depends only on $R_q(\tau)$, X, Y, q, p, and N (but *not* on the particular chosen $\{M_j\}_{j=1}^n$) such that

$$\left\|\widetilde{T}\right\|_{\mathcal{B}\left(\widetilde{L_q(X)}, \widetilde{L_q(Y)}\right)} \leq C_*\,. \tag{3.1}$$

For then $R_q(\mathcal{M})$ (resp. $R_q(\mathcal{H})$, $R_q(\mathcal{N})$) is bounded above by C_*.

Towards this, define

$$\widetilde{M}\colon \mathbb{R}^N \setminus \{0\} \to \mathcal{B}\left(\widetilde{X}, \widetilde{Y}\right)$$

by

$$\left[\widetilde{M}(\cdot)\right]\left(\{x_j\}_{j\in\mathbb{N}}\right) := \{M_j(\cdot)\, x_j\}_{j\in\mathbb{N}} \qquad \text{for} \qquad \{x_j\}_{j\in\mathbb{N}} \in \widetilde{X} .$$

Note that $\|\widetilde{M}(t)\| = R_q\left(\{M_j(t)\}_{j=1}^n\right) \leq R_q(\tau)$ for each $t \in \mathbb{R}^N \setminus \{0\}$. To show (3.1), it suffices to show that $\widetilde{M}$ is a Fourier multiplier on $L_q(\widetilde{X})$ with

$$\|T_{\widetilde{M}}\|_{\mathcal{B}(L_q(\widetilde{X}),L_q(\widetilde{Y}))} \leq C_* .$$

Indeed, for then the diagram

$$\begin{array}{ccc} \mathrm{Rad}_q\left(L_q(X)\right) & \xrightarrow{\widetilde{T}} & \mathrm{Rad}_q\left(L_q(Y)\right) \\ {\scriptstyle I_X}\uparrow & & \uparrow{\scriptstyle I_Y} \\ L_q\left(\mathrm{Rad}_q(X)\right) & \xrightarrow{T_{\widetilde{M}}} & L_q\left(\mathrm{Rad}_q(Y)\right) \end{array}$$

commutes since if $f(\cdot) = \{f_j(\cdot)\}_{j\in\mathbb{N}} \in \mathcal{S}\left(\mathbb{R}^N; \widetilde{X}\right)$ then $T_{\widetilde{M}} f = \left[\widetilde{M}(\cdot)\,\widehat{f}(\cdot)\right]^{\vee}$ and so, with the help of (2.2),

$$I_Y T_{\widetilde{M}} f = \{T_{M_j} f_j\}_{j\in\mathbb{N}} = \widetilde{T} I_X f \; ; \quad \text{hence,} \quad \|T_{\widetilde{M}}\| = \left\|\widetilde{T}\right\| .$$

First consider the set $\mathcal{M}$. Note that the set

$$\left\{|t|^{|\alpha|}\left(D^\alpha \widetilde{M}\right)(t) \in \mathcal{B}\left(\widetilde{X}, \widetilde{Y}\right) \; : t \in \mathbb{R}^N \setminus \{0\}, \alpha \in \mathbb{N}_0^N, |\alpha| \leq l\right\}$$

is R-bounded; indeed, for any fixed subsets

$$\{t_j\}_{j=1}^m \subset \mathbb{R}^N \setminus \{0\} \quad \text{and} \quad \left\{\widetilde{x}_j := \left\{x_k^j\right\}_{k\in\mathbb{N}}\right\}_{j=1}^m \subset \widetilde{X}$$

and $\alpha \in \mathbb{N}_0$ with $|\alpha| \leq l$, note that $D^\alpha \widetilde{M} = \{D^\alpha M_k\}_{k\in\mathbb{N}}$ and so for

$$N_k(t) := |t|^{|\alpha|} D^\alpha M_k(t) \qquad \text{and} \qquad \widetilde{N}(t) := \{N_k(t)\}_{k\in\mathbb{N}} ,$$

by Fact 2.6,

$$\begin{aligned}\left\|\sum_{j=1}^{m} r_j(\cdot)\,\widetilde{N}(t_j)\,\widetilde{x}_j\right\|^q_{L_q([0,1];\widetilde{Y})} &= \int\limits_{[0,1]}\int\limits_{[0,1]'}\left\|\sum_{k=1}^{n}\sum_{j=1}^{m} r_k'(u')\,r_j(u)\,N_k(t_j)\,x_k^j\right\|^q_Y du'du \\ &\le C_0^q \int\limits_{[0,1]}\int\limits_{[0,1]'}\left\|\sum_{k=1}^{n}\sum_{j=1}^{m} r_k'(u')\,r_j(u)\,x_k^j\right\|^q_X du'du \\ &\le C_0^q \left\|\sum_{j=1}^{m} r_j(\cdot)\,\widetilde{x}_j\right\|^q_{L_q([0,1];\widetilde{X})},\end{aligned} \tag{3.2}$$

where $C_0 = \alpha_q(X)\,\alpha_q(Y)\,R_q(\tau)$. Thus by Remark 2.9c) and Fact 3.1, $\widetilde{M}$ is a Fourier multiplier on $L_q\left(\widetilde{X}\right)$ with $T_{\widetilde{M}}$ bounded by some constant C_* that depends only on the items claimed.

The proofs for $\mathcal{H}$ and $\mathcal{N}$ are similar. The only difference is that for the estimate in (3.2) one considers $\alpha \in \mathbb{N}_0$ with $\alpha \le (1,1,\ldots,1)$ and uses $N_k(t) := t^\alpha (D^\alpha M_k)(t)$ in the case of $\mathcal{H}$ while one considers $\{s_j\}_{j=1}^m \subset \mathbb{R}^N \setminus \{0\}$ also and uses $N_k(s,t) := |t/s|^r [M_k(t+s) - M_k(t)]$ or $N_k(t) = M_k(t)$ in the case of $\mathcal{N}$. ■

Since $X = Y = \mathbb{C}$ have UMD, property (α), and Fourier type 2, an interesting R-boundedness result follows for scalar-valued Fourier multipliers on $L_q(\mathbb{R}^N;\mathbb{C})$.

COROLLARY 3.3. *Let $l = [N/2]+1$ and choose $r \in (1/2,1)$. Then the sets of Fourier multiplier operators $\{T_M : M \in \mathcal{M}^l\}$, $\{T_M : M \in \mathcal{H}^N\}$, and $\{T_M : M \in \mathcal{N}^r\}$ are R-bounded on $L_q(\mathbb{R}^N;\mathbb{C})$ for each $1 < q < \infty$ (of course, $N = 1$ in the latter case).*

4 CONVOLUTION OPERATORS

The theorem in the previous section gives that certain families of Fourier multiplier operators are R-bounded. The results of this section give that certain families of convolution operators are R-bounded. Recall that for a *kernel function* $k \in L_1(\mathbb{R}^N;\mathcal{B}(X,Y))$ and $q \in [1,\infty]$, one can define the corresponding *convolution operator* $K_k \in \mathcal{B}(L_q(\mathbb{R}^N;X), L_q(\mathbb{R}^N;Y))$ by

$$K_k f(t) := \int\limits_{\mathbb{R}^N} k(t-s) f(s)\,ds \qquad \text{for a.e.} \qquad t \in \mathbb{R}^N\,. \tag{4.1}$$

Note that it follows from Lemma 4.1 that $\{\widehat{k}(t) : t \in \mathbb{R}^N\}$ is R-bounded in $\mathcal{B}(X,Y)$.

LEMMA 4.1. *Let X and Y be arbitrary Banach spaces.*
a) Let $N : \Omega \to \mathcal{B}(X,Y)$ be strongly integrable on the measure space (Ω,Σ,μ). Thus

there is a constant A so that

$$\int_{\Omega} \|N(\omega)x\|_Y \, d\mu(\omega) \;\leq\; A\; \|x\|_X \qquad \textit{for each} \quad x \in X\,. \tag{4.2}$$

Then

$$\tau := \left\{ X \ni x \;\rightarrow\; \int_{\Omega} h(\omega)\, N(\omega)\, x \, d\mu(\omega) \in Y \;:\; h \in L_\infty(\Omega;\mathbb{C}) \;,\; \|h\|_\infty \leq 1 \right\}$$

is an R-bounded subset of $\mathcal{B}(X,Y)$ with $R_1(\tau) \leq 2A$.
b) If $N \in L_1(\Omega;\mathcal{B}(X,Y))$, then

$$R_1\left(\left\{\int_{\Omega} h(\omega)\, N(\omega)\, d\mu(\omega) \in \mathcal{B}(X,Y) \;:\; h \in L_\infty(\Omega;\mathbb{C}) \;,\; \|h\|_\infty \leq 1 \right\}\right) \leq 2\, \|N\|_{L_1(\Omega;\mathcal{B}(X,Y))}\,.$$

Proof. Clearly it suffices to show just part a). For $h \in L_\infty(\Omega;\mathbb{C})$, let

$$T_h x = \int_{\Omega} h(\omega)\, N(\omega)\, x d\mu(\omega)$$

so that $T_h \in \mathcal{B}(X,Y)$ with $\|T_h\|_{\mathcal{B}(X,Y)} \leq A\,\|h\|_\infty$. For $h_1,\ldots,h_n \in L_\infty(\Omega;\mathbb{C})$ with $\|h_i\|_\infty \leq 1$ and $x_1,\ldots,x_n \in X$

$$\begin{aligned}
&\int_{[0,1]} \left\| \sum\nolimits_{j=1}^{n} r_j(t) T_{h_j}(x_j) \right\|_Y dt \\
&\quad = \int_{[0,1]} \left\| \int_{\Omega} \left(\sum\nolimits_{j=1}^{n} r_j(t)\, h_j(\omega)\, N(\omega)\, x_j \right) d\mu(\omega) \right\|_Y dt \\
&\quad \leq \int_{\Omega} \int_{[0,1]} \left\| \sum\nolimits_{j=1}^{n} r_j(t)\, h_j(\omega)\, N(\omega)\, x_j \right\|_Y dt d\mu(\omega) \\
&\quad \leq 2 \int_{\Omega} \int_{[0,1]} \left\| \sum\nolimits_{j=1}^{n} r_j(t)\, N(\omega)\, x_j \right\|_Y dt\, d\mu(\omega) \\
&\quad = 2 \int_{[0,1]} \left(\int_{\Omega} \left\| N(\omega) \left[\sum\nolimits_{j=1}^{n} r_j(t)\, x_j \right] \right\|_Y d\mu(\omega) \right) dt \\
&\quad \leq 2\, A \int_{[0,1]} \left\| \sum\nolimits_{j=1}^{n} r_j(t)\, x_j \right\|_X dt
\end{aligned}$$

by Kahane's contraction principle and (4.2). ∎

Lemma 4.1 leads to an R-boundedness criterion for certain families of kernels.

PROPOSITION 4.2. *Let X and Y be UMD spaces with property (α). Let the kernel function $k\colon \mathbb{R}^N \to \mathcal{B}(X,Y)$ satisfy, for each $\alpha \in \mathbb{N}_0^N$ with $\alpha \le (1,1,\ldots,1)$,*

$$\int_{\mathbb{R}^N} |t^\alpha| \, \|D^\alpha k(t)\|_{\mathcal{B}(X,Y)} \, dt \; < \; \infty \, . \tag{4.3}$$

Then the family $\{K_\gamma \in \mathcal{B}\left(L_q\left(\mathbb{R}^N;X\right), L_q\left(\mathbb{R}^N;Y\right)\right) \; : \gamma \in \mathbb{R}\}$, defined by

$$(K_\gamma f)(t) \; := \; \int \gamma^N k(\gamma(t-s)) \, f(s) \, ds \qquad \text{for} \quad f \in \mathcal{S}\left(\mathbb{R}^N;X\right) \; ,$$

is R-bounded for each $q \in (1,\infty)$.

REMARK 4.3. For $X = Y = \mathbb{C}$, Proposition 4.2 reduces to: if $k \in \mathbb{R}^N \to \mathbb{C}$ satisfies

$$\int_{\mathbb{R}^N} |t^\alpha D^\alpha k(t)| \, dt \; < \; \infty \qquad \text{for each} \quad \alpha \in \mathbb{N}_0^N \quad \text{with} \quad \alpha \le (1,1,\ldots,1)$$

then the family $\{K_\gamma \in \mathcal{B}\left(L_q\left(\mathbb{R}^N;\mathbb{C}\right)\right) \; : \gamma \in \mathbb{R}\}$ is R-bounded for each $q \in (1,\infty)$.

REMARK 4.4. The following variant of Proposition 4.2 is also true. First replace (4.3) by

$$\int_{\mathbb{R}^N} |t^\alpha| \, \|D^\alpha k(t)\, x\|_Y \, dt \; \le A \; \|x\|_X \quad \text{for each } x \in X \, . \tag{4.3$'$}$$

Note that then k is strongly integrable and so the function $M\colon \mathbb{R}^N \to \mathcal{B}(X,Y)$, given by

$$M(t)\, x \; := \; [k(\cdot)\, x]^\wedge(t) \quad \text{for } t \in \mathbb{R}^N \quad \text{and } x \in X \; , \tag{4.4}$$

is well-defined. Assume further that the distributional derivatives $D^\alpha M$ are represented by functions for each $\alpha \in \mathbb{N}_0^N$ with $\alpha \le (1,1,\ldots,1)$. Then the conclusion of Proposition 4.2 holds.

Proof of Proposition 4.2 and Remark 4.4. Let X and Y be UMD Banach spaces with property (α). Let k satisfy either (4.3) or (4.3$'$) and M be as in (4.4). Define $k_x, M_x\colon \mathbb{R}^N \to Y$ by

$$k_x(\cdot) \; := \; k(\cdot)\, x \qquad\qquad M_x(\cdot) \; := \; M(\cdot)\, x \; = \; \widehat{k_x}(\cdot)$$

for each $x \in X$.

Fix $\alpha \in \mathbb{N}_0^N$ with $\alpha \le (1,1,\ldots,1)$. Then for each $x \in X$

$$D^\alpha(s^\alpha k_x(s)) \; = \; \sum_{\gamma \le \alpha} \tbinom{\alpha}{\gamma} D^{\alpha-\gamma}(s^\alpha)\, D^\gamma(k_x(s)) \; = \; \sum_{\gamma \le \alpha} C_{\alpha,\gamma}\, s^\gamma \, D^\gamma(k_x(s)) \tag{4.5}$$

and

$$(-1)^{|\alpha|} \left[D^\alpha(s^\alpha k_x(s))\right]^\wedge(t) \; = \; (-i)^{|\alpha|}\, t^\alpha \left[s^\alpha k_x(s)\right]^\wedge(t) \; = \; t^\alpha D^\alpha M_x(t) \; . \tag{4.6}$$

Thus if k satisfies (4.3), then $D^\alpha(s^\alpha k(s)) \in L_1(\mathcal{B}(X,Y))$ and so the distributional derivative $D^\alpha M$ is represented as a function. Thus it suffices to show just Remark 4.4; so, assume the setting of Remark 4.4.

By (4.6) and (4.5)

$$t^\alpha D^\alpha M_x(t) = (-1)^{|\alpha|} \sum_{\gamma \le \alpha} C_{\alpha,\gamma} \int_{\mathbb{R}^N} h_t(s) \left[s^\gamma D^\gamma k_x(s)\right] ds$$

where $h_t(s) := e^{-it\cdot s}$. Thus by (4.3′) and Lemma 4.1, the set

$$\tau_\alpha := \left\{t^\alpha D^\alpha M(t) : t \in \mathbb{R}^N\right\}$$

is R-bounded.

Fix $\gamma \in \mathbb{R}^N \setminus \{0\}$. Then $M_\gamma \colon \mathbb{R}^N \to \mathcal{B}(X,Y)$, defined below, satisfies

$$M_\gamma(t)(x) := \left[\gamma^N k_x(\gamma\cdot)\right]^\frown(t) = (\operatorname{sign}\gamma)^N \hat{k}_x\left(\gamma^{-1}t\right) = (\operatorname{sign}\gamma)^N M\left(\gamma^{-1}t\right)x$$

and so

$$t^\alpha D^\alpha M_\gamma(t) = (\operatorname{sign}\gamma)^N \left(\gamma^{-1}t\right)^\alpha D^\alpha M\left(\gamma^{-1}t\right) \in \pm\tau_\alpha$$

for each $t \in \mathbb{R}^N$. Hence $M_\gamma \in \mathcal{H}^N(\cup_{\alpha \le (1,\dots,1)} \pm\tau_\alpha)$ and so M_γ is a Fourier multiplier on $L_q\left(\mathbb{R}^N; X\right)$ by Fact 3.1. Furthermore, Theorem 3.2 gives that $\{K_\gamma \colon \gamma \in \mathbb{R}\}$ is R-bounded. ∎

R-boundedness of the Gauß and Poisson semigroups follows from Proposition 4.2.

EXAMPLE 4.5. Keeping with Proposition 4.2's notation, fix $\gamma > 0$. If

$$k(s) := (4\pi)^{-1/2} \exp\left(\frac{-|s|^2}{4}\right),$$

then

$$\left(K_{\gamma^{-1/2}} f\right)(t) = (4\pi\gamma)^{-N/2} \int_{\mathbb{R}^N} \exp\left(\frac{-|s|^2}{4\gamma}\right) f(t-s)ds = \left(e^{\gamma\Delta} f\right)(t) .$$

If $k(s) := C_N \left(1+|s|^2\right)^{-(N+1)/2}$ where $C_N := \Gamma\left(\frac{N+1}{2}\right) \pi^{-(N+1)/2}$ then

$$\left(K_{\gamma^{-1}} f\right)(t) = C_N \int_{\mathbb{R}^N} \frac{\gamma}{\left(\gamma^2+|s|^2\right)^{(N+1)/2}} f(t-s)\, ds = \left(e^{-\gamma\sqrt{-\Delta}} f\right)(t) .$$

Since both functions k satisfy (4.3), Remark 4.3 gives that the Gauß semigroup $\{e^{\gamma\Delta} \colon \gamma > 0\}$ and the Poisson semigroup $\{e^{-\gamma\sqrt{-\Delta}} \colon \gamma > 0\}$ are R-bounded in $\mathcal{B}\left(L_q\left(\mathbb{R}^N; \mathbb{C}\right)\right)$ for each $q \in (1,\infty)$.

A more general criterion for R-boundedness of a family of convolution operators is now formulated. For this, let τ be a R-bounded subset of $\mathcal{B}(X,Y)$ and τ_0 be the closure (in the strong operator topology) of the complex absolute convex hull of τ. Note that τ_0 is also R-bounded (indeed, $R_p(\tau_0) \le 2R_p(\tau)$). Let $\|\cdot\|_\tau : \mathcal{B}(X,Y) \to [0,\infty]$ be the Minkowski functional of τ_0, i.e.,

$$\|S\|_\tau := \inf\{\lambda \ge 0 \colon S \in \lambda\tau_0\} .$$

Define

$$\mathcal{K}^N(\tau) := \left\{ k\colon \mathbb{R}^N \to \mathcal{B}(X,Y) \ : \int_{\mathbb{R}^N} \| t^\alpha D^\alpha k(t) \|_\tau \, dt \ \le \ 1 \ \text{ for each } \alpha \le (1,\dots,1) \right\}.$$

Note that if $k \in \mathcal{K}^n(\tau)$, then k satisfies (4.3); in particular, the convolution operator K_k, as defined in (4.1), is in $\mathcal{B}\left(L_q\left(\mathbb{R}^N;X\right), L_q\left(\mathbb{R}^N;Y\right)\right)$ for each $q \in (1,\infty)$.

THEOREM 4.6. *Let X and Y be UMD spaces with property (α) and τ be an R-bounded subset of $\mathcal{B}(X,Y)$. Then the family of convolution operators*

$$\left\{ K_k \in \mathcal{B}\left(L_q\left(\mathbb{R}^N;X\right), L_q\left(\mathbb{R}^N;Y\right)\right) \ : k \in \mathcal{K}^N(\tau) \right\}$$

is R-bounded for each $q \in (1,\infty)$.

REMARK 4.7. Let $X = Y = \mathbb{C}$ (i.e. scalar-valued kernels) and τ be the unit ball of $\mathbb{C}$ (so $\|\cdot\|_\tau = |\cdot|$). Then Theorem 4.6 implies that

$$\left\{ K_k \in \mathcal{B}\left(L_q\left(\mathbb{R}^N;\mathbb{C}\right)\right) : \int_{\mathbb{R}^N} | t^\alpha D^\alpha k(t) | \, dt \ \le \ 1 \quad \text{for each } \alpha \le (1,\dots,1) \right\}$$

is R-bounded for each $q \in (1,\infty)$.

Proof of Theorem 4.6 Fix $k \in \mathcal{K}^N(\tau)$. Then, keeping with the notation of Proposition 4.2,

$$t^\alpha D^\alpha M(t) \ = \ (-1)^{|\alpha|} \sum_{\gamma \le \alpha} C_{\alpha,\gamma} \int_{\mathbb{R}^N} e^{-it\cdot s} \left[s^\gamma D^\gamma k(s) \right] ds \ .$$

Note that

$$S(t) := \int_{\mathbb{R}^N} e^{-it\cdot s} \left[s^\gamma D^\gamma k(s) \right] ds = \int e^{-it\cdot s} h(s) \tilde{k}(s) \, ds$$

where $h(\cdot) := \|(\cdot)^\gamma D^\gamma k(\cdot)\|_\tau$ is in the unit ball of $L_1\left(\mathbb{R}^N;\mathbb{R}\right)$ and

$$\tilde{k}(s) := \left[h(s)\right]^{-1} s^\gamma D^\gamma k(s) \in \tau_0 \ .$$

Therefore $S(t) \in \tau_0$. So there is a constant C_α so that $t^\alpha D^\alpha M(t) \in C_\alpha \tau_0$ for each $t \in \mathbb{R}^N \setminus \{0\}$. Now Theorem 3.2 gives the desired result. ∎

5 SMOOTH OPERATOR-VALUED FUNCTIONS

The theorem in this section gives that the range of a *sufficiently smooth* function from $\mathbb{R}^N$ into $\mathcal{B}(X,Y)$ is R-bounded. *Smoothness* is measured in terms of the *modulus of smoothness* s of a Besov space $B^s_{q,r}$. Also, the Fourier type of Y is taken into account in order to reduce the required degree of smoothness. The theorem's corollaries measure smoothness in more classical ways (e.g. in terms of derivatives or a Lipschitz-like condition) rather than in terms of Besov spaces.

There are several equivalent well-known definitions of Besov spaces $B^s_{q,r}\left(\mathbb{R}^N;Z\right)$; e.g. Besov spaces can be defined via a Paley-Littlewood decomposition (c.f., e.g., [18; Def. 2.4]), via real interpolation (c.f., e.g., [1]), or as in [31; Prop. 3.1].

THEOREM 5.1. *Let X and Y be arbitrary Banach spaces and Y have Fourier type $p \in [1,2]$. Let $M \in B_{p,1}^{N/p}\left(\mathbb{R}^N; \mathcal{B}(X,Y)\right)$ have norm A. Then the set*

$$\tau := \left\{ M(t) \in \mathcal{B}(X,Y) \; : t \in \mathbb{R}^N \right\}$$

is R-bounded and $R_1(\tau) \leq CA$ for some constant C that depends only on $\mathcal{F}_{p,N}(Y)$.

Proof. First assume in addition that $M \in \mathcal{S}(\mathcal{B}(X,Y))$. Fix $x \in X$. Note that

$$\widehat{[M(\cdot)\, x]}\,(t) = \widehat{M}(t)\, x$$

since $M \in L_1$. Thus by [18; Cor. 3.2], for each $x \in X$,

$$\int_{\mathbb{R}^N} \left\| \widehat{M}(t)\, x \right\|_Y dt \;\leq\; C_1 \left\| M(\cdot)\, x \right\|_{B_{p,1}^{N/p}(\mathbb{R}^N;Y)} \;\leq\; C_1\, A\, \|x\|_X$$

for some constant C_1 depending only on $\mathcal{F}_{p,N}(Y)$. By the Fourier inversion formula

$$M(t) \;=\; (2\pi)^{-N} \int_{\mathbb{R}^N} e^{it\cdot s}\, \widehat{M}(s)\, ds \qquad \text{for each} \quad t \in \mathbb{R}^N \;.$$

Applying Lemma 4.1 with $h_t(s) = (2\pi)^{-N} e^{it\cdot s}$ gives that

$$R_1\left(\left\{ M(t) : t \in \mathbb{R}^N \right\}\right) \leq 2 C_1 A \;.$$

This proves the claim if $M \in \mathcal{S}(\mathcal{B}(X,Y))$.

For an arbitrary function $M \in B_{p,1}^{N/p}\left(\mathbb{R}^N; \mathcal{B}(X,Y)\right)$, choose a sequence $\{M_k\}_{k\in\mathbb{N}}$ from $\mathcal{S}(\mathcal{B}(X,Y))$ so that $\sum_{k=1}^n M_k$ converges to M in $B_{p,1}^{N/p}\left(\mathbb{R}^N; \mathcal{B}(X,Y)\right)$ and

$$\sum_{k\in\mathbb{N}} \|M_k\|_{B_{p,1}^{N/p}} \leq 2\, \|M\|_{B_{p,1}^{N/p}} \;.$$

Note that $\sum_{k=1}^n M_k$ converges to M also in $L_\infty\left(\mathbb{R}^N; \mathcal{B}(X,Y)\right)$ since the formal identity mapping from $B_{p,1}^{N/p}\left(\mathbb{R}^N; Z\right)$ into $L_\infty\left(\mathbb{R}^N; Z\right)$ is continuous by the Sobolev embedding theorem. Thus

$$\begin{aligned} R_1\left(\left\{ M(t) : t \in \mathbb{R}^N \right\}\right) &\leq \sum_{k\in\mathbb{N}} R_1\left(\left\{ M_k(t) : t \in \mathbb{R}^N \right\}\right) \\ &\leq \sum_{k\in\mathbb{N}} 2\, C_1\, \|M_k\|_{B_{p,1}^{N/p}} \;\leq\; 4\, C_1\, \|M\|_{B_{p,1}^{N/p}} \end{aligned}$$

by [36; Lemma 2.4] and the above argument. ∎

REMARK 5.2. In Theorem 5.1, the assumption that $M \in B_{p,1}^{N/p}\left(\mathbb{R}^N; \mathcal{B}(X,Y)\right)$ with norm A can be replaced by a *pointwise* estimate; more precisely, it can be replaced by: M is strongly integrable and

$$\|M(\cdot)\, x\|_{B_{p,1}^{N/p}(\mathbb{R}^N;Y)} \;\leq\; A\, \|x\|_X \qquad \text{for each} \quad x \in X \;. \tag{5.1}$$

Proof. For $x \in X$, define $M_x \colon \mathbb{R}^N \to Y$ by $M_x(\cdot) := M(\cdot)\,x$. Since M is strongly integrable, the function $L \colon \mathbb{R}^N \to \mathcal{B}(X,Y)$, given by

$$[L(t)](x) := (\mathcal{F}M_x)(t) \ ,$$

is well-defined. By (5.1) and [18; Cor. 3.2], $\mathcal{F}M_x \in L_1(Y)$ with

$$\left\| \widehat{M_x} \right\|_{L_1(Y)} \leq CA \left\|x\right\|_X$$

for some constant C that depends only on $\mathcal{F}_{p,N}(Y)$. Thus by the Fourier inversion formula

$$M_x(t) \;=\; (2\pi)^{-N} \int\limits_{\mathbb{R}^N} e^{it\cdot s}\widehat{M_x}(s)\,ds \;=\; \int\limits_{\mathbb{R}^N} h_t(s)\,[L(s)](x)\,ds$$

where $h_t(s) := (2\pi)^{-N} e^{it\cdot s}$. As in the proof of Theorem 5.1, Lemma 4.1 now gives the desired result. ■

The following corollary compares to [33; Thm. 4.1].

COROLLARY 5.3. *Let X be an arbitrary Banach space and Y have Fourier type $p \in [1,2]$. Let $l = [N/p] + 1$. Assume that $t \in \mathbb{R}^N \to M(t) \in \mathcal{B}(X,Y)$ satisfies*

$$\left(\int\limits_{\mathbb{R}^N} \left\|D^\alpha M(t)\right\|^p_{\mathcal{B}(X,Y)}\,dt \right)^{1/p} \;\leq\; A$$

for each $\alpha \in \mathbb{N}_0$ with $|\alpha| \leq l$. Then $\left\{M(t) \in \mathcal{B}(X,Y) \ : t \in \mathbb{R}^N\right\}$ is R-bounded.

Proof. Since $N/p < l \in \mathbb{N}$, the formal identity map from a Sobolev space $W^l_p\left(\mathbb{R}^N; Z\right)$ into $B^{N/p}_{p,1}\left(\mathbb{R}^N; Z\right)$ is continuous. Thus there are constants K_i so that

$$\begin{aligned}\left\|M\right\|_{B^{N/p}_{p,1}(\mathbb{R}^N;\mathcal{B}(X,Y))} \;&\leq\; K_1 \left\|M\right\|_{W^l_p(\mathbb{R}^N;\mathcal{B}(X,Y))} \\ &:=\; K_1 \sum_{0\leq|\alpha|\leq l} \left\|D^\alpha M\right\|_{L_p(\mathbb{R}^N;\mathcal{B}(X,Y))} \;\leq\; K_1\,K_2\,A\ .\end{aligned}$$

Now apply Theorem 5.1. ■

Recall that for a function $\mathbb{R} \supset (a,b) \ni t \to M(t) \in \mathcal{B}(X,Y)$ with integrable derivative [36] (or more generally, for a function of bounded variation [33]) the set $\{M(t) \in \mathcal{B}(X,Y) : t \in (a,b)\}$ is R-bounded. A short proof of this in the integrable derivative case follows easily from Lemma 4.1 since

$$M(t) = M(a) + \int\limits_a^b \chi_{[a,t]}(s)\,M'(s)\,ds\ .$$

The next corollary is a variant of these results. Its proof uses the following (equivalent) definition [31; Prop. 3.1] of Besov spaces $B^{1/p}_{p,1}(\mathbb{R}; Z)$ for $p \in (1,2]$:

$$B^{1/p}_{p,1}(\mathbb{R}; Z) \;=\; \left\{ f \in L_p(\mathbb{R}; Z) \ : B^{1/p}_{p,1}(f) < \infty \right\}\ , \tag{5.2}$$

equipped with the norm $\|f\|'_{B^{1/p}_{p,1}} = \|f\|_{L_p} + B^{1/p}_{p,1}(f)$, where

$$B^{1/p}_{p,1}(f) = \int_0^\infty \sup_{|h|\le s} \|f(\cdot + h) - f(\cdot)\|_{L_p(Z)} \frac{ds}{s^{1+\frac{1}{p}}} .$$

COROLLARY 5.4. *Let X be an arbitrary Banach space and Y have Fourier type $p \in (1,2]$. Fix $\alpha \in (1/p, 1)$ and $\varepsilon \in (\alpha - 1/p, \alpha]$. If $M \in L_p(\mathbb{R}; \mathcal{B}(X,Y))$ satisfies*

$$\|M(t+s) - M(t)\|_{\mathcal{B}(X,Y)} \le A \frac{|s|^\alpha}{1+|s|^\varepsilon} (1+|t|)^{-\alpha} \qquad \text{for each} \quad s,t \in \mathbb{R}$$

for some constant A, then $\{M(t) \in \mathcal{B}(X,Y) : t \in \mathbb{R}\}$ is R-bounded.

Proof. Note that $\frac{|s|^\alpha}{1+|s|^\varepsilon}$ is increasing in $|s|$ since $\varepsilon \le \alpha$. Thus for $s \ge 0$

$$\begin{aligned}\sup_{|h|\le s}\left[\int_R \|M(t+s) - M(t)\|^p_{\mathcal{B}(X,Y)}\, dt\right]^{1/p} &\le A \frac{s^\alpha}{1+s^\varepsilon}\left[\int_R (1+|t|)^{-\alpha p}\, dt\right]^{1/p} \\ &= A\left[\frac{2p}{\alpha p - 1}\right]^{1/p} \frac{s^\alpha}{1+s^\varepsilon}\end{aligned}$$

since $1/p < \alpha$. Note that

$$\int_0^\infty \frac{s^\alpha}{1+s^\varepsilon}\frac{ds}{s^{1+\frac{1}{p}}} \le \int_0^{17} s^{\alpha-1-\frac{1}{p}}\, ds + \int_{17}^\infty s^{\alpha-1-\frac{1}{p}-\varepsilon}\, ds . \tag{5.3}$$

On the right-hand side of inequality (5.3), the first integral is finite since $1/p < \alpha$ while the second integral is finite since $\alpha - 1/p < \varepsilon$. Thus $B^{1/p}_{p,1}(M)$ is finite and so $M \in B^{1/p}_{p,1}(\mathbb{R}; \mathcal{B}(X,Y))$. Now apply Theorem 5.1. ■

Now for some applications to semigroup theory.

EXAMPLE 5.5. Let $(T_t)_{t\ge 0}$ be a c_0-semigroup on a Banach space X with generator A. Define

$$\begin{aligned}\omega(T_t) &:= \inf\left\{w \in \mathbb{R}\colon \text{there exists a constant } C_w \text{ with } \|T_t\| \le C_w e^{wt} \text{ for } t > 0\right\}\\ s_R(A) &:= \inf\{w \in \mathbb{R}\colon \{R(\lambda, A) : \Re\lambda > w\} \text{ is R-bounded}\}\\ s_N(A) &:= \inf\{w \in \mathbb{R}\colon \{R(\lambda, A) : \Re\lambda > w\} \text{ is norm bounded}\}\\ s(A) &:= \sup\{\Re\lambda\colon \lambda \in \sigma(A)\} .\end{aligned}$$

a) Then $s_R(A) \le \omega(T_t)$; however, equality need not hold in general.
b) If furthermore $(T_t)_{t\ge 0}$ is positive and X is a q-concave Banach lattice for some $q \in [1,\infty)$, then $s(A) = s_R(A)$.

QUESTION 5.6. If A generates a c_0-semigroup $(T_t)_{t\ge 0}$ on a Banach space X, then

$$s(A) \le s_N(A) \le s_R(A) \le \omega(T_t) . \tag{5.4}$$

The first (resp. last) inequality in (5.4) need not be an equality in general, as illustrated by several examples already in the literature (resp. by Example 5.5 above). It is not know whether the middle inequality in (5.4) need not be an equality in general.

Proof of Example 5.5 Towards part a), let $\Re\lambda > w_1 > \omega(T_t)$. Then for each $x \in X$

$$R(\lambda, A)\, x = \int_0^\infty e^{-(\lambda - w_1)t} \left[e^{-w_1 t}\, T_t x\right] dt$$

(cf., eg., [17; Thm. II.1.10]) and so it follows from Lemma 4.1 that

$$\{R(\lambda, A) : \Re\lambda > w_1\}$$

is R-bounded. Hence $s_R(A) \leq \omega(T_t)$.

Now assume the setting of part b). Let $\Re\lambda \geq \lambda_0 > s(A)$. Then

$$|R(\lambda, A)\, x| \leq R(\lambda_0, A)\, |x|$$

for each $x \in X$ (cf., eg., [4; C-III: Thm. 1.2, Cor. 1.3]); hence,

$$\{R(\lambda, A) : \Re\lambda > \lambda_0\}$$

is R-bounded (just consider inequality (2.1) and [28; Prop. 1.d.9]) so $s(A) = s_R(A)$.

However, there are positive c_0-semigroups on q-concave Banach lattices (q finite) for which $\omega(T_t) > s(A)$ (cf., eg., [29; Ch. 4 Ex. 4.2]). ∎

EXAMPLE 5.7. Let $-A$ be the generator of a c_0-semigroup $(T_t)_{t\geq 0}$ on a Banach space X having Fourier type $p \in (1, 2]$ and $\omega(T_t) < 0$. Denote by i_α the embedding of $D(A^\alpha)$ with its graph norm into X. Then

$$\{T_t i_\alpha : t \geq 0\} \subset \mathcal{B}(D(A^\alpha), X)$$

is R-bounded for $1/p < \alpha < 1$.

Proof. Let X_θ be the real interpolation space $(X, D(A))_{\theta,1}$. By [5; Thm. 6.7.3], the norm $\|x\|_\theta$ on X_θ is equivalent to

$$\|x\|_X + \int_0^\infty \left(t^{-\theta}\, w(x,t)\right) \frac{dt}{t} \qquad \text{where} \qquad w(x,t) := \sup_{s\leq t} \|T_s x - x\| \ .$$

Let $h_x(t) := T_{|t|}(x)$. Note that $h_x \in L_1(\mathbb{R}; X)$ since $\omega(T_t) < 0$. Furthermore, using the notation from (5.2),

$$B^{1/p}_{p,1}(h_x) \ \leq \ \left(2 \int_0^\infty \|T_u\|\, du\right) \left(\int_0^\infty t^{-1/p}\, w(x,t)\, \frac{dt}{t}\right)$$

and since $D(A^\alpha) \subset X_{1/p}$ for $\alpha > 1/p$

$$\|h_x\|_{B^{1/p}_{p,1}(X)} \ \leq \ C\, \|x\|_{X_{1/p}} \ \leq \ C_1\, \|x\|_{D(A^\alpha)} \ .$$

Now just apply Remark 5.2 with $M(t) = T_{|t|}$. ∎

BIBLIOGRAPHY

1. Herbert Amann, *Linear and quasilinear parabolic problems. Vol. I*, Birkhäuser Boston Inc., Boston, MA, 1995, Abstract linear theory.

2. W. Arendt and S. Bu, *The operator-valued Marcinkiewicz multiplier theorem and maximal regularity*, (preprint).

3. W. Arendt and S. Bu, *Tools for maximal regularity*, (preprint).

4. W. Arendt, A. Grabosch, G. Greiner, U. Groh, H. P. Lotz, U. Moustakas, R. Nagel, F. Neubrander, and U. Schlotterbeck, *One-parameter semigroups of positive operators*, Springer-Verlag, Berlin, 1986.

5. Jöran Bergh and Jörgen Löfström, *Interpolation spaces. An introduction*, Springer-Verlag, Berlin, 1976, Grundlehren der Mathematischen Wissenschaften, No. 223.

6. Earl Berkson and T. A. Gillespie, *Spectral decompositions and harmonic analysis on UMD spaces*, Studia Math. **112** (1994), no. 1, 13–49.

7. S. Blunck and P. C. Kunstmann, *Weighted norm estimates and maximal regularity*, (submitted).

8. Sönke Blunck, *Maximal regularity of discrete and continuous time evolution equations*, Studia Math. **146** (2001), no. 2, 157–176.

9. J. Bourgain, *A Hausdorff-Young inequality for B-convex Banach spaces*, Pacific J. Math. **101** (1982), no. 2, 255–262.

10. J. Bourgain, *Some remarks on Banach spaces in which martingale difference sequences are unconditional*, Ark. Mat. **21** (1983), no. 2, 163–168.

11. J. Bourgain, *Vector-valued singular integrals and the H^1-BMO duality*, Probability theory and harmonic analysis (Cleveland, Ohio, 1983), Dekker, New York, 1986, pp. 1–19.

12. J. Bourgain, *Vector-valued Hausdorff-Young inequalities and applications*, Geometric aspects of functional analysis (1986/87), Springer, Berlin, 1988, pp. 239–249.

13. P. Clément, B. de Pagter, F. A. Sukochev, and H. Witvliet, *Schauder decomposition and multiplier theorems*, Studia Math. **138** (2000), no. 2, 135–163.

14. Philippe Clément and Jan Prüss, *An operator-valued transference principle and maximal regularity on vector-valued L_p-spaces*, Evolution equations and their applications in physical and life sciences (Bad Herrenalb, 1998), Dekker, New York, 2001, pp. 67–87.

15. R. Denk, M. Hieber, and J. Prüß, *R-boundedness, Fourier multipliers and problems of elliptic and parabolic type*, (submitted).

16. Joe Diestel, Hans Jarchow, and Andrew Tonge, *Absolutely summing operators*, Cambridge University Press, Cambridge, 1995.

17. Klaus-Jochen Engel and Rainer Nagel, *One-parameter semigroups for linear evolution equations*, Springer-Verlag, New York, 2000, With contributions by S. Brendle, M. Campiti, T. Hahn, G. Metafune, G. Nickel, D. Pallara, C. Perazzoli, A. Rhandi, S. Romanelli and R. Schnaubelt.

18. Maria Girardi and Lutz Weis, *Operator-valued Fourier multiplier theorems on Besov spaces*, Mathematische Nachrichten, (to appear).

19. Maria Girardi and Lutz Weis, *Operator-valued Fourier multiplier theorems on $L_p(X)$ and geometry of Banach spaces*, (submitted).

20. Jerome A. Goldstein, *Semigroups of linear operators and applications*, The Clarendon Press Oxford University Press, New York, 1985.

21. R. Haller, H. Heck, and A. Noll, *Mikhlin's theorem for operator-valued Fourier multipliers on n-dimensional domains*, (preprint).

22. N. J. Kalton and G. Lancien, *A solution to the problem of L^p-maximal regularity*, Math. Z. **235** (2000), no. 3, 559–568.

23. N. J. Kalton and L. Weis, *The H^∞-calculus and sums of closed operators*, Math. Ann. **321** (2001), no. 2, 319–345.

24. N. J. Kalton and Lutz Weis, *Comparison and perturbation theorems for the H^∞-calculus*, (in preparation).

25. P. C. Kunstmann, *Maximal L_p-regularity for second order elliptic operators with uniformly continuous coefficients on domains*, (submitted).

26. Peer Christian Kunstmann and Lutz Weis, *Perturbation theorems for maximal L_p-regularity*, Ann. Scuola Norm. Sup. Pisa Cl. Sci. (4) **30** (2001), no. 2, 415–435.

27. S. Kwapień, *Isomorphic characterizations of inner product spaces by orthogonal series with vector valued coefficients*, Studia Math. **44** (1972), 583–595, Collection of articles honoring the completion by Antoni Zygmund of 50 years of scientific activity, VI.

28. Joram Lindenstrauss and Lior Tzafriri, *Classical Banach spaces. II*, Springer-Verlag, Berlin, 1979, Function spaces.

29. A. Pazy, *Semigroups of linear operators and applications to partial differential equations*, Springer-Verlag, New York, 1983.

30. Jaak Peetre, *Sur la transformation de Fourier des fonctions à valeurs vectorielles*, Rend. Sem. Mat. Univ. Padova **42** (1969), 15–26.

31. A. Pełczyński and M. Wojciechowski, *Molecular decompositions and embedding theorems for vector-valued Sobolev spaces with gradient norm*, Studia Math. **107** (1993), no. 1, 61–100.

32. Gilles Pisier, *Some results on Banach spaces without local unconditional structure*, Compositio Math. **37** (1978), no. 1, 3–19.

33. Ž. Štrkalj and Lutz Weis, *On operator-valued Fourier multiplier theorems*, (submitted).

34. Alberto Venni, *Marcinkiewicz and Mihlin multiplier theorems, and R-boundedness*, (preprint).

35. Lutz Weis, *A new approach to maximal L_p-regularity*, Evolution equations and their applications in physical and life sciences (Bad Herrenalb, 1998), Dekker, New York, 2001, pp. 195–214.

36. Lutz Weis, *Operator-valued Fourier multiplier theorems and maximal L_p-regularity*, Math. Ann. **319** (2001), no. 4, 735–758.

One Dimensional Hyperbolic systems and Hille-Yosida operators

RONALD GRIMMER

Department of Mathematics, Southern Illinois University
Carbondale, IL 62901, USA
e-mail: rgrimmer@math.siu.edu

EUGENIO SINESTRARI

Dipartimento di Matematica, Università di Roma "La Sapienza"
P. Aldo Moro 7, 00185 Roma Italy
e-mail: sinestrari@mat.uniroma1.it

Dedicated to Jerry Goldstein on the occasion of his 60th birthday with all the best wishes and best regards.

ABSTRACT. We shall examine one dimensional hyperbolic boundary value problems as examples of Hille-Yosida operator theory. As a motivating example we apply our results to the Timoshenko beam with various physically realizable boundary conditions.

1 INTRODUCTION

In previous work the authors have considered one-dimensional hyperbolic systems via the theory of Hille-Yosida operators and applied the results to the forced wave equation, [3], to obtain maximum norm estimates of solutions of the wave equation. More recently we used the theory of Hille-Yosida operators to obtain estimates of the classical solutions of the Timoshenko beam with fixed ends, [4]. We consider here a setting for hyperbolic systems that allows for almost all of the standard boundary conditions for a Timoshenko beam equation, cf. [2; p. 154]. In this way we obtain the existence and uniqueness of a classical solution of this equation.

2 HYPERBOLIC SYSTEMS

In this section we shall formulate the general one dimensional system with general boundary value conditions as an abstract differential equation in a space $C[0,1]^N$ and show that it is amenable to the theory of Hille-Yosida operators as originally discussed in the paper by DaPrato and Sinestrari, [1].

We write the general system as

$$\begin{aligned} D_tN(t,x) &= \Lambda_N D_x N(t,x) + B_{11}N(t,x) + B_{12}P(t,x) + f(t,x) && (2.1)\\ D_tP(t,x) &= \Lambda_P D_x P(t,x) + B_{21}N(t,x) + B_{22}P(t,x) + g(t,x) && (2.2)\\ D_tZ_0(t,x) &= B_{31}N(t,x) + B_{32}P(t,x) && (2.3)\\ D_tZ_1(t,x) &= B_{41}N(t,x) + B_{42}P(t,x) && (2.4)\\ & 0 \le x \le 1, \qquad 0 \le t < \infty. \end{aligned}$$

It is assumed that the vector valued functions N, P, Z_0, Z_1 are column vectors with dimensions q_N, q_P, r_N and r_P respectively. Λ_N is a $q_N \times q_N$ diagonal matrix with negative entries on the diagonal while Λ_P is a $q_P \times q_P$ diagonal matrix with positive entries on the diagonal. Additionally, each B_{ij} is a constant matrix of appropriate size. Further, the functions f and g are continuous vector valued functions of the appropriate dimension. Finally, we have assumed that the spatial dimension has been scaled so that $0 \le x \le 1$. We have initial conditions

$$\begin{aligned} N(0,x) &= N_0(x), & P(0,x) &= P_0(x),\\ Z_0(0,x) &= Z_{00}(x), & Z_1(0,x) &= Z_{10}(x)\\ && 0 \le x \le 1 \end{aligned}$$

and boundary conditions

$$\begin{aligned} N(t,0) &= G_P P(t,0) + H_P Z_1(t,0) && (2.5)\\ P(t,1) &= G_N N(t,1) + H_N Z_0(t,1). && (2.6) \end{aligned}$$

Here G_P, G_N, H_P, H_N, are constant matrices of appropriate size. For convenience of later computation, we also write the boundary conditions in coordinate form

$$\begin{aligned} N_i(t,0) &= \sum_{j=1}^{q_P} g^P{}_{ij} P_j(t,0) + \sum_{j=1}^{r_P} h^P{}_{ij} Z_{1j}(t,0) && (2.7)\\ i &= 1 \dots q_N && (2.8)\\ P_i(t,1) &= \sum_{j=1}^{q_N} g^N{}_{ij} N_j(t,1) + \sum_{j=1}^{r_N} h^N{}_{ij} Z_{0j}(t,1) && (2.9)\\ i &= 1 \dots q_P. && (2.10) \end{aligned}$$

For the abstract setting we will use a suitable Banach space of continuous functions. In particular, if $\|\cdot\|$ denotes the maximum norm on the space $C[0,1]$, we define

Banach spaces

$$
\begin{aligned}
X_1 =&(C[0,1])^{q_N} \text{ with norm } \|(w_1,\ldots,w_{q_N})\| = \max_k\{\|w_k\|\} \\
X_2 =&(C[0,1])^{q_P} \text{ with norm } \|(w_1,\ldots,w_{q_P})\| = \max_k\{\|w_k\|\} \\
X_3 =&(C[0,1])^{r_N} \text{ with norm } \|(z_1,\ldots,z_{r_N})\| = \max_k\{\|z_k\|\} \\
X_4 =&(C[0,1])^{r_P} \text{ with norm} \|(z_1,\ldots,z_{r_P})\| = \max_k\{\|z_k\|\}
\end{aligned}
$$

and

$$X = X_1 \times X_2 \times X_3 \times X_4$$

with norm

$$\|(N, P, Z_0, Z_1)\| = \max\{\|N\|,\ \|P\|,\ \|Z_0\|,\ \|Z_1\|\}.$$

In this setting, many of the terms on the right side of the differential equation will turn out to correspond to a bounded operator, B. In abstract notation we can write our equation as

$$U' = AU + BU + F(t) \tag{2.11}$$

where

$$D(A) = \{(N, P, Z_0, Z_1) \in X : D_x(N, P, 0, 0) \in X, BC\} \tag{2.12}$$

where BC indicates the boundary conditions

$$N(0) =e^{-b}G_P P(0) + e^{-b}H_P Z_1(0) \tag{2.13}$$

$$P(1) =e^{-c}G_N N(1) + e^{-c}H_N Z_0(1) \tag{2.14}$$

and

$$A(N, P, Z_0, Z_1)^T = (\Lambda_N D_x N, \Lambda_P D_x P, 0, 0)^T. \tag{2.15}$$

The operator B encompasses all of the terms which involve the matrices B_{ij}, $b\Lambda_P$ and $c\Lambda_N$, while $F(t) = (e^{cx}f, e^{b(1-x)}g, 0, 0)^T$.

To check the conditions for the application of the Hille-Yosida theory it is useful to rescale the unknown. It is also useful to choose an appropriate norm. We first perform a change of variables. First, choose $b,\ c \geq 0$ so that

$$\sum_{j=1}^{q_P} |g^P{}_{ij}| + \sum_{j=1}^{r_P} |h^P{}_{ij}| < e^b, \qquad i = 1, \ldots, q_N \tag{2.16}$$

$$\sum_{j=1}^{q_N} |g^N{}_{ij}| + \sum_{j=1}^{r_N} |h^N{}_{ij}| < e^c, \qquad i = 1, \ldots, q_P. \tag{2.17}$$

With these choices of b and c, take

$$\tilde{N}(t,x) =e^{cx}N(t,x) \qquad \tilde{Z}_0(t,x) =e^{cx}Z_0(t,x) \tag{2.18}$$

$$\tilde{P}(t,x) =e^{b(1-x)}P(t,x) \qquad \tilde{Z}_1(t,x) =e^{b(1-x)}Z_1(t,x) \tag{2.19}$$

Referring to our original equations (2.1) through (2.4), we obtain

$$D_t\tilde{N} = \Lambda_N D_x\tilde{N} - c\Lambda_N\tilde{N} + B_{11}\tilde{N} + e^{(b+c)x-b}B_{12}\tilde{P} + e^{cx}f \quad (2.20)$$
$$D_t\tilde{P} = \Lambda_P D_x\tilde{P} + b\Lambda_P\tilde{P} + e^{-(b+c)x+b}B_{21}\tilde{N} + B_{22}\tilde{P} + e^{b(1-x)}g \quad (2.21)$$
$$D_t\tilde{Z}_0 = B_{31}\tilde{N} + e^{(b+c)x-b}B_{32}\tilde{P} \quad (2.22)$$
$$D_t\tilde{Z}_1 = e^{-(b+c)x+b}B_{41}\tilde{N} + B_{42}\tilde{P} \quad (2.23)$$
$$0 \le x \le 1, \qquad 0 \le t < \infty$$

(We have suppressed the arguments (t, x) to aid readability.)

The initial conditions are much as before,

$$\tilde{N}(0,x) = \tilde{N}_0(x), \qquad \tilde{P}(0,x) = \tilde{P}_0(x),$$
$$\tilde{Z}_0(0,x) = \tilde{Z}_{00}(x), \qquad \tilde{Z}_1(0,x) = \tilde{Z}_{10}(x)$$
$$0 \le x \le 1$$

where $\tilde{N}_0(x) = e^{cx}N_0(x)$, $\tilde{P}_0(x) = e^{b(1-x)}P_0(x)$, $\tilde{Z}_{00}(x) = e^{cx}Z_{00}(x)$, and $\tilde{Z}_{10}(x) = e^{b(1-x)}Z_{10}(x)$. Likewise, the boundary conditions are altered but remain similar.

$$\tilde{N}(t,0) = e^{-b}G_P\tilde{P}(t,0) + e^{-b}H_P\tilde{Z}_1(t,0) \quad (2.24)$$
$$\tilde{P}(t,1) = e^{-c}G_N\tilde{N}(t,1) + e^{-c}H_N\tilde{Z}_0(t,1). \quad (2.25)$$

In this setting, many of the terms on the right side of the differential equation (2.20)- (2.23) again turn out to correspond to a bounded operator, B_1. In abstract notation we can write our equation as

$$U' = A_1U + B_1U + F_1(t) \quad (2.26)$$

where

$$D(A_1) = \{(N, P, Z_0, Z_1) \in X : D_x(N, P, 0, 0) \in X, BC\} \quad (2.27)$$

where BC indicates the boundary conditions

$$N(0) = e^{-b}G_PP(0) + e^{-b}H_PZ_1(0) \quad (2.28)$$
$$P(1) = e^{-c}G_NN(1) + e^{-c}H_NZ_0(1) \quad (2.29)$$

and

$$A_1(N, P, Z_0, Z_1)^T = (\Lambda_N D_xN, \Lambda_P D_xP, 0, 0)^T. \quad (2.30)$$

The operator B_1 encompasses all of the terms which involve the matrices B_{ij}, $b\Lambda_P$ and $c\Lambda_N$, while $F_1(t) = (e^{cx}f, e^{b(1-x)}g, 0, 0)^T$.

DEFINITION 2.1. *A linear operator C on a Banach space is a Hille-Yosida operator if there is an $\omega > 0$ such that each $\mu > \omega$ is in the resolvent set $\rho(C)$ of C and for each such μ, the resolvent operator of C satisfies $\|(\mu - \omega)(\mu I - C)^{-1}\| \le 1$.*

Our immediate goal is to show that A_1 is a Hille-Yosida operator.

THEOREM 2.2. *A_1 is a Hille-Yosida operator.*

Proof. Consider the equation

$$(\mu I - A_1)U = K$$

where $U = (N, P, Z_0, Z_1)$ and $K = (f, g, h, k)$.

If $||U|| = ||Z_{0i}||$ or $||Z_{1i}||$ then since $\mu Z_{0i} = h_i$ or $\mu Z_{1i} = k_i$ we see immediately that $||U|| \leq (1/\mu)||K||$. However, if $||U|| = ||N_i||$ or $||P_i||$ we note that $\mu N_i - \lambda_{Ni} D_x N_i = f_i$. If $||N_i|| = |N_i(1)|$, we may assume that $N_i(1) > 0$ and so in this case $D_x N_i(1) \leq 0$ and $\lambda_{Ni} < 0$ implies $\mu N_i(1) \leq f_i(1)$ and again $||N_i|| \leq (1/\mu)||K||$. Finally, if $||N_i|| = |N_i(0)|$ then we see that (2.12) and (2.16) yield

$$\begin{aligned} |N_i(0)| &\leq \sum_{j=1}^{q_P} e^{-b} |g^P{}_{ij}||P_j(0)| + \sum_{j=1}^{r_P} e^{-b} |h^P{}_{ij}||N_j(0)| \\ &\leq e^{-b} (\sum_{j=1}^{q_P} |g^P{}_{ij}| + \sum_{j=1}^{r_P} |h^P{}_{ij}|)|N_i(0)| \\ &< |N_i(0)|. \end{aligned}$$

This contradiction indicates that this last option cannot occur.

If we have $||U|| = ||P_i||$ an almost identical argument shows that $||U|| \leq (1/\mu)||K||$ in this case also.

We now address the problem of the existence of $(\mu I - A_1)^{-1}$, $\mu > 0$. Given $K \in X$ we must show there is a unique $U \in D(A)$ so that $(\mu I - A_1)U = K$. That is, we want $U = (N, P, Z_0, Z_1)$ so that

$$\mu N - \Lambda_N D_x N = f \tag{2.31}$$

$$\mu P - \Lambda_P D_x P = g \tag{2.32}$$

$$\mu Z_0 = h \tag{2.33}$$

$$\mu Z_1 = k \tag{2.34}$$

The equations (2.33) and (2.34) are clearly solvable with $Z_0 = (1/\mu)h$ and $Z_1 = (1/\mu)k$.

If we take $M_N = \mu(\Lambda_N)^{-1}$ and $M_P = \mu(\Lambda_P)^{-1}$, (2.31) and (2.32) imply

$$N(x) = e^{M_N x} N(0) - \int_0^x e^{M_N(x-s)} f(s)\, ds \tag{2.35}$$

$$P(x) = e^{M_P x} P(0) - \int_0^x e^{M_P(x-s)} g(s)\, ds. \tag{2.36}$$

The constraint that $U \in D(A_1)$ requires that the boundary conditions (2.28)and (2.29) be satisfied. Implementing the solution obtained for Z_0 and Z_1 then leads to the equations

$$N(0) = e^{-b} G_P P(0) + e^{-b} H_P k(0)/\mu \tag{2.37}$$

$$P(1) = e^{-c} G_N N(1) + e^{-c} H_N h(1)/\mu \tag{2.38}$$

Combining (2.35) and (2.37) we see that we must satisfy

$$N(x) = e^{M_N x}[e^{-b}G_P P(0) + e^{-b}H_P k(0)/\mu] - \int_0^x e^{M_N(x-s)}\Lambda_N{}^{-1} f(s)\, ds$$

and so, from (2.38)

$$\begin{aligned} P(1) =& e^{c} G_N e^{M_N} e^{-b} G_P P(0) + e^{c} G_N e^{M_N} e^{-b} H_P k(0)/\mu \\ & - e^{-c} G_N \int_0^1 e^{M_N(1-s)} \Lambda_N{}^{-1} f(s)\, ds + e^{-c} H_N h(1)/\mu. \end{aligned} \tag{2.39}$$

Also, from (2.36)

$$P(1) = e^{M_P} P(0) - \int_0^1 e^{M_P(1-s)} \Lambda_P{}^{-1} g(s)\, ds. \tag{2.40}$$

Combining (2.39) and (2.40) we obtain an equation for $P(0)$.

$$(I - e^{-c} e^{M_P} G_N e^{M_N} e^{-b} G_P) P(0) =$$

$$\begin{aligned} =& e^{-c} e^{-M_P} G_N e^{M_N} e^{-b} H_P k(0)/\mu + e^{-c} e^{-M_P} H_N h(1)/\mu \\ & + \int_0^1 e^{-M_P s} \Lambda_P{}^{-1} g(s)\, ds - e^{-M_P} e^{-c} G_N \int_0^1 e^{M_N(1-s)} \Lambda_N)^{-1} f(s)\, ds \end{aligned}$$

This equation has a unique solution as (2.16) implies that

$$(I - e^{-c} e^{M_P} G_N e^{M_N} e^{-b} G_P) \tag{2.41}$$

is invertible.

■

Because B_1 is bounded we obtain (cf. e.g. Theorem 3.4 of ([4]).)

THEOREM 2.3. *$\Lambda_1 = A_1 + B_1$ is a Hille-Yosida operator. In particular, for $\mu > \omega = \|B_1\|$ we have $\mu \in \rho(\Lambda_1)$ and*

$$\|(\mu - \Lambda_1)^{-1}\| \leq \frac{1}{\mu - \omega}$$

Proof. This follows quickly as B_1 is a bounded operator.

■

DEFINITION 2.4. *A linear operator C on a Banach space is a Hille-Yosida operator of type(M, ω) if each $\mu > \omega$ is in the resolvent set $\rho(C)$ of C and for each such μ and positive integer n the resolvent operator of C satisfies $\|(\mu - \omega)^n (\mu I - C)^{-n}\| \leq M$.*

THEOREM 2.5. *$\Lambda = A + B$ is a Hille-Yosida operator of type (M, ω) where $\omega = \|B_1\|$ and $M = \|S\|\|S^{-1}\|$ where S is the similarity transformation $A+B = S(A_1 + B_1)S^{-1}$.*

Proof. We note that S exists as a bounded operator with a bounded inverse. It is defined by the change of variables (2.18) and (2.19). ∎

3 THE TIMOSHENKO BEAM

The Timoshenko Beam model involves a function $y(x,t)$ with $0 \le x \le L$, $0 \le t$ representing the transverse or perpendicular deflection of the beam from the line $y = 0$. It is assumed that the slope of the displaced beam y_x is the sum of an angle of deflection depending solely on the bending (i.e. independent of the shear force) ψ of the beam and the angle of shear γ_0. That is, slope of the deflection curve depends not only on the bending moment as with the Euler beam but also on the shear.

In Timoshenko's development the bending moment, M, is given by

$$M/EI = -\psi_x.$$

where E is the modulus of elasticity and I is moment of inertia. The assumption for shear is that the shear force, V, is given by

$$V = K_s AG(y_x - \psi) = K_s AG\gamma_0$$

where A is the area of a cross section, G is shear modulus and K_s is a "shape factor" depending on the shape of the cross section.

This leads to the system, [2; p. 183],

$$\begin{aligned} \rho A y_{tt} &= K_s AG(y_x - \psi)_x + q. \\ \rho A r^2 \psi_{tt} &= EI\psi_{xx} + K_s AG(y_x - \psi). \end{aligned}$$

Let us now examine Timoshenko's beam with the left end of the beam constrained by a spring and the right end pinned. Many other boundary conditions are possible. We just choose this example to show how the equation and boundary conditions lead to the form of equation we have examined. See [2; p. 154] for additional information and other possible boundary conditions.

Recall that Bending Moment and Shear are given by

$$\begin{aligned} M &= -EI\psi_x \\ V &= K_s AG(y_x - \psi) \end{aligned}$$

1. In the case of a deflected spring constraint at $x = 0$, the beam and spring are constrained with only vertical motion possible at $x = 0$ so that there is no bending moment but the shear must be balanced by the spring force which obeys Hooke's law.

$$\begin{aligned} \psi_x(0,t) &= 0, \\ K_s AG(y_x - \psi)(0,t) &= K_D y(0,t) \end{aligned}$$

2. At the right end, $x = 1$, we assume that the beam is pinned. That is, the beam can have no vertical displacement at the left end but it is allowed to pivot freely. Thus, there is no bending moment. This is written as

$$y(1,t) = 0$$
$$\psi_x(1,t) = 0.$$

Let us now define new variables $p_i,\ i = 1,2,3,4$ as follows:

$$\begin{aligned} p_1 &= y_t, \\ p_2 &= V = K_s AG(y_x - \psi) = \text{shear}, \\ p_3 &= \psi_t, \\ p_4 &= -M = EI\psi_x = \text{- bending moment.} \end{aligned}$$

This leads to the system ,

$$\begin{bmatrix} \rho A & 0 & 0 & 0 \\ 0 & (K_s AG)^{-1} & 0 & 0 \\ 0 & 0 & \rho A r^2 & 0 \\ 0 & 0 & 0 & (EI)^{-1} \end{bmatrix} \begin{bmatrix} p_1 \\ p_2 \\ p_3 \\ p_4 \end{bmatrix}_t = \begin{bmatrix} 0 & 1 & 0 & 0 \\ 1 & 0 & 0 & 0 \\ 0 & 0 & 0 & 1 \\ 0 & 0 & 1 & 0 \end{bmatrix} \begin{bmatrix} p_1 \\ p_2 \\ p_3 \\ p_4 \end{bmatrix}_x + \begin{bmatrix} 0 & 0 & 0 & 0 \\ 0 & 0 & -1 & 0 \\ 0 & 1 & 0 & 0 \\ 0 & 0 & 0 & 0 \end{bmatrix} \begin{bmatrix} p_1 \\ p_2 \\ p_3 \\ p_4 \end{bmatrix}.$$

We now restate the boundary conditions for our new variables.

1. Using our new variables we see that the deflected spring boundary condition must be include the variable y. It will be convenient to do this at the other endpoint also.

$$p_2(0,t) = K_D y(0,t) \text{ Shear force balanced with spring force.}$$
$$p_4(0,t) = 0, \text{ No bending moment.}$$

2. Pinned at $x = 0$. That is, the beam can have no vertical displacement at the left end but it is allowed to pivot without any force - no bending at the left end.

$$p_1(1,t) = 0 \quad \text{Fixed position.}$$
$$p_4(1,t) = 0 \quad \text{No bending moment.}$$

Notice that the boundary condition at $x = 0$ requires information on the function y also. For this reason we shall incorporate y into our equation.

Now we diagonalize this by choosing $w_i,\ i = 1,2,3,4$ by

$$\begin{aligned} p_1 &= -\ (\rho A K_s G)^{-1/2}(w_1 - w_3) \\ p_2 &= w_1 + w_3 \\ p_3 &= -\ (\rho A r^2 EI)^{-1/2}(w_2 - w_4) \\ p_4 &= w_2 + w_4 \end{aligned}$$

If we further define

$$\begin{aligned}\alpha =&(K_sG/\rho)^{1/2}\\ \beta =&(EI/\rho Ar^2)^{1/2}\\ \gamma =&K_sAG(\rho Ar^2EI)^{-1/2}/2\\ \delta =&1/2(EI/(\rho Ar^2)^{1/2}\\ \eta =&-(\rho AK_sG)^{-1/2}\end{aligned}$$

we then get the system

$$\begin{bmatrix} w_1\\ w_2\\ w_3\\ w_4\\ y\end{bmatrix}_t = \begin{bmatrix} -\alpha & 0 & 0 & 0 & 0\\ 0 & -\beta & 0 & 0 & 0\\ 0 & 0 & \alpha & 0 & 0\\ 0 & 0 & 0 & \beta & 0\\ 0 & 0 & 0 & 0 & 0\end{bmatrix} \begin{bmatrix} w_1\\ w_2\\ w_3\\ w_4\\ y\end{bmatrix}_x + \begin{bmatrix} 0 & \gamma & 0 & -\gamma & 0\\ -\delta & 0 & -\delta & 0 & 0\\ 0 & -\gamma & 0 & \gamma & 0\\ \delta & 0 & \delta & 0 & 0\\ \eta & 0 & -\eta & 0 & 0\end{bmatrix} \begin{bmatrix} w_1\\ w_2\\ w_3\\ w_4\\ y\end{bmatrix}.$$

We thus have our system with $(w_1, w_2)^T = N$, $(w_3, w_4)^T = P$ and $y = Z_1$. The variable Z_0 is not needed in this application but could be put in as anything with a trivial differential equation and no effect on the boundary condition of P. Also, it is possible in other problems to require information about ψ on one of the boundaries.

Our boundary conditions are:

1. Using our new variables we see that the deflected spring boundary condition must be include the variable y. It will be convenient to do this at the other endpoint also.

$$w_1(0,t) = -w^3(0,t) + K_D y(0,t)$$
Shear balanced by spring force.
$$w_2(0,t) = -w_4(0,t), \quad \text{No bending moment.}$$

2. Pinned at $x = 0$. That is, the beam can have no vertical displacement at the left end but it is allowed to pivot without any force - no bending at the left end.

$$w_3(1,t) = w_1(1,t) \text{ Fixed position.}$$
$$w_4(1,t) = -w_2(1,t) \text{ No bending moment.}$$

It is easy to check now that this Timoshenko model fits into our hyperbolic system model. Because of the original boundary condition $y(1,t) = 0$ the only solutions that will correspond to the classical solutions we are concerned with will be ones that satisfy that condition. Because $y_t(1,t) = p_1(1,t) = w_1(1,t) - w_3(1,t) = 0$ we see that solutions with the correct initial data will continue to satisfy this condition.

Finally, we can state our conclusions regarding the Timoshenko Beam as a theorem. We restrict our comments to the boundary conditions we have examined here but many other boundary conditions lead to the same conclusion.

THEOREM 3.1. *The Timoshenko Beam with boundary conditions consisting of one end constrained by a spring and the other end pinned leads to a one dimensional hyperbolic system which can be reformulated as an abstract differential equation with a Hille-Yosida operator.*

BIBLIOGRAPHY

1. G. Da Prato and E. Sinestrari, *Differential operators with nondense domain*, Annali Scuola Norm. Sup. Pisa **14** (1987), 285–344.

2. Karl F. Graff, *Wave Motion in Elastic Solids*, Oxford University Press, London, 1975.

3. R. Grimmer and E. Sinestrari, *Maximum norm in one-dimensional hyperbolic problems*, Diff. Int. Equ. **5** (1992), 421–432.

4. R. Grimmer and E. Sinestrari, *Classical solutions of the Timoshenko system*, Advances in Differential Equations **7** (2002), 799–818.

On The Wave Equation Subjected to Coulomb Friction

RONALD GRIMMER
Department of Mathematics, Southern Illinois University at Carbondale
Carbondale, Illinois 62901
e-mail: rgrimmer@math.siu.edu

YUDI SOEHARYADI
Department of Mathematics, Institut Teknologi Bandung
Bandung 40132, Indonesia
e-mail: yudish@dns.math.itb.ac.id

ABSTRACT. We study the one dimensional wave equation on the unit interval subjected to Coulomb frictions on both ends. The application of this type of boundary condition introduces nonlinearity to the problem. Using the machinery of nonlinear semigroup theory we exhibit well-posedness in the context of Hilbert space of square integrable functions, also in the context of the space of continuous functions.

1 INTRODUCTION

The wave equation is an ever-inspirational source of problems. We study the problem of vibrations of a finite-length elastic string. The string is subjected to Coulomb frictions on both ends. We thus consider the wave equation on the unit interval $[0, 1]$,

$$u_{tt} = c^2 u_{xx}, \qquad \text{for} \quad 0 \leq x \leq 1, \quad t > 0. \tag{1.1}$$

Coulomb friction at the end points is given as the boundary conditions

$$u_x(0,t) \in g_0(u_t(0,t)), \; u_x(1,t) \in -g_1(u_t(1,t)), \tag{1.2}$$

for all $t \geq 0$. Here the g_i are maximal monotone graphs, given as positive multiples of the completion of the graph of signum function.

This body of work was done while YS was visiting Southern Illinois University in the year 2000-2001

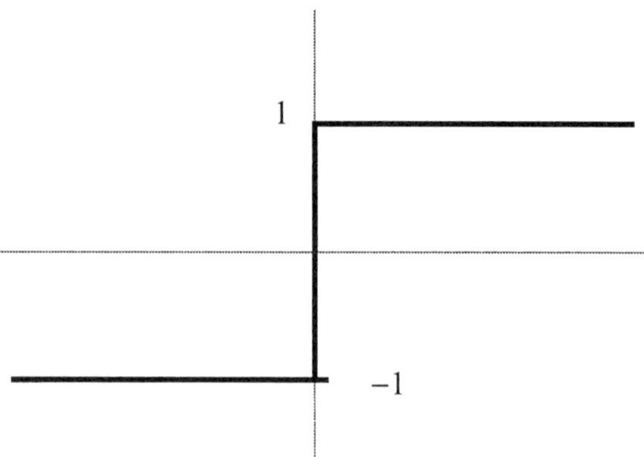

Figure 1.1: Maximal monotone graph g_i

The boundary conditions (1.2) represent tension u_x at the end points which is being balanced by dry friction mechanism.

The tension can be anything between certain bounds as long as the velocity u_t is zero. If the velocity is negative or positive then one can only have a certain amount of tension. We remark that the analysis we carry out here also applies to any monotone (and continuous) damping.

We use a semigroup theoretic approach. While our equation itself is linear, the boundary conditions we impose are not, therefore it gives rise to a nonlinear problem. Resorting to nonlinear semigroup theory, i.e., the Crandall-Liggett Theorem, we are able

to exhibit global well-posedness of the problem in two cases: the case of Hilbert space of square integrable functions and the case of the space of continuous functions; hence obtaining the existence and the uniqueness of global solutions, and continuous dependence of these solutions on initial data.

For application of this problem, for example see [4] where the authors apply this result to the viscoelastic Timoshenko beam subjected to similar boundary conditions as above.

The authors wish to take this occasion to congratulate Jerry Goldstein on the occasion of his 60th birthday and to thank him for his profound efforts in the theory of semigroups of bounded operators.

2 ABSTRACT SETTING

Let X be a Banach space. An operator A, possibly multi-valued, with domain $\mathcal{D}(A) \subset X$ into X is said to be *dissipative* if for all $\alpha > 0$, $(I - \alpha A)^{-1}$ is a (single-valued) function, and $\|(I - \alpha A)^{-1}\|_{Lip(X)} \leq 1$. A is *m-dissipative* if it is dissipative and its range satisfies $\mathcal{R}(I - \alpha A) = X$ for some $\alpha > 0$.

Let $u(t) = u(\cdot, t)$. Consider the Abstract Cauchy Problem (ACP)

$$\frac{d}{dt}u(t) \in Au(t), \qquad u(0) = u_0$$

where $u_0 \in \mathcal{D}(A)$.

In linear theory, the Hille-Yosida Theorem provides necessary and sufficient conditions for this ACP to be governed by a strongly continuous semigroup of operators $\{T(t) : t \geq 0\}$, i.e.,

$$u(t) = T(t)u_0, \quad \text{for} \quad t \geq 0.$$

Its parallel in nonlinear theory is the Crandall-Ligget Theorem, which is a complete generalization of sufficiency part of the Hille-Yosida Theorem.

THEOREM 2.1. *Let A be an operator, possibly multi-valued, on $\mathcal{D}(A) \subset X$, and ω be in $\mathbb{R}$. Suppose $A - \omega I$ is dissipative. Further we assume that there is $\alpha_0 > 0$ such that $\mathcal{R}(I - \alpha A) \supset \overline{\mathcal{D}(A)}$, for $0 < \alpha < \alpha_0$. Then for every $f \in \overline{\mathcal{D}(A)}$ the ACP*

$$\frac{d}{dt}u(t) \in Au(t), \ u(0) = f$$

is governed by a strongly continuous semigroup of operators $\{T(t) : t \geq 0\}$, and furthermore

$$T(t)f = \lim_{n \to \infty} \left(I - \frac{t}{n}A\right)^{-n} f.$$

For a more substantial discussion of the nonlinear theory of the semigroup of operators, see for example [1, 2].

For our problem we set $cv = u_t$ and $w = u_x$. Then we can rewrite (1.1) as a system

$$\frac{d}{dt}\begin{bmatrix} v \\ w \end{bmatrix} = \begin{bmatrix} 0 & c \\ c & 0 \end{bmatrix} \begin{bmatrix} v_x \\ w_x \end{bmatrix}$$

The boundary conditions transform into

$$w(0,t) \in g_0(cv(0,t)), \ w(1,t) \in -g_1(cv(1,t)), \tag{2.1}$$

for all $t \geq 0$. Let $h_0(s) = g(cs)$ and $h_1(s) = g_1(cs)$. The graphs h_0 and h_1 are also maximal monotone since c is positive. We can now transform our boundary value problem into an ACP. We let X be our ambient Banach space, which will be the Hilbert space $L^2[0,1] \times L^2[0,1]$, or the space $C[0,1] \times C[0,1]$. The ACP is

$$\frac{d}{dt}\begin{bmatrix} v \\ w \end{bmatrix} \in A\begin{bmatrix} v \\ w \end{bmatrix}, \quad (v,w) \in \overline{\mathcal{D}(A)}, \tag{2.2}$$

where

$$A = \begin{bmatrix} 0 & c\,\partial_x \\ c\,\partial_x & 0 \end{bmatrix},$$

and

$$\mathcal{D}(A) = \{(v,w) \in X' : w(i,t) \in (-1)^i h_i(v(i,t)), i = 0, 1, \text{and } t \geq 0\},$$

for some appropriate Banach space $X' \subset X$, which will be stated later.

3 WELL-POSEDNESS

We first show well-posedness of (2.2) in the Hilbert space $X = L^2[0,1] \times L^2[0,1]$. In this case

$$\mathcal{D}(A) = \{\tilde{u} = (v,w) \in H^1[0,1] \times H^1[0,1] : \\ w(i,t) \in (-1)^i h_i(v(i,t)), i = 0,1, \text{and } t \geq 0\}.$$

Here $H^1[0,1]$ denote the Sobolev space of the L^2-functions whose first order derivatives are also in $L^2[0,1]$. We must show (i) dissipativity, which in Hilbert space translates into

$$\left\langle A\tilde{u}_1 - A\tilde{u}_2, \tilde{u}_1 - \tilde{u}_2 \right\rangle_{L^2 \times L^2} \leq 0\,,$$

for $\tilde{u}_1, \tilde{u}_2 \in \mathcal{D}(A)$, and (ii) range condition, i.e., $\overline{\mathcal{R}(I - \lambda A)} = L^2[0,1] \times L^2[0,1]$, for $\lambda > 0$.

Without loss of generality we may assume $c = 1$. Then

$$\begin{aligned}
&\left\langle A\tilde{u}_1 - A\tilde{u}_2, \tilde{u}_1 - \tilde{u}_2 \right\rangle_{L^2 \times L^2} \\
&\quad = \left\langle \begin{bmatrix} \partial_x(w_1 - w_2) \\ \partial_x(v_1 - v_2) \end{bmatrix}, \begin{bmatrix} v_1 - v_2 \\ w_1 - w_2 \end{bmatrix} \right\rangle_{L^2 \times L^2} \\
&\quad = \int_0^1 ((v_1 - v_2)\partial_x(w_1 - w_2) + (w_1 - w_2)\partial_x(v_1 - v_2))dx \\
&\quad = (v_1 - v_2)(w_1 - w_2)|_{x=0}^{x=1} \\
&\quad = [(v_1(1,\cdot) - v_2(1,\cdot))(-h_1(v_1(1,\cdot)) + h_1(v_2(1,\cdot))) \\
&\qquad\quad - (v_1(0,\cdot) - v_2(0,\cdot))(h_0(v_1(0,\cdot)) - h_0(v_2(0,\cdot)))] \\
&\quad \leq 0, \text{by monotonicity of } h_0, h_1,
\end{aligned}$$

and therefore A is dissipative.

For the range condition, we let $\lambda > 0$, $f = (f_1, f_2) \in L^2[0,1] \times L^2[0,1]$. We show that the resolvent equation

$$(I - \lambda A)u = f, \tag{3.1}$$

can be solved. We rewrite (3.1) to obtain

$$\lambda \begin{bmatrix} w \\ v \end{bmatrix}_x = \begin{bmatrix} 0 & 1 \\ 1 & 0 \end{bmatrix} \begin{bmatrix} w \\ v \end{bmatrix} - \begin{bmatrix} f_1 \\ f_2 \end{bmatrix},$$

and from the variation of parameters formula

$$\begin{aligned}
\lambda \begin{bmatrix} w(x,\cdot) \\ v(x,\cdot) \end{bmatrix} &= \begin{bmatrix} \cosh x & \sinh x \\ \sinh x & \cosh x \end{bmatrix} \begin{bmatrix} w_0(\cdot) \\ v_0(\cdot) \end{bmatrix} + \\
&\int_0^1 \begin{bmatrix} \cosh(x-s) & \sinh(x-s) \\ \sinh(x-s) & \cosh(x-s) \end{bmatrix} \begin{bmatrix} f_1(s) \\ f_2(s) \end{bmatrix} ds.
\end{aligned} \tag{3.2}$$

Now we have to solve for w_0, v_0. We let

$$F_1 = \int_0^1 \cosh(x-s)f_1(s) + \sinh(x-s)f_2(s)ds,$$

$$F_2 = \int_0^1 \sinh(x-s)f_1(s) + \cosh(x-s)f_2(s)ds.$$

Note that both F_1 and F_2 are bounded. Using boundary conditions (2.1) and (3.2)

$$\begin{gathered} w_0 \in h_0(v_0) \\ w_0 \cosh 1 + v_0 \sinh 1 + F_1 \in -\lambda h_1(\frac{1}{\lambda}(w_0 \sinh 1 + v_0 \cosh 1 + F_2)), \end{gathered}$$

which can be rewritten as

$$\begin{gathered} w_0 \in h_0(v_0) \\ w_0 \cosh 1 \in -v_0 \sinh 1 - F_1 - \lambda h_1(\frac{1}{\lambda}(w_0 \sinh 1 + v_0 \cosh 1 + F_2)). \end{gathered} \tag{3.3}$$

We now think of (3.3) as two graphs on v_0w_0-plane, and by monotonicity (and continuity) of h_0, h_1, they intersect. Thus we find v_0 and w_0 which solve the resolvent equation, hence m-dissipativity, and thus well-posedness.

In the context of the space of continuous function we let $X = C[0,1] \times C[0,1]$. In this case

$$\begin{aligned} \mathcal{D}(A) = \{(\tilde{v}, \tilde{w}) \in C^1[0,1] \times C^1[0,1] :& \\ \tilde{w}(i,t) \in (-1)^i h_i(\tilde{v}(i,t)), i = 0,1, \text{ and } t \geq 0\}&. \end{aligned}$$

Here $C^1[0,1]$ denote the set of functions whose first order derivatives are in $C[0,1]$.

The proof of the range condition is similar to the one in the case of Hilbert space, so we prove dissipativity. For this purpose we diagonalize A to get $\tilde{A}$ by setting

$$v = \tilde{v} + \tilde{w}, w = \tilde{v} - \tilde{w}.$$

Dissipativity of $\tilde{A}$ will imply dissipativity of A. We have

$$\frac{d}{dt}\begin{bmatrix} v \\ w \end{bmatrix} = \begin{bmatrix} \partial_x & 0 \\ 0 & -\partial_x \end{bmatrix} \begin{bmatrix} v \\ w \end{bmatrix}.$$

The boundary conditions become

$$\begin{aligned} \frac{1}{2}(v(0,\cdot) - w(0,\cdot)) &\in h_0(\frac{1}{2}(v(0,\cdot) + w(0,\cdot))), \\ \frac{1}{2}(v(1,\cdot) - w(1,\cdot)) &\in -h_1(\frac{1}{2}(v(1,\cdot) + w(1,\cdot))). \end{aligned}$$

For simplicity we define

$$H_i(s) = 2h_i(s/2), \quad \text{for } i = 0, 1.$$

Note that maximal monotonicity of h_0, h_1 survive in H_0, H_1. Let $\tilde{u}_1 = (v_1, w_1)$ and $\tilde{u}_2 = (v_2, w_2)$ be in $\mathcal{D}(\tilde{A})$. We let also $\lambda > 0$. We shall show that

$$\|(I - \lambda\tilde{A})\tilde{u}_1 - (I - \lambda\tilde{A})\tilde{u}_2\| \geq \|\tilde{u}_1 - \tilde{u}_2\|. \tag{3.4}$$

Here we use the norm

$$\|(v, w)\| = \max(\|v\|_{C[0,1]}, \|w\|_{C[0,1]}).$$

Rewriting (3.4), what we need to show is

$$\left\|\begin{bmatrix}(v_1 - v_2) - \lambda(v_1 - v_2)_x \\ (w_1 - w_2) - \lambda(w_2 - w_1)_x\end{bmatrix}\right\| \geq \left\|\begin{bmatrix}v_1 - v_2 \\ w_1 - w_2\end{bmatrix}\right\|. \tag{3.5}$$

We have three cases.
Case 1: The maximum of $u_1(\cdot) - u_2(\cdot)$ is attained in the interior of $[0, 1]$. Suppose

$$\left\|\begin{bmatrix}v_1 - v_2 \\ w_1 - w_2\end{bmatrix}\right\| = v_1(x_0, \cdot) - v_2(x_0, \cdot) \geq 0$$

for some x_0, and $0 < x_0 < 1$. Then $(v_1 - v_2)_x(x_0, \cdot) = 0$, and thus

$$|\,(v_1 - v_2)(x_0, \cdot) - \lambda(v_1 - v_2)_x(x_0, \cdot)\,| = (v_1 - v_2)(x_0, \cdot),$$

hence (3.5) holds. We argue in a similar manner for the case $\left\|\begin{bmatrix}v_1 - v_2 \\ w_1 - w_2\end{bmatrix}\right\| = v_2(x_0, \cdot) - v_1(x_0, \cdot) \geq 0$. The case $\left\|\begin{bmatrix}v_1 - v_2 \\ w_1 - w_2\end{bmatrix}\right\| = w_1(x_0, \cdot) - w_2(x_0, \cdot)$, is handled similarly.
Case 2: The maximum is attained at the end-point $x = 0$. Suppose

$$\left\|\begin{bmatrix}v_1 - v_2 \\ w_1 - w_2\end{bmatrix}\right\| = (v_2(0, \cdot) - v_1(0, \cdot)) \geq 0. \tag{3.6}$$

Then $(v_1 - v_2)_x(0, \cdot) \leq 0$, and thus

$$|\,(v_1 - v_2)(0, \cdot) - \lambda(v_1 - v_2)_x(0, \cdot)\,| \geq (v_1 - v_2)(0, \cdot).$$

If

$$\left\|\begin{bmatrix}v_1 - v_2 \\ w_1 - w_2\end{bmatrix}\right\| = (v_1 - v_2)(0, \cdot) \geq 0.$$

then

$$|\,(v_1 - v_2)(0, \cdot) - \lambda(v_1 - v_2)_x(0, \cdot)\,| \geq |\,(v_1 - v_2)(0, \cdot)\,|,$$

so (3.6) holds.
Now suppose

$$\left\|\begin{bmatrix}v_1 - v_2 \\ w_1 - w_2\end{bmatrix}\right\| = (w_1(0, \cdot) - w_2(0, \cdot)) \geq |\,(v_1(0, \cdot) - v_2(0, \cdot))\,|\,.$$

Then $w_1(0, \cdot) - w_2(0, \cdot) \geq v_1(0, \cdot) - v_2(0, \cdot)$. Switching terms yields

$$w_1(0, \cdot) - v_1(0, \cdot) \geq w_2(0, \cdot) - v_2(0, \cdot).$$

Using the boundary condition at $x = 0$ we have

$$-H_0(v_1(0,\cdot) + w_1(0,\cdot)) \geq -H_0(v_2(0,\cdot) + w_2(0,\cdot)).$$

By monotonicity of H_0,

$$v_1(0,\cdot) + w_1(0,\cdot) \leq v_2(0,\cdot) + w_2(0,\cdot),$$

which upon switching terms becomes

$$v_1(0,\cdot) - v_2(0,\cdot) \leq w_2(0,\cdot) - w_1(0,\cdot).$$

Therefore $v_1(0,\cdot) - v_2(0,\cdot) = | w_1(0,\cdot) - w_2(0,\cdot) |$, and we are back to (3.6).
Case 3: The maximum is attained at $x = 1$. This case is similar to the Case 2. The proof is omitted.

We have shown the following:

THEOREM 3.1. *We have well-posedness of the ACP*

$$\frac{d}{dt}\begin{bmatrix} v \\ w \end{bmatrix} \in A \begin{bmatrix} v \\ w \end{bmatrix}, \quad \begin{bmatrix} v(0) \\ w(0) \end{bmatrix} = \begin{bmatrix} v_0 \\ w_0 \end{bmatrix} \in \overline{\mathcal{D}(A)} \subset X,$$

where

$$A = \begin{bmatrix} 0 & c\partial_x \\ c\partial_x & 0 \end{bmatrix},$$

and

$$\mathcal{D}(A) = \{(v,w) \in X' : w(i,t) \in (-1)^i h_i(v(i,t)), i = 0,1, \text{and}\, t \geq 0\}.$$

Here $X = L^2[0,1] \times L^2[0,1]$, *and* $X' = H^1[0,1] \times H^1[0,1]$, *or* $X = C[0,1] \times C[0,1]$, *and* $X' = C^1[0,1] \times C^1[0,1]$.

BIBLIOGRAPHY

1. V. Barbu, *Semigroups of Nonlinear Operators and Differential Equations in Banach Spaces*, Nordhoff, 1976.

2. J. A. Goldstein, *Semigroups of Nonlinear Operators and Their Applications*, (monograph in preparation).

3. K. F. Graff, *Wave Motion in Elastic Solids*, (Dover), 1991.

4. R. Grimmer and Y. Soeharyadi, On viscoelastic Timoshenko beam with Coulomb frictions, (in preparation).

Asymptotics of Perturbations to the Wave Equation

MATTHIAS HIEBER

Fachbereich Mathematik, TU Darmstadt
Schlossgartenstr. 7, D-64289 Darmstadt, Germany
e-mail: hieber@mathematik.tu-darmstadt.de

IAN WOOD

Fachbereich Mathematik, TU Darmstadt
Schlossgartenstr. 7, D-64289 Darmstadt, Germany
e-mail: wood@mathematik.tu-darmstadt.de

Dedicated to Jerry Goldstein on the occasion of his 60th birthday

1 INTRODUCTION

Determining the asymptotic behaviour of solutions to linear PDEs is often a delicate matter. Even if the solution is given by a C_0-semigroup T acting on a suitable Banach space, it can be difficult to calculate the growth bound of T. It is a well-established procedure to calculate or to estimate the spectrum of the generator A and to try to relate the location of the spectrum in the complex plane to the asymptotic behaviour of the solution. It is however well-known that the spectral bound and the growth bound of T do not coincide in general (see the counterexamples given in [1], [3], [12], [13], and [15]). Therefore, one is interested in the question of when the growth bound $\omega(T)$ and the spectral bound $s(A)$ do coincide. When this is the case, we say that the principle of linear stability holds.

The principle of linear stability holds whenever we have a suitable spectral mapping theorem for the semigroup. This is the case for a wide variety of semigroups. In [3] it is shown that the spectral bound and the growth bound coincide for eventually norm-continuous semigroups. This includes analytic semigroups which allows us to deal with parabolic PDEs. For hyperbolic equations in one dimension affirmative results are given in [10] and [11]. For higher dimensions however, there are counterexamples where the equality of bounds does not hold. In [12] such a counterexample is given which is just a first order perturbation of the wave equation in two dimensions.

Inspired by this example, we try to find conditions on the perturbation guaranteeing equality of the bounds. The well-posedness of this kind of problem is treated in the monograph by Goldstein [6]. We show that for a class of self-adjoint perturbations the equality of bounds which exists for the wave equation is preserved. Finally we show that Renardy's construction of a counterexample can be extended to higher order equations. Here, we will make use of the theory of cosine functions which was partly developed by Goldstein (cf. [5] and [6; section 2.8]).

Further results on the stability of the semigroup have recently been gained by using Fourier multiplier properties of the resolvent. For details see [7], [8], [9] and [14].

2 RENARDY'S EXAMPLE

We consider a first order perturbation of the wave equation,

$$\partial_t^2 u = \partial_x^2 u + \partial_y^2 u + e^{iy}\partial_x u, \ (x,y) \in \mathbb{R}^2$$

where u is 2π-periodic in x and y.

We rewrite the problem in $H := H^1_{per}(\Omega) \times L^2_{per}(\Omega)$ in the following way.

$$\partial_t \begin{pmatrix} u \\ v \end{pmatrix} = \mathcal{A} \begin{pmatrix} u \\ v \end{pmatrix}$$

where $\Omega = (-\pi,\pi) \times (-\pi,\pi)$,

$$\mathcal{A} = \begin{pmatrix} 0 & I \\ \Delta + e^{iy}\partial_x & 0 \end{pmatrix} \quad \text{and} \quad \mathcal{D}(\mathcal{A}) = H^2_{per}(\Omega) \times H^1_{per}(\Omega).$$

The Hilbert space H is equipped with the norm

$$||(u,v)||_H := \sqrt{||u||^2_{H^1(\Omega)} + ||v||^2_{L^2(\Omega)}}.$$

In [12], Renardy proves the following result.

THEOREM 2.1. *On $H^1_{per}(\Omega) \times L^2_{per}(\Omega)$ the operator*

$$\mathcal{A} = \begin{pmatrix} 0 & I \\ \Delta + e^{iy}\partial_x & 0 \end{pmatrix} \quad \text{with} \quad \mathcal{D}(\mathcal{A}) = H^2_{per}(\Omega) \times H^1_{per}(\Omega)$$

generates a strongly continuous semigroup $(T(t))_{t\geq 0}$ and we have $s(\mathcal{A}) = 0$, but $\omega(T) \geq \frac{1}{2}$.

3 SELF-ADJOINT PERTURBATIONS

In this section, we consider perturbations of the wave equation on a bounded domain in $\mathbb{R}^n$ with zero boundary conditions and on $\Omega = (-\pi,\pi)^2$ with periodic boundary conditions. The perturbation is chosen such that the perturbed operator remains self-adjoint. Using the theory of self-adjoint operators on Hilbert spaces, we will see that the equality $s(\mathcal{A}) = \omega(T)$ still holds.

We consider operators on H of the form

$$\mathcal{A} = \begin{pmatrix} 0 & I \\ \Delta + if(y)\partial_x & 0 \end{pmatrix},$$

where f is a bounded real-valued function and

- for the case of Dirichlet boundary conditions we have

$$H = H_0^1(\Omega) \times L^2(\Omega)\ ,\ \mathcal{D}(\mathcal{A}) = \left(H^2(\Omega) \cap H_0^1(\Omega)\right) \times H_0^1(\Omega)$$

  and

- for the periodic case

$$H = H_{per}^1(\Omega) \times L_{per}^2(\Omega) \text{ and } \mathcal{D}(\mathcal{A}) = H_{per}^2(\Omega) \times H_{per}^1(\Omega).$$

In both cases we are dealing with bounded perturbations of the wave operator, so using [6; Theorem7.8], we see that the operator $\mathcal{A}$ generates a semigroup on H. For further information on the semigroup generated by $\mathcal{A}$ we refer to [4].

We now determine the spectral bound.

For $\lambda^2 \notin \sigma(\Delta + if(y)\partial_x)$, the resolvent of $\mathcal{A}$ is given by

$$R(\lambda, \mathcal{A}) = \begin{pmatrix} \lambda & I \\ \Delta + if(y)\partial_x & \lambda \end{pmatrix} R\left(\lambda^2, \Delta + if(y)\partial_x\right) \tag{3.1}$$

If we now make the further assumption that

- $\|f\|_\infty \leq d(\Omega)^{-1}$ where $d(\Omega)$ is the diameter of Ω in the case of Dirichlet boundary conditions,
- or $\|f\|_\infty \leq 1$ in the case of periodic boundary conditions,

then by a straightforward calculation, we see that $\Delta + if(y)\partial_x$ is a negative self-adjoint operator. Therefore its spectrum lies in $]-\infty, 0]$. Then from (3.1), $\sigma(\mathcal{A}) \subseteq i\mathbb{R}$ and $s(\mathcal{A}) = 0$.

Our next task is to determine the growth bound of the semigroup.

Since H is a Hilbert space, by the Gearhart-Prüss-Theorem (see [2; Theorem 5.2.1] or, for a proof using Fourier multipliers see [7]), we have that $s_0(\mathcal{A}) = \omega(T)$ where $s_0(\mathcal{A})$ denotes the pseudo-spectral bound.

From (3.1), we obtain

$$\begin{aligned} \|R(\lambda, \mathcal{A})\|_{\mathcal{L}(H)}^2 \leq\ & \left\|\lambda\left(\lambda^2 - (\Delta + if(y)\partial_x)\right)^{-1}\right\|_{\mathcal{L}(H_0^1(\Omega))}^2 \\ & + \left\|(\lambda^2 - (\Delta + if(y)\partial_x))^{-1}\right\|_{\mathcal{L}(L^2(\Omega), H_0^1(\Omega))}^2 \\ & + \left\|(\Delta + if(y)\partial_x)(\lambda^2 - (\Delta + if(y)\partial_x))^{-1}\right\|_{\mathcal{L}(H_0^1(\Omega), L^2(\Omega))}^2 \\ & + \left\|\lambda(\lambda^2 - (\Delta + if(y)\partial_x))^{-1}\right\|_{\mathcal{L}(L^2(\Omega))}^2 . \end{aligned}$$

In order to prove $s_0(\mathcal{A}) = 0$, we have to show that all four terms are uniformly bounded for $Re\lambda > \epsilon$ for all $\epsilon > 0$.

We start with two lemmas.

LEMMA 3.1. *Let $\Omega \subseteq \mathbb{R}^n$ be a bounded domain with smooth boundary. On $H_0^1(\Omega)$ the norms*

$$\|\cdot\|_{H^1} \text{ and } \left\|(-\Delta - if(y)\partial_x + 1)^{\frac{1}{2}} \cdot\right\|_{L^2}$$

are equivalent whenever $M := \|f\|_\infty < (d(\Omega))^{-1}$.

In the periodic case we have a similar statement.

LEMMA 3.2. *Let $\Omega = (-\pi, \pi)^2 \subseteq \mathbb{R}^2$. On $H_{per}^1(\Omega)$ the norms*

$$\|\cdot\|_{H^1} \text{ and } \left\|(-\Delta - if(y)\partial_x + 1)^{\frac{1}{2}} \cdot\right\|_{L^2}$$

are equivalent whenever $M := \|f\|_\infty \leq 1$.

In both cases the proof consists of simple calculations. For Lemma 3.1 we make use of Poincaré's inequality which is why the diameter of Ω comes in.

From now on, the proof for the periodic case works in exactly the same way as for the Dirichlet case. We will therefore only give the proof for the case where Ω is a bounded domain and we have zero boundary conditions.

PROPOSITION 3.3. *For all $\epsilon > 0$,*

$$\left\|\lambda(\lambda^2 - (\Delta + if(y)\partial_x))^{-1}\right\|_{\mathcal{L}(L^2(\Omega))}$$

is uniformly bounded on $S := \{\lambda : Re\lambda > \epsilon\}$.

Proof. The idea is the following.
For any $\theta \in (0, \frac{\pi}{2})$ we can get an estimate on the resolvent in the sector $\sum_{\theta+\frac{\pi}{2}}$ using the fact that $\Delta + if(y)\partial_x$ generates a bounded analytic semigroup. However, as $\theta \to \frac{\pi}{2}$, the constants tend to infinity. Therefore, we need another estimate on the resolvent for those λ^2 that are outside the sector $\sum_{\theta+\frac{\pi}{2}}$. Here we can use that $\Delta + if(y)\partial_x$ is self-adjoint, so the norm of the resolvent at λ^2 can be estimated by $(Im\lambda^2)^{-1}$. ∎

We now use the spectral decomposition of the operator $-(\Delta + if(y)\partial_x)$.
Let $\{f_j, j \in \mathbb{N}\}$ be the set of orthonormal eigenfunctions of $-\Delta - if(y)\partial_x$ with the corresponding eigenvalues $\lambda_j \geq 0$.
Then we have for $u \in H^2(\Omega) \cap H_0^1(\Omega)$

$$(-\Delta - if(y)\partial_x)u = \sum_{n=1}^{\infty} \lambda_n \langle u, f_n\rangle_{L^2} f_n.$$

PROPOSITION 3.4. *For all $\epsilon > 0$, the terms*

$$\left\|\lambda(\lambda^2 - (\Delta + if(y)\partial_x))^{-1}\right\|_{\mathcal{L}(H_0^1(\Omega))}, \quad \left\|(\lambda^2 - (\Delta + if(y)\partial_x))^{-1}\right\|_{\mathcal{L}(L^2(\Omega), H_0^1(\Omega))}$$

and

$$\left\|(\Delta + if(y)\partial_x)(\lambda^2 - (\Delta + if(y)\partial_x))^{-1}\right\|_{\mathcal{L}(H_0^1(\Omega), L^2(\Omega))}$$

are uniformly bounded on $S := \{\lambda : Re\lambda > \epsilon\}$.

Proof. The key to the proof is given by Lemma 3.1 and the following estimates. For $\lambda = a + ib$ with $a > \epsilon > 0$ we have

$$\sup_{\lambda \in S, \lambda_n \geq 0} \left| \frac{\lambda}{\lambda^2 + \lambda_n} \right| \leq \frac{1}{\epsilon}, \quad \sup_{\lambda \in S, \lambda_n \geq 0} \left| \frac{1}{\lambda^2 + \lambda_n} \right| \leq \frac{1}{\epsilon^2}$$

and

$$\sup_{\lambda \in S, \lambda_n \geq 0} \left| \frac{\sqrt{\lambda_n}}{\lambda^2 + \lambda_n} \right| \leq \frac{1}{2\epsilon}. \tag{3.2}$$

We only consider the last term, the others can then be estimated in a similar way. Let $u \in H_0^1(\Omega)$. Then

$$\begin{aligned}
&\left\| (\Delta + if(y)\partial_x)(\lambda^2 - (\Delta + if(y)\partial_x))^{-1} u \right\|_{L^2} \\
&= \left\| \sum_n \frac{\lambda_n}{\lambda^2 + \lambda_n} \langle u, f_n \rangle_{L^2} f_n \right\|_{L^2} \\
&\leq \sup_{\lambda \in S, n \in \mathbb{N}} \left| \frac{\sqrt{\lambda_n}}{\lambda^2 + \lambda_n} \right| \left\| \sum_n \sqrt{\lambda_n} \langle u, f_n \rangle_{L^2} f_n \right\|_{L^2} \\
&\leq C \left\| \sum_n \sqrt{1 + \lambda_n} \langle u, f_n \rangle_{L^2} f_n \right\|_{L^2} \quad \text{(using estimate (3.2))} \\
&= C \left\| (-\Delta - if(y)\partial_x + 1)^{1/2} u \right\|_{L^2} \\
&\leq C \|u\|_{H^1} \quad \text{(by Lemma 3.1)}.
\end{aligned}$$

■

Collecting all our results, we obtain

THEOREM 3.5. *Let $\Omega \subseteq \mathbb{R}^n$ be a bounded domain with smooth boundary. Let $H = H_0^1(\Omega) \times L^2(\Omega)$. Then the operator*

$$\mathcal{A} = \begin{pmatrix} 0 & I \\ \Delta + if(y)\partial_x & 0 \end{pmatrix} \quad \text{with} \quad \mathcal{D}(\mathcal{A}) = \left(H^2(\Omega) \cap H_0^1(\Omega)\right) \times H_0^1(\Omega)$$

and

$$f : \mathbb{R} \to \mathbb{R} \text{ with } \|f\|_\infty < (d(\Omega))^{-1}$$

generates a strongly continuous semigroup $(T(t))_{t \geq 0}$ on H and

$$s(\mathcal{A}) = \omega(T) = 0.$$

For the periodic case, we obtain

THEOREM 3.6. *Let $\Omega = (-\pi, \pi)^2 \subseteq \mathbb{R}^2$. On $H = H_{per}^1(\Omega) \times L_{per}^2(\Omega)$ the operator*

$$\mathcal{A} = \begin{pmatrix} 0 & I \\ \Delta + if(y)\partial_x & 0 \end{pmatrix} \quad \text{with} \quad \mathcal{D}(\mathcal{A}) = H_{per}^2(\Omega) \times H_{per}^1(\Omega)$$

and

$$f : (-\pi, \pi) \to \mathbb{R} \text{ with } \|f\|_\infty \leq 1$$

generates a strongly continuous semigroup $(T(t))_{t\geq 0}$ *and*

$$s(\mathcal{A}) = \omega(T) = 0.$$

Choosing $f(y) = \sin y$, we see that in Renardy's example, it is the term $\cos y\partial_x$ which destroys the equality of growth and spectral bound.

Our results suggest that symmetric lower order perturbations of the Laplacian might guarantee that the bounds remain equal. This could be the case even if the bounds do not stay equal to zero.

4 A HIGHER ORDER EQUATION

We now show that, by following Renardy's example, we can construct an operator for a fourth order differential equation where the growth bound of the generated semigroup and the spectral bound of the generator differ.

We consider the equation

$$\partial_t^2 u = -\partial_x^4 u - \partial_y^4 u - ie^{iy}\partial_x^2 u \quad , \quad (x,y) \in \mathbb{R}^2$$

where u is 2π-periodic in both x and y.

Again, we rewrite the problem in the following way.

$$\partial_t \begin{pmatrix} u \\ v \end{pmatrix} = \tilde{\mathcal{A}} \begin{pmatrix} u \\ v \end{pmatrix}$$

where $\Omega = (-\pi,\pi) \times (-\pi,\pi)$,

$$\tilde{\mathcal{A}} = \begin{pmatrix} 0 & I \\ -\partial_x^4 - \partial_y^4 - ie^{iy}\partial_x^2 & 0 \end{pmatrix}, \quad \text{and} \quad \mathcal{D}(\tilde{\mathcal{A}}) = H^4_{per}(\Omega) \times H^2_{per}(\Omega).$$

The underlying Hilbert space is $H := H^2_{per}(\Omega) \times L^2_{per}(\Omega)$ with norm

$$||(u,v)||_H := \sqrt{||u||^2_{H^2(\Omega)} + ||v||^2_{L^2(\Omega)}}.$$

4.1 Generation of a Semigroup

We first want to show that $\tilde{\mathcal{A}}$ generates a strongly continuous semigroup . Let

$$\mathcal{A} = \begin{pmatrix} 0 & I \\ -\partial_x^4 - \partial_y^4 & 0 \end{pmatrix} \text{ and } \mathcal{C} = \begin{pmatrix} 0 & 0 \\ -ie^{iy}\partial_x^2 & 0 \end{pmatrix}$$

Then $\mathcal{C}$ is bounded on H, so it is enough to show that $\mathcal{A}$ with $\mathcal{D}(\mathcal{A}) = H^4_{per}(\Omega) \times H^2_{per}(\Omega)$ generates a strongly continuous semigroup on H.

We introduce the following operators.

- Define A by $Au = -\partial_x^4 u - \partial_y^4 u$, $\mathcal{D}(A) = H^4_{per}(\Omega)$.
- Let $f(x,y) = \sum_{m,n} f_{m,n} e^{imx} e^{iny} \in L^2_{per}(\Omega)$.
  We define $Cos, Sin : \mathbb{R}^+ \to \mathcal{L}(L^2_{per}(\Omega))$ by

$$\begin{aligned} Cos(t)f(x,y) &:= \sum_{m,n} f_{m,n} \cos(t\sqrt{m^4+n^4}) e^{imx} e^{iny}, \\ Sin(t)f(x,y) &:= tf_{0,0} + \\ &\quad \sum_{m^2+n^2\neq 0} \frac{1}{\sqrt{m^4+n^4}} f_{m,n} \sin(t\sqrt{m^4+n^4}) e^{imx} e^{iny}. \end{aligned}$$

PROPOSITION 4.1. $(T(t))_{t\geq 0} = \begin{pmatrix} Cos(t) & Sin(t) \\ ASin(t) & Cos(t) \end{pmatrix}_{t\geq 0}$ *is a strongly continuous semigroup on* H *with generator* $\mathcal{A}$.

Proof. The proof is done in the following steps.

I. $(T(t))_{t\geq 0}$ is strongly continuous.

II. $(T(t))_{t\geq 0}$ is exponentially bounded.

III. The generator of the semigroup is $\mathcal{A}$.

I.$(T(t))_{t\geq 0}$ is strongly continuous.
Let $f \in H^2_{per}(\Omega)$. For the first entry in the matrix we obtain

$$\begin{aligned} &||Cos(t)f - Cos(s)f||^2_{H^2} \\ &= \left\| \sum_{m,n} \left(\cos(t\sqrt{m^4+n^4}) - \cos(s\sqrt{m^4+n^4})\right) f_{m,n}e^{imx}e^{iny} \right\|^2_{H^2} \\ &\leq C\sum_{m,n} \left|\cos(t\sqrt{m^4+n^4}) - \cos(s\sqrt{m^4+n^4})\right|^2 (1+m^4+n^4)|f_{m,n}|^2. \end{aligned}$$

Given $\epsilon > 0$, choose N_0 sufficiently large such that

$$4C \sum_{|m|+|n|>N_0} (1+m^4+n^4)|f_{m,n}|^2 < \frac{\epsilon^2}{2}.$$

For m, n with $|m|+|n| \leq N_0$, by continuity of the cosine function, there exists $\delta_{m,n}$ such that $|t-s| < \delta_{m,n}$ implies

$$\left|\cos(t\sqrt{m^4+n^4}) - \cos(s\sqrt{m^4+n^4})\right|^2 < \frac{\epsilon^2}{8C(1+m^4+n^4)|f_{m,n}|^2 N_0^2}.$$

Let $\delta := \min \delta_{m,n}$. Then for $|t-s| < \delta$,

$$||Cos(t)f - Cos(s)f||^2_{H^2} \leq \sum_{|m|+|n|\leq N_0} \frac{\epsilon^2}{8N_0^2} + \frac{\epsilon^2}{2} \leq \epsilon^2.$$

The other entries of the matrix can be treated in the same way. This proves that $(T(t))_{t\geq 0}$ is strongly continuous.

II. $(T(t))_{t\geq 0}$ is exponentially bounded. It is easy to check that all terms are exponentially bounded.

III. The generator of the semigroup is $\mathcal{A}$.
By [2; Theorem 3.1.7][1] it remains to show that

$$R(\lambda, \mathcal{A}) = \int_0^\infty e^{-\lambda t} \begin{pmatrix} Cos(t) & Sin(t) \\ ASin(t) & Cos(t) \end{pmatrix} dt \tag{4.1}$$

[1]cf. also [2; Definition 3.1.8.]

for $Re\lambda > \omega$, where ω is the growth bound of the semigroup.
By explicitly calculating the right hand side of equation (4.1), we see that the equation holds for $Re\lambda > 0$. Thus, $\mathcal{A}$ is the generator of the semigroup $(T(t))_{t\geq 0}$, which ends the proof of the proposition. ∎

We have therefore shown that the perturbed operator $\tilde{\mathcal{A}}$ also generates a strongly continuous semigroup.

4.2 Calculating the Growth and the Spectral Bound

PROPOSITION 4.2. *For the spectral bound of $\tilde{\mathcal{A}}$, we have $s(\tilde{\mathcal{A}}) = 0$.*

Proof. Because Ω is bounded, $\tilde{\mathcal{A}}$ has compact resolvent and its spectrum consists only of eigenvalues.
Let $(u, v) \in H^4_{per}(\Omega) \times H^2_{per}(\Omega)$ and

$$\tilde{\mathcal{A}} \begin{pmatrix} u \\ v \end{pmatrix} = \lambda \begin{pmatrix} u \\ v \end{pmatrix},$$

then $v = \lambda u$ and it is enough to solve

$$\lambda^2 u + \partial_x^4 u + \partial_y^4 u + ie^{iy}\partial_x^2 u = 0.$$

With $u = \sum_{m,n} u_{m,n} e^{imx} e^{iny} \neq 0$, this implies

$$(\lambda^2 + m^4 + n^4)u_{m,n} - im^2 u_{m,n-1} = 0 \tag{4.2}$$

for all $m, n \in \mathbb{Z}$. If for $m = 0, u_{m,n} \neq 0$, then by equation (4.2), $\lambda = \pm in^2$. Else, there exists $m \neq 0$ such that $u_{m,n} \neq 0$ and

$$u_{m,n-1} = \frac{\lambda^2 + m^4 + n^4}{im^2} u_{m,n}.$$

But since $lim_{|n|\to\infty} u_{m,n} = 0$, there must exist m, n such that $\lambda^2 + m^4 + n^4 = 0$, so $\lambda \in i\mathbb{R}$.
Therefore $\sigma(\tilde{\mathcal{A}}) \subseteq i\mathbb{R}$ and $s(\tilde{\mathcal{A}}) = 0$. ∎

PROPOSITION 4.3. *The growth bound of $(T(t))_{t\geq 0}$ satisfies $\omega(T) \geq \frac{1}{2}$.*

Proof. We use the Gearhart-Prüss Theorem and show that $s_0(\tilde{\mathcal{A}}) \geq \frac{1}{2}$.
To this end, we construct a sequence λ_n such that $Re\lambda_n \to \frac{1}{2}$ and $\left\| R(\lambda_n, \tilde{\mathcal{A}}) \right\|_H \to \infty$ as $n \to \infty$.
We set

$$\lambda_n = \sqrt{-n^4 + in^2} = \sqrt[4]{n^8 + n^4}\, e^{\frac{i}{2}(\pi - arctan\frac{1}{n^2})}.$$

Then $Re\lambda_n \to \frac{1}{2}$.
Let ϵ_n be a sequence such that

- $\epsilon_n \in (0, \pi)$,
- $\epsilon_n \to 0$ for $n \to \infty$,

- $n\epsilon_n^2 \to \infty$ for $n \to \infty$.

Let $\Phi \in C_c^\infty(-1,1), \Phi \neq 0$ and define $u_n(x,y) := e^{inx}\Phi(\frac{y}{\epsilon_n})$ and $v_n(x,y) := \lambda_n u_n$. We show $\frac{\|(\lambda_n - \tilde{A})(u_n,v_n)\|_H}{\|(u_n,v_n)\|_H} \to 0$ for $n \to \infty$. This implies the desired result, since then we have that

$$\left\|R(\lambda_n, \tilde{A})\right\|_H = \sup_{y \in \mathcal{D}(\tilde{A})} \frac{\|y\|_H}{\left\|(\lambda_n - \tilde{A})y\right\|_H} \to \infty \text{ as } n \to \infty.$$

We get the following estimates.

$$\|u_n\|_{H^2}^2 \geq C_1 n^4 \epsilon_n, \tag{4.3}$$

$$\lambda_n^2 u_n + (\partial_x^4 + \partial_y^4 + ie^{iy}\partial_x^2)u_n = in^2(1 - e^{iy})u_n + e^{inx}\partial_y^4\Phi(\frac{y}{\epsilon_n}), \tag{4.4}$$

$$\left\|in^2(1 - e^{iy})u_n\right\|_{L^2}^2 \leq C_2 n^4 \epsilon_n^3, \tag{4.5}$$

$$\left\|e^{inx}\partial_y^4\Phi(\frac{y}{\epsilon_n})\right\|_{L^2}^2 = C_3 \epsilon_n^{-7}. \tag{4.6}$$

Combining (4.3), (4.4), (4.5) and (4.6), we obtain

$$\begin{aligned}
\frac{\left\|(\lambda_n - \tilde{A})(u_n, v_n)\right\|_H^2}{\|(u_n, v_n)\|_H^2} &= \frac{\left\|\lambda_n^2 u_n + (\partial_x^4 + \partial_y^4 + ie^{iy}\partial_x^2)u_n\right\|_{L^2}^2}{\|u_n\|_{H^2}^2 + \|v_n\|_{L^2}^2} \\
&\leq \frac{C_2 n^4 \epsilon_n^3 + C_3 \epsilon_n^{-7}}{C_1 n^4 \epsilon_n} \longrightarrow 0, \text{ as } n \to \infty.
\end{aligned}$$

■

We summarise our results in the following theorem.

THEOREM 4.4. *Let* $\Omega = (-\pi, \pi)^2 \subseteq \mathbb{R}^2$ *and* $H = H_{per}^2(\Omega) \times L_{per}^2(\Omega)$. *Then the operator*

$$\tilde{A} = \begin{pmatrix} 0 & I \\ -\partial_x^4 - \partial_y^4 - ie^{iy}\partial_x^2 & 0 \end{pmatrix}$$

with

$$\mathcal{D}(\tilde{A}) = H_{per}^4(\Omega) \times H_{per}^2(\Omega)$$

generates a strongly continuous semigroup $(T(t))_{t \geq 0}$ *and*

$$\omega(T) \geq \frac{1}{2} > s(\tilde{A}) = 0.$$

This gives us another example arising from a partial differential equation where the growth and spectral bound differ.

BIBLIOGRAPHY

1. W. Arendt, *Spectrum and growth of positive semigroups*, Evolution Equations (G. Ferreyra, G. R. Goldstein, and F. Neubrander, eds.), Lecture Notes in Pure and Appl. Math., vol. 168, Marcel Dekker, 1995, pp. 21–28.

2. W. Arendt, C. J. K. Batty, M. Hieber, and F. Neubrander, *Vector valued Laplace transforms and Cauchy problems*, Birkhäuser, 2001.

3. K.-J. Engel and R. Nagel, *One-parameter semigroups for linear evolution equations*, Graduate Text in Mathematics, vol. 194, Springer, New York, 2000.

4. J. A. Goldstein, *Semigroups and second-order differential equations*, J. Func. Anal. **4** (1969), 50–70.

5. J. A. Goldstein, *On the convergence and approximation of cosine functions*, Aeq. Math. **10** (1974), 201–205.

6. J. A. Goldstein, *Semigroups of linear operators and applications*, Oxford University Press, Oxford, 1985.

7. M. Hieber, *A characterization of the growth bound of a semigroup via Fourier multipliers*, Evolution Equations and Their Applications in Physical and Life Sciences (G. Lumer and L. Weis, eds.), Lect. Notes in Pure and Appl. Math., vol. 215, Marcel Dekker, 2001, pp. 121–124.

8. Y. Latushkin and F. Räbiger, *Fourier multipliers in stability and control theory*, preprint, 2000.

9. Y. Latushkin and R. Shvydkoy, *Hyperbolicity of semigroups and Fourier multipliers*, preprint, 2000.

10. A. F. Neves, H. de Sousa Ribeiro, and O. Lopez, *On the spectrum of evolution operators generated by hyperbolic systems*, J. Func. Anal. **67** (1986), 320–344.

11. M. Renardy, *On the type of certain C_0-semigroups*, Comm. Part. Diff. Eq. **18** (1993), 1299–1307.

12. M. Renardy, *On the linear stability of hyperbolic PDEs and viscoelastic flows*, Z. Angew. Math. Phys. **45** (1994), 854–865.

13. J. M. A. M. van Neerven, *The asymptotic behaviour of a semigroup of linear operators*, Birkhäuser, 1996.

14. L. Weis, *Stability theorems for semigroups via multiplier theorems*, Differential Equations, Asymptotic Analysis and Math. Physics (M. Demuth and B.-W. Schulze, eds.), Akademie Verlag, 1997, pp. 407–411.

15. J. Zabczyk, *A note on C_0-semigroups*, Bull. Acad. Polon. Sci. Ser. Math. Astr. Phys. **23** (1975), 895–898.

A class of ordinary differential operators with jump boundary conditions

ROBERT M. KAUFFMAN

Department of Mathematics, University of Alabama at Birmingham
e-mail: kauffman@math.uab.edu

HONGKUN ZHANG

Department of Mathematics, University of Alabama at Birmingham
e-mail: Zhang@math.uab.edu

ABSTRACT. In this paper, we study a class of differential operators which are second order, matrix coefficient Schrödinger operators with infinitely many jump conditions. Our operators are more general than those occurring in dispersive billiards, in other words, dispersive geodesic flow on Riemannian manifold with boundary, but include these as a special case. In this case, the jump conditions correspond to the reflections. A derivation of the jump conditions for dispersive billiards on m dimensional Riemannian manifold with boundary is included.

1 INTRODUCTION AND MOTIVATIONS

A very important problem in ergodic theory concerns the ergodicity of geodesic flow. This paper is motivated by the question of ergodicity for geodesic flow on manifolds where the trajectories reflect off the boundary. However, this is a paper not in ergodic theory, but about the spectral theory of a class of ordinary differential operators which comes up naturally in this context. The class of operators we study is a general class of second order differential operators with jump boundary conditions.

For studying geodesic flow, a very important question concerns **hyperbolicity;** roughly speaking, this is the situation where trajectories close to a given trajectory either diverge exponentially in time from it or converge in the same way. (This actually occurs in the universal cover, not in the underlying manifold, which we shall assume to be compact.) The differential operator L we study is associated with the linearization of the flow, hence we ask whether solutions to $Lf = 0$ have the corresponding property. Also the L of this paper correspond only to good trajectories. It turns out in application that almost every trajectory is good.

Studying the flow by using differential operator theory in the case of a manifold of nonpositive curvature, where the sectional curvatures at one point are all negative, was done by Kauffman [4]; the purpose of this study is to try to obtain an analytic perspective on the critical questions of ergodic theory such as absolute continuity. In the situation we consider here, there is a boundary, and the flow reflects through the normal at the boundary. The manifold and the boundary are linked in such a way that the shape operator is positive definite at any point on the boundary. Hence, the boundary is allowed to be concave outward, if the curvature of the manifold is sufficiently negative to counteract this effect.

Our assumptions guarantee that the boundary is curved in such a fashion that the reflections cause the flow to disperse; the Jacobi fields decay (or grow) exponentially in time. The novelty of our approach is that we study this phenomenon using only differential operator theory; no dynamical information is involved. As a consequence, of course, we only obtain the basic results of exponential decay and growth of the solutions to the Jacobi equation; the subtle aspects of the flow which carry the dynamical information are not analyzed.

In particular, ergodicity of geodesic flow in this situation depends upon the fine structure of the weakly stable manifolds associated with singular trajectories. The differential operators associated with these trajectories are not studied in this paper; our hypotheses do not allow "grazing" trajectories, or trajectories which hit a nonsmooth portion of the boundary. Hence we are far from being able to understand the problem of ergodicity. A bit of the history of this problem is in order, though it is very sketchy, as we are not experts in this area.

Sinai, in 1970, [9] showed ergodicity in $\mathbb{R}^2$, for flat regions with a concave boundary (so that the center of curvature lies outside the region). The proof of this result was thought for some time to extend to $\mathbb{R}^m$, but it was recently found that a technical lemma used in the proof fails even for $m = 3$, which makes the proof of the general result incomplete. The proof was completed for manifold with algebraic boundaries in the recent work of Balint, Chernov, Szasz and I. P. Toth. A list of references to the fundamental papers we know about is included, though an exact analysis of the contributions of the many leading mathematicians who have worked on these problems is beyond our competence.

Our class of differential operators are second order, matrix coefficient Schrödinger operators with infinitely many jump conditions. Our operators are more general than those occurring in geodesic flow, but include these as a special case. In this case, the jump conditions correspond to the reflections.

The derivation of the jump conditions for the differential operator is included. This information is not new. For example, the two dimensional Euclidian version, in a slightly different form, appeared in 1997 in [11]. It seems clear that all the information in our derivation is implicit in the fundamental papers in the area, such as that of Chernov and Sinai [2], [3]. After we finished the first version of this paper, we found out that Wojtkowski gave an elegant proof in his paper [12] which is to appear. But for completeness we nevertheless give a self-contained elementary version here, since our purpose is to motivate the class of ordinary differential operators we study.

As differential operators, the operators we study are a bit remarkable. For example, though in the flat case the differential expression is just

$$l(Y) = -Y'',$$

where Y is a vector valued function with range in $\mathbb{R}^m$ and satisfies the jump boundary conditions, nevertheless there is an $m-$dimensional L^2 solution space where all solutions decay exponentially, and a complementary space of the same dimension where all the solutions grow exponentially. (This operator corresponds to variations of the trajectories which are perpendicular to the flow.)

In summary, this is a paper in differential operator theory, motivated by ergodic theory. While the results of this paper are preliminary, and they give some hope that the structure of differential operator theory may be used later to analyze some of the problems of interest in dynamics.

2 ORDINARY DIFFERENTIAL OPERATORS WITH JUMP BOUNDARY CONDITIONS

- Throughout this section, we will let $0 = t_0 < t_1 < ... < t_i < ...$ be a partition of $[0, \infty)$. Denote $I_i = [t_i, t_{i+1})$, for all $i \geq 0$. Assume that there exists a constant $0 < b < +\infty$, such that

$$b = \sup_{i \geq 0} (t_i - t_{i-1}). \tag{2.1}$$

  Denote $u \cdot v$ as the inner product of vectors $u, v \in \mathbb{C}^m$. If I is a real interval, denote by $L^2(I)$ the Hilbert space of all measurable functions f from I into $\mathbb{C}^m$, such that

$$\int_I |f|^2 \, dt < \infty,$$

  with inner product

$$[f, g] = \int_I (f \cdot g) \, dt.$$

- In the application to billiards, the hypothesis that the t_i go to infinity with i rules out trajectories which get stuck in a corner (these have measure zero in applications). The assumption about b rules out trajectories which take a long time to hit the boundary; this is a hypothesis which might not hold in some reasonable applications, but which holds for example when the manifold with boundary (the billiard table) is compact.

DEFINITION 2.1. Let l be the differential expression on an interval I defined by

$$lf = -\frac{d^2 f}{dt^2} + A(t)f, \tag{2.2}$$

where $A(t)$ is a positive semi-definite symmetric real $m \times m$ matrix for each $t \in \mathbb{R}^+$ and a continuous function of t.

DEFINITION 2.2. Let L be a differential operator defined by

$$Lf(t) = lf(t), \text{ for all } t \in I_i \cap I, i \geq 0 \tag{2.3}$$

satisfying the following boundary conditions : for any $i \in \mathbb{N}$

$$\begin{cases} f(t_i) = A_i f(t_i - 0), \\ f'(t_i) = C_i f(t_i - 0) + D_i f'(t_i - 0)), \end{cases} \tag{2.4}$$

where f is a function from I into $\mathbb{C}^m$, and A_i, C_i are real $m \times m$ matrices such that for each $i \in \mathbb{N}$ the following holds:

(2.3.1).$D_i^* A_i = I$, where I is a m by m identity matrix;
(2.3.2).$A_i^* C_i$ is positive definite on the real vectors; and
there exists a universal constant c, such that
for any unit vector $u \in \mathbb{R}^m$, $(A_i^* C_i u, u) \geq c$.

- Note that a function f from I into $\mathbb{C}^m$ satisfies these boundary conditions if and only if its real and imaginary part do. Note furthermore that any solution to $Lf = 0$ can therefore be broken up into real and imaginary parts, both of which are solutions.

- The hypotheses above can be improved in the sense that they only hold for a subsequence $\{i_k\} \subset \{i\}$. However , the cost of this is to complicate the arguments. It seems best to leave the study of the best possible version of these hypotheses for later work.

- (**Hypothesis A**) In what follows, we assume the following three alternatives, in the definition of L above:

  1. for all i,

  $$||A_i|| \leq 1, \text{ and } 0 < \lim_{n\to\infty} \sup_{i \geq n} (t_i - t_{i-1}) = b_1 < \infty;$$

  2. for all i,

  $$||A_i^{-1}|| \leq 1, \text{ and } 0 < \lim_{n\to\infty} \sup_{i \geq n} (t_i - t_{i-1}) = b_1 < \infty;$$

  3. for all i,

  $$A_i \text{ is unitary in } \mathbb{C}^m.$$

- In the application to billiards, Alternative (3) of Hypothesis A above holds.

THEOREM 2.3. *Let $I = [0, \infty)$.*

(1)Assume that Alternative (1) or (3) of Hypothesis A above holds. Then a real solution to $Lf = 0$ is bounded on $\mathbb{R}^+$ if and only if its magnitude is non-increasing. If f is a bounded solution whose amplitude is not a constant, then there exists a universal constant κ, which depends only upon c and b, such that for all $\lambda \in (0, \frac{\kappa}{2})$ and for all $t \in [t_1, \infty)$:

$$|f(t)| \leq e^{-\lambda(t-t_1)} |f(t_1)|.$$

(2)Assume that Alternative (2) or (3) of Hypothesis A above holds. If f is a solution to $Lf = 0$ whose amplitude is not bounded, then there exists a universal constant τ, which depends only upon c and b, such that for all $\lambda \in (0, \frac{\tau}{2})$, there exists a positive real number T, such that for all $t \in (T, \infty)$:

$$|f(t)| \geq e^{\lambda(t-T)}|f(T)|.$$

Proof. Let $t \in I_i, i \geq 0$, unless otherwise specified.

Assume that Alternative (1) or (3) of Hypothesis A above holds. Since for any real solution f to $Lf = 0$,

$$(f \cdot f)'(t) = 2f'(t) \cdot f(t);$$

$$(f \cdot f)''(t) = 2|f'|^2(t) + 2(A(t)f(t) \cdot f(t)) \geq 0.$$

By assumption

$$(f \cdot f')(t_i) - (f \cdot f')(t_i - 0) = (A_i^* C_i f \cdot f)(t_i - 0) \geq c|f|^2(t_i - 0),$$

so

$$(f \cdot f)'(t_i) \geq (f \cdot f)'(t_i - 0).$$

By the preceding, we see that $(f \cdot f)'$ is nondecreasing on $[0, \infty)$. If Alternative (3) of Hypothesis A holds, the nondecreasing nature of $(f \cdot f')$ shows directly that f is bounded if and only if its magnitude is non-increasing, since $|f|$ is continuous on $\mathbb{R}^+$. We need only consider the situation of Alternative (1).

Suppose that for some $s > 0$,

$$(f \cdot f)'(s) = \varepsilon > 0.$$

Then for any $i > 1$, on each interval I_{i-1} with length greater than (say) $b_1/2$ and $t_i > s$,

$$|f|^2(t_i - 0) \geq \varepsilon b_1.$$

Then, by the preceding, it follows that for $t > t_i$

$$(f \cdot f')(t_i) - (f \cdot f')(t_i - 0) \geq c\varepsilon^2 b_1^2.$$

Iterating this argument, we see that

$$\lim_{t \to \infty} f \cdot f'(t) = \infty.$$

But then, it follows that $|f|^2$ cannot be bounded, because for every $n \in \mathbb{N}$, on all intervals $I_{i_n - 1}$ where $f \cdot f'(t) \geq n$, with length greater than δ,

$$|f|^2(t_{i_n} - 0) \geq n\delta.$$

We have seen that, if f is bounded, then for all $t > 0$, $f \cdot f'(t) \leq 0$.

Since $(f \cdot f)(t_i) = (A_i^* A_i f \cdot f)(t_i - 0) \leq (f \cdot f)(t_i - 0)$, therefore f is bounded if and only if its magnitude is non-increasing.

Note that

$$\begin{aligned} |f|^2(t_{i+1}-0) &= |f|^2(t_i) + 2\int_{t_i}^{t_{i+1}} f\cdot f'(t)\,dt \\ &\geq |f|^2(t_i) + 2(t_{i+1}-t_i)\, f\cdot f'(t_i). \end{aligned}$$

Moreover, again for f bounded, it now follows that for i large,

$$\begin{aligned} 0 &\geq (f\cdot f)'(t_{i+1}-0) + c|f|^2(t_{i+1}-0) \\ &\geq (f\cdot f)'(t_i) + 2bc(f\cdot f)'(t_i) + c|f|^2(t_i) \\ &= (1+2bc)(f\cdot f)'(t_i) + c|f|^2(t_i). \end{aligned}$$

This implies that

$$(f\cdot f)'(t_i) \leq -\frac{c}{1+2bc}|f|^2(t_i).$$

Let

$$\kappa_1 = \frac{c}{1+2bc}$$

then we have proved that for any $i \in \mathbb{N}$

$$(f\cdot f)'(t_i) \leq -\kappa_1 |f|^2(t_i),$$

and

$$(f\cdot f)'(t_i-0) \leq -\kappa_1 |f|^2(t_i-0).$$

For $t \in I_i$,

$$\begin{aligned} |(f\cdot f)'(t)| &\geq |(f\cdot f)'|(t_{i+1}-0) \\ &\geq \kappa_1 |f|^2(t_{i+1}-0) \\ &= \kappa_1\Big[|f|^2(t) + \int_t^{t_{i+1}} (f\cdot f)'(s)ds\Big] \\ &= \kappa_1\Big[\frac{1}{2}|f|^2(t) + \frac{1}{2}|f|^2(t) + \int_t^{t_{i+1}} (f\cdot f)'(s)ds\Big]. \end{aligned}$$

If

$$\frac{1}{2}|f|^2(t) + \int_t^{t_{i+1}} (f\cdot f)'(s)ds \geq 0,$$

then

$$|(f\cdot f)'(t)| \geq -\frac{\kappa_1}{2}|f|^2(t).$$

But on the other hand, if

$$\frac{1}{2}|f|^2(t) + \int_t^{t_{i+1}} (f\cdot f)'(s)ds \leq 0,$$

since $t_{i+1} - t \leq b$, so we have

$$|(f \cdot f)'(t)| \geq -\frac{1}{2b}|f|^2(t).$$

Let $\kappa = min\{\kappa_1/2, 1/2b\}$, then we get for all $t \in I_i$, $i > 0$:

$$(f \cdot f)'(t) \leq -\kappa|f|^2(t).$$

By Gronwall's inequality, we know that on each I_i, $i \in \mathbb{N}$:

$$|f(t)|^2 \leq e^{-\lambda(t-t_i)}|f(t_i)|^2$$

hold for any $\lambda \in (0, \kappa)$. So for any $t > t_1$, there exists $n \in \mathbb{N}$, such that $t_n \leq t < t_{n+1}$. Thus we have

$$|f|^2(t) \leq e^{-\lambda(t-t_i)}|f(t_i)|^2 \leq e^{-\lambda(t-t_i)}|f(t_i - 0)|^2 \leq e^{-\lambda(t-t_1)}|f(t_1)|^2.$$

Therefore $|f(t)| \leq e^{-\lambda(t-t_1)}|f(t_1)|$ hold for any $\lambda \in (0, \kappa/2)$and any $t \in [t_1, +\infty)$. The conclusion about unbounded solutions follows from a completely parallel argument. In this case, of course, $\left(|f|^2\right)'$ is eventually positive. But T may not be the same as t_1, since the unbounded solutions may not be nondecreasing from the beginning. ■

- Below, $AC_{loc}(I)$ denotes the set of locally absolutely continuous complex valued functions on I.

- (**Hypotheses B**) We now need some additional hypotheses on the matrices A_i, C_i, in order to be able to calculate the dimension of the square integrable solution space to the equation

  $$Lf = 0.$$

  These hypotheses are needed to be able to make a self-adjoint operator out of L. They are as follows: for any $i \in \mathbb{N}$

  1. There exists a universal constant $M > 0$, such that

     $$||A_i|| \leq M$$

  2. $$A_i^* C_i = C_i^* A_i.$$

DEFINITION 2.4. Let L be defined as in (2.2) and (2.3) on interval $I = [0, \infty)$ or $I = [\alpha, \beta]$. Let D_M be the set of all $f \in L^2(I)$, such that $Lf \in L^2(I)$, $f' \in AC(I_i) \cap I$, and f satisfies (2.4), for all $i \geq 0$. Then the maximal operator L_M associated with L is defined as Lf for all f in this domain. Let $\hat{D}$ to be the set of all $f \in D_M$, such that $f \in C^\infty(I_i) \cap I$, for all $i \geq 0$ and f vanishes in a neighborhood of both "end points" of I and supp(f) is compact. Denote $\hat{L}$ to be L restricting on $\hat{D}$, then we define the minimal operator L_0 to be the smallest closed operator in $L^2(I)$ which extends $\hat{L}$. (It is clear that this operator exists).

- Let [,] denote the inner product in $L^2(I)$.

LEMMA 2.5. *If L is defined as in Definition 2.2 on* $I = [0, \infty)$, *then*

$$L_0 \subseteq L_M$$

and for any f in the domain of L_0 , *we have* $f(0) = 0, f'(0) = 0$. *If I is compact, say* $I = [0, T] \subset [0, \infty)$, *then also* $f(T) = f'(T) = 0$.

Proof. Let $\{f_n\}$ be a Cauchy sequence of functions from $\hat{D}$ converging to f with $L_0 f_n$ converging to $L_0 f$ in $L^2(I)$. It follows from the variation of parameters formula in the theory of ordinary linear differential equations that, on any compact interval $[\alpha, \beta] \subseteq I_i, t \geq 0$, f_n and f_n' both converge uniformly. Therefore f_n'' also converges in each $L^2(I_i)$, so f_n'' converges in $L^2(I)$. It follows that f' is absolutely continuous on each I_i and $f'(0) = 0$. If $I = [0, T]$, then also $f'(T) = 0$. since $f_n^{(i)}$ converges to $f^{(i)}$ for $0 \leq i \leq 2$, we see that $L_0 f = Lf$, Lf is in L^2 and so $f \in D_M$. ∎

DEFINITION 2.6. A densely defined operator T on a Hilbert space is said to be symmetric if $[Tf, g] = [f, Tg]$ for all f, g in the domain of T.

LEMMA 2.7. *If Hypothesis B holds,* L_0 *is a symmetric operator.*

Proof. For any f, g in the domain of L_0, by Lemma 2.5, we know that $f(0) = f'(0) = 0$, and $g(0) = g'(0) = 0$. Without loss of generality, we may assume that f and g are real vector valued functions. By hypotheses on the boundary conditions (2.3), we get from a somewhat tedious but elementary calculation

$$\begin{aligned}[Lf, g] - [f, Lg] &= -\int_0^\infty (f'' \cdot g - f \cdot g'')(t)dt + [Af, g] - [f, Ag] \\ &= \sum_{i=1}^{\infty} [(f' \cdot g - f \cdot g')(t_i) - (f' \cdot g - f \cdot g')(t_i - 0)] \\ &= 0.\end{aligned}$$

Thus by definition, L_0 is symmetric. ∎

THEOREM 2.8. *If L is defined as in Definition 2.2, for all f in the domain of* L_0, $[Lf, f] \geq \lambda \|f\|^2$, *where* λ *is a universal positive constant depending only on b and c, where b and c are defined as in (2.1) and (2.3.2).*

Proof. Since L is symmetric, and takes real vector valued functions to real vector valued functions, if f and g are real we have

$$[L(f + ig), f + ig] = [Lf, f] + [Lg, g].$$

Hence, we may restrict ourselves to real vector valued functions f.

Let $d = \max\{M^2/c, b\}$, where M is defined as in Hypotheses B. Notice that for all $t \geq 0$, there exists a $i \in \mathbb{N}$, such that $t \in [t_i, t_{i+1}]$.

$$f(t) = f(t_i) + \int_{t_i}^{t} f'(s)ds$$

so

$$\frac{1}{2}|f(t)|^2 \leq |f(t_i)|^2 + |\int_{t_i}^{t} f'(s)ds|^2$$

$$\leq |f(t_i)|^2 + (t-t_i)\int_{t_i}^{t} |f'(s)|^2 ds$$

$$\leq |f(t_i)|^2 + d\int_{t_i}^{t_{i+1}} |f'(s)|^2 ds.$$

Thus

$$\frac{1}{2d^2}\int_{t_i}^{t_{i+1}} |f(s)|^2 ds \leq \frac{1}{d}|f(t_i)|^2 + \int_{t_i}^{t_{i+1}} |f'(s)|^2 ds$$

$$= \frac{1}{d}|f(t_i)|^2 + \int_{t_i}^{t_{i+1}} |f'(s)|^2 ds.$$

By assumption

$$\begin{aligned}&(f\cdot f')(t_i) - (f\cdot f')(t_i - 0)\\ &= (A_i^* C_i f(t_i - 0)\cdot f\,(t_i - 0))\\ &\geq c|f|^2(t_i - 0).\end{aligned}$$

Combining all these facts together, we get:

$$[L_0 f, f] = [f', f'] + \sum_{i=1}^{\infty}(f(t_i)\cdot f'(t_i) - f(t_i - 0)\cdot f'(t_i - 0)) + [Af, f]$$

$$\geq [f', f'] + \sum_{i=1}^{\infty} c|f(t_i - 0)|^2$$

$$\geq \sum_{i=1}^{\infty}(\int_{t_i}^{t_{i+1}} |f'(s)|^2 ds + 1/d|f(t_i)|^2) + \sum_{i=1}^{\infty}(c - M^2/d)|f(t_i - 0)|^2$$

$$\geq 1/2d^2[f,f] + \sum_{i=1}^{\infty}(c - M^2/d)|f(t_i - 0)|^2.$$

By assumption, since $c \geq M^2/d$, let $\lambda = 1/2d^2$, then we get the desired result. ∎

- In the following three Lemmas, we will let L to be defined as in Definition 2.2, but choose $I = [\alpha, \beta] \subset [0, \infty)$,

LEMMA 2.9. *For $h \in L^2([\alpha, \beta])$, the equation $Lf = h$ has a solution f in the domain of L_0 if and only if h is orthogonal to all solutions of $Lg = 0$.*

Proof. If for $h \in L^2(I)$, the equation $Lf = h$ has a solution $f \in D_0$, then there exists $k, n \in \mathbb{N}$, such that

$$t_n \leq \beta < t_{n+1} \text{ and } t_k < \alpha \leq t_{k+1}.$$

For any $g \in D_M$, such that $Lg = 0$, the following hold:

$$\begin{aligned}
[h, g] &= [L_0 f, g] \\
&= \int_\alpha^\beta -f'' \cdot g + A(t) f \cdot g dt \\
&= [f', g'] + [f, Ag] + \sum_{i=k+1}^{n} (f(t_i) \cdot g'(t_i) - f(t_{i-0}) \cdot g'(t_i - 0)) + f(\beta) \cdot g'(\beta) \\
&= [f, Lg] \\
&= 0.
\end{aligned}$$

Conversely, if h is orthogonal to all solutions of $Lg = 0$, choose f such that $Lf = h$ and $f(\alpha) = f'(\alpha) = 0$. We need to show that $f(\beta) = f'(\beta) = 0$. Since using the same calculation as in the proof of Lemma 2.7,

$$[Lf, g] - [f, Lg] = f(\beta) \cdot g'(\beta) - f'(\beta) \cdot g(\beta).$$

On the other hand,

$$[Lf, g] - [f, Lg] = [Lf, g] = [h, g] = 0.$$

Thus we get that $f(\beta) \cdot g'(\beta) - f'(\beta) \cdot g(\beta) = 0$ for any g, such that $Lg = 0$.
Choose g, s.t. $g(\beta) \neq 0$, but $g'(\beta) = 0$, then we get $f'(\beta) = 0$. Similarly, we can get $f(\beta) = 0$. ∎

LEMMA 2.10. *If L is defined as in Definition 2.2, and $I = [\alpha, \beta] \subset [0, \infty)$. Let T denote the restriction of L_M to the functions which vanish along with their first derivative at α and β. Then $T = L_0$.*

Proof. By Lemma 2.9, the range of T is orthogonal complement in $L^2[\alpha, \beta]$ of the m dimensional solution space of $Lf = 0$. So $\hat{D} \cap$ range T is dense in range T. However, if $Tf = g$ and g is in $\hat{D}$ then f vanishes in a neighborhood of each end point of I. since $g \in C^\infty(I_i)$, it follows that f is in $\hat{D}$. For $Tf = g$, then there is a sequence g_n in $\hat{D} \cap$ range T with g_n converging to g in $L^2(I)$. Let $Tf_n = g_n$, then f_n is in $\hat{D}$, and it is clear that f_n converges to f and f'_n converges to f'. Thus $T \subseteq L_0$. By Lemma 2.5, we also have $T \supset L_0$. This complete the proof. ∎

LEMMA 2.11. *If L is defined as in Definition 2.2, and $I = [\alpha, \beta] \subset [0, \infty)$. Then*

$$L_0^* = L_M$$

Proof. It follows from Lemma 2.5, and integration by parts that $(L_0)^* \supseteq L_M$.
Let f be in the domain of L_0^*. From the variation formula it follows that L_M is surjective since I is compact, so there is a vector g in D_M such that

$$L_M g = L_0^* f.$$

Thus

$$L_0^*(f-g)=0,$$

so that if u is in the range of L_0 with $L_0 v = u$, we have

$$[f-g,u]=[f-g,L_0 v]=[L_0^*(f-g),v]=0.$$

Therefore $f-g$ is orthogonal to the range of L_0 which is the same as range T. But by Lemma 2.9, $f-g$ is in the null space of L_M. Hence $f \in D_M$ and $L_M f = L_0^* f$. ∎

THEOREM 2.12. *Suppose for* L *defined as in Definition 2.2,* $I=[0,\infty)$. *Then*

$$L_0^* = L_M$$

Proof. Since L_0 is the smallest closed extension of $\hat{L}$, it follows that

$$L_0^* = \ddot{L}^*.$$

Since for any $f \in \hat{D}$ and $g \in D_M$,

$$[\hat{L}f,g]=[f,L_M g].$$

It follows that

$$L_M \subseteq \hat{L}^* = L_0^*.$$

If g is in the domain of L_0^*, then on any compact subinterval $[\alpha,\beta]$ of I, we have

$$[Lf,g]=[f,L_0^* g], \forall f \in \hat{D}.$$

Thus the restriction of g to $[\alpha,\beta]$ is in the domain of R^*, where R is the restriction of $\hat{L}$ to $L^2[\alpha,\beta]$. But by Lemma 2.11, the restriction of $L_0^* g$ to $[\alpha,\beta]$ must agree with $L_M g$ on $[\alpha,\beta]$. Since $[\alpha,\beta]$ is arbitrary, the theorem is proved. ∎

DEFINITION 2.13. With L as in Definition 2.2, let $Q(f,g) = [Lf,g]$, for all $f,g \in \hat{D}$.

- It follows from operator theory that the semi-bounded symmetric operator L_0 has equal deficiency indices, and therefore by Von Neumann's theorem, such an operator always has self-adjoint extensions. There is a distinguished extension H, called the **Friedrichs extension,** which is obtained from the quadratic form associated to L_0.

LEMMA 2.14. *Q is a closable quadratic form and its closure $\hat{Q}$ is the quadratic form of a unique self-adjoint operator H defined by*

$$D(H)=\left\{f \in L^2([0,\infty)) \,:\, |Q(f,g)| \le C_f[g,g],\ \forall g \in \hat{D},\ \textit{where } C_f \textit{ is a constant}\right\}$$

and

$$Hf = L_M f,\ \forall f \in D(H).$$

Furthermore,

$$[Hf,f] \ge \lambda[f,f],\ \forall f \in D(H),$$

where λ is defined as in Theorem 2.8.

Proof. This is a form of the definition of the Friedrichs extension: for the semi-bounded symmetric operator L_0 and the closable quadratic form Q, the restriction of L_0^* to the domain of the closure of Q is in fact the Friedrichs extension. This form is clearly closable. The last inequality follows from the construction of the Friedrichs extension H, together with Lemma 2.8. ∎

THEOREM 2.15. *There exist exactly m linearly independent bounded solutions to the equation $Lf = 0$, where m is the dimension of the underlying vector space.*

Proof. It is sufficient to consider real solutions.

Let f be any real solution of $Lf = 0$. For any $i \in \mathbb{N}$,

$$(f \cdot f)'(t_i) = (f \cdot f)'(t_i - 0) + c(f \cdot f)(t_i - 0) \geq (f \cdot f)'(t_i - 0),$$

and

$$(f \cdot f)''(t) = 2|f'|^2(t) + 2(A(t)f(t) \cdot f(t)) \geq 0.$$

So $(f \cdot f)'$ is nondecreasing on $[0, \infty)$.

Choose f, such that $f(0) = 0, f'(0) \neq 0$. Then

$$(f \cdot f)''(0) \geq 2(f' \cdot f')(0) > 0,$$

and for any $i \in \mathbb{N}$,

$$(f \cdot f)'(t_i) \geq c(f \cdot f)(t_i - 0) > 0.$$

So

$$(f \cdot f)'(t) > 0, \forall t \in (0, \infty).$$

Thus we know that those solutions of $Lf = 0$ satisfying $f(0) = 0$ and $f'(0) \neq 0$ are unbounded solutions.

Let $Ker(L)$ be the null space of L, then $\dim(Ker(L)) \leq 2m$. The set

$$\{f \in Ker(L) | f(0) = 0, f'(0) \neq 0\}$$

is m dimensional, which means that there are at most m linearly independent bounded solutions for $Lf = 0$.

If H is the Friedrichs extension of L_0, then by Lemma 2.14, H is self-adjoint and positive definite. Also for any f in $D(H)$, $f(0) = 0$. If we choose m compactly supported piecewise smooth functions f_k, $k = 1, ..., m$, such that the space spanned by $\{f_k(0) | k = 1, 2, ..., m\}$ has dimension m, then $\{f_k | k = 1, 2, ..., m\}$ are linearly independent mod $D(H)$. It is clear that $f_k \in D_M$ for $k = 1, 2, ..., m$. This forces the Fredholm index to increase by m. Since H is already surjective, this produces an m dimensional L^2 solution space to $Lf = 0$. So $\dim(Ker(L_M)) \geq m$. But we already showed that $\dim(Ker(L_M)) \leq m$, so $\dim(Ker(L_M)) = m$. ∎

THEOREM 2.16. *The Friedrichs extension H above is the restriction T of the maximal operator L_M to*

$$\{f \in D_M \ : \ f(0) = 0\}.$$

Proof. By Lemma 2.14, H is positive definite and self-adjoint , therefore it is surjective. It is clearly contained in T. But the elementary theory of Fredholm operators shows that an m dimensional extension of H has an m dimensional null space. Thus by the above theorem, L_M has exactly an m dimensional null space. Hence every element of D_M differs from an element of the domain of H by a smooth compactly supported function which does not vanish at zero. In particular, since

$$[Hf, f] = [f', f'] + [Af, f] + \sum_{i=1}^{\infty}((f \cdot f')(t_i) - (f \cdot f')(t_i - 0)) + (f \cdot f')(0),$$

it follows from integration by parts that the same is true for T. Since T contains H, the two operators are the same. ■

3 THE JACOBI FIELD AT THE REFLECTION POINTS

- As an application, we can deal with the Jacobi field along a piecewise smooth geodesic using the above theorems. Since this is basically a paper in differential equations, we review some basic differential geometry.

  Let M be a complete, connected smooth m-dimensional Riemannian manifold with all sectional curvatures non-positive. Let Q be a bounded, smooth, open submanifold in M with boundary ∂Q. Assume ∂Q is a finite union of smooth compact submanifolds of codimension one, $\partial Q = \Gamma_1 \cup \cdots \cup \Gamma_r$, $r \geq 1$. Let the set

  $$\Gamma^* = \cup_{i \neq j}(\Gamma_i \cap \Gamma_j)$$

  be a finite union of smooth compact submanifolds of codimension ≥ 2. This set includes all the corner points of the boundary ∂Q. We will call it the singular part of ∂Q. Denote $B = (\partial Q) \setminus \Gamma^*$. The points $q \in B$ are said to be regular. If a geodesic hits the singular set Γ^* (a corner point of the boundary), its further trajectory is not defined. If a geodesic hits the boundary at a regular point, it is continued by reflection through the normal.

  For each point $p \in M$, the tangent space to M at p will be denoted by T_pM; and the tangent bundle, that is the union of all the tangent spaces of M endowed with its natural differentiable structure, will be denoted by

  $$TM = \{(p, v) : p \in M, v \in T_pM\}.$$

  For $u, v \in T_pM$, we also use the notation $\langle u, v\rangle$ for the inner product. Then we may consider the norm $|u|$ of a tangent vector u.

  The set $U_pM = \{u \in T_pM; |u| = 1\}$ of unit tangent vectors at p forms the unit sphere in T_pM, and the set

  $$UM = \bigcup_{p \in M} U_pM$$

  of all unit tangent vectors is a submanifold of TM of codimension 1, called the unit tangent bundle of M.

DEFINITION 3.1. A vector field V on M is a smooth function that assigns to each point $p \in M$ a tangent vector $V(p)$ to M at p.

- Let $\aleph(M)$ denote the set of all vector fields on M.

DEFINITION 3.2. A connection D on a smooth manifold M is a function $D : \aleph(M) \times \aleph(M) \mapsto \aleph(M)$ such that

(1) $D_V W$ is $C^\infty(M)$ linear in V,
(2) $D_V W$ is $\mathbb{R}$ linear in W,
(3) $D_V(fW) = (Vf)W + fD_V W$ for $f \in C^\infty(M)$.

$D_V W$ is called the covariant derivative of W with respect to V for the connection D.

LEMMA 3.3. *On a Riemannian manifold M there is a unique connection D such that*

(4) $[V, W] = D_V W - D_W V$, *and*
(5) $X\langle V, W\rangle = \langle D_X V, W\rangle + \langle V, D_X W\rangle$,

for all $X, V, W \in \aleph(M)$. D is called the Levi-Civita connection of M, and is characterized by the Koszul formula

$$2\langle D_V W, X\rangle = V\langle W, X\rangle + W\langle X, V\rangle - X\langle V, W\rangle - \langle V, [W, X]\rangle + \langle W, [X, V]\rangle + \langle X, [V, W]\rangle.$$

DEFINITION 3.4. The curvature R of a Riemannian manifold M is a correspondence that associates to every pair X, Y in $\aleph(M)$ a mapping $R(X, Y) : \aleph(M) \mapsto \aleph(M)$ given by

$$R(X, Y)Z = D_Y D_X Z - D_X D_Y Z + D_{[X,Y]}Z, \; Z \in \aleph(M),$$

where D is the Riemannian connection of M.

- By definition, we know that $R(X, Y)Z$ is a multi-linear transformation taking the ordered triples in the tangent space $T_p(M)$ into $T_p(M)$ for each p.

- Let $\gamma : I \mapsto M$ be a parameterized curve in M, with $V : I \mapsto \mathbb{R}^m$ a vector field along γ tangent to M. We say a vector field V along γ is *parallel* if $D_{\gamma'} V \equiv 0$. The fundamental existence and uniqueness theorem for the linear ordinary differential equations gives:

LEMMA 3.5. *For a curve $\gamma : I \mapsto M$, let $a \in I$ and $y \in T_{\gamma(a)}M$. Then there is a unique parallel vector field Y on γ such that $Y(a) = y$.*

- In the notation of the above Lemma, if $b \in I$ then the function

$$P = P_a^b(\gamma) : T_p M \mapsto T_q M$$

sending each $Y(a)$ to $Y(b)$ is called *parallel translation* along γ from $p = \gamma(a)$ to $q = \gamma(q)$. We can also express the covariant derivative in terms of the parallel translation. Namely, for $V, W \in \aleph(M)$ and a curve $\gamma(t)$ with $\gamma(0) = p$, $\gamma'(0) = V(p)$, then

$$D_{V(p)} W = \lim_{t \mapsto 0} \frac{1}{t}(P_t^0(\gamma) W(\gamma(t)) - W(p)).$$

LEMMA 3.6. *Parallel translation is a linear isometry.*

Proof. With notation as above, let $v, w \in T_pM$ correspond to the parallel vector fields V, W. Since $V + W$ is also parallel,

$$P(v+w) = (V+W)(b) = V(b) + W(b) = P(v) + P(w).$$

Similarly, $P(cv) = cP(v)$. Thus P is linear.

If $P(v) = 0$ then by the uniqueness of the parallel vector field, V can only be the identically zero vector field on γ. Hence $v = V(\gamma) = 0$. Thus P is one-to-one, and since tangent spaces to M have the same dimension, P is a linear isomorphism.

Finally, for V, W as above,

$$D_{\gamma'}\langle V, W\rangle = \langle D_{\gamma'}V, W\rangle + \langle V, D_{\gamma'}W\rangle = 0.$$

Hence $\langle V, W\rangle$ is constant, so

$$\langle P(v), P(w)\rangle = \langle V(s), W(s)\rangle = \langle V(t), W(t)\rangle = \langle v, w\rangle.$$

In general, parallel translation from p to q depends on the particular curve joining p to q. ■

DEFINITION 3.7. A smooth parameterized curve $\gamma : I \mapsto M$ on an interval I is a geodesic at $s \in I$ if $D_{\gamma'}(\gamma') = 0$ at the point s; if γ is a geodesic at t, for all $t \in I$, we say that γ is a geodesic. A broken geodesic is a piecewise smooth curve segment whose smooth subsegments are geodesics.

DEFINITION 3.8. A geodesic variation of a broken geodesic $\gamma : [0, \infty)$ on $Q \cup B$ is a two-parameter mapping

$$\psi : [0, \infty) \times (-\varepsilon, \varepsilon) \longrightarrow Q \cup B,$$

such that $\gamma(t) = \psi(t, 0)$ for all $t \geq 0$, and for any $s \in (-\varepsilon, \varepsilon)$, $\psi(t, s)$ is a broken geodesic on $Q \cup \partial B$. The variation vector field J on γ given by $J(t) = \psi_s(t, s)$ is called the Jacobi field along γ.

LEMMA 3.9. *The Jacobi field along a broken geodesic γ satisfies the Jacobi equation along any smooth segment of γ:*

$$D_{\gamma'}D_{\gamma'}J + R(\gamma'(t), J(t))\gamma'(t) = 0. \tag{3.1}$$

- A Jacobi field is determined by its initial conditions $J(0)$, $D_{\gamma'}J(0)$. Indeed, let $e_1(t), ..., e_m(t)$ be parallel, orthonormal vector fields along γ. We shall write:

$$J(t) = \sum_i f_i(t) e_i(t),$$

$$a_{i,j} = \langle R(\gamma'(t), e_i(t))\gamma'(t), e_j(t)\rangle, \; i, j = 1, ..., m.$$

Then the Jacobi equation is equivalent to the system

$$f_j''(t) + \sum_i a_{i,j}(t) f_i(t) = 0, \; j = 1, ..., m,$$

which is a linear system of the second order. Hence, given the initial conditions $J(0)$, $D_{\gamma'}J(0)$, there exist $2m$ linearly independent Jacobi fields along γ.

- It is worth noting that $\gamma'(t)$ and $t\gamma'(t)$ are Jacobi fields along γ. The first field has derivative zero and vanishes nowhere; the second field is zero if and only if $t = 0$. A Jacobi field tangent to a geodesic γ is of scant importance since it is the infinitesimal model for a family of geodesics that merely reparametrize γ. Thus in considering the jump condition of the Jacobi field, we only emphasize the case $J \perp \gamma'$. By Lemma 3.1, if we choose a parallel frame fields along γ, the Jacobi fields along γ which are perpendicular to γ' satisfies the following equation on each smooth segment of γ:

$$Y''(t) + A(t)Y(t) = 0, \tag{3.2}$$

  where $A(t)$ is the $(m-1) \times (m-1)$ matrix corresponding to reduced curvature operator, since we rule out the Jacobi fields which are parallel to γ'. Note that since M has non-positive curvature , $A(t)$ turns out to be positive semi-definite.

- For each $p \in B$, the tangent space T_pB is a nondegenerate subspace of T_pM. Hence the inner product on T_pM splits T_pM into the direct sum

$$T_pM = T_pB \oplus (T_pB)^{\perp},$$

  where $(T_pB)^{\perp}$ is the orthogonal complement of T_pB in T_pM,which is also nondegenerate and has codimension 1. If $v \in T_pM$, $p \in B$, we can write

$$v = v^T + v^N, \ v^T \in T_pB, \ v^N \in (T_pB)^{\perp}.$$

  We call v^T the tangential component of v and v^N the normal component of v. In the following of this paper, the Riemannian connection on M and B will be denoted by $\bar{D}$ and D respectively. If X,Y are local vector fields on B, and $\bar{X}, \bar{Y}$ are local extensions to M, define

$$D_XY = (\bar{D}_{\bar{X}}\bar{Y})^T,$$

  and

$$\Pi(X,Y) = (\bar{D}_{\bar{X}}\bar{Y})^N = \bar{D}_{\bar{X}}\bar{Y} - D_XY,$$

  Π is called the shape tensor (or second fundamental form tensor) of $B \subset M$. Since

$$\Pi(V,W) - \Pi(W,V) = (\bar{D}_VW - \bar{D}_WV)^N = [V,W]^N = 0,$$

  so Π is symmetric.

DEFINITION 3.10. Let N be the unit outward normal vector field on the boundary $B \subset M$. Let $p \in B$, and $v, u \in T_pB$. The operator $S_p : T_pB \to T_pB$ defined by :

$$\langle S_p(v), u\rangle = \langle \Pi(v,u), N(p)\rangle,$$

is called the shape operator of the boundary at p.

LEMMA 3.11. *If S_p is the shape operator at $p \in B$, then $S_p(v) = -D_vN(p)$, and at each point the linear operator S_p on T_pB is self-adjoint.*

Proof. Since $\langle N, N\rangle$ is constant, $\langle \bar{D}_V N, N\rangle = 0$. Hence $\bar{D}_V N$ is tangent to B for all $V \in \aleph(M)$. But if $W \in \aleph(M)$ then

$$\langle S_p(V), W\rangle = \langle \Pi(V,W), N(p)\rangle = \langle \bar{D}_V W, N(p)\rangle = -\langle \bar{D}_V N(p), W\rangle.$$

Hence $S_p(V) = -\bar{D}_V N(p) = D_V N(p)$. The symmetry of Π implies that S_p is self-adjoint. ∎

- This characterization of S_p shows how it describes the shape of B in M: S_p measures the M rate of change of N in all tangent directions, and since $\{N(p)\}^{\perp} = T_pB$, it thereby records the turning of T_pB in M as p traverses B. Given a point $p \in M$ and a two-dimensional subspace $\Gamma \subset T_pM$, let $v, w \in \Gamma$ be orthonormal vectors. Let $K_p(v,w)$ denote the sectional curvature of Γ at p on M. The relation between the sectional curvatures $\bar{K}$ of M and K of B is given by the following famous *Gauss equation*:

LEMMA 3.12. *Let $p \in B$ and let v, w be orthonormal vectors in T_pB. Then*

$$K_p(v,w) = \bar{K}_p(v,w) + \langle S_p v, v\rangle\langle S_p w, w\rangle - \langle S_p v, w\rangle^2, \tag{3.3}$$

where S_p is the shape operator of $B \subset M$.

The proof of this Lemma can be found in O'Neill [7].

(**Hypothesis H**) In what follows, we make the following assumptions:

(H1) There exist a constant $c > 0$, such that for any $u \in T_pB$,

$$\langle S_p u, u\rangle \geq c\langle u, u\rangle.$$

(H2) Let $\gamma : [0,\infty) \longrightarrow Q \cup B$ be a unit speed broken geodesic, such that γ is smooth in Q and it only reaches the boundary at regular points at a sequence of times $\{t_i\}$ converging to infinity, obeying the law of elastic reflection at each reflection point.

(H3) There exists a positive $b < \infty$, such that:

$$b = \sup_{i\in\mathbb{N}} (t_i - t_{i-1}).$$

- Note that from (H1) and Lemma 3.12, if the manifold M is nonpositively curved, we only assume that the boundary is more positively curved, though it maybe negatively curved. This hypothesis can be improved in the sense that we may only assume that the shape operator S_p is positive semidefinite for any $p \in B$, and there exists one point $q \in B$, such that S_q is positive definite. However, the cost of this is to complicate the arguments. It seems better to leave the study of the best possible version of this hypothesis for later work.

- The (H2) and (H3) of this Hypothesis corresponds physically to trajectories always escaping corners, then hitting regular points of the boundary after a length of time which is bounded above. In applications to billiards, under usual hypotheses, this assumption holds for almost every trajectory. When the geodesic reaches the boundary, it is reflected by the rule of elastic collisions.

This means as mentioned earlier that the geodesic is continued by reflection through the normal.

For $p \in M$, $v \in T_pM$, denote the reflection operator $R_v : T_pM \longrightarrow T_pM$ by

$$R_v w = w - 2\langle v, w\rangle v, \quad \forall w \in T_pM. \tag{3.4}$$

Let Y be a Jacobi field along γ. The Jacobi field is the variation vector field for a geodesic variation of γ. The variation is continuous, but not smooth. In fact the first derivative of the Jacobi field has certain jumps at the reflection points of γ. This effect is investigated in detail for m-dimensional systems in the following theorems:

THEOREM 3.13. *Assume that the broken geodesic $\gamma : [0,\infty) \mapsto Q \cup \partial B$ reflects at point $p = \gamma(T) \in B$, $T \in (0,\infty)$, and γ is smooth on $[0,T)$, satisfying the Hypothesis H. Let Y be a Jacobi field along γ which perpendicular to γ', let $\Pi(T) = \{u \in T_pM | \langle u, \gamma'(T-0)\rangle = 0\}$. Let $\hat{N}$ be the unit normal vector field on B, $n = \hat{N}(p)$. Then the following formulas give the jump of the Jacobi field and its first derivative at p.*

$$\begin{aligned} Y(T) &= R_n(Y(T-0)), \\ Y'(T) &= R_n Y'(T-0) + 2\langle \gamma'(T-0), n\rangle C(T) Y(T-0), \end{aligned}$$

where $R_n : T_pM \mapsto T_pM$ is the reflection operator defined as in (3.4), $R_nC(T)$ is a self-adjoint operator on $\Pi(T)$. Also for any unit vector $u \in \Pi(T)$,

$$\langle R_n C(T) u, u\rangle \geq c,$$

where c is defined as in (H1).

Proof. We will use the notation that for any $v \in T_pM$,$|v| = \langle v, v\rangle^{\frac{1}{2}}$.

Since $Y(T-0) \in T_pM$, so it is clear that

$$Y(T) = R_n Y(T-0).$$

Let $\psi(t,s) : [0,\infty) \times (-\varepsilon, \varepsilon) \mapsto Q \cup B$ be the geodesic variation of γ, where $\varepsilon > 0$ sufficiently small. Choose a sequence $\{s_i\} \subset (-\varepsilon,\varepsilon)$ which converges to zero, such that for each s_i, $\psi(t, s_i)$ is a broken geodesic on $Q \cup B$ and there exists $T_i \in [0,\infty)$, such that $\psi(t,s_i)$ is smooth on $[0,T_i)$ and $\psi(T_i, s_i) \in B$. Let $q_i = \psi(T_i, s_i)$. Since $p \in B$, so we can choose $i_0 \in \mathbb{N}$ big enough, such that there exists a smooth parametrized curve $c(s) : [0, s_{i_0}] \mapsto B$ connecting p and q_{i_0} in B, such that $c(0) = p$, and $c(s_i) = q_i$, for $i \geq i_0$. Let $v \in U_p(B)$ be the unit tangent vector in the plane spanned by $\gamma'(T-0)$ and $\gamma'(T)$. Let Γ be the plane spanned by n and $c'(0)$. Denote by Λ the tangent plane of B at point p spanned by v and $c'(0)$; in the case where v and $c'(0)$ are parallel, we simply let Λ denote the tangent line. Then by geometry, we have

$$c'(0) = Y(T-0) - \frac{\langle n, Y(T-0)\rangle}{\langle n, \gamma'(T-0)\rangle}\gamma'(T-0). \tag{3.5}$$

In fact we can introduce an operator $P_p : \Pi(T) \to T_pB$, such that for any $w \in \Pi(T)$,

$$P_p w = w - \frac{\langle n, w\rangle}{\langle n, \gamma'(T-0)\rangle}\gamma'(T-0). \tag{3.6}$$

Then $c'(0) = PY(T-0)$. Also $\|P_p\| \geq 1$, since for any $w \in \Pi(T)$,

$$\langle w, \gamma'(T-0)\rangle = 0.$$

Let $N(c(t))$ be the vector field along c induced by the projection of the unit normal vector field $\hat{N}$ on plane Γ.

Since γ is a geodesic, $D_{\gamma'(T-0)}\gamma'(T-0) = 0$, so by (3.5) we get

$$D_{c'(0)}\gamma'(T-0) = D_{Y(T-0)}\gamma'(T-0).$$

Thus

$$Y'(T-0) = D_{Y(T-0)}\gamma'(T-0) = D_{c'(0)}\gamma'(T-0).$$

By Lemma 3.5 and its remark, for any $i \geq i_0$,

$$s_i Y'(T-0) = P_s^0(c)\psi_t'(s_i, T-0) - \gamma'(T-0) + O(s_i^2)u_1, \tag{3.7}$$

where $u_1 \in U_pM$.

Similarly from

$$Y'(T) = D_{c'(0)}\gamma'(T),$$

we get

$$s_i Y'(T) = P_{s_i}^0(c)\psi_t'(s_i, T) - \gamma'(T) + O(s_i^2)u_2. \tag{3.8}$$

where $u_2 \in U_pM$.

Also

$$D_{c'(0)}N = \lim_{s\mapsto 0}\frac{1}{s}(P_s^0(c)N(c(s)) - n), \tag{3.9}$$

so

$$P_{s_i}^0(c)N(q_i) - n = s_i D_{c'(0)}N + O(s_i^2)u_3,$$

where u_3 is a unit tangent vector at p.

On the other hand, by Lemma 3.11 we know that

$$D_{c'(0)}N = -S_p(c'(0)).$$

Now let us try to find the jump condition for the first derivative of the Jacobi field at the reflection point. Let G_p be the operator defined on T_pB, for any $v \in T_pB$:

$$G_p v = v - \frac{\langle \gamma'(T), v\rangle}{\langle \gamma'(T), n\rangle} n. \tag{3.10}$$

By (3.7), we get

$$\begin{aligned}
&P_{s_i}^0(c)R_{N(q_i)}\psi_t'(s_i, T-0) - R_n P_{s_i}^0(c)\psi_t'(s_i, T-0)) \\
&= R_{P_{s_i}^0(c)N(q_i)}\gamma'(T-0) - R_n\gamma'(T-0) + O(s_i^2)u_4 \\
&= 2\langle n, \gamma'(T-0)\rangle n - 2\langle P_{s_i}^0(c)N(q_i), \gamma'(T-0)\rangle P_{s_i}^0(c)N(q_i) + O(s_i^2)u_4 \\
&= -2\langle s_i D_{c'(0)}N, \gamma'(T-0)\rangle n - 2\langle n, \gamma'(T-0)\rangle s_i D_{c'(0)}N + O(s_i^2)u_5 \\
&= -2\langle s_i D_{c'(0)}N, \gamma'(T)\rangle n + 2\langle n, \gamma'(T)\rangle s_i D_{c'(0)}N + O(s_i^2)u_5 \\
&= 2s_i\langle n, \gamma'(T-0)\rangle G_p S_p P_p(Y(T-0)) + O(s_i^2)u_5,
\end{aligned}$$

where $u_4, u_5 \in U_pM$. Combine all these facts together, and by (3.8), we get

$$\begin{aligned} & s_i Y'(T) \\ &= P^0_{s_i}(c) R_{N(q_i)} \psi'_t(s_i, T-0) - R_n \gamma'(T-0) + O(s_i^2) u_2 \\ &= s_i R_n Y'(T-0) + P^0_{s_i}(c) R_{N(q_i)} \psi'_t(s_i, T-0) - R_n P^0_{s_i}(c) \psi'_t(s_i, T-0) + O(s_i^2) u_6 \\ &= s_i R_n (Y'(T-0) + 2\langle n, \gamma'(T-0)\rangle G_p S_p P_p(Y(T-0)) + O(s_i^2) u_7, \end{aligned}$$

where $u_6, u_7 \in U_pM$. Let

$$C(T) = G_p S_p P_p.$$

Since for any unit vector $u \in \Pi(T)$,

$$G_p^* u = u - \frac{\langle u, n\rangle}{\langle \gamma'(T), n\rangle} \gamma'(T),$$

we have

$$G_p^* R_n u = P_p u.$$

Thus $R_n C(T)$ is a self-adjoint operator on $\Pi(T)$.

Since the shape operator S_p is positive definite on T_pB, so for any unit vector $u \in \Pi(T)$, we have

$$\langle R_n G_p S_p P_p u, u\rangle = \langle S_p P_p u, G_p^* R_n u\rangle = \langle S_p P_p u, P_p u\rangle \geq c\|P_p u\|^2 \geq c\|u\|^2.$$

Thus we get the desired result. ∎

THEOREM 3.14. *Assume that a geodesic $\gamma : [0, \infty) \mapsto Q \cup B$ reflects at points $\gamma(t_i) \in B$, for $t_i \in (0, \infty)$, $i \in \mathbb{N}$, satisfying the Hypothesis H. Let Y be a Jacobi field along γ which perpendicular to γ', let $\Pi_i = \{u \in T_{\gamma(t_i)}M | \langle u, \gamma'(t_i - 0)\rangle = 0\}$. Let N be the unit normal vector field on B, let $n_i = N(\gamma(t_i))$ for all $i \in \mathbb{N}$. Then the following formulas give the jump conditions of the Jacobi field and its first derivative at $\gamma(t_i)$.*

$$\begin{aligned} Y(t_i) &= R_i(Y(t_i - 0)), \\ Y'(t_i) &= R_i Y'(t_i - 0) + 2\langle \gamma'(t_i - 0), n_i\rangle C(t_i) Y(t_i - 0), \end{aligned} \tag{3.11}$$

where $R_i = R_{n_i}$ is the reflection operator defined as in (3.4), and $C(t_i) = G_i S_i P_i$. S_i is the shape operator at $\gamma(t_i)$, and P_i, G_i are the operators defined as in (3.6) and (3.10). For any $i \in \mathbb{N}$, $R_i C(t_i)$ is a self-adjoint operator on Π_i, and for any unit vector $u \in \Pi_i$,

$$\langle R_i C(t_i) u, u\rangle \geq c,$$

where c is given in the assumption (H1).

THEOREM 3.15. *Assume that a geodesic $\gamma : [0, \infty) \mapsto Q \cup B$, obeys the Hypothesis H. Let N be the unit normal vector field on B, and $n_i = N(\gamma(t_i))$ for all $i \in \mathbb{N}$. Let $\Pi_i = \{u \in T_{\gamma(t_i)}M | \langle u, \gamma'(t_i - 0)\rangle = 0\}$. Assume there exists a constant $d > 0$, such that*

$$\langle \gamma'(t_i - 0), n_i\rangle \geq d, \forall i \in \mathbb{N}.$$

Then the following statements are true:

(1) The Jacobi field Y along γ is bounded if only if its magnitude is non-increasing.

(2) There exist exactly m linearly independent bounded Jacobi fields along γ, in which one of them has constant magnitude.

(3) Let Y be a bounded Jacobi field along γ which perpendicular to γ', then there exists universal constant κ, which depends only upon b, c and d, such that for all $\lambda \in (0, \frac{\kappa}{2})$ and for all $t \in [t_1, \infty)$:

$$|Y(t)| \leq e^{-\lambda(t-t_1)}|Y(t_1)|.$$

(4) Let Y be a Jacobi field along γ which perpendicular to γ' whose amplitude does not go to zero at infinity, then there exists a universal constant τ, which depends only upon b, c and d, and there exists $T > 0$, for all $\lambda \in (0, \frac{\tau}{2})$ and for all $t \in [T, \infty)$:

$$|Y(t)| \geq e^{\lambda(t-T)}|Y(T)|.$$

Proof. Note that in this case, Alternative (3) of Hypothesis Λ holds. The Jacobi field Y along γ which is perpendicular to γ' is the solution of $LY = 0$, where L is defined as in (2.2) and (2.3), $A(t)$ is defined as in (3.2), $A_i = R_i$, $C_i = 2\langle\gamma'(t_i - 0), n_i\rangle C(t_i)$, $D_i = R_i$. The condition that A_iC_i is positive definite on the real vectors follows from the fact that $R_iC(T_i)$ is positive definite on Π_i which is a $m-1$ dimensional vector space. Notice that γ' itself is a bounded Jacobi field along γ, so by Theorem 2.15, we know that there are m linearly independent bounded Jacobi fields along γ. Now by Theorem 2.3, we get the result. ■

- From this Theorem, we can simply study the differential operator L to get the properties of the Jacobi field along a broken geodesic.

BIBLIOGRAPHY

1. Manfredo Perdigão do Carmo, *Riemannian Geometry*, Birkhäuser, 1992.
2. N. I. Chernov, Ya. G. Sinai, *Entropy of a Gas of Hard Spheres with respect to the Group of Space-time Shifts*, Trudy Sem. Petrovsk. **8** (1982), 218–238.
3. N. I. Chernov, Ya. G. Sinai, *Ergodic Properties of Certain Systems of 2-D Disks and 3-D Balls* Russian Math. Surveys (3) **42** (1987), 181–207.
4. Robert M. Kauffman, *On the Ergodicity of geodesic flow*, to appear.
5. Robert M. Kauffman, Thomas T. Read, and Anton Zettl, *The Deficiency Index Problem for Powers of Ordinary Differential Expressions*, Springer, 1977.
6. A. Krámli, N. Simányi, and D. Szász, *Dispersing Billiards without Focal Points on Surfaces are Ergodic*, Comm. Math. Phys. **125** (1989), no. 3, 439–457.
7. Barrett O'Neill, *Semi-Riemannian Geometry*, Academic Press, 1983.

8. Takashi Sakai, *Riemannian Geometry*, American Mathematical Society, **149**, 1996.

9. Ya. G. Sinai, *Dynamical Systems with Elastic Reflections*, Russian Math. Surveys **25:2** (1970), 137–189.

10. Ya. G. Sinai, *Developments of Krylov's Ideas* in N.S. Krylov, *Works on Foundations of Statistical Physics*, trans. A.B. Migdal, Ya.G. Sinai and Y.U. Zeeman, Princeton: Princeton Universitty Press, 1979.

11. Tamás Tasnádi, *The Behavior of Nearby Trajectaries in Magnetic Billiards*, J. Math. Phys. **37** (1996), 5577–5598.

12. M. Wojtkowski, *W- Flow on Weyl Manifolds and Gaussian Thermostats*, to appear.

An alternate proof of Kato's inequality

M. K. KWONG

Lucent Technologies, Lisle, Illinois, U.S.A.
e-mail: mkkwong@lucent.com

ANTON ZETTL

Mathematics Department, Northern Illinois University,
DeKalb, Illinois, 60115
e mail: zettl@math.niu.edu

This paper is dedicated to Jerry Goldstein on the occasion of his 60th birthday.

ABSTRACT. We give an alternate proof of the well known Kato inequality for dissipative operators on Hilbert space.

1 KATO'S INEQUALITY

Norms of a function y and its derivative y' can be arbitrary i.e. they are, in general, not related to each other. But the norms of y, y', y'' cannot be arbitrary. They must satisfy certain inequalities. Such inequalities are well known and have become associated with the names of Landau, Hadamard, Hardy and Littlewood, Berdishev, Ditzian and others. Higher order versions of these inequalities are associated with the names of Kolmogorov, Schonberg and Cavaretta, Ljubic, Kupcov etc. See [7] for details.

In 1970 Kallman and Rota [4] published an abstract version of the second order inequalities for operators in Banach spaces. In 1971 Kato [3] showed that the best constant in the Kallman-Rota inequality can be improved if the Banach space setting is specialized to a Hilbert space.

This paper contains an alternate proof of this Kato inequality [3] for dissipative or accretive operators in Hilbert space. This proof, like Kato's, is short and elementary. We believe it yields insights which are not transparent from [3]. For example, it can be extended to higher orders [5] while the proof in [3], being based primarily on a clever use of the Schwarz inequality, does not seem to have such an extension. Using methods based on deep results from the theory of semi-groups of operators Chernoff [1] and Phong [8] proved the higher order version of Kato's inequality. This higher order version was also established by Kwong and Zettl in [5] using only methods from finite dimensional linear algebra. Our presentation is based on [5].

THEOREM 1.1. *Let A be a linear operator on an inner product space H; and let $D(A)$ denote its domain. If*

$$\mathrm{Re}(Af,\, f) \le 0 \textit{ for all } f \textit{ in } D(A), \tag{1.1}$$

then

$$||Af||^2 \le 2\,||f||\,||A^2 f||, \quad \textit{for all } f \in D(A^2). \tag{1.2}$$

Furthermore, equality holds for $f \neq 0$ if and only if

$$\mathrm{Re}(A^2 f,\, f) = 0 \tag{1.3}$$

and for some $d \ge 0$

$$A^2 f + d\,Af + d^2\, f = 0. \tag{1.4}$$

Proof. Let $f \in D(A^2)$, $c > 0$. Then, using (1.1), we have that

$$\mathrm{Re}(cA(cAf + f),\, cAf + f) \le 0, \tag{1.5}$$

implies

$$(c^2 A^2 f + cAf,\, cAf + f) + (\, cAf + f,\, c^2 A^2 f + cAf) \le 0. \tag{1.6}$$

Adding

$$(c^2 A^2 f,\, c^2 A^2 f) + (f, f) \tag{1.7}$$

to both sides of (1.6) we obtain

$$(cAf, cAf) + (c^2 A^2 f + cAf + f,\, c^2 A^2 f + cAf + f) \le (c^2 A^2 f,\, c^2 A^2 f) + (f, f). \tag{1.8}$$

Since the second term on the left of (1.8) is nonnegative we may conclude that

$$c^2 ||Af||^2 = (cAf,\, cAf) \le (c^2 A^2 f,\, c^2 A^2 f) + (f, f) = c^4 ||A^2 f||^2 + ||f||^2. \tag{1.9}$$

Dividing (1.9) by c^2 and then letting $c \to \infty$, we see that if $||A^2 f|| = 0$, then $||Af|| = 0$ and the inequality (1.2) holds. Otherwise let $c^2 = ||f||/||A^2 f||$, then the inequality (1.9) reduces to (1.2), i.e.

$$||Af||^2 \le 2\,||f||\,||A^2 f||.$$

To prove the furthermore part of Theorem 1.1 assume that for some $f \in D(A^2)$, $f \neq 0$, we have

$$||Af||^2 = 2\,||f||\,||A^2 f||. \tag{1.10}$$

If $A^2 f = 0$ then, as above, $Af = 0$ and (1.3) holds as well as (1.4) with $d = 0$. If $A^2 f \neq 0$, let $c^2 = ||f||/||A^2 f||$, and substitution in (1.8) yields

$$c^2 A^2 f + cAf + f = 0, \tag{1.11}$$

now divide by c^2 to obtain (1.4). Working backward from (1.8) to (1.5) we get

$$\begin{aligned}0 &= \mathrm{Re}(cA(cAf + f), cAf + f) = \mathrm{Re}(c^2A^2f + cAf, cAf + f)\\ &= \mathrm{Re}(-f, c^2A^2f) = -c^2\,\mathrm{Re}(f, A^2f),\end{aligned}$$

and this implies (1.3).

Conversely, suppose (1.3) and (1.4) hold. If $d = 0$, then $A^2f = 0$. This implies, as above, that $Af = 0$ and (1.2) holds. If $d > 0$, then (1.11) holds with $c = 1/d > 0$. Noting that equality holds in (1.9) and using the elementary inequality

$$2ab \le \frac{1}{d^2}a^2 + d^2b^2,$$

we obtain from (1.2)

$$\frac{1}{d^2}||A^2f||^2 + d^2||f||^2 = ||Af||^2 \le 2\,||f||\,||A^2f|| \le \frac{1}{d^2}||A^2|| + d^2||f||^2;$$

and hence $||Af||^2 = 2\,||f||\,||A^2f||$.This concludes the proof of Theorem 1.1.

REMARK 1.2. The constant 2 in inequality (1.2) is universal; the inequality holds for any linear operator A satisfying (1.1) in any inner product space. For a particular operator A there may be a smaller constant, e.g. if A is symmetric then $||Af||^2 \le ||f||\,||A^2f||$. We don't know of any case where the best constant, say k, satisfies $1 < k < 2$ and all cases of equality are known.

Our example below is the celebrated Hardy-Littlewood inequality mentioned in the introduction which, together with the Kallman-Rota inequality in Banach space, motivated Kato to discover and prove Theorem 1.1.

EXAMPLE 1.3. Let $J = (a, \infty)$, $-\infty < a < \infty$, $H = L^2(J)$. Define

$$\begin{aligned}D(A) &= \{f \in H : f \in AC_{loc}(J),\ f' \in H\}\\ Af &= f'\ (f \in D(A))\end{aligned}$$

where $AC_{loc}(J)$ denote the complex valued functions on J which are absolutely continuous on all compact subintervals of J.

To show that A satisfies (1.1), let $f = u + iv \in D(A)$. Since

$$\int_a^t u'u = \frac{1}{2}[u^2(t) - u^2(a)]$$

and the integral is convergent it follows that

$$\lim_{t\to\infty} u^2(t) = L,\ 0 \le L < \infty$$

and $L = 0$ since $L > 0$ contradicts $u \in H$. Thus

$$\mathrm{Re}(Af, f) = \int_a^\infty (u'u + v'v) = \frac{-1}{2}[u^2(a) + v^2(a)] \le 0.$$

Hence, by Theorem 1.1,

$$||f'||^2 \leq 2\,||f||\,||f''|| \tag{1.12}$$

for all

$$\begin{aligned} f \in D(A^2) &= \{f \in D(A) : Af = f' \in D(A)\} \\ &= \{f \in H : f,\ f' \in AC_{loc}(J),\ f',\ f'' \in H\}. \end{aligned}$$

Note that (1.12) holds for all $f \in H$, f, $f' \in AC_{loc}(J)$, $f'' \in H$ since this implies that $f' \in H$. Using the furthermore part of Theorem 1.1 to characterize all nontrivial cases of equality in (1.12) note that conditions (1.3), and (1.4) reduce to the requirements that f satisfy the constant coefficient linear differential equation

$$f'' + df' + d^2 f = 0$$

for some $d > 0$ and $\mathrm{Re}(f'',\, f) = 0$. A tedious but straightforward computation then shows that all nontrivial cases of equality are given by

$$f(t) = z\,e^{(-t/(2d))}\sin(t\sqrt{3}/(2d) - \pi/3),\ t \in J,\ z \in \mathbb{C}.$$

REMARK 1.4. Does (1.12) hold when $H = L^2(J)$ is replaced by $H = L^2(J, w) = \{f : \int_J |f|^2 w < \infty\}$? That the answer is no can be seen from the simple example: $J = (0, \infty)$, $w(t) = e^{-t}$, $f(t) = t$. Using Theorem 1.1 Kwong and Zettl [6] showed that

$$\Big(\int_J |f'|^2 w\Big)^2 \leq 4 \int_J |f|^2 w \int_J |f''|^2 w \tag{1.13}$$

holds for any $f \in L^2(J, w)$ with $f' \in AC_{loc}(J)$ and $f'' \in L^2(J, w)$ under the condition that w is a positive nondecreasing function on J; this for both $J = (0, \infty)$ or $J = (-\infty, \infty)$. Characterizing the class of weight functions w for which (1.13) holds seems to be a difficult open problem.

REMARK 1.5. In [2] Goldstein, Kwong and Zettl studied the more general inequality

$$\Big(\int_J |f'|^p w\Big)^2 \leq K \int_J |f|^p w \int_J |f''|^p w \tag{1.14}$$

for $1 \leq p < \infty$, $J = (0, \infty)$ or $J = (-\infty, \infty)$; here the question is for what 'weight functions' w does there exist a constant K, $0 < K < \infty$, such that (1.14) holds for all functions f for which the two integrals on the right are finite; the finiteness of the integral on the left is part of the conclusion. For instance these authors show that the class of weight functions $w > 0\,a.e.$, $w \in L_{loc}(J)$ for which (1.14) holds for a given K form a positive cone. Also in [2] sufficient conditions on 'weight functions' w with zeros are found for 1.14 to hold.

BIBLIOGRAPHY

1. P. Chernoff, *Optimal Landau-Kolmogorov inequalities for dissipative operators in Hilbert and Banach spaces*, Adv. in Math. **34** (1979), 134–147.

2. J. A. Goldstein, M. K. Kwong and A. Zettl, *Weighted Landau Inequalities*, J. Math. Anal. and Appl. **95** (1983), 20–28.

3. T. Kato, *On an inequality of Hardy, Littlewood and Polya*, Adv. in Math. 7 (1971), 217-218.

4. R.R. Kallman and G.-C. Rota, *On the inequality* $||f'||^2 \leq 4\,||f||\,||f''||'$, Inequalities II, Academic Press, New York, 1970.

5. M. K. Kwong and A. Zettl, *Norm inequalities for dissipative operators in inner product spaces*, Houston J. Math. **5** (1979), 543–547.

6. M. K. Kwong and A. Zettl, *An extension of the Hardy-Littlewood inequality*, Proc. Amer. Math. Soc. **77** (1979), 117–118.

7. M. K. Kwong and A. Zettl, *Norm inequalities for derivatives and differences*, Lecture Notes in Mathematics, vol. 1536, Springer Verlag, 1992.

8. V. Q. Phong, *On inequalities for powers of linear operators and for quadratic forms*, Proc. Roy. Soc. Edinburgh A **89** (1981), 25–50.

On a Continuous Coagulation and Fragmentation Equation with a Singular Fragmentation Kernel

WILSON LAMB

Department of Mathematics, University of Strathclyde
Glasgow, U.K.
e-mail: wl@maths.strath.ac.uk

ADAM C. MCBRIDE

Department of Mathematics, University of Strathclyde
Glasgow, U.K.
e-mail: acmcb@maths.strath.ac.uk

Dedicated to Jerry Goldstein on the occasion of his 60th birthday.

ABSTRACT. An integro–differential equation modelling coagulation and multiple–fragmentation is investigated. Under the assumptions that the coagulation kernel is constant and the fragmentation kernel takes the form $(\nu + 2)(y/x)^{\nu} x^{-1}$, where $\nu > -1$, global existence and uniqueness of mass–conserving solutions are established. This extends similar results obtained elsewhere, such as [1,4,8,10,12,13] since the fragmentation kernel considered here is singular at the origin. In the case of pure fragmentation, when no coagulation occurs, semigroup methods are used to derive a closed–form solution of the resulting linear equation. This solution agrees with that obtained in [9] by means of a non–rigorous limiting argument.

1 INTRODUCTION

The study of coagulation–fragmentation processes is of relevance to many diverse areas of pure and applied science, such as colloidal chemistry, astrophysics and polymer science. In modelling the dynamical behaviour of such processes, it is often assumed that individual elements in the system can be uniquely characterised by a non–negative scalar quantity. For example, this dynamical variable could represent polymer mass or size in the case of a system of reacting polymers. If it is also assumed that the individual dynamical variable distribution varies only as a result of two processes, namely coalescence and fragmentation, with reaction rates that depend on the sizes of the individual elements involved, then the dynamical behaviour of the system may be described mathematically by the continuous coagulation–fragmentation equation

$$\frac{\partial}{\partial t}u(x,t) = (\mathcal{F}u)(x,t) + (\mathcal{C}u)(x,t)\,, \tag{1.1}$$

where

$$(\mathcal{F}u)(x,t) = -u(x,t)\int_0^x \frac{y}{x}\gamma(x,y)\,dy \;+\; \int_x^\infty \gamma(y,x)u(y,t)\,dy\,, \tag{1.2}$$

and

$$(\mathcal{C}u)(x,t) = \frac{1}{2}\int_0^x K(x-y,y)u(x-y,t)u(y,t)\,dy - u(x,t)\int_0^\infty K(x,y)u(y,t)\,dy \tag{1.3}$$

model fragmentation and coagulation respectively; see [11,12].

If, for convenience, we concentrate on the case of a system of reacting particles that are distinguished by their mass, then the function u appearing in the integro–differential equation (1.1) is interpreted as a density function, with $u(x,t)$ denoting the density of particles with mass x at time t. Implicit in this interpretation is the assumption that the number of particles in the system is large enough to justify the use of a density function. The coagulation kernel $K(x,y)$ then represents the rate at which particles of mass x coalesce with particles of mass y, while the fragmentation kernel $\gamma(x,y)$ represents the rate at which particles of mass y are produced from fragmenting particles of mass x. An explanation of each of the terms in (1.2) and (1.3) is given in [11].

Equation (1.1) is referred to as the continuous coagulation and multiple–fragmentation equation as fragmenting particles can split into more than two pieces. However, the continuous coagulation and binary–fragmentation equation, investigated, for example, in [13,14], can be obtained as a special case of (1.1) by setting

$$\gamma(x,y) \;=\; F(x-y,y) \quad 0 \le y \le x < \infty\,, \tag{1.4}$$

where F is assumed to be symmetric, i.e. $F(x,y) = F(y,x)$ for all $x, y > 0$. In this binary–fragmentation model, the function F represents the rate at which particles of mass $x-y$ and y are obtained from a fragmenting particle of mass x. If we use the relation

$$\int_0^x \frac{y}{x}F(x-y,y)\,dy \;=\; \frac{1}{2}\int_0^x F(x-y,y)\,dy\,,$$

then (1.2) becomes

$$(\mathcal{F}u)(x,t) = -\frac{1}{2}u(x,t)\int_0^x F(x-y,y)\,dy + \int_x^\infty F(x,y-x)u(y,t)\,dy\,. \tag{1.5}$$

Similarly, (1.1) with

$$K(x,y) \equiv 0, \quad a(x) := \int_0^x \frac{y}{x}\gamma(x,y)\,dy\,, \quad b(x,y) = \frac{\gamma(y,x)}{a(y)} \tag{1.6}$$

reduces to the fragmentation equation

$$\frac{\partial}{\partial t}u(x,t) = -a(x)u(x,t) + \int_x^\infty a(y)b(x,y)u(y,t)\,dy\,, \tag{1.7}$$

studied by McGrady and Ziff [9]. In (1.7), $a(x)$ is the rate at which particles of size x fragment and $b(x,y)$ represents the rate of production of particles of size x from a fragmenting particle of size y.

Many results on the existence and uniqueness of solutions to the different forms of the coagulation–fragmentation equation have already been established by various authors. For example, Aizenman and Bak [1] use semigroup methods to analyse the case when both the coagulation kernel K and binary–fragmentation kernel F are constant. Using a similar approach, McLaughlin *et al* [11] establish existence and uniqueness of non–negative, mass–conserving solutions to the multiple–fragmentation equation under the condition that

$$\int_0^x \frac{y}{x}\gamma(x,y)\,dy \ \le C_n \ < \infty \ \text{ for all } x \in (0,n],\ n > 0\,, \tag{1.8}$$

where the sequence $\{C_n\}$ may be unbounded. This is extended in [10] to the combined coagulation and multiple–fragmentation equation (1.1) under the assumptions that K is constant and

$$\gamma \in L^\infty((0,\infty) \times (0,\infty))\,. \tag{1.9}$$

An alternative strategy in which existence of solutions to the combined coagulation and binary–fragmentation equation is proved via a weak compactness argument is found in the work of Stewart [13] and Dubovskiĭ and Stewart [4]. In the latter paper, existence and uniqueness of a non–negative solution are established for continuous, non–negative kernels K and F satisfying

$$K(x,y) \ \le \ k(1+x+y)\,, \tag{1.10}$$

$$\int_0^x F(x-y,y)\,dy \ \le \ b(1+x^{m_1}), \quad x \in [0,\infty)\,, \tag{1.11}$$

$$F(x-y',y) \ \le \ b(1+x^m), \quad 0 \le y \le y' \le x < \infty, \tag{1.12}$$

where k, m, m_1 and b are positive constants with $m_1 \le 1$. Under an additional constraint, the solution is shown to be mass–conserving.

More recently, a similar analysis of the combined coagulation and multiple–fragmentation equation has been carried out by Laurençot [8]. In this case existence of mass–conserving solutions is established for a class of coagulation and fragmentation kernels. Specifically, K takes the form

$$K(x,y) = r(x)r(y) + \alpha(x,y) \tag{1.13}$$

where $r \in C[0,\infty)$, $\alpha \in C([0,\infty) \times [0,\infty))$ and

$$0 \le \alpha(x,y) = \alpha(y,x) \le Ar(x)r(y) \quad \forall (x,y) \in [1,\infty) \times [1,\infty).$$

The function r is the dominant term in the coagulation–fragmentation processes considered in [8] since the kernel $\gamma \in C([0,\infty) \times [0,\infty))$ is assumed to satisfy

$$\int_0^x \gamma(x,y)\,dy \le \omega(x)\max(x, r(x)),\ x \ge 0 \tag{1.14}$$

$$\gamma(x,y) \le B(1 + \max(x, r(x))),\ x, y \ge 0, \tag{1.15}$$

where B is a positive real number and $\omega : [0,\infty) \to [0,\infty)$ is such that $\omega(x) \to 0$ as $x \to \infty$.

In [12] Melzak deals with certain bounded kernels by assuming that solutions are of the form $u(x,t) = \sum_0^\infty a_k(x)t^k$. Similarly, for the specific equation with $K \equiv 0$ and

$$\gamma(x,y) = (\nu+2)(y/x)^\nu x^\alpha, \quad -2 < \nu \le 0, \tag{1.16}$$

McGrady and Ziff [9] use a series expansion to obtain an explicit solution in terms of the confluent hypergeometric function for each $\alpha \ne -1$. By formally allowing $\alpha \to -1$, a solution, now involving a modified Bessel function of the first kind, is also produced for the case $\alpha = -1$. Note that the functions a and b, related to the kernel γ in (1.16) via (1.6), are given by

$$a(x) = x^{\alpha+1}, \quad b(x,y) = (\nu+2)(x/y)^\nu y^{-1}. \tag{1.17}$$

The restrictions on ν stated in (1.16) then follow from the conditions

$$\int_0^y x\,b(x,y)\,dx = y,$$

$$\int_0^z x\,b(x,y)\,dx \ge \int_{y-z}^y (y-x)\,b(x,y)\,dx, \quad z < y/2,$$

which reflect, respectively, mass–conservation and the requirement that the breakup process is a single event with no arrangement of the mass allowed; see [9] for details.

Here our aim is to establish the existence and uniqueness of solutions to (1.1) in the case of a constant coagulation kernel

$$K(x,y) \equiv k, \quad k > 0,$$

and the unbounded fragmentation kernel

$$\gamma(x,y) = (\nu+2)(y/x)^\nu x^{-1}, \quad -1 < \nu \le 0. \tag{1.18}$$

Thus equation (1.1) becomes

$$\frac{\partial}{\partial t}u(x,t) = (\mathcal{F}u)(x,t) + (\mathcal{C}u)(x,t),$$

where

$$(\mathcal{F}u)(x,t) = -u(x,t) + (\nu+2)\,x^\nu \int_x^\infty y^{-\nu-1}u(y,t)\,dy, \tag{1.19}$$

and

$$(\mathcal{C}u)(x,t) = \frac{k}{2}\int_0^x u(x-y,t)u(y,t)\,dy - k\,u(x,t)\int_0^\infty u(y,t)\,dy\,. \tag{1.20}$$

Clearly, when $k = 0$, we obtain the fragmentation equation studied by McGrady and Ziff in the case $\alpha = -1$. However, the results we present here extend those in [9] since we consider *non-zero* constant coagulation kernels. Note also that the various existence and uniqueness results discussed above do not apply since the fragmentation kernel γ, given by (1.18), fails to satisfy the required conditions. For example, $\nu = 0$ leads to the binary–fragmentation kernel $F(x,y) = 2(x+y)^{-1}$ which does not satisfy (1.12).

To establish existence and uniqueness of solutions we follow the strategy used in [1, 10, 11]. In Section 2, we deal with the pure fragmentation equation

$$\frac{\partial}{\partial t}u(x,t) = -u(x,t) + (\nu+2)x^\nu \int_x^\infty y^{-\nu-1}u(y,t)\,dy\,, \tag{1.21}$$

where $\nu > -2$. The associated initial–value problem is expressed as an abstract Cauchy problem (ACP) of the form

$$\frac{d}{dt}\mathrm{u}(t) = A[\mathrm{u}(t)]\,,\ t > 0\,;\quad \mathrm{u}(0) = u_0 \tag{1.22}$$

where the operator A is defined on suitable functions f by

$$(Af)(x) := -f(x) + (\nu+2)x^\nu \int_x^\infty y^{-\nu-1}f(y)\,dy\,,\quad x > 0. \tag{1.23}$$

Both the Kato-Voigt perturbation approach used by Banasiak in [2] and the strategy adopted in [11] of initially considering a truncated version of (1.21) can be applied to (1.22). However, the fact that A is bounded on an appropriate Banach space simplifies matters considerably, as a strong solution is obtained immediately in the form

$$\mathrm{u}(t) = S(t)u_0 := e^{At}u_0 := \left(\sum_{n=0}^\infty \frac{(At)^n}{n!}\right)u_0. \tag{1.24}$$

This leads to the existence and uniqueness of mass–conserving solutions to (1.21). In addition, (1.24) can be used to derive the solution obtained in [9] in a simple, but mathematically rigorous manner.

In Section 3, the initial–value problem for the combined coagulation and fragmentation equation (1.1) is expressed as the autonomous semilinear ACP

$$\frac{d}{dt}\mathrm{u}(t) = A[\mathrm{u}(t)] + N[\mathrm{u}(t)]\,,\ t > 0;\quad \mathrm{u}(0) = u_0\,, \tag{1.25}$$

where the nonlinear operator N is given by

$$(Nf)(x) := \frac{k}{2}\int_0^x f(x-y)f(y)\,dy - kf(x)\int_0^\infty f(y)\,dy\,,\quad x > 0\,. \tag{1.26}$$

Relevant properties of N are given in [10] where the existence and uniqueness of a global (in time) mild solution are established for the case of bounded fragmentation and constant coagulation kernels. The arguments used in [10] require boundedness of γ since this enables a crucial estimate on the fragmentation semigroup of the form

$$||S(t)|| \leq 1 + Mt\,, \quad M > 0, \tag{1.27}$$

to be obtained. For the unbounded fragmentation kernel investigated here, (1.27) no longer holds. However, the alternative estimate

$$||S(t)|| \leq e^{t/(\nu+1)} \tag{1.28}$$

can be obtained provided that $\nu > -1$, and, since most of the arguments used in [10] remain valid when (1.27) is replaced by (1.28), the existence and uniqueness of a mild, local, mass–conserving solution follow routinely. One aspect not considered in [10] is the existence of a strong solution to the related semilinear ACP. Here, we prove that the nonlinear operator N is Fréchet differentiable, and therefore, on applying [3; Theorem 3.30], we can deduce that the mild solution is also a strong solution. Finally, an application of Gronwall's inequality establishes that this strong solution is defined for all $t \geq 0$.

2 THE FRAGMENTATION EQUATION

In the absence of coagulation, the initial–value problem associated with (1.1) takes the form

$$\frac{\partial u}{\partial t}(x,t) = (\mathcal{F}u)(x,t)\,, \tag{2.1}$$

$$u(x,0) = u_0(x)\,, \tag{2.2}$$

where u_0 represents the initial mass distribution. We consider the specific problem in which $\mathcal{F}$ is given by (1.19). As indicated earlier, this problem can be expressed as the ACP

$$\frac{d}{dt}\mathrm{u}(t) = A[\mathrm{u}(t)]\,,\ t > 0; \quad \mathrm{u}(0) = u_0\,, \tag{2.3}$$

where A is defined by (1.23). As we expect the mass in the system to be a conserved quantity, an appropriate space in which to study (2.3) is the Banach space Y consisting of equivalence classes of measurable, real-valued functions f which satisfy

$$||f||_Y := \int_0^\infty x\,|f(x)|\,dx \ < \ \infty\,. \tag{2.4}$$

THEOREM 2.1. *If* $\nu > -2$ *then* (2.3) *has a unique, strongly continuously differentiable solution* $\mathrm{u} : [0,\infty) \to Y$ *for each* $u_0 \in Y$.

Proof. A simple calculation shows that $A \in B(Y)$ with $||A|| \leq 2$. Consequently, the unique, strongly continuously differentiable solution to (2.3) is given by

$$\mathrm{u}(t) = e^{At}u_0\,,\ t \geq 0\,, \tag{2.5}$$

where $e^{At} \in B(Y)$ is defined by the exponential series

$$e^{At} = \sum_{n=0}^{\infty} \frac{(At)^n}{n!} . \tag{2.6}$$

■

Our next aim is to obtain an explicit formula for e^{At} which can be used to establish that the solution $e^{At}u_0$ is non–negative and conserves mass for each $u_0 \in Y^+$, where

$$Y^+ := \{f \in Y : f(x) \geq 0 \text{ for a.a. } x > 0\}.$$

To this end, let $S(t)$ be the operator

$$S(t) := e^{-t}(I + G(t)) , \; t \geq 0 , \tag{2.7}$$

where I is the identity operator and $G(t)$ is a Mellin convolution operator

$$[G(t)f](x) = (g_t \star f)(x) := \int_0^\infty g_t(x/y)\, f(y)\, y^{-1}\, dy , \tag{2.8}$$

with kernel

$$g_t(x) := \sqrt{(\nu+2)t}\, x^\nu\, I_1\left(2\sqrt{-(\nu+2)t\ln x}\right) H(1-x) \,/\sqrt{-\ln x} . \tag{2.9}$$

In (2.9), H is the Heaviside function and I_1 is the modified Bessel function of the first kind, given by

$$I_1(x) = \sum_{n=0}^{\infty} \frac{(x/2)^{2n+1}}{n!\,(n+1)!} .$$

LEMMA 2.2. *If $t \geq 0$ and $\nu > -2$ then $G(t) \in B(Y)$ and $||G(t)|| = e^t - 1$.*

Proof. The case $t = 0$ is trivial, so let $t > 0$. Then

$$\int_0^\infty x^{s-1}\, |\, [G(t)f](x)\, |\, dx$$

$$\leq \int_0^\infty x^{s-1} \int_0^\infty g_t(x/y)|f(y)|y^{-1}\, dy\, dx \tag{2.10}$$

$$= \int_0^\infty \int_0^\infty x^{s-1} g_t(x/y)|f(y)|y^{-1}\, dx\, dy . \tag{2.11}$$

The substitution $\omega = \ln(y/x)$ reduces (2.11) to

$$\sqrt{(\nu+2)t} \int_0^\infty \int_0^\infty y^{s-1}\, e^{-(\nu+s)\omega}\, \omega^{-1/2}\, I_1(2\sqrt{(\nu+2)\omega t})\, |f(y)|\, d\omega\, dy ,$$

which, from [5; page 197, (16)], becomes

$$\int_0^\infty y^{s-1} \left[e^{(\nu+2)t/(\nu+s)} - 1\right] |f(y)| \, dy \tag{2.12}$$

provided that $\Re(\nu + s) > 0$. If we now let $s = 2$ then we obtain

$$||G(t)f||_Y \leq (e^t - 1) \, ||f||_Y \, , \quad \forall f \in Y \, .$$

Moreover, for $f \in Y^+$, the inequality in (2.10) becomes equality, and so

$$||G(t)f||_Y = (e^t - 1) \, ||f||_Y \, , \quad \forall f \in Y^+ \, . \tag{2.13}$$

Hence $||G(t)|| = e^t - 1$ as required. ■

THEOREM 2.3. *Let A and $S(t)$ be defined by (1.23) and (2.7) respectively. Then, for each $t \geq 0$ and $\nu > -2$, $e^{At} = S(t)$ in $B(Y)$.*

Proof. Let B be defined by

$$(Bf)(x) := (\nu + 2) \int_x^\infty (x/y)^\nu f(y) \, y^{-1} \, dy \, .$$

Then

$$\begin{aligned} e^{At} &= e^{(-I+B)t} = e^{-t} \, e^{Bt} \\ &= e^{-t} \left(I + \sum_{n=1}^\infty B^n t^n / n! \right) . \end{aligned}$$

It is a routine matter to show that

$$(B^n f)(x) = \frac{(\nu+2)^n}{(n-1)!} \int_x^\infty (x/y)^\nu \left[\ln(y/x)\right]^{n-1} f(y) \, y^{-1} \, dy$$

and

$$||B^n f||_Y \leq ||f||_Y$$

for all $f \in Y$ and $n = 1, 2, \ldots$ Consequently, for each $f \in Y$

$$\begin{aligned} & \left\| G(t)f - \left(\sum_{n=1}^N B^n \, t^n / n! \right) f \right\|_Y \\ \leq \; & \frac{[(\nu+2)t]^n}{n! \, (n-1)!} \int_0^\infty \int_x^\infty \sum_{n=N+1}^\infty (x/y)^{\nu-1} \left[\ln(y/x)\right]^{n-1} |f(y)| \, dy \, dx \\ = \; & \left\| \left(\sum_{n=N+1}^\infty t^n B^n / n! \right) |f| \right\|_Y \\ \leq \; & \left(\sum_{n=N+1}^\infty t^n / n! \right) ||f||_Y \to 0 \text{ as } N \to \infty \, . \end{aligned}$$

■

LEMMA 2.4. *If $\nu > -2$ then $\{e^{At}\}_{t\geq 0}$ is a C_0- semigroup of non-negative isometries on Y.*

Proof. Let $f \in Y^+$. Since $e^{At} = e^{-t}(I + G(t))$ and $G(t)$ has a non-negative kernel, it follows that $e^{At}f \in Y^+$. Moreover, from (2.13),

$$\begin{aligned} \|e^{At}f\|_Y &= e^{-t}\left(\int_0^\infty xf(x)\,dx + \int_0^\infty x[G(t)f](x)\,dx\right) \\ &= e^{-t}\left(\|f\|_Y + (e^t - 1)\|f\|_Y\right) = \|f\|_Y\,. \end{aligned}$$

To establish that the last result also holds for a general $f \in Y$ we simply split f into its positive and negative parts. ∎

THEOREM 2.5. *If $\nu > -2$ then the ACP (2.3) has a unique, strongly continuously differentiable, non-negative, mass-conserving solution for all $u_0 \in Y^+$. This solution is given by $\mathrm{u}(t) = S(t)u_0$, where $S(t)$ is given by (2.7).*

Proof. This follows immediately from the results above. ∎

THEOREM 2.6. *Let $\nu > -2$ and $u_0 \in Y^+$, and define u by*

$$u(x,t) := \left[e^{At}u_0\right](x)\,.$$

Then

(i) u is measurable on $(0,\infty) \times [0,\infty)$;

(ii) $u(x,t) \geq 0$ for all $t \geq 0$ and a.a. $x > 0$;

(iii) $u(x,0) = u_0(x)$ for a.a. $x > 0$;

(iv) $\|u(\cdot,t)\|_Y = \|u_0\|_Y$ for all $t \geq 0$;

(v) u satisfies the fragmentation equation (1.21) a.e.

Moreover, u is the unique solution satisfying (1.21) and the initial condition $u(x,0) = u_0(x)$ such that $\mathrm{u}(t) := u(\cdot,t)$ is a strong solution of (2.3) in the Banach space Y.

Proof. The above results follow on repeating the arguments used to establish [11; Theorem 7.1 & Corollary 7.2]. ∎

It follows immediately that a solution of (1.21) is given by

$$u(x,t) = e^{-t}\left(u_0(x) + \int_0^\infty g_t(x/y)\,u_0(y)\,y^{-1}\,dy\right), \quad t \geq 0 \text{ and a.a. } x > 0\,,$$

for each $u_0 \in Y^+$. This coincides with the solution obtained in [9] via a formal limiting argument.

3 THE COAGULATION AND FRAGMENTATION EQUATION

We now turn our attention to the combined coagulation and fragmentation equation (1.1) and consider the initial–value problem

$$\frac{\partial}{\partial t}u(x,t) = (\mathcal{F}u)(x,t) + (\mathcal{C}u)(x,t)\,, \tag{3.1}$$

$$u(x,0) = u_0(x)\,, \tag{3.2}$$

where $\mathcal{F}$ and $\mathcal{C}$ are given by (1.19) and (1.20) respectively, and $u_0(x)$ is assumed to be non–negative for almost all $x \in (0\,,\infty)$. In this case, the corresponding semilinear ACP takes the form

$$\frac{d}{dt}\mathrm{u}(t) = A[\mathrm{u}(t)] + N[\mathrm{u}(t)]\,,\; t > 0; \quad \mathrm{u}(0) = u_0\,, \tag{3.3}$$

where A and N are defined by (1.23) and (1.26). Although our treatment of the fragmentation equation was carried out in the space Y, we now choose the Banach space X defined by

$$X := \{f \in Y \cap L_1 \,:\, ||f||_X := ||f||_Y + ||f||_1 < \infty\}$$

for our analysis of the combined equation. This space provides a more convenient setting for establishing that the nonlinear operator N is locally Lipschitz and Fréchet differentiable. We begin by showing that the fragmentation semigroup $\{S(t)\}_{t\geq 0}$ introduced in Section 2 also forms a C_0– semigroup in X under the more restrictive condition that $\nu > -1$.

THEOREM 3.1. *Let A be defined by (1.23) and let $\nu > -1$. Then*

(i) $A \in B(X)$*;*

(ii) $e^{At} = S(t),\ t \geq 0,$ *where $S(t)$ is given by (2.7);*

(iii) $||S(t)||_{B(X)} \leq \exp(t/(\nu+1))\ \forall t \geq 0$.

Proof. (i) If $f \in X$ then

$$\begin{aligned} ||Af||_X &= ||Af||_Y + ||Af||_1 \\ &\leq 2||f||_Y + ||f||_1 + \frac{\nu+2}{\nu+1}||f||_1 \\ &\leq \frac{2\nu+3}{\nu+1}||f||_X\,. \end{aligned}$$

(ii) The proof is identical to that of Theorem 2.3.

(iii) For each $f \in X$ and $t \geq 0$,

$$\begin{aligned} ||S(t)f||_1 &= e^{-t}\,||\,e^{tB}f\,||_1 \\ &\leq e^{-t}\,e^{t\,||B||}\,||f||_1 \\ &\leq e^{-t}\,e^{(\nu+2)t/(\nu+1)}\,||f||_1\,, \end{aligned}$$

and therefore

$$\begin{aligned}||S(t)f||_X &= ||S(t)f||_Y + ||S(t)f||_1 \\ &\le ||f||_Y + e^{t/(\nu+1)}\,||f||_1 \\ &\le e^{t/(\nu+1)}\,||f||_X .\end{aligned}$$

■

The arguments used in [10; Section 3] could now be used (with the estimate $||S(t)|| \le (1 + t\,C_F)$ replaced by $||S(t)|| \le \exp[t/(\nu+1)]$) to prove the existence of a unique, mild solution to (3.3). However, a more direct, and profitable, approach is to apply [3; Theorem 3.30] which gives sufficient conditions for the existence of strong solutions to semilinear ACPs. To establish that the operator N satisfies these conditions it is convenient to introduce an associated operator $\mathcal{N}$ defined by

$$\mathcal{N}[f,g](x) := \frac{k}{2}\int_0^x f(x-y)g(y)\,dy - k f(x)\int_0^\infty g(y)\,dy\,, \quad x > 0\,.$$

LEMMA 3.2. *The operator $\mathcal{N} : X \times X \to X$ is bilinear and*

$$||\,\mathcal{N}[f,g]\,||_X \le 2k\,||f||_X\,||g||_X \quad \forall f, g \in X\,. \tag{3.4}$$

Proof. The operator $\mathcal{N}$ is clearly bilinear and (3.4) follows on routine application of Fubini's theorem and the simple results that $||f||_1 \le ||f||_X$ and $||f||_Y \le ||f||_X$ for all $f \in X$. ■

THEOREM 3.3. *Let N be defined by (1.26). Then*

(i) $N : X \to X$;

(ii) N is locally Lipschitz on X ;

(iii) N is Fréchet differentiable on X .

Proof. (i), (ii). On applying Lemma 3.2, we obtain

$$||Nf||_X = ||\,\mathcal{N}[f,f]\,||_X \le 2k\,||f||_X^2 \quad \forall f \in X$$

and

$$\begin{aligned}||Nf - Ng||_X &= ||\,\mathcal{N}[f,f] - \mathcal{N}[g,g]\,||_X \\ &= ||\,\mathcal{N}[f-g,f] + \mathcal{N}[g,f-g]\,||_X \\ &\le 2k\,||f-g||_X\,(\,||f||_X + ||g||_X\,)\,.\end{aligned}$$

Consequently, if $u_0 \in X$ is fixed then

$$||Nf - Ng||_X \le C(r,u_0)\,||f-g||_X \quad \forall f, g \in \bar{B}(u_0, r)\,,$$

where

$$\begin{aligned} \bar{B}(u_0, r) &:= \{ f \in X : \|f - u_0\|_X \le r \} , \text{ and} \\ C(r, u_0) &:= 4k \, (r + \|u_0\|_X) . \end{aligned} \tag{3.5}$$

(iii) Let $f, \delta \in X$. Then

$$\begin{aligned} N[f+\delta] &= \mathcal{N}[f+\delta, f+\delta] \\ &= \mathcal{N}[f,f] + \mathcal{N}[f,\delta] + \mathcal{N}[\delta,f] + \mathcal{N}[\delta,\delta] \\ &= Nf + \mathcal{N}[f,\delta] + \mathcal{N}[\delta,f] + N\delta . \end{aligned}$$

For fixed f, $\mathcal{N}[f,\cdot] + \mathcal{N}[\cdot,f]$ is in $B(X)$ with

$$\| \mathcal{N}[f,\delta] + \mathcal{N}[\delta,f] \|_X \le 4k \|f\|_X \|\delta\|_X \quad \forall \delta \in X .$$

Also

$$\frac{\|N\delta\|_X}{\|\delta\|_X} \le 2k \|\delta\|_X \to 0 \text{ as } \|\delta\|_X \to 0 .$$

Hence N is Fréchet differentiable at each $f \in X$ and the Fréchet derivative N_f at f is given by

$$N_f g := \mathcal{N}[f,g] + \mathcal{N}[g,f] \quad \forall g \in X .$$

■

THEOREM 3.4. *The semilinear ACP (3.3) has a unique, local (in time), strongly differentiable solution for each* $u_0 \in X$ *and* $\nu > -1$.

Proof. We have already established that $A \in \mathcal{G}(1, 1/(\nu+1), X)$ and also that N satisfies a local Lipschitz condition and is Fréchet differentiable. If we now examine the Fréchet derivative, then it is easily seen that

$$\|N_f g\|_X \le C(r, u_0) \|g\|_X \quad \forall g \in X, \ f \in B(u_0, r) ,$$

where the constant $C(r, u_0)$, given by (3.5), is independent of f. In addition,

$$\begin{aligned} \|N_{f_1} g - N_{f_2} g\|_X &= \| \mathcal{N}[f_1, g] + \mathcal{N}[g, f_1] - \mathcal{N}[f_2, g] - \mathcal{N}[g, f_2] \|_X \\ &= \| \mathcal{N}[f_1 - f_2, g] + \mathcal{N}[g, f_1 - f_2] \|_X \\ &\le 4k \|f_1 - f_2\|_X \|g\|_X \to 0 \text{ as } \|f_1 - f_2\|_X \to 0 . \end{aligned}$$

The conditions of [3; Theorem 3.30] are therefore satisfied and the existence of a unique, local, strongly differentiable solution follows. ■

Let the maximal interval of existence of the solution $\mathfrak{u}$ to (3.3) be $[0, T)$, where $0 < T \le \infty$, and, for each $\tau \in (0, T)$, let $Z(\tau)$ denote the Banach space $C([0,\tau], X)$ equipped with the supremum norm

$$\|\mathfrak{g}\|_{Z(\tau)} := \sup_{0 \le t \le \tau} \|\mathfrak{g}(t)\|_X \quad \mathfrak{g} \in Z(\tau) .$$

Since $\mathfrak{u}$ is also a mild solution to (3.3) on $(0,T)$, it follows that $\mathfrak{u}$ satisfies the integral equation

$$\mathfrak{u}(t) = S(t)u_0 + \int_0^t S(t-s)N[\mathfrak{u}(s)]\,ds \tag{3.6}$$

in $Z(\tau)$ for each $\tau < T$. Furthermore,

$$\mathfrak{u} \in \bar{B}_{Z(\tau)}(u_0, r) \subset Z(\tau)\,, \tag{3.7}$$

for some $r > 0$, where

$$\begin{aligned}\bar{B}_{Z(\tau)}(u_0, r) &= \{\mathfrak{g} \in Z(\tau) : \mathfrak{g}(t) \in \bar{B}(u_0, r) \quad \forall t \in [0,\tau]\,\} \\ &= \{\mathfrak{g} \in Z(\tau) : \|\mathfrak{g}(t) - u_0\|_X \le r \quad \forall t \in [0,\tau]\,\}\,.\end{aligned}$$

By using (3.6) and (3.7) we shall now deduce that $\mathfrak{u}$ is mass–conserving and is in the positive cone $X^+ := \{f \in X : f \ge 0 \ \text{ a.e.}\}$ for all $t \in [0,T)$ and $u_0 \in X^+$.

THEOREM 3.5. *Let $\mathfrak{u}$ be the unique strong solution on $[0,T)$ to the semilinear ACP (3.3). Then*

$$\int_0^\infty x\,[\mathfrak{u}(t)\,](x)\,dx = \int_0^\infty x\,u_0(x)\,dx \qquad \textit{for all } t \in [0,T)\,.$$

Proof. From (3.6), we can write

$$\int_0^\infty x\,[\mathfrak{u}(t)\,](x)\,dx = \int_0^\infty x\,u_0(x)\,dx + I(t)\,,$$

where

$$I(t) = \int_0^\infty x \left[\int_0^t S(t-s)\,N[\mathfrak{u}(s)]\,ds\right](x)\,dx\,.$$

An identical argument to that used in the proof of [10; Theorem 3.3] now establishes that $I(t) = 0$ for all $t \in [0,T)$. ∎

THEOREM 3.6. *Let $\mathfrak{u}$ be the unique strong solution on $[0,T)$ to the semilinear ACP (3.3). If $u_0 \in X^+$ then $\mathfrak{u}(t) \in X^+$ for all $t \in [0,T)$.*

Proof. The proof is essentially that given in [10; Section 4] but is presented in a slightly different form. Let J denote the continuous linear functional defined on X by

$$Jf := -k \int_0^\infty f(x)\,dx\,,$$

and express the operator $\mathcal{N}$ as the sum $\mathcal{N} = \mathcal{N}_1 + \mathcal{N}_2$, where, for $f, g \in X$,

$$\mathcal{N}_1[f,g](x) := \frac{k}{2}\int_0^x f(x-y)g(y)\,dy\,, \quad x>0\,,$$

$$\mathcal{N}_2[f,g] := (Jg)f\,.$$

Since the strong solution $\mathfrak{u}$ to (3.3) satisfies the integral equation (3.6), it follows that, for each $\tau \in (0,T)$, $\mathfrak{u}$ is also a fixed point of the map $W : Z(\tau) \to Z(\tau)$ given by

$$(W\mathfrak{g})(t) := S(t)u_0 + \int_0^t S(t-s)\,(\mathcal{N}_1[\mathfrak{u}(s),\mathfrak{u}(s)] + \mathcal{N}_2[\mathfrak{g}(s),\mathfrak{u}(s)])\,ds$$

$$= S(t)u_0 + \int_0^t S(t-s)\,\mathcal{N}_1[\mathfrak{u}(s),\mathfrak{u}(s)]\,ds + \int_0^t S(t-s)\,J[\mathfrak{u}(s)]\,\mathfrak{g}(s)\,ds\,.$$

Moreover, a simple calculation using Theorem 3.1(iii), the inequalities

$$\|\mathcal{N}_i[f,g]\|_X \le k\,\|f\|_X\,\|g\|_X \quad \forall f,g \in X\,,\ i=1,2\,,$$

and (3.7) shows that

$$\|W\mathfrak{g} - u_0\|_{Z(\tau)} \le q(\tau) \text{ and}$$

$$\|W\mathfrak{g}_1 - W\mathfrak{g}_2\|_{Z(\tau)} \le p(\tau)\,\|\mathfrak{g}_1 - \mathfrak{g}_2\|_{Z(\tau)}\,,$$

where

$$q(\tau) = \max_{0\le t\le\tau} \|S(t)u_0 - u_0\|_X + 2k\tau\, e^{\tau/(\nu+1)}\,(r + \|u_0\|_X)^2 \quad \text{and}$$

$$p(\tau) = k\tau\, e^{\tau/(\nu+1)}\,(r + \|u_0\|_X).$$

Hence, for sufficiently small τ, $\mathfrak{u}$ is the unique fixed point of W in $\bar{B}_{Z(\tau)}(u_0,r)$, and so is the limit in $Z(\tau)$ of the sequence $(\mathfrak{u}_n)$ generated by the iterative formula $\mathfrak{u}_{n+1} = W\mathfrak{u}_n$ for any choice of initial estimate $\mathfrak{u}_1 \in \bar{B}_{Z(\tau)}(u_0,r)$. In particular, if we take

$$\mathfrak{u}_1 = S(t)u_0 + \int_0^t S(t-s)\,\mathcal{N}_1[\mathfrak{u}(s),\mathfrak{u}(s)]\,ds\,,\ 0\le t\le\tau\,,$$

and use the formula

$$\frac{1}{j!}\int_a^b \phi(t_1)\left(\int_a^{t_1}\phi(t_2)\,dt_2\right)^j dt_1 = \frac{1}{(j+1)!}\left(\int_a^b \phi(t_1)\,dt_1\right)^{j+1},$$

then an inductive proof establishes that

$$\mathfrak{u}_n(t) = \left[\sum_{j=0}^{n-1} \frac{1}{j!}\left(\int_0^t J[u(\sigma)]\,d\sigma\right)^j\right] S(t)u_0$$

$$+\int_0^t \left[\sum_{j=0}^{n-1} \frac{1}{j!}\left(\int_s^t J[u(\sigma)]\,d\sigma\right)^j\right] S(t-s)\,\mathcal{N}_1\left[\,\mathfrak{u}(s),\mathfrak{u}(s)\,\right] ds\,,$$

for each $n = 1, 2, \ldots$. On allowing $n \to \infty$, we deduce that $\mathfrak{u}$ is a fixed point of the map $V : Z(\tau) \to Z(\tau)$ defined by

$$(V\mathfrak{y})(t) := \exp\left(\int_0^t J[\mathfrak{u}(\sigma)]\,d\sigma\right) S(t)u_0$$

$$+\int_0^t \exp\left(\int_s^t J[\mathfrak{u}(\sigma)]\,d\sigma\right) S(t-s)\,\mathcal{N}_1\left[\,\mathfrak{g}(s),\mathfrak{g}(s)\,\right] ds\,.$$

Arguing as before, it can be shown that V is a contraction on $\bar{B}_{Z(\tau)}(u_0, r)$ for τ sufficiently small, and therefore $\mathfrak{u} = \lim \mathfrak{g}_n$ where $\mathfrak{g}_{n+1} = V\mathfrak{g}_n$ and $\mathfrak{g}_1(t) = u_0$ for all $t \in [0, \tau]$. Since $u_0 \in X^+$ and V preserves non–negativity, we deduce that $\mathfrak{u}(t) \in X^+$ for $0 \leq t \leq \tau$ provided τ is sufficiently small. Finally, to show that $\mathfrak{u}(t) \in X^+$ for $t \in [0, T)$, let $T_0 \in (0, T)$ be arbitrarily fixed and define T_{max} by

$$T_{max} := \sup\{0 < \tau \leq T_0 : \mathfrak{u}(t) \in X^+ \ \forall t \in [0, \tau]\}\,.$$

Suppose that $T_{max} < T_0$ and consider the semilinear ACP

$$\frac{d}{dt}\mathfrak{g}(t) = A[\mathfrak{g}(t)] + N[\mathfrak{g}(t)]\,,\ t > 0;\quad \mathfrak{g}(0) = \mathfrak{u}(T_{max})\,,$$

which has solution on $[0, T_0 - T_{max}]$ given by $\mathfrak{g}(t) = \mathfrak{u}(t + T_{max})$. Since X^+ is closed, $u(T_{max}) \in X^+$ and the previous analysis shows that $\mathfrak{u}(t + T_{max}) \in X^+$ for sufficiently small t. This contradicts the definition of T_{max} and therefore $u(t) \in X^+$ for all $t \in [0, T_0]$. ■

THEOREM 3.7. *The semilinear ACP (3.3) has a unique, globally–defined strong solution for each* $u_0 \in X^+$ *and* $\nu > -1$.

Proof. We show that the solution $\mathfrak{u}$ has maximal interval of existence $[0, \infty)$ by proving that $||\mathfrak{u}(t)||_X$ cannot blow up in finite time; see [6; Theorem 2.6, pp.90–91]. Since $\mathfrak{u}$ is a strong solution and X is a Banach space of type L [7; pp 69-70], it follows that

$$[\mathfrak{u}(t)](x) = u_0(x) + \int_0^t (\,A[\mathfrak{u}(s)] + N[\mathfrak{u}(s)]\,)\,(x)\,ds \quad \text{for } a.a.\ x \in (0, \infty)\,. \qquad (3.8)$$

Consequently, on setting $u(x,t) = [\mathfrak{u}(t)](x)$, we obtain

$$||\mathfrak{u}(t)||_1 = ||u_0||_1 + \int_0^\infty \int_0^t \left(\int_x^\infty (\nu+2)\,(x/y)^\nu y^{-1}\, u(y,s)\, dy \; - u(x,s) \right) ds\, dx$$

$$+\frac{k}{2} \int_0^\infty \int_0^t \int_0^x u(x-y,s)\, u(y,s)\, dy\, ds\, dx - k \int_0^\infty \int_0^t \int_0^\infty u(x,s)\, u(y,s)\, dy\, ds\, dx$$

$$= ||u_0||_1 + \int_0^\infty \int_0^t \left(\int_x^\infty (\nu+2)\,(x/y)^\nu y^{-1}\, u(y,s)\, dy \; - u(x,s) \right) ds\, dx$$

$$-\frac{k}{2} \int_0^t \int_0^\infty \int_0^\infty u(x,s)\, u(y,s)\, dy\, dx\, ds$$

$$\leq ||u_0||_1 + \int_0^t \int_0^\infty \int_0^y (\nu+2)\,(x/y)^\nu\, y^{-1}\, u(y,s)\, dx\, dy\, ds\,.$$

It follows that

$$||\mathfrak{u}(t)||_1 \leq ||u_0||_1 + \frac{\nu+2}{\nu+1} \int_0^t ||\mathfrak{u}(s)||_1\, ds$$

and therefore, on applying Gronwall's inequality,

$$||\mathfrak{u}(t)||_1 \leq ||u_0||_1 \exp[(\nu+2)t/(\nu+1)]\,.$$

Hence, from Theorem 3.5,

$$\begin{aligned} ||\mathfrak{u}(t)||_X \;&\leq\; ||u_0||_Y + ||u_0||_1 \exp[(\nu+2)t/(\nu+1)] \\ &\leq\; ||u_0||_X \exp[(\nu+2)t/(\nu+1)]\,, \end{aligned}$$

from which we deduce that $||\mathfrak{u}(t)||_X$ does not blow up in finite time. ■

On combining the above results we obtain the following existence and uniqueness theorem for solutions of equation (1.1).

THEOREM 3.8. *The coagulation–fragmentation equation (1.1), with $\mathcal{F}$ and $\mathcal{C}$ given by (1.19) and (1.20) respectively, has a non-negative, mass-conserving solution u for each $u_0 \in X^+$ and $\nu > -1$. This solution is given by*

$$u(x,t) = [\mathfrak{u}(t)](x) \qquad \textit{for a.a. } (x,t) \in (0,\infty) \times [0,\infty)\,,$$

where $\mathfrak{u}$ is the unique strong solution of the semilinear ACP (3.3). Moreover, this is the only solution of (1.1) such that the vector-valued function $u(\cdot,t)\,, t \geq 0$ is a strong solution of (3.3).

Proof. By (3.8), the scalar–valued function u, defined by

$$u(x,t) := [\mathrm{u}(t)](x) \quad \text{for all } t \geq 0 \text{ and } a.\,a.\ x \in (0,\infty)\,,$$

is absolutely continuous and satisfies the coagulation and fragmentation equation (1.1), with kernels given by (1.19) and (1.20), for $a.\,a.\ (x,t) \in (0,\infty) \times [0,\infty)$. The scalar–valued function inherits the non–negativity and mass conservation from the vector–valued function. Finally, the uniqueness of the strong solution of (3.3) leads to the uniqueness statement in the theorem. ∎

4 CONCLUSION

By expressing the initial–value problem (3.1) and (3.2) as a semilinear abstract Cauchy problem, we have established the existence and uniqueness of a strongly differentiable solution $\mathrm{u} : [0,\infty) \to X^+$ under the assumptions that $\mathrm{u}(0) \in X^+$, the coagulation kernel is constant, and the fragmentation kernel takes the form

$$\gamma(x,y) = (\nu+2)(y/x)^{\nu} x^{-1},\ \nu > -1.$$

Clearly there are a number of ways in which the work presented here could be developed further. For example, it is expected that only minor modifications would be required to deal with bounded, but non–constant, coagulation kernels. Also, initial investigations into the more general fragmentation kernel

$$\gamma(x,y) = (\nu+2)(y/x)^{\nu} x^{\alpha}, \quad \nu > -1\,,\ \alpha \geq -1\,,$$

indicate that it may be possible to conduct a similar analysis of the abstract formulation of the related coagulation–fragmentation equation.

BIBLIOGRAPHY

1. M. Aizenman and T.A. Bak, *Convergence to equilibrium in a system of reacting polymers*, Comm. Math. Phys. **65** (1979), 203–230.

2. J. Banasiak, *On an extension of the Kato–Voigt perturbation theorem for substochastic semigroups and its application*, Taiwanese J. Math. **5** (2001), 169–191.

3. A. Belleni-Morante and A. C. McBride, *Applied nonlinear semigroups*, Wiley, Chichester, 1998.

4. P. B. Dubovskiĭ and I. W. Stewart, *Existence, uniqueness and mass conservation for the coagulation–fragmentation equation*, Math. Meth. Appl. Sci. **19** (1996), 571–591.

5. A. Erdélyi et al., *Tables of Integral Transforms*, vol. 1, McGraw–Hill, New York, 1954.

6. J. A. Goldstein, *Semigroups of linear operators and applications*, Oxford University Press, Oxford, 1985.

7. E. Hille and R. S. Phillips, *Functional analysis and semigroups*, AMS Colloquium Publications, AMS, Providence, 1957.

8. P. Laurençot, *On a class of continuous coagulation–fragmentation equations*, J. Diff. Eq. **167** (2000), 245–274.

9. E. D. McGrady and R. M. Ziff, *"Shattering" transition in fragmentation*, Phys. Review Lett. **58** (1987), 892–895.

10. D. J. McLaughlin, W. Lamb, and A. C. McBride, *An existence and uniqueness theorem for a coagulation and multiple-fragmentation equation*, SIAM J. Math. Anal. **28** (1997), 1173–1190.

11. D. J. McLaughlin, W. Lamb, and A. C. McBride, *A semigroup approach to fragmentation models*, SIAM J. Math. Anal. **28** (1997), 1158–1172.

12. Z. A. Melzak, *A scalar transport equation*, Trans. Amer. Math. Soc. **85** (1957), 547–560.

13. I. W. Stewart, *A global existence theorem for the general coagulation–fragmentation equation with unbounded kernels*, Math. Meth. Appl. Sci. **11** (1989), 627–648.

14. I. W. Stewart, *A uniqueness theorem for the coagulation–fragmentation equation*, Math. Proc. Camb. Philos. Soc. **107** (1990), 573–578.

Almost periodicity of inhomogeneous parabolic evolution equations

LAHCEN MANIAR

Université Cadi Ayyad, Faculté des Sciences Semlalia
B.P. 2390 40000 Marrakech, Morocco
e-mail: maniar@ucam.ac.ma

ROLAND SCHNAUBELT

FB Mathematik und Informatik, Martin–Luther–Universität
06099 Halle, Germany
e-mail: schnaubelt@mathematik.uni-halle.de

Dedicated to Jerry Goldstein on the occasion of his 60th birthday

ABSTRACT. We show the (asymptotic) almost periodicity of the bounded solution to the parabolic evolution equation $u'(t) = A(t)u(t) + f(t)$ on $\mathbb{R}$ (on $\mathbb{R}_+$) assuming that the linear operators $A(t)$ satisfy the 'Acquistapace–Terreni' conditions, that the evolution family generated by $A(\cdot)$ has an exponential dichotomy, and that $R(\omega, A(\cdot))$ and f are (asymptotically) almost periodic.

1 INTRODUCTION

In the present work we investigate the almost periodicity of the solutions to the parabolic inhomogeneous evolution equations

$$u'(t) = A(t)u(t) + f(t), \quad t \in \mathbb{R}, \tag{1.1}$$

$$u'(t) = A(t)u(t) + f(t), \quad t > 0, \qquad u(0) = x, \tag{1.2}$$

in a Banach space X. It is assumed that the linear operators $A(t)$ satisfy the 'Acquistapace–Terreni' conditions, that the evolution family U solving the homogeneous problem has an exponential dichotomy, and that the functions $t \mapsto R(\omega, A(t))$, for an $\omega \geq 0$, and f are almost periodic (compare Section 2). In Theorem 4.5 we show that then the unique bounded mild solution $u : \mathbb{R} \to X$ of (1.1) is almost periodic. This fact is a consequence of the almost periodicity of Green's function corresponding to U, established also in Theorem 4.5. In Theorem 5.4 we prove the analogous results for (1.2) on $\mathbb{R}_+$ in the context of asymptotic almost periodicity imposing a necessary compatibility condition on x and f.

The almost periodicity of inhomogeneous problems has been studied by many authors in the autonomous case, where $A(t) = A$, and in the periodic case, where $A(t) = A(t+p)$, see [3], [4], [6], [11], [12], [16], [21], and the references therein. Equations with almost periodic $A(\cdot)$ are treated in, e.g., [8] and [10] for $X = \mathbb{C}^n$ and in [13] for a certain class of parabolic problems, see also [5], [15], [17]. For general evolution families U (but subject to an extra condition not assumed here), it is shown in [14] that U has an exponential dichotomy *with* an almost periodic Green's function if and only if there is a unique almost periodic mild solution u of (1.1) for each almost periodic f, see also [8; Prop.8.3]. Our main theorems extend [8; Prop.8.4], [10; Thm.7.7], and [13; p.240], and complement [14; Thm.5.3] in the case of parabolic evolution equations. The initial value problem (1.2) was not studied in [8], [10], [13], and [14] in the context of asymptotic almost periodicity.

Our strategy is similar to Henry's approach in [13; §7.6] who derived the almost periodicity of Green's function Γ corresponding to U (and thus of the dichotomy projections $P(\cdot)$) from a formula for $\Gamma(t+\tau, s+\tau) - \Gamma(t,s)$ (compare (4.2)). In the context of [13], this equation allows to verify Bohr's definition of almost periodicity by straightforward estimates in operator norm if τ is a pseudo period of $A(\cdot)$. However, in the present more general situation one obtains such a formula only on a subspace of X so that one cannot proceed in this way. To overcome this difficulty, we employ Yosida approximations. This requires some preparations given in Section 3 and a somewhat delicate limiting process presented in Section 4. We point out that we can *not* estimate the relevant quantities for the approximating problems independently of n, cf. Lemma 4.1. The almost periodicity of the mild solution u of (1.1) then follows from the standard formula (2.11) which expresses u in terms of Γ, see Theorem 4.5. We also give a straightforward application to a second order parabolic equation. The initial value problem (1.2) is treated in Section 5 by essentially the same methods.

In the next section we collect several concepts and preliminary results. We refer to [9] for unexplained notation. By $c = c(\alpha, \beta, \cdots)$ we denote a generic constant only depending on the constants in the hypotheses involved and the quantities $\alpha, \beta, \cdots$.

2 PREREQUISITES

Let X be a Banach space. A set $U = \{U(t,s) : t \geq s,\ t,s \in \mathbb{R}\}$ of bounded linear operators on X is called an *evolution family* if

(E1) $U(t,s) = U(t,r)U(r,s)$ and $U(s,s) = I$ for $t \geq r \geq s$ and

(E2) $(t,s) \mapsto U(t,s)$ is strongly continuous for $t > s$.

We say that an evolution family U has an *exponential dichotomy* (or is *hyperbolic*) if there are projections $P(t)$, $t \in \mathbb{R}$, being uniformly bounded and strongly continuous in t and constants $\delta > 0$ and $N \geq 1$ such that

(a) $U(t,s)P(s) = P(t)U(t,s)$,

(b) the restriction $U_Q(t,s) : Q(s)X \to Q(t)X$ of $U(t,s)$ is invertible (and we set $U_Q(s,t) := U_Q(t,s)^{-1}$),

(c) $\|U(t,s)P(s)\| \le Ne^{-\delta(t-s)}$ and $\|U_Q(s,t)Q(t)\| \le Ne^{-\delta(t-s)}$

for $t \ge s$ and $t, s \in \mathbb{R}$. Here and below we let $Q = I - P$ for a projection P. Exponential dichotomy is a classical concept in the study of the long–term behaviour of evolution equations, see e.g. [7], [8], [9], [13], [15]. If U is hyperbolic, then the operator family

$$\Gamma(t,s) := \begin{cases} U(t,s)P(s), & t \ge s,\ t,s \in \mathbb{R}, \\ -U_Q(t,s)Q(s), & t < s,\ t,s \in \mathbb{R}, \end{cases}$$

is called *Green's function* corresponding to U and $P(\cdot)$. If $P(t) = I$ for $t \in \mathbb{R}$, then U is *exponentially stable.* The evolution family is called *exponentially bounded* if there are constants $M > 0$ and $\gamma \in \mathbb{R}$ such that $\|U(t,s)\| \le Me^{\gamma(t-s)}$ for $t \ge s$. For computations involving Green's function it is useful to observe that

$$\begin{aligned} &(t,s) \mapsto U_Q(t,s)Q(s) \qquad \text{is strongly continuous for } t,s \in \mathbb{R},\ t \ne s, \\ &U_Q(t,s)Q(s) = U_Q(t,r)U_Q(r,s)Q(s) \quad \text{for } t,r,s \in \mathbb{R}, \end{aligned}$$

cf. [9; Lemma VI.9.17].

Let U be an exponentially bounded evolution family on X. It is a well known and important fact that the exponential dichotomy of U persists under small perturbations, see e.g. [7], [8], [9; §VI.9], [13], [18]. Our approach also relies on this property. More precisely, we will use Proposition 2.1 below which is a refinement of [20; Prop.2.3]. This result is based on a characterization of exponential dichotomy by means of the *evolution semigroup*

$$(T_U(t)f)(s) := U(s,s-t)f(s-t), \qquad t \ge 0,\ s \in \mathbb{R},\ f \in C_0(\mathbb{R},X),$$

on $C_0(\mathbb{R},X)$ (the space of continuous functions vanishing at infinity). Note that T_U is an exponentially bounded semigroup being strongly continuous on $(0,\infty)$, but not necessarily at $t = 0$. We then have

$$U \text{ has exponential dichotomy} \iff I - T_U(1) \text{ is invertible}, \tag{2.1}$$

see the equivalence (a)⇔(b) in [9; Thm.VI.9.18] or [7; §3.2.3] and the references therein. There it is also shown that then T_U is hyperbolic and that the dichotomy projections of U are given by

$$P(\cdot) = \frac{1}{2\pi i} \int\limits_{|\lambda|=1} R(\lambda, T_U(1))\, d\lambda. \tag{2.2}$$

To be precise, in these works it is assumed that $(t,s) \mapsto U(t,s)$ is strongly continuous for $t \ge s$ (and not only for $t > s$ as in our paper), but for the proof of the above mentioned facts this does not matter, see [18].

PROPOSITION 2.1. *Let U and V be evolution families with*

$$\|U(t,s)\|, \|V(t,s)\| \le Me^{\gamma(t-s)}$$

for $t \geq s$. Assume that U has an exponential dichotomy with projections $P_U(s)$ and constants $N, \delta > 0$ and that

$$q := \sup_{s \in \mathbb{R}} \|U(s+1,s) - V(s+1,s)\| \leq \frac{(1-e^{-\delta})^2}{8N^2}. \tag{2.3}$$

Then V has an exponential dichotomy with projections $P_V(s)$ and constants $0 < \delta' < -\log(2qN + e^{-\delta})$ and N', where N' only depends on $M, \gamma, N, \delta, \delta'$. Moreover,

$$\|P_U(t) - P_V(t)\| \leq q \, \frac{16N^2}{3(1-e^{-\delta})^2}\,. \tag{2.4}$$

Proof. The result is a consequence of [20; Prop.2.3] and its proof except for the uniformity of N'. Equations (2.6) and (2.7) of [20] combined with (2.3) yield

$$\|R(1, T_V(1))\| \leq \frac{8N}{3(1-e^{-\delta})}\,.$$

The uniformity of N' thus follows from the next lemma (a variant of [19; Lem.4]). ∎

LEMMA 2.2. *Let U be an evolution family on a Banach space X such that $\|U(t,s)\| \leq Me^{\gamma(t-s)}$, $t \geq s$, and $\|R(1, T_U(1))\| \leq C$ for the evolution semigroup $T_U(\cdot)$ on $C_0(\mathbb{R}, X)$. Then U has an exponential dichotomy with constants $N, \delta > 0$ depending only on M, γ, C.*

Proof. We first recall that if $\lambda \in \rho(T_U(1))$, then $\lambda e^{i\xi} \in \rho(T_U(1))$ and

$$\|R(\lambda e^{i\xi}, T_U(1)\| = \|R(\lambda, T_U(1))\|,$$

where $\xi \in \mathbb{R}$ and $\lambda \in \mathbb{C}$, see [7; (3.8)] or [9; p.483]. This fact implies

$$\|R(\lambda, T_U(1))\| \leq \tilde{C} := \frac{C}{1-(e^{\delta}-1)C} \tag{2.5}$$

for $|\lambda| = e^{\alpha}$ and $0 \leq |\alpha| \leq \delta < \log(1+\frac{1}{C})$. By (2.1) and a simple rescaling argument, we obtain the exponential dichotomy of U for every exponent $0 < \delta < \log(1+\frac{1}{C})$. Fix such a δ. If the result were false, then there would exist evolution families U_n on Banach spaces X_n satisfying the asumptions, real numbers t_n and s_n, and elements $x_n \in X_n$ such that $\|x_n\| = 1$ and

$$e^{\delta|t_n - s_n|} \|\Gamma_n(t_n, s_n)x_n\| \longrightarrow \infty \qquad \text{as } n \to \infty, \tag{2.6}$$

where Γ_n is Green's function of U_n. By assumption and (2.2), the projections $P_n(t)$ are uniformly bounded for $t \in \mathbb{R}$ and $n \in \mathbb{N}$. Hence the operators $\Gamma_n(t,s)$, $s \leq t \leq s+1$, $n \in \mathbb{N}$, are also uniformly bounded. Thus we have either $t_n > s_n + 1$ or $t_n < s_n$ in (2.6).

In the first case, we write $t_n = s_n + k_n + \tau_n$ for $k_n \in \mathbb{N}$ and $\tau_n \in (0,1]$. Otherwise, $t_n = s_n - k_n + \tau_n$ for $k_n \in \mathbb{N}$ and $\tau_n \in (0,1]$. Take continuous functions φ_n with $0 \leq \varphi_n \leq 1$, $\operatorname{supp}\varphi_n \subset (s_n + \frac{\tau_n}{2}, s_n + \frac{3\tau_n}{2})$, and $\varphi_n(t_n \mp k_n) = 1$ (here $t_n - k_n$ is used in the first case, and $t_n + k_n$ in the second). Set $\lambda = e^{\mp\delta}$ and

$f_n(s) = e^{\pm\delta(s-s_n)} \varphi_n(s) U_n(s, s_n) x_n$. (We let $U(t,s) := 0$ for $t < s$.) Using [9; (VI.9.4)] in the second inequality, we compute

$$\begin{aligned}
[R(\lambda, T_{U_n}(1)) f_n](t_n) &= \sum_{k=0}^{\infty} \lambda^{-(k+1)} [T_{U_n}(k) P_n(\cdot) f_n](t_n) \\
&\quad - \sum_{k=1}^{\infty} \lambda^{k-1} [T_{U_n,Q}(k)^{-1} Q_n(\cdot) f_n](t_n) \\
&= \sum_{k=0}^{\infty} \lambda^{-(k+1)} U_n(t_n, t_n - k) P_n(t_n - k) f_n(t_n - k) \\
&\quad - \sum_{k=1}^{\infty} \lambda^{k-1} U_{n,Q}(t_n, t_n + k) Q_n(t_n + k) f(t_n + k) \\
&= \sum_{k=0}^{\infty} e^{\pm\delta(k+1)} e^{\pm\delta(t_n - k - s_n)} \varphi_n(t_n - k) U_n(t_n, s_n) P_n(s_n) x_n \\
&\quad - \sum_{k=1}^{\infty} e^{\mp\delta(k-1)} e^{\pm\delta(t_n + k - s_n)} \varphi_n(t_n + k) U_{n,Q}(t_n, s_n) Q_n(s_n) x_n.
\end{aligned}$$

Here exactly one term does not vanish, namely $k = k_n$ in the first sum if $t_n > s_n$ and $k = k_n$ in the second sum if $t_n < s_n$. Therefore

$$R(\lambda, T_{U_n}(1)) f_n(t_n) = e^{\pm\delta} e^{\delta|t_n - s_n|} \Gamma_n(t_n, s_n) x_n ,$$

and the assumptions and (2.5) imply

$$\|e^{-\delta} e^{\delta|t_n - s_n|} \Gamma_n(t_n, s_n) x_n\| \le \|R(\lambda, T_{U_n}(1)) f_n\|_\infty \le \tilde{C} M e^{2(\gamma+\delta)}.$$

This estimate contradicts (2.6). ■

In the present work we study operators $A(t)$ on X subject to the following hypotheses.

(H1) There is an $\omega \ge 0$ such that the operators $A(t)$, $t \in \mathbb{R}$, satisfy $\Sigma_\phi \cup \{0\} \subseteq \rho(A(t) - \omega)$, $\|R(\lambda, A(t) - \omega)\| \le \frac{K}{1+|\lambda|}$, and

$$\|(A(t) - \omega) R(\lambda, A(t) - \omega) \, [R(\omega, A(t)) - R(\omega, A(s))]\| \le L \, |t - s|^\mu \, |\lambda|^{-\nu}$$

for $t, s \in \mathbb{R}$, $\lambda \in \Sigma_\phi := \{\lambda \in \mathbb{C} \setminus \{0\} : |\arg \lambda| \le \phi\}$, and constants $\phi \in (\frac{\pi}{2}, \pi)$, $L, K \ge 0$, and $\mu, \nu \in (0, 1]$ with $\mu + \nu > 1$.

This assumption was introduced by P. Acquistapace and B. Terreni in [2] (for $\omega = 0$). It implies that there exists a unique evolution family U on X such that $(t,s) \mapsto U(t,s) \in \mathcal{L}(X)$ is continuous for $t > s$, $U(\cdot, s) \in C^1((s, \infty), \mathcal{L}(X))$, $\partial_t U(t,s) = A(t) U(t,s)$, and

$$\|A(t)^k U(t,s)\| \le C \, (t-s)^{-k}, \tag{2.7}$$

$$\|A(t) U(t,s) R(w, A(s))\| \le C, \tag{2.8}$$

$$\|U(t,s)(\omega - A(s))^\alpha x\| \le C \, (\mu - \alpha)^{-1} \, (t-s)^{-\alpha} \, \|x\| \tag{2.9}$$

for $0 < t - s \leq 1$, $k = 0, 1$, $0 \leq \alpha < \mu$, $x \in D((\omega - A(s))^\alpha)$, and a constant C depending only on the constants in (H1). Moreover, $\partial_s^+ U(t,s)x = -U(t,s)A(s)x$ for $t > s$ and $x \in D(A(s))$ with $A(s)x \in \overline{D(A(s))}$. This follows by an obvious rescaling from [1; Thm.2.3] and [23; Thm.2.1], see also [2], [22]. We say that $A(\cdot)$ *generates* U. Note that U is exponentially bounded by (2.7) with $k = 0$. We further suppose that

(H2) the evolution family U generated by $A(\cdot)$ has an exponential dichotomy with constants $N, \delta > 0$, dichotomy projections $P(t)$, $t \in \mathbb{R}$, and Green's function Γ.

We point out that it is quite difficult to find conditions on $A(\cdot)$ implying (H2). Such results are usually based on (2.1) and variants of it, see [8], [10], [13], [15], [18], [20], [19].

Assume that (H1) and (H2) hold and that $f : \mathbb{R} \to X$ is bounded and continuous. A *classical solution* of (1.1) is a function $u \in C^1(\mathbb{R}, X)$ satisfying $u(t) \in D(A(t))$ for $t \in \mathbb{R}$ and (1.1). It is known that then

$$u(t) = U(t,s)u(s) + \int_s^t U(t,\tau)f(\tau)\,d\tau \qquad \text{for all } t \geq s, \tag{2.10}$$

see [1; Prop.3.2, 5.1]. On the other hand, there is a unique bounded continuous function satisfying (2.10), namely

$$u(t) = \int_{\mathbb{R}} \Gamma(t,\tau)f(\tau)\,d\tau, \qquad t \in \mathbb{R}, \tag{2.11}$$

see e.g. (the proof of) [7; Thm.4.28]. We call the function u given by (2.11) the *mild solution* of (1.1). Observe that the mild solution is a classical one if, for instance, f is Hölder continuous due to [2; Thm.6.3] and (2.10).

It remains to introduce the concept of almost periodicity, see e.g. [4], [10], [15], [16]. The next definition due to H. Bohr is the most convenient one for our purposes.

DEFINITION 2.3. Let Y be a Banach space and $g : \mathbb{R} \to Y$ be continuous. A number $\tau \in \mathbb{R}$ is an ϵ*–almost period* of g if $\|g(t+\tau) - g(t)\| \leq \epsilon$ holds for all $t \in \mathbb{R}$ and some $\epsilon > 0$. The function g is called *almost periodic* if for every $\epsilon > 0$ there exist a set $P(\epsilon) \subseteq \mathbb{R}$ of ϵ–pseudo periods of g and a number $\ell(\epsilon) > 0$ such that each interval $(a, a + \ell(\epsilon))$, $a \in \mathbb{R}$, contains a number $\tau = \tau_\epsilon \in P(\epsilon)$. The space of almost periodic functions is denoted by $AP(\mathbb{R}, Y)$.

We recall that $AP(\mathbb{R}, Y)$ is a closed subspace of the space of bounded and uniformly continuous functions, see [15; Chap.1]. Our last assumption reads as follows.

(H3) $R(\omega, A(\cdot)) \in AP(\mathbb{R}, \mathcal{L}(X))$ with pseudo periods $\tau = \tau_\epsilon$ belonging to sets $P(\epsilon, A)$.

It is not difficult to verify that then $R(\omega', A(\cdot)) \in AP(\mathbb{R}, \mathcal{L}(X))$ if $\omega' \geq \omega$.

3 YOSIDA APPROXIMATIONS

Assume that (H1) holds and define the Yosida opproximations

$$A_n(t) = nA(t)R(n, A(t))$$

of $A(t)$ for $n > \omega$ and $t \in \mathbb{R}$. These operators generate an evolution family U_n on X. We want to show that $A_n(\cdot)$ satisfies the same assumptions as $A(\cdot)$. The elementary (but tedious) proof of the next result is omitted, cf. [2; Lem.4.2] or [22; Prop.2.1].

LEMMA 3.1. *Assume that (H1) holds. Fix $\omega' > \omega$ and $\phi' \in (\frac{\pi}{2}, \phi)$. Then there are constants $n_0 > \omega$, $L' \geq L$, and $K' \geq K$ (only depending on the constants in (H1), ω', and ϕ') such that the operators $A_n(t)$, $t \in \mathbb{R}$, satisfy (H1) with constants $K', \phi', \omega', L', \mu, \nu$ for all $n \geq n_0$.*

Since (H1), (H2), and (H3) still hold for $A(\cdot)$ with constants K', L', ω', ϕ', we can assume that $A(\cdot)$ and $A_n(\cdot)$ satisfy (H1) with the same constants, denoted by K, L, ω, ϕ.

LEMMA 3.2. *If (H1) and (H3) hold, then there is a number $n_1 \geq n_0$ such that $R(\omega, A_n(\cdot)) \in AP(\mathbb{R}, \mathcal{L}(X))$ for $n \geq n_1$, with pseudo periods belonging to $P(\epsilon/\kappa, A)$, where $\kappa := 2 + 4K$.*

Proof. Let $\tau > 0$ and $t \in \mathbb{R}$. We first observe that

$$R(\lambda, A_n(t)) = \tfrac{1}{\lambda+n}(n - A(t))R(\tfrac{\lambda n}{\lambda+n}, A(t)) = \tfrac{n^2}{(\lambda+n)^2}R(\tfrac{\lambda n}{\lambda+n}, A(t)) + \tfrac{1}{\lambda+n} \tag{3.1}$$

if $n \geq n_0$ and $|\arg(\lambda - \omega)| \leq \phi$. Equation (3.1) yields

$$\begin{aligned} &R(\omega, A_n(t+\tau)) - R(\omega, A_n(t)) \\ &\quad = \frac{n^2}{(\omega+n)^2}\Big(R\Big(\frac{\omega n}{\omega+n}, A(t+\tau)\Big) - R\Big(\frac{\omega n}{\omega+n}, A(t)\Big)\Big) \\ &\quad = \frac{n^2}{(\omega+n)^2}R(\omega, A(t+\tau))\Big[1 - \frac{\omega^2}{\omega+n}R(\omega, A(t+\tau))\Big]^{-1} \\ &\quad\quad - \frac{n^2}{(\omega+n)^2}R(\omega, A(t))\Big[1 - \frac{\omega^2}{\omega+n}R(\omega, A(t))\Big]^{-1}, \end{aligned}$$

where we have used that

$$\Big\|\frac{\omega^2}{\omega+n}R(\omega, A(s))\Big\| \leq \frac{\omega^2}{\omega+n}\frac{K}{1+\omega} \leq \frac{\omega K}{n} \leq \frac{1}{2}$$

for $n \geq n_1 := \max\{n_0, 2\omega K\}$ and $s \in \mathbb{R}$. In particular,

$$\Big\|\Big[1 - \frac{\omega^2}{\omega+n}R(\omega, A(s))\Big]^{-1}\Big\| \leq 2. \tag{3.2}$$

Hence, (3.2) implies

$$\begin{aligned} &\|R(\omega, A_n(t+\tau)) - R(\omega, A_n(t))\| \\ &\quad \leq 2\,\|R(\omega, A(t+\tau)) - R(\omega, A(t))\| \\ &\quad + \frac{K}{1+\omega}\,\Big\|\Big[1 - \frac{\omega^2}{\omega+n}R(\omega, A(t+\tau))\Big]^{-1} - \Big[1 - \frac{\omega^2}{(\omega+n)^2}R(\omega, A(t))\Big]^{-1}\Big\|. \end{aligned}$$

Employing (3.2) again, we obtain

$$\begin{aligned}&\Big\|\Big[1-\frac{\omega^2}{\omega+n}R(\omega,A(t+\tau))\Big]^{-1}-\Big[1-\frac{\omega^2}{\omega+n}R(\omega,A(t))\Big]^{-1}\Big\|\\&\quad\le 4\Big\|\Big[1-\frac{\omega^2}{\omega+n}R(\omega,A(t+\tau))\Big]-\Big[1-\frac{\omega^2}{\omega+n}R(\omega,A(t))\Big]\Big\|\\&\quad\le 4\omega\,\|R(\omega,A(t+\tau))-R(\omega,A(t))\|.\end{aligned}$$

Putting everything together, we arrive at

$$\|R(\omega,A_n(t+\tau))-R(\omega,A_n(t))\|\le(2+4K)\|R(\omega,A(t+\tau))-R(\omega,A(t))\| \tag{3.3}$$

for $n\ge n_1$ and $t\in\mathbb{R}$. The assertion thus follows from (H3). ■

In order to see that $A_n(\cdot)$ satisfies also (H2), we need the following result which is of interest in itself. For $\omega=0$ it is shown in [5; Prop.4.4] (note that our result does not simply follow by rescaling). We give here a different, more elementary proof leading to a different rate of convergence.

PROPOSITION 3.3. *Let (H1) hold and fix $0<t_0<t_1$. Then $\|U(t,s)-U_n(t,s)\|\le c(t_1,\theta)\,n^{-\theta}$ for $0<t_0\le t-s\le t_1$, $n\ge n_2(t_0):=\max\{n_0,t_0^{-2/\mu}\}$, and any $0<\theta<\min\{\mu/2,1-\mu/2,\mu(\mu+\nu-1)/2\}$. Moreover, $\|(U(t,s)-U_n(t,s))R(\omega,A(s))\|\le c(t_1,\alpha)\,n^{-\alpha}$ for $0\le t-s\le t_1$, $n\ge n_0$, and $\alpha\in(0,\mu)$.*

Proof. Let $0<h<t_0$ and $0<t_0\le t-s\le t_1$. Then we have

$$\begin{aligned}U(t,s)-U_n(t,s)&=(U(t,s+h)-U_n(t,s+h))U(s+h,s)\\&\quad-U_n(t,s+h)[U(s+h,s)-e^{hA(s)}+e^{hA(s)}-e^{hA_n(s)}+e^{hA_n(s)}-U_n(s+h,s)]\\&=:S_1-S_2.\end{aligned}$$

Due to Lemma 4.3 in [2] and equation (2.6) and Lemma 2.2 in [1], we obtain

$$\|S_2\|\le c(t_1)\,(h^{\mu+\nu-1}+(hn)^{-1}). \tag{3.4}$$

The other term can be transformed into

$$\begin{aligned}S_1&=\int_{s+h}^{t}U_n(t,\sigma)(A(\sigma)-A_n(\sigma))U(\sigma,s)\,d\sigma\\&=\int_{s+h}^{t}U_n(t,\sigma)(\omega-A_n(\sigma))^{\alpha}(\omega-A_n(\sigma))^{1-\alpha}((A(\sigma)-\omega)^{-1}-(A_n(\sigma)-\omega)^{-1})\\&\qquad\cdot(A(\sigma)-\omega)U(\sigma,s)\,d\sigma.\end{aligned}$$

where $\alpha\in(0,\mu)$. The estimates (2.7), (2.9), and

$$\begin{aligned}\|(A_n(\sigma)-\omega)^{-1}-(A(\sigma)-\omega)^{-1}\|&=\|\tfrac{1}{\omega+n}A(\sigma)R(\tfrac{\omega n}{\omega+n},A(\sigma))A(\sigma)R(\omega,A(\sigma))\|\le\frac{c}{n},\\\|(\omega-A_n(\sigma))^{1-\alpha}\|&\le c\,\|\omega-A_n(\sigma)\|^{1-\alpha}\le cn^{1-\alpha}\end{aligned} \tag{3.5}$$

(use the moment inequality [9; Thm.II.5.38] in the second line) lead to

$$\|S_1\| \le c(t_1, \alpha)\, h^{-1} n^{-\alpha}. \tag{3.6}$$

Combining (3.4), (3.4), and (3.6), we deduce

$$\|U(t,s) - U_n(t,s)\| \le c(t_1, \alpha)\, ((nh)^{-1} + h^{\mu+\nu-1} + n^{-\alpha} h^{-1}).$$

The first assertion now follows if we set $h := n^{-\mu/2}$. The second one can be shown using the formula

$$(U(t,s) - U_n(t,s))R(\omega, A(s)) = \int_s^t U_n(t,\sigma)(\omega - A_n(\sigma))^{\alpha}(\omega - A_n(\sigma))^{1-\alpha}$$
$$\cdot ((A(\sigma) - \omega)^{-1} - (A_n(\sigma) - \omega)^{-1})(A(\sigma) - \omega)U(\sigma, s)R(\omega, A(s))\, d\sigma,$$

and the estimates (2.9), (3.5), and (2.8). ∎

Propositions 2.1 and 3.3 immediately imply that $A_n(\cdot)$ fulfills also (H2).

COROLLARY 3.4. *Let (H1) and (H2) hold. Then there is a number* $n_3 \ge n_2(1)$ *such that* U_n *has an exponential dichotomy for* $n \ge n_3$ *with constants* $\delta' \in (0, \delta)$ *and* $N' = N'(\delta')$ *independent of* n. *Moreover, the dichotomy projections* $P_n(t)$ *of* U_n *satisfy* $\|P_n(t) - P(t)\| \le c(\theta)\, n^{-\theta}$ *for* $t \in \mathbb{R}$, *where* $\theta \in (0,1)$ *and* $n_2(1)$ *are given by Proposition 3.3.*

4 MAIN RESULTS FOR EQUATIONS ON $\mathbb{R}$

We first establish the almost periodicity of Green's function Γ_n for the evolution family U_n generated by the Yosida approximation $A_n(\cdot)$.

LEMMA 4.1. *Assume that (H1), (H2), and (H3) hold. Let* $n \ge \max\{n_1, n_3\}$, $\eta > 0$, *and* $\tau \in P(\eta/\kappa, A)$, *where* n_1, n_3, *and* κ *were given in Section 3. We then have*

$$\|\Gamma_n(t+\tau, s+\tau) - \Gamma_n(t,s)\| \le c\,\eta\, n^2\, e^{-\frac{\delta}{2}|t-s|}, \quad t, s \in \mathbb{R}.$$

Proof. The operators $\Gamma_n(t,s)$ exist by Corollary 3.4. It is easy to see that

$$\begin{aligned} g_n(\sigma) &:= \frac{d}{d\sigma}\Big(\Gamma_n(t,\sigma)\Gamma_n(\sigma+\tau, s+\tau)\Big) \\ &= \Gamma_n(t,\sigma)(A_n(\sigma+\tau) - A_n(\sigma))\Gamma_n(\sigma+\tau, s+\tau) \\ &= \Gamma_n(t,\sigma)(A_n(\sigma) - \omega)((A_n(\sigma) - \omega)^{-1} - (A_n(\sigma+\tau) - \omega)^{-1}) \\ &\quad \cdot (A_n(\sigma+\tau) - \omega)\Gamma_n(\sigma+\tau, s+\tau), \end{aligned}$$

for $\sigma \neq t, s$, $\tau \geq 0$, and $n \geq n_3$. Hence,

$$\int_{\mathbb{R}} g_n(\sigma)\,d\sigma = \begin{cases} \int_{-\infty}^{s} g_n(\sigma)d\sigma + \int_s^t g_n(\sigma)d\sigma + \int_t^\infty g_n(\sigma)d\sigma, & t \geq s, \\ \int_{-\infty}^{t} g_n(\sigma)d\sigma + \int_t^s g_n(\sigma)d\sigma + \int_s^\infty g_n(\sigma)d\sigma, & t < s, \end{cases}$$

$$= \begin{cases} -\Gamma_n(t,s)Q_n(s+\tau) + P_n(t)\Gamma_n(t+\tau, s+\tau) \\ \qquad -\Gamma_n(t,s)P_n(s+\tau) + Q_n(t)\Gamma_n(t+\tau, s+\tau), \\ P_n(t)\Gamma_n(t+\tau, s+\tau) - \Gamma_n(t,s)Q_n(s+\tau) \\ \qquad +Q_n(t)\Gamma_n(t+\tau, s+\tau) - \Gamma_n(t,s)P_n(s+\tau), \end{cases}$$

$$= \Gamma_n(t+\tau, s+\tau) - \Gamma_n(t,s).$$

We have shown that

$$\Gamma_n(t+\tau, s+\tau) - \Gamma_n(t,s) = \int_{\mathbb{R}} \Gamma_n(t,\sigma)(A_n(\sigma) - \omega)\,[R(\omega, A_n(\sigma+\tau)) - R(\omega, A_n(\sigma))] \cdot (A_n(\sigma+\tau) - \omega)\Gamma_n(\sigma+\tau, s+\tau)\,d\sigma \tag{4.1}$$

for $s, t \in \mathbb{R}$ and $n \geq n_3$. Lemma 3.2 and Corollary 3.4 now yield

$$\|\Gamma_n(t+\tau, s+\tau) - \Gamma_n(t,s)\| \leq c\eta n^2 \int_{\mathbb{R}} e^{-\frac{3\delta}{4}|t-\sigma|}\, e^{-\frac{3\delta}{4}|\sigma-s|}\,d\sigma$$

if also $n \geq n_1$ and $\tau \in P(\eta/\kappa, A)$, which gives the asserted estimate. ∎

LEMMA 4.2. *Assume that (H1) and (H2) hold. Fix $0 < t_0 < t_1$ and let $\theta > 0$, $n_2(t_0)$, and n_3 be given by Proposition 3.3 and Corollary 3.4. Then $\|\Gamma(t,s) - \Gamma_n(t,s)\| \leq c(t_1,\theta)\, n^{-\theta}$ holds for $t_0 \leq |t-s| \leq t_1$ and $n \geq \max\{n_3, n_2(t_0)\}$.*

Proof. If $t_0 \leq t - s \leq t_1$, we write

$$\Gamma_n(t,s) - \Gamma(t,s) = (U_n(t,s) - U(t,s))P_n(s) + U(t,s)(P_n(s) - P(s)).$$

So the assertion is a consequence of Proposition 3.3 and Corollary 3.4. For $-t_1 \leq t - s \leq -t_0$, we have

$$\begin{aligned} \Gamma(t,s) - \Gamma_n(t,s) &= U_{n,Q}(t,s)Q_n(s) - U_Q(t,s)Q(s) \\ &= U_Q(t,s)Q(s)(Q_n(s) - Q(s)) + (Q_n(t) - Q(t))U_{n,Q}(t,s)Q_n(s) \\ &\quad - U_Q(t,s)Q(s)(U_n(s,t) - U(s,t))U_{n,Q}(t,s)Q_n(s). \end{aligned}$$

Again the asserted estimate follows from Proposition 3.3 and Corollary 3.4. ∎

Employing (4.1), we extend a formula given on [13; p.240] to the present setting.

COROLLARY 4.3. *Let (H1) and (H2) hold, $t, s, \tau \in \mathbb{R}$, and $x \in D((\omega - A(s))^\beta)$ for some $\beta > 0$ (or x contained in a suitable interpolation space). We then have*

$$\Gamma(t+\tau, s+\tau)x - \Gamma(t,s)x = \int_{\mathbb{R}} \Gamma(t,\sigma)(A(\sigma) - \omega)\,[R(\omega, A(\sigma+\tau)) - R(\omega, A(\sigma))] \cdot (A(\sigma+\tau) - \omega)\Gamma(\sigma+\tau, s+\tau)x\,d\sigma. \tag{4.2}$$

Proof. We want to obtain (4.2) by taking the limit as $n \to \infty$ in (4.1) evaluated at $x \in D((\omega - A(s))^\beta)$. The left hand side converges as required due to Lemma 4.2 and Corollary 3.4. For $r \neq \rho$, $n \geq n_3$, and $\alpha \in (1-\nu, \mu)$, we write

$$\Gamma_n(r,\rho)(\omega - A_n(\rho))^\alpha = \begin{cases} P_n(r)U_n(r,\rho)(\omega - A_n(\rho))^\alpha, & \rho < r \leq \rho + 1, \\ U_n(r,\rho+1)P_n(\rho+1) \\ \qquad \cdot U_n(\rho+1,\rho)(\omega - A_n(\rho))^\alpha, & \rho + 1 \leq r, \\ -U_{n,Q}(r,\rho+1)Q_n(\rho+1) \\ \qquad \cdot U_n(\rho+1,\rho)(\omega - A_n(\rho))^\alpha, & r < \rho, \end{cases}$$

$$A_n(r)\Gamma_n(r,\rho) = \begin{cases} A_n(r)U_n(r,\rho)P_n(\rho), & \rho < r \leq \rho+1, \\ A_n(r)U_n(r,r-1)U_n(r-1,\rho)P_n(\rho), & \rho+1 \leq r, \\ -A_n(r)U_n(r,r-1)U_{n,Q}(r-1,\rho)Q_n(\rho)x, & r < \rho. \end{cases}$$

By (the proof of) [22; Prop.3.1] and Lemma 4.2 these terms converge strongly to the analogous terms without n as $n \to \infty$. Using also [22; Prop.2.1], one deduces the pointwise convergence of the integrand in (4.1). Moreover, we observe that

$$\begin{aligned} A_n(r)U_n(r,\rho)P_n(\rho)x = A_n(r)U_n(r,\rho)(\omega - A_n(\rho))^{-\beta}\Big((\omega - A_n(\rho))^\beta x \\ - (\omega - A_n(\rho))^\beta U_n(\rho,\rho-1)U_{n,Q}(\rho-1,\rho)Q_n(\rho)x\Big). \end{aligned}$$

Due to (2.7) and [23; Thm.2.1], the norm of the right hand side is smaller than

$$c\,(r-\rho)^{\beta-1}(\|(\omega - A_n(\rho))^\beta x\| + \|x\|).$$

The moment inequality, see e.g. [9; Thm.II.5.38], further yields

$$\|(\omega - A_n(\rho))^\beta x\| = \|\left[(\omega - A_n(\rho))(\omega - A(\rho))^{-1}\right]^\beta (\omega - A(\rho))^\beta x\| \leq c\,\|(\omega - A(\rho))^\beta x\|.$$

Combining these estimates with (2.9), [22; Prop.2.1], and (2.7), we obtain an integrable, n–independent bound of the integrand in (4.1) evaluated at $x \in D((\omega - A(s))^\beta)$. The assertion thus follows from the theorem of dominated convergence. ∎

Though the above formula is quite interesting, the proofs of our main results only use the two preceding lemmas.

PROPOSITION 4.4. *Assume that (H1), (H2), and (H3) hold. Let $\epsilon > 0$ and $h > 0$. Then*

$$\|\Gamma(t+\tau, s+\tau) - \Gamma(t,s)\| \leq \epsilon\, e^{-\frac{\delta}{2}|t-s|}$$

holds for $|t-s| \geq h$ and $\tau \in P(\eta/\kappa, A)$, where $\kappa = 2 + 4K$ and $\eta = \eta(\epsilon, h) \to 0$ as $\epsilon \to 0$ and h is fixed.

Proof. Let $\epsilon > 0$ and $h > 0$ be fixed. Then there is a $t_\epsilon > h$ such that

$$\|\Gamma(t+\tau, s+\tau) - \Gamma(t,s)\| \leq \epsilon\, e^{-\frac{\delta}{2}|t-s|}$$

for $|t-s| \geq t_\epsilon$. Let $h \leq |t-s| \leq t_\epsilon$ and $\tau \in P(\eta/\kappa, A)$. Lemmas 4.1 and 4.2 show that

$$\|\Gamma(t+\tau, s+\tau) - \Gamma(t,s)\| \leq (c(t_\epsilon)e^{\frac{\delta}{2}t_\epsilon} n^{-\theta} + c\eta n^2)\, e^{-\frac{\delta}{2}|t-s|}$$

for $n \geq \max\{n_1, n_2(h), n_3\}$. We now choose first a large n and then a small $\eta > 0$ (depending on ϵ and h) in order to obtain the assertion. ∎

THEOREM 4.5. *Assume that (H1), (H2), and (H3) hold. Then* $r \mapsto \Gamma(t+r, s+r)$ *belongs to* $AP(\mathbb{R}, \mathcal{L}(X))$ *for* $t, s \in \mathbb{R}$, *where we may take the same pseudo periods for* t, s *with* $|t-s| \geq h > 0$. *If* $f \in AP(\mathbb{R}, X)$, *then the unique bounded mild solution* $u = \int_{\mathbb{R}} \Gamma(\cdot, s) f(s)\, ds$ *of (1.1) is almost periodic.*

Proof. In Lemma 4.1 we have seen that $P_n(\cdot) \in AP(\mathbb{R}, \mathcal{L}(X))$. Corollary 3.4 then shows that $P(\cdot) \in AP(\mathbb{R}, \mathcal{L}(X))$. Thus the first assertion follows from Proposition 4.4. Further, for $\tau, h > 0$ and $t \in \mathbb{R}$, we write

$$
\begin{aligned}
u(t+\tau) - u(t) &= \int_{\mathbb{R}} \Gamma(t+\tau, s+\tau) f(s+\tau)\, ds - \int_{\mathbb{R}} \Gamma(t,s) f(s)\, ds \\
&= \int_{\mathbb{R}} \Gamma(t+\tau, s+\tau)(f(s+\tau) - f(s))\, ds \\
&\quad + \int_{|t-s| \geq h} (\Gamma(t+\tau, s+\tau) - \Gamma(t,s)) f(s)\, ds \\
&\quad + \int_{|t-s| \leq h} (\Gamma(t+\tau, s+\tau) - \Gamma(t,s)) f(s)\, ds.
\end{aligned}
$$

For $\bar{\epsilon} > 0$ let $\eta = \eta(\bar{\epsilon}, h)$ be given by Proposition 4.4. Let $P(\epsilon, A, f)$ be the set of pseudo periods for the almost periodic function $t \mapsto (f(t), R(\omega, A(t)))$, cf. [15; p.6]. Taking $\tau \in P(\eta/\kappa, A, f)$, we deduce from Proposition 4.4 and (H2) that

$$
\|u(t+\tau) - u(t)\| \leq \tfrac{2N}{\delta\kappa}\, \eta(\bar{\epsilon}, h) + (\tfrac{4}{\delta}\, \bar{\epsilon} + 4Nh)\|f\|_\infty.
$$

for $t \in \mathbb{R}$. Given an $\epsilon > 0$, we can take first a small $h > 0$ and then a small $\bar{\epsilon} > 0$ such that $\|u(t+\tau) - u(t)\| \leq \epsilon$ for $t \in \mathbb{R}$ and $\tau \in P(\eta/\kappa, A, f) =: P(\epsilon)$. ∎

REMARK 4.6. For $g \in AP(\mathbb{R}, Y)$ and $\lambda \in \mathbb{R}$ the means

$$
\lim_{t\to\infty} \frac{1}{2t} \int_{-t}^{t} e^{-i\lambda s} g(s)\, ds
$$

exist. They are different from zero for at most countable many λ, which are called the *frequencies* of g, see e.g. [4; §4.5], [15; §2.3]. The *module* of g is the smallest additive subgroup of $\mathbb{R}$ containing all frequencies of g. By [15; p.44] (see also [10; Thm.4.5]) and the proof of Theorem 4.5 the module of the solution u to (1.1) is contained in the joint module of f and $R(\omega, A(\cdot))$ (which is the smallest additive subgroup of $\mathbb{R}$ containing the frequencies of the function $t \mapsto (f(t), R(\omega, A(t)))$). Similarly, the modules of $\Gamma(t+\cdot, s+\cdot)$, $t \neq s$, are contained in that of $R(\omega, A(\cdot))$.

EXAMPLE 4.7. Consider the parabolic problem

$$
\begin{aligned}
&D_t\, u(t,x) = \sum_{k,l=1}^{n} D_k\, a_{kl}(t,x)\, D_l u(t,x) + a_0(t,x) u(t,x) + f(t,x), \quad t \in \mathbb{R},\ x \in \Omega, \\
&\sum_{k,l=1}^{n} n_k(x)\, a_{kl}(t,x)\, D_l u(t,x) = 0, \quad t \in \mathbb{R},\ x \in \partial\Omega,
\end{aligned}
\tag{4.3}
$$

on the time interval $\mathbb{R}$. Here $\Omega \subseteq \mathbb{R}^n$ is a bounded domain with C^2–boundary $\partial\Omega$ being locally on one side of Ω, $D_t = d/dt$, $D_k = d/dx_k$, and $n(x)$ is the outer unit normal vector. We assume that the coefficients satisfy

$$a_{kl} \in C_b^\mu(\mathbb{R}, C(\overline{\Omega})) \cap C_b(\mathbb{R}, C^1(\overline{\Omega})) \cap AP(\mathbb{R}, L^n(\Omega)), \quad k, l = 1, \cdots, n,$$
$$a_0 \in C_b^\mu(\mathbb{R}, L^n(\Omega)) \cap C_b(\mathbb{R}, C(\overline{\Omega})) \cap AP(\mathbb{R}, L^{n/2}(\Omega))$$

for some $\frac{1}{2} < \mu \leq 1$, where $n/2$ is replaced by 1 if $n = 1$. Moreover, (a_{kl}) is supposed to be symmetric and real and to satisfy $\sum_{k,l} a_{kl}(t,x)\, y_k\, y_l \geq \eta\, |y|^2$ for a constant $\eta > 0$, $x \in \overline{\Omega}$, $t \in \mathbb{R}$, $y \in \mathbb{R}^n$. Let

$$A(t,x,D) = \sum_{k,l=1}^{n} D_k\, a_{kl}(t,x)\, D_l \; + a_0(t,x).$$

We then define on $X = L^p(\Omega)$, $1 < p < \infty$, the operator $A(t)\varphi = A(t,\cdot,D)\varphi$ with domain

$$D(A(t)) = \{\varphi \in W^{2,p}(\Omega) : \sum_{k,l=1}^{n} n_k(\cdot)\, a_{kl}(t,\cdot)\, D_l\varphi = 0 \text{ on } \partial\Omega\},$$

where the boundary condition is understood in the sense of traces if necessary. It is shown in [22; §4] that $A(t)$, $t \in \mathbb{R}$, fulfill (H1) for μ and each $\nu \in (0, \frac{1}{2})$. Thus there exists an evolution family U on X solving the Cauchy problem corresponding to (4.3) for $f = 0$. In the same way, one shows that (H3) holds (with the same pseudo periods as the coefficients). Therefore (4.3) has a unique mild solution $u \in AP(\mathbb{R}, X)$ provided that $f \in AP(\mathbb{R}, X)$ and U has an exponential dichotomy. The solution is classical if, e.g., f is also Hölder continuous in time.

5 MAIN RESULTS FOR EQUATIONS ON $\mathbb{R}_+$

The arguments and results of the previous sections can be extended to the problem (1.2) on $\mathbb{R}_+$. Here we will concentrate on the necessary modifications for the sake of brevity. We first introduce the space

$$AP(\mathbb{R}_+, Y) := \{g : \mathbb{R}_+ \to Y : \exists \tilde{g} \in AP(\mathbb{R}, Y) \text{ s.t. } \tilde{g}|\mathbb{R}_+ = g\}$$

of almost periodic functions on $\mathbb{R}_+$. We remark that the function $\tilde{g}$ in the above definition is uniquely determined, cf. [4; Prop.4.7.1]. The following concept is more important for our investigations.

DEFINITION 5.1. A continuous function $g : \mathbb{R}_+ \to Y$ is called *asymptotically almost periodic* if for every $\epsilon > 0$ there exists a set $P(\epsilon) \subseteq \mathbb{R}_+$ and numbers $s(\epsilon), \ell(\epsilon) > 0$ such that each interval $(a, a + \ell(\epsilon))$, $a \geq 0$, contains a number $\tau = \tau_\epsilon \in P(\epsilon)$ and the estimate $\|g(t+\tau) - g(t)\| \leq \epsilon$ holds for all $t \geq s(\epsilon)$ and $\tau \in P(\epsilon)$. The space of asymptotically almost periodic functions is denoted by $AAP(\mathbb{R}_+, Y)$.

Due to e.g. [4; Thm.4.7.5], these spaces are related by the equality

$$AAP(\mathbb{R}_+, Y) = AP(\mathbb{R}_+, Y) \oplus C_0(\mathbb{R}_+, Y). \tag{5.1}$$

Evolution families and exponential dichotomy on time intervals $[a,\infty)$ are defined by restricting the definitions on $\mathbb{R}$ to parameters $t,s \geq a$. So we can make the following assumptions.

(H1') The operators $A(t)$, $t \geq -1$, satisfy (H1) for $t,s \geq -1$.

(H2') The evolution family U generated by $A(\cdot)$ has an exponential dichotomy on $[-1,\infty)$ with projections $P(t)$, $t \geq -1$, constants $N,\delta > 0$, and Green's function Γ.

(H3') $R(\omega, A(\cdot)) \in AAP(\mathbb{R}_+, \mathcal{L}(X))$ with constants $s(\epsilon, A)$ and sets $P(\epsilon, A)$.

(We have to involve the interval $[-1,\infty)$ for technical reasons, see (5.6). Each interval $[b,\infty)$ with $b < 0$ would do the job.) Assume that (H1') and (H2') hold and let $f : \mathbb{R}_+ \to X$ be bounded and continuous. Then the *mild solution* of (1.2) is given by

$$u(t) := U(t,0)x + \int_0^t U(t,s)f(s)\,ds, \qquad t \geq 0.$$

This function is a classical solution (i.e., $u \in C^1((0,\infty),X) \cap C(\mathbb{R}_+,X)$, $u(t) \in D(A(t))$, and (1.2) holds for $t > 0$) if $x \in \overline{D(A(0))}$ and, e.g., f is Hölder continuous, see [2; Thm.6.3]. By [1; Prop.3.2,5.1], each classical solution is a mild one if $x \in \overline{D(A(0))}$. Writing $f(s) = P(s)f(s) + Q(s)f(s)$, one sees that a mild solution u satisfies

$$u(t) = U(t,0)\Big(x + \int_0^\infty U_Q(0,s)Q(s)f(s)\,ds\Big) + \int_0^\infty \Gamma(t,s)f(s)\,ds, \quad t \geq 0. \tag{5.2}$$

Thus u is bounded if and only if the term in brackets belongs to $P(0)X$ if and only if

$$Q(0)x = -\int_0^\infty U_Q(0,s)Q(s)f(s)\,ds. \tag{5.3}$$

In this case the mild solution of (1.2) is given by

$$u(t) = U(t,0)P(0)x + \int_0^\infty \Gamma(t,s)f(s)\,ds, \qquad t \geq 0. \tag{5.4}$$

Again we consider the Yosida approximations $A_n(t)$, $t \geq 0$, and the evolution family U_n generated by $A_n(\cdot)$. Lemma 3.1 and Proposition 3.3 clearly hold on $\mathbb{R}_+$ if we replace (H1) by (H1'). Since also the estimate (3.3) remains valid for $t,\tau \geq 0$, Lemma 3.2 is still true if we replace (H1) and (H3) by (H1') and (H3') and $AP(\mathbb{R},\mathcal{L}(X))$ by $AAP(\mathbb{R}_+,\mathcal{L}(X))$. Moreover, we can take $s(\epsilon, A_n) = s(\epsilon, A)$.

However, it is not immediate that Corollary 3.4 can be extended to the half line case because the proof of the perturbation result Proposition 2.1 only works on $\mathbb{R}$. But one can overcome this obstacle using a suitable extension of U.

LEMMA 5.2. *Assume that (H1') and (H2') hold. Then there is a number* $n_3' \geq n_0$ *such that, for* $n \geq n_3'$, *the evolution family* U_n *generated by* $A_n(\cdot)$ *has an exponential dichotomy on* $\mathbb{R}_+$ *with dichotomy projections* $P_n(t)$ *and constants* $\delta' \in (0,\delta)$ *and* $N' = N'(\delta')$ *independent of* n. *Moreover,* $\|P_n(t) - P(t)\| \leq cn^{-\theta}$ *for* $t \geq 0$, *where* $\theta \in (0,1)$ *is a fixed number given by Proposition 3.3.*

Proof. Let $d \geq \delta$, $n \geq n_0$, and set $R = \delta Q(0) - dP(0)$. We define

$$U^d(t,s) := \begin{cases} U(t,s), & t \geq s \geq 0, \\ U(t,0)e^{-sR}, & t \geq 0 \geq s, \\ e^{(t-s)R}, & 0 \geq t \geq s, \end{cases} \qquad U_n^d(t,s) := \begin{cases} U_n(t,s), & t \geq s \geq 0, \\ U_n(t,0)e^{-sR}, & t \geq 0 \geq s, \\ e^{(t-s)R}, & 0 \geq t \geq s. \end{cases}$$

Clearly, U^d and U_n^d satisfy (E1). Observing that $e^{rR} = e^{-rd}P(0) + e^{r\delta}Q(0)$, it is easy to see that U^d and U_n^d are exponentially bounded independent of d and n and that U^d has an exponential dichotomy on $\mathbb{R}$ with constants N, δ and projections $P^d(t) = P(t)$ for $t \geq 0$ and $P^d(t) = P(0)$ for $t \leq 0$. Moreover, $(t,s) \mapsto U_n^d(t,s)$ is norm continuous for $t \geq s$, $s \mapsto U^d(s+t,s)$ is norm continuous from the left for each $t > 0$, and $t \mapsto U^d(t,s)$ is norm continuous from the left for $t > s$ ((E2) only holds for U^d if $D(A(0))$ is dense). As shown in [18], Proposition 2.1 remains valid under these conditions. To apply this result, we want to find $n_3' \geq n_0$ and $d_n \geq \delta$ such that

$$\|U^{d_n}(s+1,s) - U_n^{d_n}(s+1,s)\| \leq cn^{-\theta} \tag{5.5}$$

for $s \in \mathbb{R}$, $n \geq n_3'$, and some $\theta > 0$. Due to Proposition 3.3, we have

$$\|U^d(s+1,s) - U_n^d(s+1,s)\| \leq \begin{cases} cn^{-\theta}, & s \geq -\frac{1}{2}, \\ 0, & -1 \geq s, \end{cases}$$

for a fixed $\theta \in (0,1)$ and $n \geq n_2(1/2)$. If $s \in (-1,-1/2)$, then Proposition 3.3 and (2.7) yield

$$\begin{aligned} &\|U^d(s+1,s) - U_n^d(s+1,s)\| \qquad\qquad (5.6)\\ &\leq \|(U(s+1,0) - U_n(s+1,0))(R(\omega,A(0))(\omega - A(0))U(0,-1)U_Q(-1,0)Q(0)e^{-s\delta}\| \\ &\qquad + \|(U(s+1,0) - U_n(s+1,0))P(0)e^{sd}\| \\ &\leq c(n^{-\theta} + e^{-d/2}) \leq 2cn^{-\theta} \end{aligned}$$

if we choose a sufficiently large $d =: d_n$. Thus (5.5) holds for $n \geq n_2(1/2) =: n_3'$ and these d_n. The assertions then follow from Proposition 2.1 by restricting U_n^d to $\mathbb{R}_+$. ■

We can now proceed almost as in the previous section; we only have to take care of certain additional exponentially decaying terms.

PROPOSITION 5.3. *Assume that (H1'), (H2'), and (H3') hold. Let* $\epsilon > 0$, $h > 0$, *and* $|t-s| \geq h$. *Then there are numbers* $\eta = \eta(\epsilon,h)$ *and* $\tilde{s}(\epsilon) \geq s(\eta, A)$ *such that*

$$\|\Gamma(t+\tau, s+\tau) - \Gamma(t,s)\| \leq \epsilon e^{-\frac{\delta}{2}|t-s|}$$

for $\tau \in P(\eta/\kappa, A)$ *and* $t,s \geq \tilde{s}(\epsilon)$, *where* $\kappa = 2 + 4K$ *and* $\eta = \eta(\epsilon,h) \to 0$ *as* $\epsilon \to 0$ *and* h *is fixed. Moreover,* $P(\cdot) \in AAP(\mathbb{R}_+, \mathcal{L}(X))$.

Proof. Let $\epsilon > 0$ and $h > 0$ be fixed. Let $t, s \geq 0$. Then there is a $t_\epsilon > h$ such that

$$\|\Gamma(t+\tau, s+\tau) - \Gamma(t,s)\| \leq \epsilon\, e^{-\frac{\delta}{2}|t-s|}$$

for $|t-s| \geq t_\epsilon$. For $h \leq |t-s| \leq t_\epsilon$ we deduce as in Lemma 4.2 from Proposition 3.3 and Lemma 5.2 that

$$\|\Gamma(t,s) - \Gamma_n(t,s)\| \leq c(t_\epsilon, \theta)\, n^{-\theta} e^{-\frac{\delta}{2}|t-s|} \tag{5.7}$$

if $n \geq \max\{n_3', n_2(h)\}$. Using the same function g_n, the arguments given in the proof of Lemma 4.1 lead to

$$\Gamma_n(t+\tau, s+\tau) - \Gamma_n(t,s) = \Gamma_n(t,a)\Gamma_n(a+\tau, s+\tau) + \int_a^\infty g_n(\sigma)\, d\sigma$$

for $t, s \geq a \geq 0$ and $\tau \geq 0$. Taking $\eta > 0$, $\tau \in P(\eta/\kappa, A)$, and $t, s \geq b \geq a := s(\eta, A)$, this equality yields as in Lemma 4.1

$$\begin{aligned}\|\Gamma_n(t+\tau, s+\tau) - \Gamma_n(t,s)\| &\leq c\eta n^2 \int_a^\infty e^{-\frac{3\delta}{4}|t-\sigma|}\, e^{-\frac{3\delta}{4}|\sigma-s|}\, d\sigma + c e^{-\frac{3\delta}{4}(t-a)}\, e^{-\frac{3\delta}{4}(s-a)} \\ &\leq c e^{-\frac{\delta}{2}|t-s|}(\eta n^2 + e^{-\frac{3\delta}{2}(b-a)}),\end{aligned} \tag{5.8}$$

where we use Lemma 5.2. We first choose a sufficiently large $n = n(\epsilon, h)$, then a small $\eta = \eta(\epsilon, h)$, and finally a large $b =: \tilde{s}(\epsilon) \geq s(\eta, A)$ in order to obtain the asserted estimate for $h \leq |t-s| \leq t_\epsilon$ from (5.7) and (5.8). The last claim is shown as in Theorem 4.5. ∎

THEOREM 5.4. *Assume that (H1'), (H2'), and (H3') hold and that $x \in X$ and $f \in AAP(\mathbb{R}_+, \mathcal{L}(X))$ satisfy (5.3). Then the mild solution u of (1.2) is asymptotically almost periodic.*

Proof. Using (5.4), we write

$$\begin{aligned} u(t+\tau) - u(t) &= (U(t+\tau, t) - I)U(t,0)P(0)x \\ &\quad + U(t+\tau,\tau)P(\tau)\int_0^\tau U(\tau,s)P(s)f(s)\,ds \\ &\quad + \int_0^\infty \Gamma(t+\tau, s+\tau)(f(s+\tau) - f(s))\,ds \\ &\quad + \int_0^\infty (\Gamma(t+\tau, s+\tau) - \Gamma(t,s))f(s)\,ds =: S_1 + S_2 + S_3 + S_4 \end{aligned}$$

for $t, \tau \geq 0$. Clearly, $\|S_1\| + \|S_2\| \leq c e^{-\delta t}$. Let $\epsilon > 0$ be given and set $a := s(\epsilon, f)$.

For $t \geq a$ and $\tau \in P(\epsilon, f)$, the asymptotic almost periodicity of f yields

$$S_3 = U(t+\tau, a+\tau)P(a+\tau) \int_0^a U(a+\tau, s)P(s)(f(s+\tau) - f(s))\, ds$$

$$+ \int_a^\infty \Gamma(t+\tau, s+\tau)(f(s+\tau) - f(s))\, ds,$$

$$\|S_3\| \leq \frac{2N^2}{\delta} \|f\|_\infty e^{-\delta(t-a)} + \frac{2N}{\delta} \epsilon.$$

For $\bar{\epsilon} > 0$ and $h > 0$, let $\eta = \eta(\bar{\epsilon}, h)$ and $b := \max\{\tilde{s}(\bar{\epsilon}), s(\epsilon, f)\}$ be given by Proposition 5.3. Choosing $t \geq b$ and $\tau \in P(\eta/\kappa, A)$, we deduce from Proposition 5.3 that

$$S_4 = U(t+\tau, b+\tau)P(b+\tau) \int_0^b U(b+\tau, s+\tau)P(s)f(s)\, ds$$

$$- U(t,b)P(b) \int_0^b U(b,s)P(s)f(s)\, ds$$

$$+ \int_b^{t-h} (\Gamma(t+\tau, s+\tau) - \Gamma(t,s))f(s)\, ds + \int_{t-h}^{t+h} \cdots\, ds + \int_{t+h}^{\infty} \cdots\, ds,$$

$$\|S_4\| \leq \Big(\frac{2N^2}{\delta} e^{-\delta(t-b)} + \frac{4\bar{\epsilon}}{\delta} + 4Nh\Big)\|f\|_\infty .$$

We now take first a small $h > 0$ and $\bar{\epsilon} > 0$ and then a large $\hat{s}(\epsilon) \geq b$ such that $\|u(t+\tau) - u(t)\| \leq c\epsilon$ for $t \geq \hat{s}(\epsilon)$ and $\tau \in P(\eta/\kappa, A, f)$, the joint set of pseudo periods of f and $R(\omega, A(\cdot))$. ■

We conclude with some remarks concerning solutions of (1.2) in $AP(\mathbb{R}_+, X)$. Assume that (H1), (H2), and (H3) hold and $f \in AP(\mathbb{R}_+, X)$. Let $\tilde{f} \in AP(\mathbb{R}, X)$ be the extension of f. By Theorem 4.5 the function

$$\tilde{u}(t) = \int_{\mathbb{R}} \Gamma(t,s)\tilde{f}(s)\, ds, \qquad t \in \mathbb{R}, \tag{5.9}$$

belongs to $AP(\mathbb{R}, X)$. Its restriction $u \in AP(\mathbb{R}_+, X)$ satisfies

$$u(t) = U(t,0) \int_{-\infty}^0 U(0,s)P(s)\tilde{f}(s)\, ds + \int_0^\infty \Gamma(t,s)f(s)\, ds, \quad t \geq 0.$$

In view of (5.2) this function is a mild solution of (1.2) with the initial value

$$x = \tilde{u}(0) = \int_{\mathbb{R}} \Gamma(0,s)\tilde{f}(s)\, ds. \tag{5.10}$$

(Note that (5.10) implies (5.3).)

Conversely, if $u \in AP(\mathbb{R}_+, X)$ is a mild solution of (1.2), then it is given by (5.4). The formula (5.4) gives

$$u(t) = U(t,0)\Big(P(0)x - \int_{-\infty}^{0} U(0,s)P(s)\tilde{f}(s)\,ds\Big) + \int_{\mathbb{R}} \Gamma(t,s)\tilde{f}(s)\,ds, \quad t \geq 0.$$

Since the second summand is almost periodic, the decomposition (5.1) shows that

$$P(0)x = \int_{-\infty}^{0} U(0,s)P(s)\tilde{f}(s)\,ds.$$

On the other hand, (5.3) must hold so that x has to satisfy (5.10).

THEOREM 5.5. *Assume that (H1), (H2), and (H3) hold and that* $f \in AP(\mathbb{R}_+, X)$*. Then (1.2) has a mild solution* $u \in AP(\mathbb{R}_+, X)$ *if and only if the initial value* x *is given by (5.10). In this case* u *is the restriction of* $\tilde{u}$ *defined in (5.9).*

BIBLIOGRAPHY

1. P. Acquistapace, *Evolution operators and strong solutions of abstract linear parabolic equations*, Differential Integral Equations **1** (1988), 433–457.

2. P. Acquistapace and B. Terreni, *A unified approach to abstract linear nonautonomous parabolic equations*, Rend. Sem. Mat. Univ. Padova **78** (1987), 47–107.

3. B. Amir and L. Maniar, *Asymptotic behaviour of hyperbolic Cauchy problems for Hille–Yosida operators with an application to retarded differential equations*, Quaest. Math. **23** (2000), 343–357.

4. W. Arendt, C. J. K. Batty, M. Hieber, and F. Neubrander, *Vector Valued Laplace Transforms and Cauchy Problems*, Birkhäuser, 2001.

5. C. J. K. Batty and R. Chill, *Approximation and asymptotic behaviour of evolution families*, Differential Integral Equations **15** (2002), 477–512.

6. C. J. K. Batty, W. Hutter, and F. Räbiger, *Almost periodicity of mild solutions of inhomogeneous Cauchy problems*, J. Differential Equations **156** (1999), 309–327.

7. C. Chicone and Y. Latushkin, *Evolution Semigroups in Dynamical Systems and Differential Equations*, Amer. Math. Soc., 1999.

8. W. A. Coppel, *Dichotomies in Stability Theory*, Springer–Verlag, 1978.

9. K.-J. Engel and R. Nagel, *One-parameter Semigroups for Linear Evolution Equations*, Graduate Text in Mathematics, vol. 194, Springer–Verlag, 2000.

10. A. M. Fink, *Almost Periodic Differential Equations*, Springer–Verlag, 1974.

11. G. Gühring and F. Räbiger, *Asymptotic properties of mild solutions of nonautonomous evolution equations with applications to retarded differential equations*, Abstr. Appl. Anal. **4** (1999), 169–194.

12. G. Gühring, F. Räbiger, and R. Schnaubelt, *A characteristic equation for nonautonomous partial functional differential equations*, J. Differential Equations **181** (2002), 439–462.

13. D. Henry, *Geometric Theory of Semilinear Parabolic Equations*, Springer–Verlag, 1981.

14. W. Hutter and F. Räbiger, *Spectral mapping theorems for evolution semigroups on spaces of almost periodic functions*, preprint.

15. B. M. Levitan and V. V. Zhikov, *Almost Periodic Functions and Differential Equations*, Cambridge University Press, 1982.

16. J. Prüss, *Evolutionary Integral Equations and Applications*, Birkhäuser, 1993.

17. W. M. Ruess and W. H. Summers, *Weak almost periodicity and the strong ergodic limit theorem for periodic evolution systems*, J. Funct. Anal. **94** (1990), 177–195.

18. R. Schnaubelt, *Asymptotic behaviour of parabolic nonautonomous evolution equations*, preprint.

19. R. Schnaubelt, *A sufficient condition for exponential dichotomy of parabolic evolution equations*, Evolution Equations and their Applications in Physical and Life Sciences (Proceedings Bad Herrenalb, 1998) (G. Lumer and L. Weis, eds.), Marcel Dekker, 2000, pp. 149–158.

20. R. Schnaubelt, *Asymptotically autonomous parabolic evolution equations*, J. Evol. Equ. **1** (2001), 19–37.

21. Vũ Quôc Phóng, *Stability and almost periodicity of trajectories of periodic processes*, J. Differential Equations **115** (1995), 402–415.

22. A. Yagi, *Parabolic equations in which the coefficients are generators of infinitely differentiable semigroups II*, Funkcial. Ekvac. **33** (1990), 139–150.

23. A. Yagi, *Abstract quasilinear evolution equations of parabolic type in Banach spaces*, Boll. Un. Mat. Ital. B **5** (1991), no. 7, 241–368.

Linear delay equations in the L_p-context

LAHCEN MANIAR
Cadi Ayyad University, Faculty of Sciences Semlalia
B.P. 2390, Marrakesh, Morocco
e-mail: maniar@ucam.ac.ma

JÜRGEN VOIGT
Fachrichtung Mathematik, Technische Universität Dresden
01062 Dresden, Germany
e-mail: voigt@math.tu-dresden.de

Dedicated to Jerry Goldstein on the occasion of his 60th birthday

ABSTRACT. We show well-posedness of the Cauchy problem for the delay equation $u'(t) = Au(t) + \Phi u_t$, with initial values in $X \times L_p(-h, 0; X)$; X a Banach space, A the generator of a C_0-semigroup on X, $h \in \{1, \infty\}$, $1 \leqslant p < \infty$. In the first result, Φ is defined by a Stieltjes integral, and $p = 1$. In the second result, Φ is a continuous linear mapping from $W_p^1(-h, 0; X)$ to the Favard class of A.

INTRODUCTION

We study linear delay equations of the form

$$\begin{cases} u'(t) = Au(t) + \Phi u_t \quad (t \geqslant 0), \\ u(0) = x, \quad u_0 = f \end{cases} \tag{DE}$$

on a Banach space X, in the L_p-context, for $1 \leqslant p < \infty$. Here, A is the generator of a C_0-semigroup on X, and Φ is a linear operator from $W_p^1(-h, 0; X)$ to X, where $h = 1$ or $h = \infty$. The aim is to show that equation (DE) is solvable for initial values

$$(x, f) \in D(A) \times W_p^1(-h, 0; X)$$

satisfying $x = f(0)$, and that the solution depends continuously on (x, f) with respect to the topology of $X \times L_p(-h, 0; X)$.

It was shown in [1; Sec.2] that (DE) is equivalent to an abstract Cauchy problem

$$\begin{cases} \mathcal{U}'(t) = \mathcal{A}_\Phi \mathcal{U}(t) \quad (t \geqslant 0), \\ \mathcal{U}(0) = \binom{x}{f} \end{cases}$$

on the space $\mathcal{E} := X \times L_p(-h, 0; X)$, where $\mathcal{A}_\Phi$ is given by

$$\mathcal{A}_\Phi := \begin{pmatrix} A & \Phi \\ 0 & \frac{d}{d\tau} \end{pmatrix},$$

with domain

$$D(\mathcal{A}_\Phi) := \left\{(x, f) \in D(A) \times W_p^1(-h, 0; X);\, f(0) = x\right\}.$$

Formulated in this context, the aim of the present paper is to show that $\mathcal{A}_\Phi$ is a generator.

We treat two types of mappings Φ, and accordingly, the paper is divided into two parts.

In the first part we treat continuous linear mappings $\Phi\colon C([-h, 0]; X) \to X$ (of a special form; see Section 1). Here we consider only the case $p = 1$, the case $1 < p < \infty$ having been treated in [1]. (This is not strictly true since in [1] only the case $h = 1$ of finite delay is considered. The proofs, however, cover also the case $h = \infty$ of infinite delay.) In this sense, this part should be considered as a supplement of [1; Example 3.4(b)].

In the second part we show that for continuous linear mappings

$$\Phi\colon W_p^1(-h, 0; X) \to D(A)$$

the operator $\mathcal{A}_\Phi$ defined above is a generator. (In fact, the result is slightly more general.) This generalizes results of [3], [4].

In Section 1 we will provide some notation, state the first result and recall the Miyadera perturbation theorem.

Section 2 contains the proof of the first result.

In Section 3 we state and prove our second result. In the proof we use a theorem of Desch and Schappacher [4].

In the Appendix, the last mentioned result is stated and provided with a new proof.

ACKNOWLEDGEMENT. This paper was started during a stay of J. Voigt at Marrakesh, Morocco, which was supported by the DAAD as well as the Cadi Ayyad University, Marrakesh.

1 NOTATION AND FIRST RESULT

Let X be a Banach space, and let A be the generator of a C_0-semigroup $(S(t);\, t \geqslant 0)$ on X. Let $1 \leqslant p < \infty$. We define a C_0-semigroup on the Banach space

$$\mathcal{E} := X \times L_p(-h, 0; X)$$

(norm given by $\|(x, f)\| := \|x\| + \|f\|_p$) by

$$\mathcal{T}(t) := \begin{pmatrix} S(t) & 0 \\ S_t & T_0(t) \end{pmatrix},$$

where $(T_0(t);\, t \geqslant 0)$, given by

$$T_0(t)f := f(\cdot + t),$$

is the C_0-semigroup of left translation on $L_p(-h, 0; X)$, and

$$S_t x(\tau) := \begin{cases} S(t+\tau)x & \text{for } -t \leqslant \tau < 0, \\ 0 & \text{for } -h < \tau < -t; \end{cases}$$

cf. [1]. The generator $\mathcal{A}$ of $\mathcal{T}(\cdot)$ is then given by

$$\mathcal{A} = \begin{pmatrix} A & 0 \\ 0 & \frac{d}{d\tau} \end{pmatrix},$$

with domain

$$D(\mathcal{A}) = \left\{(x, f) \in D(A) \times W_p^1(-h, 0; X);\, f(0) = x\right\}.$$

(We will write vectors of a product space as row vectors; only if an operator matrix is applied to such a vector we will write it as a column vector.)

The operator Φ which we will treat here is described in the following way. Let

$$\eta\colon [-h, 0] \to L(X)$$

be of bounded variation. Let

$$\Phi\colon C([-h, 0]; X) \to X$$

be the bounded linear operator given by the Stieltjes integral

$$\Phi(f) := \int_{[-h,0]} d\eta(\tau) f(\tau). \tag{1.1}$$

(Here we have used the notation

$$C([-\infty, 0]; X) := \{f \in C((-\infty, 0]; X);\, f(-\infty) := \lim_{\tau \to -\infty} f(\tau) \text{ exists}\}.)$$

Note that, by Sobolev's embedding theorem, $W_p^1(-h, 0; X) \subseteq C([-h, 0]; X)$ with continuous embedding. (In fact, for $h = \infty$ one has

$$W_p^1(-\infty, 0; X) \subseteq \{f \in C([-\infty, 0]; X);\, f(-\infty) = 0\}$$

with continuous embedding.) Thus, the operator

$$\mathcal{B} := \begin{pmatrix} 0 & \Phi \\ 0 & 0 \end{pmatrix}$$

is defined on $D(\mathcal{A})$, and is $\mathcal{A}$-bounded.

In order to explain the terminology in (DE) completely we recall that, for a function $u\colon (-h,\infty) \to X$, the function $u_t\colon (-h,0) \to X$ is defined by

$$u_t(\tau) := u(t+\tau),$$

for $t \geqslant 0$, $-h < \tau < 0$.

It was shown in [1; Theorem 2.8] that (DE) is well-posed if and only if

$$\mathcal{A}_\Phi := \begin{pmatrix} A & \Phi \\ 0 & \frac{d}{d\tau} \end{pmatrix},$$

with $D(\mathcal{A}_\Phi) = D(\mathcal{A})$, is a generator. In this way, the problem of solvability of (DE) is transformed into the question to show that $\mathcal{A}_\Phi$ is a generator.

It was further shown in [1; Example 3.4 (b)] that, for $1 < p < \infty$ and $h = 1$, $\mathcal{A}_\Phi$ is a generator. This was shown by proving that $\mathcal{B}$ defined above is an infinitesimally small Miyadera perturbation of $\mathcal{A}$ (see the terminology below). The same result, with the same proof, holds for $h = \infty$. Also, for $p = 1$ (and $h = 1$), it was shown that $\mathcal{B}$ is a small Miyadera perturbation of $\mathcal{A}$ if the total variation $|\eta|([-h,0])$ of η is small enough, and therefore $\mathcal{A}_\Phi = \mathcal{A} + \mathcal{B}$ is a generator.

The first result of this paper is the following.

THEOREM 1.1. *Let A, Φ be as above, and let $p = 1$. Then $\mathcal{A}_\Phi = \mathcal{A} + \mathcal{B}$ is a generator in $\mathcal{E}$. As a consequence, (DE) has a unique solution for all $(x,f) \in D(\mathcal{A}_\Phi) = D(\mathcal{A})$, and the solution depends continuously on (x,f).*

REMARKS 1.2. (a) For the case that A is bounded Theorem 1.1 has been proved in [3] (for $\dim X < \infty$) and [4] (see also [6]), even for more general Φ. In this case, the operator A can be included in Φ. (This, in a way, is in strong contrast to our method since we will assume that η has no contribution at zero; see Section 2.) The method of proof in [3], [4], [6] is different from our method.

(b) For the case that $h = 1$ and η is given by a density $\phi \in L_1(-1,0;L(X))$, it is shown in [2; Lemma 4.1] that Φ is an infinitesimally small Miyadera perturbation of $\mathcal{A}$ (and thus $\mathcal{A} + \mathcal{B}$ is a generator).

(c) For a variety of existence results for differential equations with delay, in Hilbert space, we refer to [12].

An essential tool in the proof of Theorem 1.1 will be the Miyadera perturbation theorem. Let $(T(t);\, t \geqslant 0)$ be a C_0-semigroup on the Banach space X, and let A be its generator. Let B be an A-bounded operator. Then B is called a Miyadera perturbation of A if there exist $\alpha > 0$, $\gamma \geqslant 0$ such that

$$\int_0^\alpha \|BT(t)x\|\, dt \leqslant \gamma \|x\|$$

for all $x \in D(A)$. The operator B is called a *small* Miyadera perturbation if α, γ can be chosen such that $\gamma < 1$, and *infinitesimally small* if the infimum of all possible γ is zero.

The Miyadera perturbation theorem states that, for a small Miyadera perturbation B, the operator $A + B$ is a generator; cf. [7; III, Theorem 3.14], [8], [9], [11].

2 PROOF OF THE FIRST RESULT.

Let A, η, Φ, $\mathcal{A}$, $\mathcal{A}_\Phi$ be as in Section 1, with $p = 1$.

We will assume that η has no mass at zero, i.e.

$$\eta(0) = \lim_{\tau \to 0} \eta(\tau). \tag{2.1}$$

This can be done without restriction because otherwise one can add $\eta_0 := \eta(0) - \lim_{\tau \to 0-} \eta(\tau)$ to A; note that the operators $\mathcal{A}$ and $\begin{pmatrix} A + \eta_0 & \Phi - \eta_0 \delta_0 \\ 0 & \frac{d}{d\tau} \end{pmatrix}$ coincide on $D(\mathcal{A})$. For convenience, we will henceforth assume that η is continuous from the left. This can be done without restriction since η has right and left limits at every point in $[-h, 0]$, and replacing η by its left-continuous version does not affect the definition of Φ.

As a matter of notation, we will denote by $|\eta|\colon [-h, 0] \to \mathbb{R}$ the variation of η,

$$|\eta|(\tau) := \sup \left\{ \sum_{j=1}^{n} \|\eta(\tau_j) - \eta(\tau_{j-1})\|; -h = \tau_0 < \ldots < \tau_n = \tau, n \in \mathbb{N} \right\},$$

but also the measure associated with that monotone function.

The main idea of our proof is to add $\mathcal{B}$ to $\mathcal{A}$ in two steps. For $0 < \alpha \leqslant 1$ we define

$$\eta_\alpha(\tau) := \begin{cases} \eta(\tau) & \text{for} \quad -h \leqslant \tau < -\alpha, \\ \eta(-\alpha) & \text{for} \quad -\alpha \leqslant \tau \leqslant 0, \end{cases}$$

and $\tilde{\eta}_\alpha := \eta - \eta_\alpha$. Then

$$|\eta_\alpha|([-h, 0]) = |\eta|([-h, -\alpha)) \leqslant |\eta|([-h, 0]) \quad (-h \leqslant \alpha < 0),$$

and $|\tilde{\eta}_\alpha|([-h, 0]) = |\eta|([-\alpha, 0]) \to 0$ as $\alpha \to 0$, according to the requirement (2.1). As in (1.1), we define Φ_α, $\tilde{\Phi}_\alpha$ corresponding to η_α, $\tilde{\eta}_\alpha$, and $\mathcal{B}_\alpha = \begin{pmatrix} 0 & \Phi_\alpha \\ 0 & 0 \end{pmatrix}$, $\tilde{\mathcal{B}}_\alpha = \begin{pmatrix} 0 & \tilde{\Phi}_\alpha \\ 0 & 0 \end{pmatrix}$, respectively. Then we first show that $\mathcal{A} + \mathcal{B}_\alpha$ is a generator, and then that $\tilde{\mathcal{B}}_\alpha$ is a small Miyadera perturbation of $\mathcal{A} + \mathcal{B}_\alpha$, if α is small.

LEMMA 2.1. *For $0 < \alpha \leqslant 1$, $\mathcal{A} + \mathcal{B}_\alpha$ is the generator of a C_0-semigroup $\mathcal{T}_\alpha(\cdot)$, and*

$$c := \sup\{\|\pi_2 \mathcal{T}_\alpha(t)\|; 0 < t \leqslant \alpha \leqslant 1\} < \infty.$$

(Here, π_2 denotes the canonical projection onto the second component of $X \times L_1(-h, 0; X)$.)

Proof. (i) First we show that $\mathcal{A} + \mathcal{B}_\alpha$ is a generator.

For $(x, f) \in D(\mathcal{A})$, the second component of $\mathcal{T}(t)\binom{x}{f}$ is given by u_t, with a function $u\colon (-h, \infty) \to X$, $u|_{(-h,0)} = f$. Therefore we obtain

$$\begin{aligned} \int_0^\alpha \left\| \mathcal{B}_\alpha \mathcal{T}(t) \binom{x}{f} \right\| dt &= \int_0^\alpha \|\Phi_\alpha u_t\| \, dt = \int_0^\alpha \left\| \int_{[-h,-\alpha)} d\eta(\tau) u(t + \tau) \right\| dt \\ &\leqslant \int_{[-h,-\alpha)} \int_0^\alpha \|f(t + \tau)\| \, dt \, d|\eta|(\tau) \leqslant |\eta|([-h, -\alpha)) \, \|f\|_1 \, . \end{aligned} \tag{2.2}$$

Choose $n \in \mathbb{N}$ such that $|\eta|([-h,-\alpha)) < n$. Then (2.2) shows that $\frac{1}{n}\mathcal{B}_\alpha$ is a small Miyadera perturbation of $\mathcal{A}$, and therefore $\mathcal{A}+\frac{1}{n}\mathcal{B}_\alpha$ is the generator of a C_0-semigroup $\mathcal{T}_{\alpha,1}(\cdot)$. It is clear that $\pi_2\mathcal{T}_{\alpha,1}(t)\binom{x}{f}$ and $\pi_2\mathcal{T}(t)\binom{x}{f}$ coincide on $(-h,-\alpha)$, for $0 \leqslant t \leqslant \alpha$. From the fact that η_α has no mass in $[-\alpha, 0]$ it follows that $\mathcal{B}_\alpha\mathcal{T}_{\alpha,1}(t)\binom{x}{f} = \mathcal{B}_\alpha\mathcal{T}(t)\binom{x}{f}$ for $0 \leqslant t \leqslant \alpha$, and therefore one obtains the same estimate as (2.2) for $\mathcal{T}_{\alpha,1}$,

$$\int_0^\alpha \left\| \mathcal{B}_\alpha\mathcal{T}_{\alpha,1}(t)\begin{pmatrix} x \\ f \end{pmatrix} \right\| dt \leqslant |\eta|([-h,-\alpha))\, \|f\|_1\,,$$

for all $(x,f) \in D(\mathcal{A}) = D(\mathcal{A}+\frac{1}{n}\mathcal{B}_\alpha)$. As before we conclude that $(\mathcal{A}+\frac{1}{n}\mathcal{B}_\alpha)+\frac{1}{n}\mathcal{B}_\alpha = \mathcal{A}+\frac{2}{n}\mathcal{B}_\alpha$ is a generator. Repeating this argument n times we obtain that $\mathcal{A}+\mathcal{B}_\alpha$ is a generator.

(ii) Let $(x,f) \in D(\mathcal{A})$. For $0 \leqslant t \leqslant \alpha$ we have $\mathcal{B}_\alpha\mathcal{T}_\alpha(t)\binom{x}{f} = \mathcal{B}_\alpha\mathcal{T}(t)\binom{x}{f}$, and therefore

$$\begin{aligned} \mathcal{T}_\alpha(t)\begin{pmatrix} x \\ f \end{pmatrix} &= \mathcal{T}(t)\begin{pmatrix} x \\ f \end{pmatrix} + \int_0^t \mathcal{T}(t-s)\mathcal{B}_\alpha\mathcal{T}_\alpha(s)\begin{pmatrix} x \\ f \end{pmatrix} ds \\ &= \mathcal{T}(t)\begin{pmatrix} x \\ f \end{pmatrix} + \int_0^t \mathcal{T}(t-s)\mathcal{B}_\alpha\mathcal{T}(s)\begin{pmatrix} x \\ f \end{pmatrix} ds. \end{aligned}$$

For the first component $u(t)$ of $\mathcal{T}_\alpha(t)\binom{x}{f}$ this means

$$u(t) = S(t)x + \int_0^t S(t-s) \int_{[-h,-\alpha)} d\eta(\tau) f(s+\tau)\, ds.$$

With this function we have

$$\mathcal{T}_\alpha(t)\begin{pmatrix} x \\ f \end{pmatrix} = \begin{pmatrix} u(t) \\ u_t \end{pmatrix}$$

(where $u|_{(-h,0)} = f$), and therefore, with $c_S := \sup\limits_{0 \leqslant t \leqslant 1} \|S(t)\|$,

$$\begin{aligned} &\left\| \pi_2\mathcal{T}_\alpha(t)\begin{pmatrix} x \\ f \end{pmatrix} \right\| = \|u_t\|_1 \\ &\leqslant \|f\|_1 + \alpha c_S\left(\|x\| + \int_0^\alpha \int_{[-h,-\alpha)} \|f(s+\tau)\|\, d|\eta|(\tau)\, ds \right) \\ &\leqslant \|f\|_1 + \alpha c_S\left(\|x\| + \int_{[-h,-\alpha)} d|\eta|(\tau)\, \|f\|_1 \right) \\ &= \alpha c_S\|x\| + (1 + \alpha c_S|\eta|([-h,-\alpha)))\, \|f\|_1\,. \end{aligned}$$

■

LEMMA 2.2. *With the constant c from Lemma 2.1 one has*

$$\int_0^\alpha \left\| \tilde{\mathcal{B}}_\alpha \mathcal{T}_\alpha(t) \binom{x}{f} \right\| dt \leqslant c\,|\eta|([-\alpha,0])\;\|(x,f)\|$$

for all $0 < \alpha \leqslant 1$, $(x,f) \in D(\mathcal{A})$.

Proof. With the estimate from Lemma 2.1 we obtain (denoting $u_t = \pi_2 \mathcal{T}_\alpha(t)\binom{x}{f}$ as before)

$$\int_0^\alpha \left\| \tilde{\mathcal{B}}_\alpha \mathcal{T}_\alpha(t) \binom{x}{f} \right\| dt = \int_0^\alpha \left\| \int_{[-\alpha,0]} d\eta(\tau) u_t(\tau) \right\| dt$$

$$\leqslant \int_0^\alpha \int_{[-\alpha,0]} \|u(t+\tau)\|\, d|\eta|(\tau)\, dt = \int_{[-\alpha,0]} \int_0^\alpha \|u(t+\tau)\|\, dt\, d|\eta|(\tau)$$

$$\leqslant \int_{[-\alpha,0]} \|u_{\tau+\alpha}\|_1\; d|\eta|(\tau) \leqslant c\,|\eta|([-\alpha,0])\;\|(x,f)\|\,.$$

■

Proof of Theorem 1.1. For the constant c from Lemma 2.1, we choose $\alpha > 0$ small enough to obtain $c\,|\eta|([-\alpha,0]) < 1$. From Lemma 2.1 we know that $\mathcal{A} + \mathcal{B}_\alpha$ is a generator. Then Lemma 2.2 states that $\tilde{\mathcal{B}}_\alpha$ is a small Miyadera perturbation of $\mathcal{A} + \mathcal{B}_\alpha$, and this implies that $\mathcal{A} + \mathcal{B}_\alpha + \tilde{\mathcal{B}}_\alpha = \mathcal{A} + \mathcal{B} = \mathcal{A}_\Phi$ is a generator. ■

3 THE SECOND RESULT

Let $1 \leqslant p < \infty$. In this section we treat the case where

$$\Phi\colon \{f \in W_p^1(-h,0;X);\; f(0) \in D(A)\} \to F_A \tag{3.1}$$

is linear and continuous. Here, $\{f \in W_p^1(-h,0;X);\; f(0) \in D(A)\}$ is endowed with the norm coming from the graph norm of $D(\mathcal{A})$, $\|f\|_{\mathcal{A}} := \|f\|_{W_p^1} + \|f(0)\| + \|Af(0)\|$, and F_A is the *Favard class* of A defined by

$$F_A := \{x \in X;\; \|x\|_{F_A} := \|x\| + \sup_{0<t\leqslant 1} \frac{1}{t}\|T(t)x - x\| < \infty\}.$$

Recall that D_A is continuously embedded in F_A, where $D_A = (D(A), \|\cdot\|_A)$, with the graph norm $\|x\|_A = \|x\| + \|Ax\|$.

THEOREM 3.1. *The operator*

$$\mathcal{A}_\Phi := \begin{pmatrix} A & \Phi \\ 0 & \frac{d}{d\tau} \end{pmatrix},$$

with $D(\mathcal{A}_\Phi) = \{(x,f) \in D(A) \times W_p^1(-h,0;X);\; f(0) = x\}$ *is a generator.*

REMARKS 3.2. (a) If $A \in L(X)$, then A can be included in Φ, and therefore this case enters the case $A = 0$ treated in [4], [6]; see also Remark 1.2 (a).

(b) Our method of proof is essentially the proof in [4]; we only have to choose the auxiliary space Z differently.

Proof of Theorem 3.1. We will apply [4; Theorem on p. 330], which we recall and prove in the Appendix.

We are going to show that Theorem A.1 is applicable with $Z := F_A \times \{0\}$, for the C_0-semigroup $\mathcal{T}(\cdot)$ generated by $\mathcal{A} := \begin{pmatrix} A & 0 \\ 0 & \frac{d}{d\tau} \end{pmatrix}$ with $D(\mathcal{A}) = D(\mathcal{A}_\Phi)$ as above, and the perturbation $\mathcal{B} := \begin{pmatrix} 0 & \Phi \\ 0 & 0 \end{pmatrix}$. Recall that $\mathcal{T}(\cdot)$ is given as follows: For $(x, f) \in \mathcal{E}$ let $u\colon (-h, \infty) \to X$,

$$u(t) := \begin{cases} f(t) & \text{for} \quad -h < t < 0, \\ S(t)x & \text{for} \quad 0 \leqslant t. \end{cases}$$

Then $\mathcal{T}(t)(x, f) = (u(t), u_t)$ $(t \geqslant 0)$.

Now we show that Z satisfies (Z1), (Z2) of Theorem A.1. (Z1) is evident. In order to show (Z2) we take $\varphi\colon [0,1] \to Z$ continuous, $\varphi(t) = (\varphi_1(t), 0)$. Then

$$\mathcal{T}(t-s)\varphi(s) = (S(t-s)\varphi_1(s), \tau \mapsto S(t-s+\tau)\varphi_1(s))$$

for $0 \leqslant s \leqslant t$ (where $S(t) := 0$ for $t < 0$),

$$\int_0^t \mathcal{T}(t-s)\varphi(s)\,ds = \left(\int_0^t S(t-s)\varphi_1(s)\,ds, \tau \mapsto \int_0^{t+\tau} S(t+\tau-s)\varphi_1(s)\,ds \right). \tag{3.2}$$

We recall that F_A satisfies properties (Z1) and (Z2) of Theorem A.1 (for $Z = F_A$); cf. [4; p.336]. In fact, there exists $c_A \geqslant 0$ such that

$$\left\| A \int_0^t S(t-s)\varphi_1(s)\,ds \right\| \leqslant c_A t \sup_{0 \leqslant s \leqslant t} \|\varphi_1(s)\|_{F_A}$$

for $0 \leqslant t \leqslant 1$. From (Z1) for F_A we obtain $\int_0^t S(t-s)\varphi_1(s)\,ds \in D(A)$, and this is also the value of the second component in (3.2) for $\tau = 0$. Using standard methods and properties (Z1), (Z2) (for $Z = F_A$, as above) one shows that the second component

$$\psi_1(\tau) := \int_0^{t+\tau} S(t+\tau-s)\varphi_1(s)\,ds$$

is right differentiable on $[-t, 0)$,

$$\left(\frac{d}{d\tau}\right)_{\text{right}} \psi_1(\tau) = \varphi_1(t+\tau) + A \int_0^{t+\tau} S(t+\tau-s)\varphi_1(s)\,ds.$$

Using again (Z2) one shows that the last expression depends continuously on τ, and therefore ψ_1 is continuously differentiable on $[-t, 0]$.

Noting $\psi_1(-t) = 0$ we obtain $\int_0^t \mathcal{T}(t-s)\varphi(s)\,ds \in D(\mathcal{A})$. Moreover

$$\begin{aligned}
&\left\| \mathcal{A} \int_0^t \mathcal{T}(t-s)\varphi(s)\,ds \right\| \\
&= \Bigg\| \Bigg(A \int_0^t S(t-s)\varphi_1(s)\,ds, \\
&\qquad\qquad \tau \mapsto \Big(\varphi_1(t+\tau) + A \int_0^{t+\tau} S(t+\tau-s)\varphi_1(s)\,ds\Big) \Bigg) \Bigg\| \\
&\leqslant \left\| A \int_0^t S(t-s)\varphi_1(s)\,ds \right\| \\
&\qquad + \|\varphi_1(t+\cdot)\|_{L_p} + \left\| \tau \mapsto A \int_0^{t+\tau} S(t+\tau-s)\varphi_1(s)\,ds \right\|_{L_p} \\
&\leqslant c_A t \sup_{0\leqslant s\leqslant t} \|\varphi_1(s)\|_{F_A} \\
&\qquad + \|\varphi_1(t+\cdot)\|_{L_p} + c_A \left(\int_{-t}^0 (t+\tau)^p\,d\tau \right)^{1/p} \sup_{0\leqslant s\leqslant t} \|\varphi_1(s)\|_{F_A} \\
&\leqslant c_A t \sup_{0\leqslant s\leqslant t} \|\varphi_1(s)\|_{F_A} \\
&\qquad + t^{1/p} \sup_{0\leqslant s\leqslant t} \|\varphi_1(s)\| + c_A \left(\frac{1}{p+1} t^{p+1} \right)^{1/p} \sup_{0\leqslant s\leqslant t} \|\varphi_1(s)\|_{F_A}.
\end{aligned}$$

This shows the inequality needed in (Z2).

Assumption (3.1) just means that $\mathcal{B}\colon D_{\mathcal{A}} \to Z$ is continuous. ∎

REMARK 3.3. Note that the proof given above also implies that the space Z satisfies condition (Z_p) of [5; see p.134].

APPENDIX

In this appendix we recall the result of Desch and Schappacher used in Section 3. We also provide a new proof of this result.

THEOREM A.1. *(Desch, Schappacher [4; p. 330]) Let A be the generator of a C_0-semigroup $T(\cdot)$ on a Banach space E. Let $(Z, \|\cdot\|_Z)$ be a Banach space satisfying*

(Z1) Z is continuously embedded in X;

(Z2) there exist $t_0 > 0$, $\gamma\colon (0,t_0] \to [0,\infty)$, $\lim_{t\to 0} \gamma(t) = 0$, such that for all continuous functions $\varphi\colon [0,t_0] \to Z$ one has

$$\int_0^t T(t-s)\varphi(s)\,ds \in D(A),$$

$$\left\| A \int_0^t T(t-s)\varphi(s)\,ds \right\| \leqslant \gamma(t) \sup_{0 \leqslant s \leqslant t} \|\varphi(s)\|_Z \,.$$

Let $B\colon D_A \to Z$ be linear and continuous, where $D_A := (D(A), \|\cdot\|_A)$. Then $A+B$ is a generator.

Proof. We are going to show that (i) there exists $\hat{t} > 0$ such that for all $x \in D(A+B) = D(A)$ the Cauchy problem

$$u'(t) = (A+B)u(t), \quad u(0) = x \tag{A.1}$$

has a unique solution on $[0, \hat{t}]$, (ii) $\rho(A+B) \neq \emptyset$. These properties imply that $A+B$ is a generator; cf. [10; Ch. 4, Theorem 1.3].

(i) There exists $\hat{t} > 0$ such that

$$c := \left(\hat{t}\, c_T \,\|id\|_{L(Z,X)} + \sup_{0<t\leqslant \hat{t}} \gamma(t) \right) \|B\|_{L(D_A,Z)} < 1,$$

where $c_T := \sup_{0\leqslant t\leqslant 1} \|T(t)\|$. Let $x \in D(A)$. We want to show that there exists a unique solution $f \in C([0,\hat{t}]; D_A)$ of the integral equation

$$f(t) = T(t)x + \int_0^t T(t-s)Bf(s)\,ds. \tag{A.2}$$

In order to show this we note first that for $f \in C([0,\hat{t}]; D_A)$ the function Ψf given by

$$\Psi f(t) := T(t)x + \int_0^t T(t-s)Bf(s)\,ds$$

is again in $C([0,\hat{t}]; D_A)$; cf. [4; p. 331]. Moreover, it follows from the choice of $\hat{t}$ that Ψ is a contraction with contraction constant $c < 1$. Therefore the contraction mapping principle implies the existence of a unique fixed point f of Ψ, i.e., of a solution f of (A.2).

Next we show that $t \mapsto \int_0^t T(t-s)Bf(s)\,ds$ is differentiable as an X-valued function. In

$$\frac{1}{h}\left(\int_0^{t+h} T(t+h-s)Bf(s)\,ds - \int_0^t T(t-s)Bf(s)\,ds \right)$$
$$= \frac{1}{h} \int_t^{t+h} T(t+h-s)Bf(s)\,ds + \frac{1}{h}(T(h)-I) \int_0^t T(t-s)Bf(s)\,ds$$

we let $h \to 0+$ and obtain

$$Bf(t) + A \int_0^t T(t-s)Bf(s)\,ds.$$

Since this function is continuous we obtain the differentiability, and as a consequence, f is a solution of (A.1).

Since every solution of (A.1) is also a solution of (A.2) the uniqueness of solutions of (A.2) implies the same for (A.1).

(ii) In [4; p. 332] it is shown that there exists $\lambda \in \rho(A)$ such that $(\lambda - A)^{-1}B$ is contractive as a mapping in D_A. This implies that $I_{D_A} - (\lambda - A)^{-1}B$ is invertible in $L(D_A)$; therefore

$$\lambda - A - B = (\lambda - A)(I_{D_A} - (\lambda - A)^{-1}B)$$

has the inverse

$$\left(I_{D_A} - (\lambda - A)^{-1}B\right)^{-1}(\lambda - A)^{-1} \in L(X, D_A) \subseteq L(X),$$

i.e., $\lambda \in \rho(A + B)$. ■

BIBLIOGRAPHY

1. A. Bátkai and S. Piazzera, *Semigroups and linear partial differential equations with delay*, J. Math. Anal. Appl. **264** (2001), 1–20.

2. S. Brendle, *On the asymptotic behavior of perturbed strongly continuous semigroups*, Math. Nachr. **226** (2001), 35–47.

3. M. C. Delfour, *The largest class of hereditary systems defining a C_0 semigroup on the product space*, Can. J. Math. **32** (1980), 969–978.

4. W. Desch and W. Schappacher, *On relatively bounded perturbations of linear C_0-semigroups*, Ann. Sc. Norm. Super. Pisa., Cl. Sci. **XI** (1984), 327–341.

5. W. Desch and W. Schappacher, *Some generation results for perturbed semigroups*, Semigroup Theory and Applications (Clément, Invernizzi, Mitidieri, and Vrabie, eds.), Lect. Notes Pure Appl. Math., no. 116, 1989, pp. 125–152.

6. K.-J. Engel, *Spectral theory and generator property of one-sided coupled operator matrices*, Semigroup Forum **58** (1999), 267–295.

7. K.-J. Engel and R. Nagel, *One-parameter Semigroups for Linear Evolution Equations*, Graduate Text in Mathematics, vol. 194, Springer, New York, 2000.

8. I. Miyadera, *On perturbation theory for semi-groups of operators*, Tôhoku Math. J. **18** (1966), 299–310.

9. I. Miyadera, *On perturbation for semigroups of linear operators*, Scientific Researches, School of Education, Waseda Univ. **21** (1972), 21–24 (japanese).

10. A. Pazy, *Semigroups of Linear Operators and Applications to Partial Differential Equations*, Springer, New York, 1983.

11. J. Voigt, *On the perturbation theory for strongly continuous semigroups*, Math. Ann. **229** (1977), 163–171.

12. G. F. Webb, *Functional differential equations and nonlinear semigroups in L^p-spaces*, J. Differ. Equations **20** (1976), 71–89.

Integrated Form of Continuous Newton's Method

J. W. NEUBERGER
Department of Mathematics, University of North Texas
Denton, TX 76203
e-mail: jwn@unt.edu

To Jerry Goldstein

ABSTRACT. An integrated form of continuous Newton's method is defined. Under rather minimal conditions the method is shown to lead to a zero of the given function. The result is applied to recover a recent Nash-Moser type inverse function theorem.

1 INTRODUCTION

Many problems in PDE may be expressed as a problem of finding u so that

$$F(u) = 0 \tag{1.1}$$

where F is a function from a Banach space H to a Banach space K. A prospective method for solving (1.1) is called here Integrated Continuous Newton's Method: Given $z_0 \in H$, find a continuous function $z : [0, \infty) \to H$ such that

$$z(0) = z_0,\ (F(z))'(t) = -F(z(t)),\ t \geq 0. \tag{1.2}$$

If F is C^1, $F'(x)^{-1}$ exists for $x \in H$ and z is differentiable, then (1.2) takes the form

$$z(0) = z_0,\ z'(t) = -F'(z(t))^{-1}F(z(t)),\ t \geq 0. \tag{1.3}$$

If we interpret invertibility in the set theoretic sense

$$F'(x)^{-1}y = \{w \in H : F'(x)w = y\}$$

then a substitute for (1.3) takes the form

$$z(0) = z_0,\ z'(t) \in -F'(z(t))^{-1}F(z(t)),\ t \geq 0, \tag{1.4}$$

what may be called a differential inclusion.

However, even in the case of ordinary complex polynomials [5], (1.2) may have meaning where neither (1.3),(1.4) does. Accordingly we concentrate on (1.2).

We attempt some perspective on PDE and the present work. Denote $[0,1]^2$ by Ω, $H = H^{1,2}(\Omega)$, $K = L_2(\Omega)$ and define F so that

$$F(u) \;=\; u_1,\; u \in H,$$

where u_1 indicates the derivative of u in its first argument. Given $g \in K$, we have the problem of finding $u \in H$ such that

$$F(u) \;=\; g. \tag{1.5}$$

Some reflection yields that this problem has no solution if g lacks a certain amount of smoothness in its second argument. It is this lack of proper inverses that makes discrete Newton's method difficult (see [3]). Earlier related work on continuous Newton's method is found in [4].

Recent papers [1], [2], [6] have used continuous methods to overcome loss of derivative problems which seem unavoidable in conventional Newton's method. The present note seeks to relate this line of development to (1.2).

2 STATEMENT OF MAIN RESULT

Suppose that each of H, J, K is a Banach space with H compactly embedded in J in the sense that the points of H form a dense linear subspace of J and that every bounded sequence $\{x_k\}_{k=1}^{\infty}$ in H has a subsequence convergent in J to a member $x \in H$ so that

$$\|x\|_H \leq \limsup_{k\to\infty} \|x_k\|_H.$$

If $x \in H, s > 0$, then

$$B_s(x) \;=\; \{y \in H : \|y - x\|_H \leq s\}$$

and

$$b_s(x) \;=\; \{y \in H : \|y - x\|_H < s\}.$$

Fix $v \in H,\ r > 0$. Suppose that F is a function with range in K so that the domain of F contains $B_r(v)$. Suppose also that F is continuous as a function on J into K.

THEOREM 2.1. *Suppose that for each $x \in b_r(v)$ there is $h \in B_r(0)$ so that*

$$F'(x)h \;=\; -F(v). \tag{2.1}$$

Then there is a function $z : [0,\infty) \to H$, continuous as a function into J so that

$$z(0) = v,\; (F(z))'(t) \;=\; -F(z(t)),\; t \geq 0.$$

Moreover,

$$u \;=\; J - \lim_{t\to\infty} z(t) \textit{ exists },u \in H \textit{ and } F(u) \;=\; 0.$$

The derivative in (2.1) is defined simply as

$$F'(x)h = \lim_{t \to 0+} \frac{1}{t}(F(x+th) - F(x)).$$

No Fréchet differentiability is assumed.

An argument is based on the following:

LEMMA 2.2. *Under the hypothesis of Theorem 2.1, suppose that*

$$t \in [0,1), \ y \in b_{tr}(v), \ s \in (0, 1-t].$$

Then if $\epsilon > 0$, *there is* $w \in B_{sr}(y)$ *so that*

$$||F(w) - F(y) + sF(v)||_K \le \epsilon s.$$

3 PROOFS

Lemma 2.2 Suppose $0 \le t < 1$, $y \in b_{tr}(v), \epsilon > 0$. Pick $h \in B_r(0)$ so that

$$F'(y)h = -F(v).$$

There is $\delta \in (0, 1-t]$ such that if $0 < s \le \delta$, then

$$||F(y+sh) - F(y) + sF(v)||_K \le s\epsilon.$$

Note that $y + sh \in B_{sr}(y)$. Denote by $M_{y,\epsilon}$ the collection of all $\delta \in (0, 1-t]$ such that if $0 < s \le \delta$ then there is $w \in B_{sr}(y)$ so that

$$||F(w) - F(y) + sF(v)||_K \le s\epsilon.$$

Clearly $M_{y,\epsilon}$ is connected. The lemma follows if it can be shown that $M_{y,\epsilon}$ contains $1-t$.

Denote $\sup M_{y,\epsilon}$ by q. Denote by $\{w_k\}_{k=1}^{\infty}$ a sequence in H and $\{q_k\}_{k=1}^{\infty}$ an increasing sequence of positive numbers convergent to q such that

$$w_k \in B_{rq_k}(y) \text{ and } ||F(w_k) - F(y) + q_k F(v)||_K \le q_k \epsilon. \ k = 1, 2, \ldots.$$

Choose an increasing sequence of positive integers $\{k_i\}_{i=1}^{\infty}$ so that

$$\{w_{k_i}\}_{i=1}^{\infty}$$

converges in $||\cdot||_J$ to some $w \in H$ and

$$||w||_H \le \limsup_{k \to \infty} ||w_k||_H.$$

Note then that $w \in B_{qr}(y)$ and

$$||F(w) - F(y) + qF(v)||_K \le q\epsilon. \tag{3.1}$$

Hence $q \in M_{y,\epsilon}$. If $q = 1-t$ the argument is finished.

Suppose then that $q < 1 - t$. Define $h_0 \in B_r(0)$ so that

$$F'(w)h_0 = -F(v)$$

Denote by d a positive number so that $d < 1 - (t+q)$ and so that if $0 < s \leq d$ then

$$\|F(w + sh_0) - F(w) + sF(v)\|_K \leq s\epsilon. \tag{3.2}$$

Note that for such a choice of s, $w + sh_0 \in B_{(q+s)r}(y)$. Combining (3.1), (3.2), if $0 < s \leq d$, then

$$\|F(w + sh_0) - F(y) + (s+q)F(v)\|_K \leq (s+q)\epsilon.$$

This places $d + q \in M_{y,\epsilon}$, contradicting the proposition that $q = \sup M_{y,\epsilon}$. Thus the assumption that $q < 1 - t$ is false and the argument is finished. ■

Theorem 2.1 Denote by $\{p^{(k)}\}_{k=1}^{\infty}$ a sequence of partitions of $[0,1]$ so that $p^{(k+1)}$ is a refinement of $p^{(k)}$, $k = 1, 2, \ldots$, and so that the mesh of $p^{(k)} \to 0$ as $k \to \infty$. Suppose that k is a positive integer and that

$$p^{(k)} = t_0^{(k)}, \ldots, t_{n_k}^{(k)}.$$

Denote by z_k a continuous function on $[0,1]$ which is linear on each of $[t_{i-1}^{(k)}, t_i^{(k)}]$, $i = 1, \ldots, n_k$, and so that $z_k(0) = v$,

$$\|z_k(t_i^{(k)}) - z_k(t_{i-1}^{(k)})\|_H \leq (t_i^{(k)} - t_{i-1}^{(k)})r,$$

and

$$\|F(z_k(t_i^{(k)})) - F(z_k(t_{i-1}^{(k)})) + (t_i^{(k)} - t_{i-1}^{(k)})F(v)\|_K \leq (t_i^{(k)} - t_{i-1}^{(k)})/k,$$

$i = 1, \ldots, n_k$, $k = 1, 2, \ldots$. Summing the above inequalities from $i = 1$ to $i = j \leq n_k$ and using the triangle inequality we see that

$$\|F(z_k(t_j^{(k)}) - F(v) + t_j^{(k)}F(v)\|_K \leq t_j^{(k)}/k \leq 1/k. \tag{3.3}$$

(The preceding lemma permits this construction.)

Note that $\{z_k\}_{k=1}^{\infty}$ is uniformly bounded relative to $\|\cdot\|_H$ and equicontinuous (actually uniformly lipschitz with constant r). Hence there is an increasing sequence $\{k_j\}_{j=1}^{\infty}$ of positive integers so that

$$\{z_{k_j}\}_{j=1}^{\infty}$$

converges uniformly using $\|\cdot\|_J$ to a J continuous $H-$ valued function z on $[0,1]$. Since F is continuous as a function on J, $\{F(z_{k_j})\}_{j=1}^{\infty}$ converges uniformly to $F(z)$ on $[0,1]$.

Suppose now that $s \in [0,1]$. Then s is a sequential limit of a sequence $\{s_k\}_{k=1}^{\infty}$ where $s_k \in p^{(k)}$, $k = 1, 2, \ldots$. Using (3.3) it follows that

$$F(z(s)) = (1-s)F(v). \tag{3.4}$$

Hence

$$(F(z))'(s) = -F(v),\ s \in [0,1]. \tag{3.5}$$

It is clear that $F(z(1)) = 0$. Conclusions (3.4), (3.5) are at least as significant as the stated conclusion to the theorem.

To reach our final conclusion, define $\alpha : [0,\infty) \to [0,1)$ by

$$\alpha(t) = 1 - \exp(-t),\ t \geq 0.$$

Define $w : [0,\infty) \to H$ by

$$w(t) = z(\alpha(t)),\ t \geq 0$$

and note that

$$(F(w))'(t) = ((F(z))(\alpha))'(t) = (F(z))'(\alpha(t))\alpha'(t) = \\ -\exp(-t)F(v) = -F(w(t)),\ t \geq 0$$

since

$$F(w(t)) = F(z(\alpha(t))) = (1-\alpha(t))F(v) = \exp(-t)F(v).$$

As noted above, $F(z(1)) = 0$ so it follows that

$$u = J - \lim_{t\to\infty} w(t) = z(1) \in H,$$

and so

$$F(u) = 0.$$

■

4 APPLICATION TO A NASH-MOSER INVERSE FUNCTION THEOREM

The above theorem can be used to recover an inverse function theorem in [6]:

THEOREM 4.1. *Suppose H, J, K are as in Theorem 1, $r, M > 0$ and G is a function with domain $B_r(0)$ so that $G(0) = 0$ and G is continuous as a function on J. Suppose furthermore that $g \in K$ and if $y \in b_r(0)$, there is $h \in B_M(0)$ so that*

$$G'(y)h = g.$$

Then there is $\lambda > 0$ so that

$$G(u) = \lambda g \text{ for some } u \in B_r(0).$$

Proof. Pick $\lambda > 0$ so that $\lambda M \leq r$. Take $v = 0$ in Theorem 1. Define F on $B_r(0)$ so that

$$F(x) = G(x) - \lambda g,\ x \in B_r(0).$$

Suppose $x \in b_r(0)$. Then there is $h_0 \in B_M(0)$ so that

$$G'(x)h_0 = g$$

and so with $h = \lambda h_0$,

$$G'(x)h \;=\; \lambda g,$$

i.e.,

$$F'(x)h \;=\; G'(x)h \;=\; \lambda g \;=\; -F(0),\ h \in B_r(0).$$

Applying Theorem 1 we get the existence of $u \in B_r(0)$ so that $F(u) = 0$, i.e.,

$$G(u) \;=\; \lambda g.$$

∎

BIBLIOGRAPHY

1. A. Castro and J. W. Neuberger, *A Local Inversion Principle of the Nash-Moser Type and Applications to Symmetric Positive Systems*, SIAM J. Math. Anal. **33** (2001), 989–993.

2. A. Castro and J. W. Neuberger, *An Inverse Function Theorem via Continuous Newton's Method*, Nonlinear Analysis **47** (2001), 3223–3229.

3. J. Moser, *A Rapidly Convergent Iteration Method and Nonlinear Differential Equations*, Ann. Scuola Normal Sup. Pisa **20** (1966), 265–315.

4. J. W. Neuberger, *Sobolev Gradients and Differential Equations*, Lecture Notes in Mathematics, vol. 1670, Springer, 1997.

5. J. W. Neuberger, *Continuous Newton's Method for Polynomials*, Math. Intelligencer **21** (1999), 18–23.

6. J. W. Neuberger, *A Nash-Moser Theorem with Near-minimal Hypothesis*, Int. J. Pure Appl. Math, to appear.

Effects of a variable step-size in some abstract product formulas

MICHEL PIERRE

Antenne de Bretagne de l'ENS Cachan
et Institut de Recherche Mathématique de Rennes
Campus de Ker Lann, 35170 - BRUZ, France
e-mail: pierre@bretagne.ens-cachan.fr

MOUNIR RIHANI

Université Hassan II - Mohammédia, Faculté des Sciences Ben M'Sik
B. P. 7955, Casablanca, Maroc
e-mail: rihani@bretagne.ens-cachan.fr

Cet article est dédié à Jerry dont nous saluons l'enthousiasme et l'influence scientifiques

ABSTRACT. A continuous semigroup of linear contractions $\big(S(t)\big)_{t>0}$ on a Banach space X can be generated through the product formula

$$\forall x \in X, \quad \forall t \geq 0, \quad S(t)x = \lim_{n\to\infty} U\left(\frac{t}{n}\right)^n x, \tag{0.1}$$

where $(U(h))_{h>0}$ is any family of contractions on X whose right-derivative at $h = 0$ coincides with the infinitesimal generator $-A$ of $\big(S(t)\big)$, that is

$$\forall x \in D(A) \quad \lim_{h\downarrow 0} \frac{x - U(h)x}{h} = Ax \left(= \lim_{t\downarrow 0} \frac{x - S(t)x}{t}\right). \tag{0.2}$$

Actually formula (0.1) extends to the case when $\big(S(t)\big)$ is a semigroup of *nonlinear* contractions "generated" by an m-accretive operator A and condition (0.2) may also be weakened to

$$\frac{I - U(h)}{h} \xrightarrow{h\downarrow 0} A \quad \text{in the sense of graphs.} \tag{0.3}$$

We look here at the same formula (0.1) when the regular step-size t/n is replaced by a variable step-size, namely

$$\lim_{n\to\infty} U(h^n_{N^n})U(h^n_{N^n-1})\dots U(h^n_1)x = S(t)x \quad \forall x \in \overline{D(A)}, \tag{0.4}$$

$$\text{where} \quad \sum_{1\le i\le N^n} h^n_i = t \quad \text{and} \quad \lim_{n\to\infty}(\max_{1\le i\le N^n} h^n_i) = 0.$$

Surprisingly, it turns out that (0.4) fails under assumption (0.3): there may be lack of stability and even lack of consistency. We exhibit examples in this direction. However, we prove that (0.4) holds true under the stronger assumption (0.2).

1 INTRODUCTION

Given an abstract evolution equation of the form

$$\frac{du}{dt}(t) + Au(t) = 0 \quad \text{on} \quad (0,T)$$
$$u(0) = x,$$

where A is a suitable operator on a Banach space X and x is given in X, many numerical schemes are based on abstract product formulas of the form

$$u(t) = \lim_{n\to\infty} U\left(\frac{t}{n}\right)^n x, \tag{1.1}$$

where $(U(h))_{h>0}$ is a family of mappings whose right-derivative at $h=0$ coincides with $-A$, say, for instance,

$$\forall x \in D(A) \quad \lim_{h\downarrow 0} \frac{x - U(h)x}{h} = Ax.$$

Formula (1.1) is a natural extension of the exponential formula

$$u(t) = \lim_{n\to\infty}\left(I + \frac{t}{n}A\right)^{-n} x.$$

It provides an extensively used tool for Lie-Trotter formulas insuring the convergence of various alternating direction methods to approximate solutions of evolution equations.

In view of numerical applications, it is important to relax the restriction that the time step-size be constant in the above schemes when, for instance, coupled with path control technics or even for stability requirements.

The goal of this paper is to decide whether formula (1.1) remains valid under similar assumptions when the step-size is variable, namely whether

$$\left.\begin{array}{lcl} \lim_{n\to\infty} U(h^n_{N_n})U(h^n_{N_n-1})\dots U(h^n_1)x & = & S(T)x, \\ \text{when} \sum_{i=1}^{N_n} h^n_i = T,\ \lim_{n\to\infty}(\max_{1\le i\le N_n} h^n_i) & = & 0\ . \end{array}\right\} \tag{1.2}$$

As proved in this paper, it turns out that, in a natural abstract framework, instability, lack of convergence as well as lack of consistency may appear as soon as the step-size is slightly perturbed!

Let us describe now the situation more precisely.

Given $(S(t))_{t\geq 0}$ a continuous semigroup of **linear** contractions in a Banach space X, it can be generated via the exponential formula:

$$\forall t \geq 0, \forall x \in X, \quad S(t)x = \lim_{n\to\infty}\left(I + \frac{t}{n}A\right)^{-n} x, \tag{1.3}$$

where $-A$ denotes its infinitesimal generator. More generally, one can replace the resolvents $(I + hA)^{-1}$ in (1.3) by a family of contractions $(U(h))_{h\geq 0}$ from X into X satisfying

$$\forall x \in D(A) \quad \lim_{h\downarrow 0} \frac{x - U(h)x}{h} = Ax. \tag{1.4}$$

Thus, it has been proved (see [6], [19]) that (1.4) implies

$$\forall t \geq 0, \forall x \in X, \quad \lim_{n\to\infty} U\left(\frac{t}{n}\right)^n x = S(t)x. \tag{1.5}$$

Actually, as proved in [7], [12], the condition (1.4) can be weakened to assuming that $\frac{I-U(h)}{h}$ converges to A in the sense of the resolvents, namely

$$\forall x \in X, \forall \lambda > 0, \quad \lim_{h\downarrow 0}\left(I + \lambda\frac{I - U(h)}{h}\right)^{-1} x = (I + \lambda A)^{-1}x, \tag{1.6}$$

which is exactly equivalent to saying that $\frac{I-U(h)}{h}$ *converges to* A *in the sense of graphs*, that is

$$\forall (x, y) \in A, \ \exists\, x_h \text{ with } \lim_{h\to 0} x_h = x, \ \lim_{h\to 0} \frac{x_h - U(h)x_h}{h} = y. \tag{1.7}$$

This convergence is more natural in many situations and, certainly, more adapted to extensions to a nonlinear framework.

Indeed, the implication (1.6) $\Rightarrow$ (1.5) (or (1.7) $\Rightarrow$ (1.5)) turns out to have a nonlinear version as proved by Brezis-Pazy [5]. Namely, let A be an *m-accretive* operator in X, that is $A \subset X \times X$ and

$$\forall \lambda > 0, \quad (I + \lambda A)^{-1} \text{ is a contraction everywhere defined on } X. \tag{1.8}$$

Let $(S^A(t))_{t\geq 0}$ be the nonlinear semigroup generated by A in the sense of Crandall-Liggett [9], that is

$$\forall t \geq 0, \forall x \in \overline{D(A)}, \quad S^A(t)x = \lim_{n\to\infty}\left(I + \frac{t}{n}A\right)^{-n} x. \tag{1.9}$$

Then, if $(U(h))_{h\geq 0}$ is a family of contractions from X into X satisfying (1.6) or (1.7), we have the product formula

$$\forall t \geq 0, \forall x \in \overline{D(A)}, \quad \lim_{n\to\infty} U\left(\frac{t}{n}\right)^n x = S^A(t)x. \tag{1.10}$$

The purpose of this paper is to look at the formula (1.10) when the regular step-size t/n is replaced by a *variable step-size.* In particular, under the same assumptions as above, does the product formula (1.2) hold for all $x \in \overline{D(A)}$?

We give here a negative and a positive answer to this question:

α) *The condition (1.6) (or (1.7)) does not imply* (1.2): perturbing a regular path, even slightly, may create strong instability, even for *linear* semi-groups in a *Hilbert* space. Moreover, one may even observe a lack of consistency (i.e. the product formula converges, but not towards the semigroup).

β) On the other hand, if A is a single-valued m-accretive operator and if $U(h)$ satisfies the stronger assumption (1.4), then, curiously enough, (1.2) holds in a general nonlinear setting.

Note that, since the first papers about Lie-Trotter-Chernoff formulas (see [6, 7, 19]), there has been lots of contributions about approximating linear or nonlinear evolutions by various abstract product formulas, mainly in the case of regular path. Actually, the question of deciding whether, for instance, $U(h) = S^A(h)S^B(h)$ is a good mapping for approximating the evolution generated by $A + B$ turns out to be difficult as shown by counterexamples in various directions (see e.g. [3, 11, 13]). Further comments may also be found in the last section of this paper, and for $J_h^A J_h^B$ as well. See also [8] for recent applications. Let us also mention [18] where a general result of convergence for (1.2) is given under adequate stability assumptions on $U(h)$.

Part of the results here had been announced in the report [16], but were left unpublished. The goal of this paper is to make them available together with new counterexamples proving possibility of non consistency and instability even in a Hilbert space.

2 THE COUNTEREXAMPLES

2.1 The first counterexample: lack of stability

On the Hilbert space $X = L^2(\mathbb{R})$, with its usual norm denoted by $\|\cdot\|$, we define the operator

$$D(A) = \{u \in L^2(\mathbb{R}); u' \in L^2(\mathbb{R})\}, \quad \forall u \in D(A), \quad Au = u'.$$

Then, $-A$ is the infinitesimal generator of the semigroup $\big(S(t)\big)$ on X defined by

$$\forall u \in L^2(\mathbb{R}), \quad S(t)u(x) = u(x - t).$$

Given $u_0 \in X$, $\alpha :\ (0,1) \to \mathbb{R}$ with $\lim_{h\downarrow 0} \alpha(h) = 0$ and $\theta : (0,1) \to \mathbb{R}$, we define

$$\forall h \in (0,1),\ \forall u \in X, \quad U(h)u = S(h)u + \alpha(h)(u_0^h - S(h)u_0^h), \tag{2.1}$$

where $u_0^h(\cdot) = u_0(\cdot + \theta(h))$. Then we have

THEOREM 2.1. *i)* $U(h)$ *is nonexpansive from* X *into* X *and* $\frac{I-U(h)}{h}$ *converges to* A *in the sense of graphs as* h *tends to* 0 *(see(1.7)).*

ii) One can choose u_0, α, θ as above and a sequence $\{h_1^n, \ldots, h_n^n\}$ such that

$$\sum_{i=1}^{n} h_i^n = T, \ \lim_{n\to\infty} \max_{1\le i\le n} h_i^n = 0,$$

$$\lim_{n\to\infty} \|U(h_n^n)\, U(h_{n-1}^n) \ldots U(h_1^n)\, 0\| = +\infty.$$

Moreover, one may choose the sequence so that $\max_{i\neq j} |h_i^n - h_j^n|$ *is as small as one wants.*

REMARK 2.2. As it will be clear from the proof, the only requirement on the sequence is that h_i^n be different from h_j^m for $i \neq j$ and all n, m . This shows that a very small perturbation of a regular path may lead to very strong instability.

Proof. The first statement of $i)$ is immediate from the definition (2.1) and the fact that $S(h)$ is nonexpansive in X.

Now let $u \in D(A)$, $u_h = u + \alpha(h)u_0^h$. Then u_h converges to u in X as h tends to 0 since $\alpha(h)$ tends to 0 and u_0^h is bounded in X. Moreover, we have

$$h^{-1}(I - U(h))u_h = h^{-1}(u_h - S(h)u_h - \alpha(h)(u_0^h - S(h)u_0^h)) = h^{-1}(u - S(h)u),$$

so that, since $-A$ is the infinitesimal generator of $\big(S(h)\big)$,

$$\lim_{h\to 0} h^{-1}(I - U(h))u_h = A\,u,$$

which proves $i)$ (see (1.7)).

Now we choose

$$\alpha(h) = h^p, 0 < p < 1/2,\ u_0 := \chi_{[0,1]},$$

and we choose a sequence of numbers $(h_i^n)_{1\le i\le n}$ in $(0,1)$ satisfying

$$\left.\begin{array}{rcl} \sum_{i=1}^n h_i^n & = & T > 0, \\ \forall n,\quad i \neq j & \Rightarrow & h_i^n \neq h_j^n, \\ \frac{T}{n+1/2} < \min_{1\le i\le n} h_i^n & < & \max_{1\le i\le n} h_i^n < \frac{T}{n-1/2}. \end{array}\right\} \tag{2.2}$$

(This may be obtained by slightly perturbing a regular subdivision). The conditions (2.2) ensure that all the numbers $\{h_i^n,\ 1 \le i \le n, n \ge 1\}$ are different. This will allow to assign the values $\theta(h_i^n)$ as we wish. We do it as follows:

Set $t_i^n = \sum_{k=1}^{i} h_k^n$, $t_0^n = 0$, and choose:

$$\forall\, 1 \le i \le n,\ \theta(h_i^n) := t_n^n - t_{i-1}^n, \tag{2.3}$$

and $\theta(h) = 0$ otherwise, (i.e. for all $h \neq h_i^n$, $1 \le i \le n \le +\infty$). We claim that with this choice, if we set

$$f_k^n = U(h_k^n)U(h_{k-1}^n)\ldots U(h_1^n)0 \text{ for } 1 \le k \le n, \quad \text{and} \quad f_0^n \equiv 0,$$

then

$$\lim_{n\to\infty} \|f_n^n\| = \infty. \tag{2.4}$$

Indeed, by definition (2.1), we have

$$f_k^n = S(h_k^n)f_{k-1}^n + \alpha(h_k^n)\left(u_0^{h_k^n} - S(h_k^n)u_0^{h_k^n}\right).$$

By induction, we check that

$$f_k^n = \sum_{i=1}^{k} \alpha(h_i^n)\left(S(t_k^n - t_i^n)u_0^{h_i^n} - S(t_k^n - t_{i-1}^n)u_0^{h_i^n}\right),$$

or also

$$f_k^n = \sum_{i=1}^{k} \alpha(h_i^n)\left(u_0(\cdot + \theta(h_i^n) - t_k^n + t_i^n) - u_0(\cdot + \theta(h_i^n) - t_k^n + t_{i-1}^n)\right).$$

>From the choice of θ and u_0, it follows that

$$f_n^n = \sum_{i=1}^{n} \alpha(h_i^n)\left(u_0(\cdot + h_i^n) - u_0(\cdot)\right) = \sum_{i=1}^{n} \alpha(h_i^n)(\chi_{[-h_i^n,0]} - \chi_{[1-h_i^n,1]}). \tag{2.5}$$

Since the intervals $[-h_i^n, 0]$ and $[1 - h_i^n, 1]$ are disjoint, we have

$$||f_n^n||^2 \geq ||\sum_{i=1}^{n} \alpha(h_i^n)\chi_{[-h_i^n,0]}||^2.$$

By the choice of the h_i^n (see 2.2), this is bounded from below as follows

$$||f_n^n||^2 \geq \int_{\frac{-T}{n+1/2}}^{0} \left(\sum_{i=1}^{n} \alpha(h_i^n)\right)^2 \geq \frac{T}{n+1/2}\left(n\left\{\frac{T}{n+1/2}\right\}^p\right)^2 = T^{2p+1}n^2/(n+1/2)^{2p+1},$$

and this tends to $+\infty$ by the choice of p. ■

2.2 The second counterexample: lack of consistency

Let X be the space of continuous functions from $[0,1]$ into $\mathbb{R}$ equipped with the uniform norm $||\cdot||_\infty$. Let Y be the subspace of constant functions of X and J_λ the family of mappings from X into X defined by

$$\forall \lambda > 0, \forall f \in X, J_\lambda f := \max_{x\in[0,1]} f(x). \tag{2.6}$$

It is obvious that the J_λ are nonexpansive mappings and, more precisely, it is easy to verify that they are the resolvents of the nonlinear *m-accretive* operator C defined on X by

$$D(C) = Y,\ \forall g \in Y,\ C\,g = \{w \in X; \max w = 0\}. \tag{2.7}$$

Thus $J_h^C = (I + hC)^{-1} = J_h$. Moreover, the semigroup $\left(S^C(t)\right)$ generated by C is defined on Y and satisfies $S^C(t) = I$ for all $t \geq 0$.

We choose $\alpha \in X$ as $\alpha(x) = 1 - 2x$ and we define two new operators A, B by

$$D(A) = D(C) = Y, \;\; D(B) = X,$$

$$\forall f \in X, Af = Cf - \alpha, \; \forall f \in X, Bf = \alpha.$$

These two operators are also m-accretive and $A+B = C$. We denote their resolvents by J_h^A, J_h^B and we introduce $U(h) = J_h^B J_h^A$. Note that we have

$$\forall u \in X, \; U(h)u = J_h^A u - h\alpha = J_h^C(u + h\alpha) - h\alpha. \tag{2.8}$$

For any subdivision σ of the form $0 = t_0 < t_1 < ... < t_N \leq T$, we set

$$\forall i = 1, ...N, h_i = t_i - t_{i-1}, \; |\sigma| = \max_{1 \leq i \leq N} h_i,$$

and for $u_0 \in Y$, $\forall i = 1, ...N$, $\forall t \in (t_{i-1}, t_i]$,

$$U(\sigma, t, u_0) = U(h_i)U(h_{i-1})...U(h_1)u_0, \; U(\sigma, 0, u_0) = u_0.$$

THEOREM 2.3. *(i) The operators $h^{-1}(I - U(h))$ converge to $A + B = C$ in the sense of graphs as h tends to 0.*
(ii) For $u_0 \in Y$ and for a sequence of subdivisions σ_n with

$$\lim_{n\to\infty} \sum_{i=1}^{N_n} h_i^n = T, \; \lim_{n\to\infty} |\sigma_n| - 0, \tag{2.9}$$

$U(\sigma_n, t, u_0)$ converges uniformly on $[0, T]$ to $S^C(t)u_0 (\equiv u_0)$ if and only if

$$\lim_{n\to\infty} \sum_{i=2}^{N_n} |h_i^n - h_{i-1}^n| = 0.$$

(iii) Moreover, one can choose σ_n so that $U(\sigma_n, t, u_0)$ converges uniformly on $[0, T]$ towards a continuous function $u_\infty(t) \neq S^C(t)u_0$.

REMARK 2.4. The third assertion says that the approximate scheme is not consistent. Note that, here, $h^{-1}(I - U(h))$ does converge *in the strong sense* to a section of the multivalued operator C. This explains why it is necessary to assume that the limit operator is single-valued in the positive convergence results given in next section.

Proof of Theorem 2.3. Proof of (i): The convergence of $h^{-1}(I - U(h))$ in the sense of graphs to C may be obtained as follows (see 1.7). Let $(g, w) \in C$; we check that $f_h = g + h(w - \alpha)$ converges to g and $h^{-1}(I - U(h))f_h = w$.
Proof of (ii): for $i \geq 1$, set $u_i^n = U(\sigma, t_i^n, u_0), v_i^n = u_i^n + h_i^n \alpha$ and $u_0^n = v_0^n = u_0$. We have

$$u_i^n = U(h_i^n)u_{i-1}^n = J_{h_i^n}^C(u_{i-1}^n + h_i^n \alpha) - h_i^n \alpha = max(u_{i-1}^n + h_i^n \alpha) - h_i^n \alpha.$$

Thus,

$$v_i^n = max(v_{i-1}^n + (h_i^n - h_{i-1}^n)\alpha) = v_{i-1}^n + \max\{\alpha(h_i^n - h_{i-1}^n)\}$$

since v_{i-1}^n is constant. By the choice of α, for all $i \geq 2$,

$$v_i^n = v_{i-1}^n + |h_i^n - h_{i-1}^n|.$$

By induction

$$v_i^n = u_0 + \sum_{k=2}^{i} |h_k^n - h_{k-1}^n| + h_1.$$

We see that the convergence of the u_i^n to $S(t)u_0 \equiv u_0$ is equivalent to the convergence to 0 of $\sum_{k=2}^{i} |h_k^n - h_{k-1}^n|$.
Proof of (iii): Let us choose $h^n = T/3n, N_n = 2n$ and

$$\forall\, 1 \leq j \leq n,\ h_{2j-1}^n = h^n, h_{2j}^n = 2h^n.$$

Then

$$v_{2n}^n = u_0 + 2nh^n, U(\sigma_n, T, u_0) = v_{2n}^n - 2h^n\alpha,$$

so that

$$\lim_{n\to\infty} U(\sigma_n, T, u_0) = u_0 + \frac{2T}{3} \neq u_0.$$

■

3 THE POSITIVE RESULT

Let X be a Banach space with its norm denoted by $|\cdot|$ and let A be a *single-valued, accretive* operator on X, that is

$$\forall\lambda > 0, \forall x, \hat{x} \in D(A), \quad |x - \hat{x}| \leq |x - \hat{x} + \lambda(Ax - A\hat{x})|. \tag{3.1}$$

For $h \in (0, h_0)$, let $U(h) : \overline{D(A)} \to \overline{D(A)}$ be nonexpansive and satisfy

$$\forall x \in D(A), \quad \lim_{h\downarrow 0} \frac{x - U(h)x}{h} = Ax. \tag{3.2}$$

Given $T > 0$ and $\sigma : 0 = t_0 < t_1 < \cdots < t_{N-1} < T \leq t_N$, a subdivision of $(0, T)$, we denote:

$$|\sigma| = \max_{1\leq i\leq N} h_i \textit{ where } h_i = t_i - t_{i-1}.$$

For x_0 given in X, let us consider, for all $i = 1, ..., N$, $x_i \in \overline{D(A)}, z_i \in X$ such that

$$x_i - U(h_i)x_{i-1} = h_i z_i. \tag{3.3}$$

Next, define z and v_{σ,z,x_0} by

$$\forall i = 1, \ldots, N,\ \forall t \in]t_{i-1}, t_i], \quad z(t) = z_i, \quad v_{\sigma,z,x_0}(t) = x_i, \tag{3.4}$$

and $v_{\sigma,z,x_0}(0) = x_0$.

REMARK 3.1. If $z \equiv 0$, the relation (3.3) always defines a sequence (x_i) such that $x_i = U(h_i)U(h_{i-1})\ldots U(h_1)x_0$ which are the expressions we are mainly interested in. However, many situations involve perturbed schemes like (3.3) where z is not zero but of small norm in $L^1(0, T;\ X)$ and our method applies as well to this more general situation.

Our goal is to describe the behavior of v_{σ^n,z^n,x^n} when

$$\lim_{n\to\infty} |\sigma^n| = \lim_{n\to\infty} \int_0^T |z^n(t)|dt = \lim_{n\to\infty} |x^n - x_0| = 0.$$

We expect it to tend to the solution (if it exists) of

$$\frac{du}{dt} + Au = 0 \quad \text{on} \quad (0,T), \quad u(0) = x_0 \in \overline{D(A)}. \tag{3.5}$$

By solution of (3.5), we mean a *mild solution* in the sense of ([1]) or "bonne solution" as in ([2]), i.e. a function $u \in C([0,T];\ X)$ such that there exists a sequence of step-functions $u^p : [0,T] \to X$ satisfying

$$\forall j = 1,\ldots,N^p, \forall s \in]s^p_{j-1}, s^p_j], \quad u^p(s) = u^p_j, \tag{3.6}$$

where

$$0 = s^p_0 < s^p_1 < \cdots < s^p_{N^p-1} < T \leq s^p_{N^p}, \quad l^p_j = s^p_j - s^p_{j-1}, \tag{3.7}$$

$$\frac{u^p_j - u^p_{j-1}}{l^p_j} + Au^p_j = w^p_j \quad (u^p_0 = u^p(0)), \tag{3.8}$$

$$0 = \lim_{p\to\infty} \sup_{0\leq s\leq T} |u^p(s) - u(s)| = \lim_{p\to\infty} \sum_{j=1}^{N^p} l^p_j |w^p_j| = \lim_{p\to\infty} \max_{1\leq j\leq N^n} l^p_j. \tag{3.9}$$

THEOREM 3.2. *Assume that A is a single-valued accretive operator in X and that $(U(h))$ is a family of nonexpansive mappings from $\overline{D(A)}$ into itself satisfying (3.2). Let u be a mild solution of* (3.5). *Let $v^n = v_{\sigma^n,z^n,x^n}$ be a sequence of functions defined by* (3.3),(3.4) *where*

$$\lim_{n\to\infty} |\sigma^n| = \lim_{n\to\infty} \int_0^T |z^n(t)|dt = \lim_{n\to\infty} |x^n - x_0| = 0.$$

Then,

$$\lim_{n\to\infty} \sup_{0\leq t\leq T} |u(t) - v^n(t)| = 0.$$

In the case when the range of $I + \lambda A$ contains $\overline{D(A)}$ for all $\lambda > 0$, thanks to Crandall-Liggett's theorem [9], one knows that

$$\forall x \in \overline{D(A)}, \forall t \geq 0, \quad S^A(t)x = \lim_{n\to\infty} \left(I + \frac{t}{n}A\right)^{-n} x \quad \text{exists}, \tag{3.10}$$

and $(S^A(t))$ is a semigroup of nonlinear contractions from $\overline{D(A)}$ into itself. Moreover, by definition, $t \to S^A(t)x_0$ is a mild solution of (3.5) for all $T > 0$.

In that situation, as an immediate corollary of Theorem 3.2, we have

PROPOSITION 3.3. *Assume A is a single-valued accretive operator satisfying* (3.10) *and assume that* (3.2) *holds. Then if $(h_i^n)_{1\le i\le N^n}$ is a sequence such that*

$$\lim_{n\to\infty}\left(\max_{1\le i\le N^n} h_i^n\right)=0,\qquad \lim_{n\to\infty}\sum_{i=1}^{N^n} h_i^n=t, \tag{3.11}$$

we have for all $x_0\in\overline{D(A)}$

$$\lim_{n\to\infty} U(h_{N^n}^n)U(h_{N^n-1}^n)\dots U(h_1^n)x_0=S^A(t)x_0 \tag{3.12}$$

and the convergence is uniform in t on bounded intervals.

More consequences of this theorem will be given in the next section.

The proof of Theorem 3.2 is based on the next lemma which is an adaptation of technics used in [10] and [14].

LEMMA 3.4. *Let $u^p:[0,T]\to X$ satisfying* (3.6)-(3.8). *Then, for all $\varepsilon>0$ there exists $\eta=\eta(\varepsilon,p)>0$ such that for all functions v_{σ,z,x_0} satisfying* (3.3)-(3.4) *with $|\sigma|<\eta$ we have*

$$\forall y\in D(A),\forall i=0,\dots,N,\forall j=0,\dots,N^p,$$

$$|x_i-u_j^p|\le|x_0-y|+|u_0^p-y|+\{(t_i-s_j^p)^2+|\sigma|t_i+|\sigma^p|s_j^p\}^{1/2}M_\sigma(y)$$
$$+\sum_{k=1}^{i}h_k\left(|z_k|+\varepsilon\right)+\sum_{k=1}^{j}l_k^p|w_k^p|, \tag{3.13}$$

where

$$|\sigma^p|=\max_{1\le j\le N^p} l_j^p\quad \textit{and}\quad M_\sigma(y)=\max\{|Ay|,\max_{1\le i\le N} h_i^{-1}|y-U(h_i)y|\}. \tag{3.14}$$

Proof of lemma 3.4. Let us denote $A^h=h^{-1}(I-U(h))$. By (3.2), for $\varepsilon>0$ and p fixed, there exists $\eta=\eta(\varepsilon,p)>0$ such that

$$h\in]0,\eta]\Rightarrow\forall j=1,\dots,N^p,\qquad |A^h u_j^p-Au_j^p|<\varepsilon. \tag{3.15}$$

Now, we will assume $|\sigma|<\eta$ and, since p is fixed we will drop the indexation by p.

Let us denote $a_{i,j}=|x_i-u_j|$ and let us establish (3.13) in four steps. We fix $y\in D(A)$.

i) estimate of $a_{0,j}$: since A is accretive and by (3.8)

$$|u_j-y|\le|u_j-y+l_j(Au_j-Ay)|\le|u_{j-1}-y|+l_jw_j+l_j|Ay|.$$

By induction, for all $j\ge 1$

$$|u_j-y|\le|u_0-y|+\sum_{k=1}^{j}l_kw_k+s_j|Ay|.$$

Hence, for all $j\ge 0$

$$a_{0,j}\le|x_0-y|+|u_0-y|+s_j|Ay|+\sum_{k=1}^{j}l_kw_k \tag{3.16}$$

which implies (3.13) for $i = 0$ and $j = 0, \dots, N^p$.

ii) <u>estimate of</u> $a_{i,0}$:

$$\begin{aligned}|x_i - y| &\le |x_i - U(h_i)x_{i-1}| + |U(h_i)x_{i-1} - U(h_i)y| + |U(h_i)y - y| \\ &\le h_i|z_i| + |x_{i-1} - y| + h_i M_\sigma(y).\end{aligned}$$

By induction, for all $i \ge 1$

$$|x_i - y| \le |x_0 - y| + \sum_{k=1}^{i} h_k|z_k| + t_i M_\sigma(y).$$

Hence for all $i \ge 0$

$$a_{i,0} \le |u_0 - y| + |x_0 - y| + \sum_{k-1}^{i} h_k|z_k| + t_i M_\sigma(y). \tag{3.17}$$

iii) <u>estimate of</u> $a_{i,j}$ <u>in terms of</u> $a_{i-1,j}$ <u>and</u> $a_{i,j-1}$: for all $i, j \ge 1$

$$\left(1 + l_j h_i^{-1}\right) a_{i,j} = |x_i - u_j + l_j h_i^{-1} \left(x_i - U(h_i)u_j + U(h_i)u_j - u_j\right)|. \tag{3.18}$$

Using (3.3), we have

$$|x_i - U(h_i)u_j| \le a_{i-1,j} + h_i|z_i|.$$

Plugging this inequality in (3.18) together with (3.15) applied with $h = h_i < \eta$ leads to

$$\left(1 + l_j h_i^{-1}\right) a_{i,j} \le l_j h_i^{-1} a_{i-1,j} + l_j|z_i| + l_j\varepsilon + |x_i - u_j - l_j A u_j|. \tag{3.19}$$

Now by (3.8)

$$|x_i - u_j - l_j A u_j| \le a_{i,j-1} + l_j|w_j|. \tag{3.20}$$

Finally, (3.19) and (3.20) yield

$$a_{i,j} \le \frac{l_j}{h_i + l_j} a_{i-1,j} + \frac{h_i}{h_i + l_j} a_{i,j-1} + \frac{l_j h_i}{h_i + l_j}(|z_i| + |w_j| + \varepsilon). \tag{3.21}$$

iv) <u>induction procedure</u>: we assume that (3.13) holds with (i, j) replaced by $(i - 1, j)$ and $(i, j - 1)$. Plugging it into (3.21) gives (3.13) after an easy computation. Thanks to the first two steps, we deduce that (3.13) holds for all (i, j). ∎

Proof of Theorem 3.2. Let $v^n = v_{\sigma^n, z^n, x^n}$ be a sequence of functions defined by (3.3)-(3.4) where

$$\lim_{n\to\infty} |\sigma^n| = \lim_{n\to\infty} \int_0^T |z^n(t)|dt = \lim_{n\to\infty} |x^n - x_0| = 0.$$

Let u be a mild solution of (3.5) and let u^p be an approximated sequence as in (3.6) - (3.9). Let us denote $|\hat{\sigma}^p| = \max_{1\le j\le N^p} l_j^p$. By Lemma 3.4, for all $p \ge 1$ and $\varepsilon > 0$, there exists $N(\epsilon, p)$ such that

$$\begin{aligned}&\forall n \ge N(\epsilon, p), \forall 1 \le i \le N^n, \forall 1 \le j \le N^p, \forall y \in D(A), \\ |x_i^n - u_j^p| &\le |x^n - y| + |u_0^p - y| + \{(t_i^n - s_j^p)^2 + |\sigma^n|t_i^n + |\hat{\sigma}^p|s_j^p\}^{1/2} M_\infty(y) \\ &+ \textstyle\sum_{k=1}^{i} h_k^n(|z_k^n| + \varepsilon) + \sum_{k=1}^{j} l_k^p|w_k^p|,\end{aligned}$$

where $M_\infty(y) = \sup_{h\in(0,1)} |h^{-1}(y - U(h)y)|$. Using

$$|v^n(t) - u(t)| \le |v^n(t) - u^p(t)| + |u^p(t) - u(t)|,$$

we deduce that for all $p \ge 0$ and $\varepsilon > 0$:

$$\begin{aligned}\limsup_{n\to\infty}\left(\sup_{0\le t\le T}|v^n(t)-u(t)|\right) \quad &\le \quad \sup_{0\le t\le T}|u^p(t)-u(t)| + |x_0 - y| + |u_0^p - y| \\ +\{|\hat{\sigma}^p|^2 + T|\hat{\sigma}^p|\}^{1/2} M_\infty(y) + \quad \varepsilon T \quad &+ \sum_{k=1}^{N^p} l_k^p |w_k^p|.\end{aligned}$$

Now letting first p tend to ∞, then ε tend to 0 (and according to (3.9)), we deduce

$$\limsup_{n\to\infty}\left(\sup_{0\le t\le T}|v^n(t)-u(t)|\right) \le 2|x_0 - y|.$$

Since $x_0 \in \overline{D(A)}$ and y is arbitrary in $D(A)$, the conclusion of the theorem follows. ■

4 MORE COMMENTS

It is important to notice that Theorem 3.2 is stated for *mild solutions* of a (single-valued) accretive operator A. We remember that, by definition, a mild solution for the closure $\overline{A}$ of A is also a mild solution for A. Therefore, the conclusion of Theorem 3.2 is valid as soon as (3.2) is satisfied for the accretive operator A itself, even if $\overline{A}$ is multivalued. This remark will be used below.

If $-B, -C$ and $-\overline{B+C}$ are infinitesimal generators of linear semigroups of contractions and if one sets $A = B + C$ and

$$U(h) = (I + hB)^{-1}(I + hC)^{-1} \text{ or } U(h) = S^B(h)S^C(h),$$

one easily verifies that assumption (3.2) holds. Since $S^{\overline{B+C}}(t)u_0$ is a mild solution of $\frac{du}{dt} + Au = 0$, $u(0) = u_0$ for all u_0 (see [1]), for both choices one has, as a consequence of Theorem 3.2

$$\lim_{n\to\infty} U\left(h_{N^n}^n\right) U\left(h_{N^n-1}^n\right) \ldots U\left(h_1^n\right) u_0 = S^{\overline{B+C}}(t)u_0.$$

Note that we only need to check (3.2) in $D(B + C)$ and not in $D(\overline{B+C})$.

Let A be an m-accretive operator in a *Hilbert space* (or equivalently maximal monotone), possibly multivalued, and let $\big(U(h)\big)$ be a family of contractions from $\overline{D(A)}$ into itself satisfying

$$\forall x \in D(A) = D(A^o),\ \lim_{h\to 0} \frac{I - U(h)}{h} x = A^o x,$$

where A^o denotes the minimal section of A (i.e. $|A^o x| = min\{|y|, y \in Ax\}$). Then the product formula with variable step-size converges to the semi-group generated by A. The point here is that A is not assumed to be single-valued (in which case, the result would fit directly into Theorem 3.2). The idea is that any mild solution for A is also a mild solution for A^o. This may be seen from a classical regularity result for mild solutions of m-accretive operators in a Hilbert space (see e.g. [4]). Indeed, if $u_0 \in D(A)$, then $u(t) = S^A(t)u_0$ is a strong solution of $u'(t) + A^o u(t) = 0$ and therefore also a mild solution (see e.g. [4]). We finish for any $u_0 \in \overline{D(A)}$ by using the closure property of mild solutions.

Theorem 3.2 may be applied to more concrete nonlinear operators like for instance $A = -\Delta\beta$, where Δ is the Laplacian operator on a regular bounded open subset Ω of $\mathbb{R}^N$ with "good", possibly nonlinear, boundary conditions, and β is a continuous nondecreasing function from $\mathbb{R}$ into $\mathbb{R}$. The right space is then $L^1(\Omega)$. The idea here is that, although the operator is nonlinear, he does have good enough regularity properties so that natural approximations $U(h)$ will converge in the strong sense when converging in the weak sense. Details may be found in [17].

Let now A be an m-accretive operator in a Banach space X and B a continuous accretive operator on X. Then, $A+B$ is also m-accretive and generates a semigroup S^{A+B} on $\overline{D(A)}$. We saw in the second counterexample that, for $U(h) = J_h^B J_h^A$, the product formula with variable step-size is generally not an approximation of the semigroup, even if B is constant (although it is so for a regular path). We may wonder what happens for $U(h) = J_h^A J_h^B$. When the step-size is constant, the approximate formula is essentially the same as for $U(h) = J_h^B J_h^A$ (upon replacing u_0 by $J_h^B u_0$) and both converge appropriately. The formulas are now quite different when the step-size is variable. What is somehow surprising is that, in the second situation $U(h) = J_h^A J_h^B$, *the variable step-size formula does always converge to the semigroup*, and this happens, although the convergence of $h^{-1}(I - U(h))$ holds only in the sense of graphs and not in the strong sense in general. This last result requires a specific proof which may be found in ([17]). The idea is to start with the case when B is constant: then, $J_h^A J_h^B = J_h^{A+B}$ and the result follows from the Crandall-Liggett theorem. The case of a general continuous B is also based on this identity.

Finally, let us mention another counterexample to the convergence of the product formula with variable step-size, due to T.G. Kurtz (see [17]), which is constructed for a linear infinitesimal generator in $C_0(\mathbb{R}^2)$.

BIBLIOGRAPHY

1. Ph. Bénilan, M. G. Crandall, and A. Pazy, *Evolution equations governed by m-accretive operators*, to appear.

2. Ph. Bénilan, M. G. Crandall, and A. Pazy, *Bonnes solutions d'un problème d'évolution semi-linéaire*, C.R.A.S. Paris **306** (1988), no. I, 527–530.

3. Ph. Bénilan and S. Ismaïl, *Générateurs de semi-groupes non linéaires et formule de Lie-Trotter*, Ann. Fac. Sc. Toulouse **7** (1985), 151–160.

4. H. Brezis, *Opérateurs maximaux monotones et semi-groupes de contractions dans les espaces de Hilbert*, vol. 5, Amsterdam, 1973.

5. H. Brezis and A. Pazy, *Convergence and approximation of semigroups of nonlinear operators in Banach spaces*, J. Funct. Anal. **9** (1972), 63–74.

6. P. R. Chernoff, *Note on product formulas for operator semigroups*, J. Funct. Anal. **2** (1968), 238–242.

7. P. R. Chernoff, *Product formulas, nonlinear semigroups, and addition of unbounded operators*, Mem. Amer. Math. Soc. **140** (1974).

8. M. Cliff, J. A. Goldstein, and M. Wacker, *Positivity, Trotter products and blow-up*, to appear.

9. M. G. Crandall and T. M. Liggett, *Generation of semigroups of nonlinear transformation on general Banach spaces*, Amer. J. Math. **93** (1971).

10. Y. Kobayashi, *Difference approximation of Cauchy problems for quasi-dissipative operators and generation of nonlinear semigroups*, J. Math. Soc. Japan **27** (1975), 640–665.

11. F. Kühnemund and M. Wacker, *The Lie-Trotter product formula does not hold for arbitrary sums of generators*, Semigroup Forum **60** (2000), 478–485.

12. T. G. Kurtz, *Extensions of Trotter's operator semigroup approximation theorem*, J. Funct. Anal. **3** (1969), 354–375.

13. T. G. Kurtz and M. Pierre, *A counterexample for the Trotter product formula*, J. Diff. Equ. **52** (1984), 407–414.

14. I. Miyadera and Y. Kobayashi, *Convergence and approximation of nonlinear semigroups*, Functional Analysis and Numerical Analysis, Japan-France Seminar, Tokyo and Kyoto, (1976) (H. Fujila, ed.), Japan Society for the Promotion of Science, 1978.

15. A. Pazy, *Semigroups of linear operators and applications to partial differential equations*, Applied Mathematical Sciences, vol. 44, Springer-Verlag, New York, 1983.

16. M. Pierre and M. Rihani, *About product formula with variable step-size*, Tech. Report 2783, MRC, Madison, WI, USA, 1985.

17. M. Rihani, *Discrétisation à pas variable d'équations d'évolution abstraites*, thèse université de Nancy 1 (1985).

18. E. Schechter, *Stability conditions for nonlinear products and semigroups*, Pac. J. Math. **85** (1979), 179–200.

19. H. F. Trotter, *Approximation of semigroups of operators*, Pac. J. Math. **8** (1958), 887–919.

Evolution Operators in Stochastic Processes and Inference

M. M. RAO

Department of Mathematics. University of California
Riverside, CA 92521
e-mail: rao@ucrmath.ucr.edu

To Jerome A. Goldstein on his 60th birthday

ABSTRACT. In this paper the role of the evolution operator family in the analysis of (higher order) stochastic differential equations and in the theory of likelihood ratios in stochastic processes is studied. Although the existence and unicity of solutions of SDEs can be established with methods of semi-group theory, further work including the sample path properties of solutions demand a (full) use of evolution families. Moreover, in inference theories of stochastic processes, involving testing composite hypotheses, a detailed study of evolution families is indispensable. Some related and several new problems arising in these studies are included.

1 INTRODUCTION

To present an intelligible discussion of the problems and their analysis, it is desirable to recall a general definition of stochastic integrals, using an extended boundedness principle originally proposed by Bochner [1]. Let (Ω, Σ, P) be a probability space and $\mathcal{F}_t \subset \Sigma, t \geq a$, be a *standard filtering family* of σ-algebras to mean that $\mathcal{F}_t \subset \mathcal{F}_{t'}$ for $t \leq t', \cap_{t>s}\mathcal{F}_t = \mathcal{F}_{s+0} = \mathcal{F}_s, s \geq a$, (the last is also called right continuity of the filtration). A process $\{X_t, \mathcal{F}_t, t \geq a\}$ is an adapted family of random variables (i.e., X_t is $\mathcal{F}_t$-measurable for each $t \geq a$) and is an L^p-process if $X_t \in L^p(\mathcal{F}_t) \subset L^p(\Sigma), t \geq a$ in what follows. Let $f = \sum_{i=0}^{n-1} a_i \chi_{[t_i, t_{i+1})}, a_i \in \mathbb{R}$, a simple function on $I = [a, b] \subset \mathbb{R}$ with $a = t_0 < t_1 < \cdots < t_n = b$. If one sets

$$\tau : f \mapsto \tau_{ab}(f) = \sum_{i=0}^{n-1} a_i (X_{t_{i+1}} - X_{t_i}),$$

then τ is unambiguously defined and is an element of $L^p(\Sigma)$, and if $\mathcal{S} = \mathcal{S}(I, \mathcal{B})$ is the set of simple functions on the Borelian space $(I, \mathcal{B})$, it follows that $\tau : \mathcal{S} \to L^p(\Sigma)$ is a linear operator. This leads to the following concept which is essentially due to Bochner [1]:

DEFINITION 1.1. A process $\{X_t, \mathcal{F}_t, t \in I = [a, b]\}$ is $L^{p,q}$ – *bounded* if there is a constant $C_1 > 0$ and a measure $\alpha : \mathcal{B} \to \mathbb{R}^+$ such that for each $f \in \mathcal{S}$, one has

$$E(|\tau f|^q) \leq C_1 \int_I |\tau f|^p \, d\alpha(t), \quad 0 < p, q < \infty. \tag{1.1}$$

In fact (1.1) implies that the X_t-process defines a vector measure $Z : \mathcal{B} \to L^p(\Sigma)$. For, if $Z([a, t)) = X_t - X_a$ for an interval $[a, t) \subset I$, and since by (1.1) τ is a bounded linear mapping on the dense subset $\mathcal{S}$ of $L^q(I, \mathcal{B}, \alpha)$ into $L^p(\Omega, \Sigma, P)$ so that it has a unique bound preserving extension, denoted by the same symbol, to all of $L^q(I, \alpha)$, then it may be represented as $\tau(f) = \int_I f \, dZ$ for a unique vector measure $Z : \mathcal{B} \to L^p(P)$ when $1 < p < \infty$ by a classical form of the Riesz representation theorem. It is not difficult by specializing f to see that $Z([a, t))$ is of the form given above, except for an additive constant (taken zero here). Writing $dZ = dX$ and taking $p = q = 2$, the process X_t as Brownian Motion (BM), and setting $\tau(f) = \int_I f \, dX$, one has:

$$E(|\int_I f(t) \, dX_t|^2) = \int_I |f(t)|^2 \, dt, \tag{1.2}$$

so that the Wiener stochastic integral is included in this definition, with equality in (1.1), $\alpha = Leb.meas.$, $C_1 = 1$ and that it is even an isometry. Moreover, in this case it is possible to express $\tau(f)$ as a Bochner integral which gives an alternative representation of the classical Paley-Wiener-Zygmund definition. This is seen as follows:

Let $g : \mathbb{R} \to \mathbb{R}$ be defined as $g(t) = f(t), t \in I; = 0$, for $t \in \mathbb{R} - I$ so that

$$||\tau(f)||_2^2 = ||\int_I f \, dX||_2^2 = \int_{\mathbb{R}} |g(t)|^2 \, dt = \int_{\mathbb{R}} |\hat{f}(t)|^2 \, dt, \tag{1.3}$$

by Plancherel's theorem, where $\hat{f}$ is the Fourier transform on $L^2(\mathbb{R}, dt)$ onto itself and is an isometry there. Let $T : \hat{f} \mapsto \tau(f)$ so that from (1.2) and (1.3), $T : L^2(\mathbb{R}, dt) \to L^2(P)$ is linear and bounded. Hence by Riesz's theorem again there is a unique vector measure $\tilde{Z}$ such that

$$T(\hat{g}) = \int_{\mathbb{R}} \hat{g}(u) \, d\tilde{Z}(u).$$

Consequently,

$$\int_I f(t)\, dX_t = \tau(f) = T(\hat{f}) = \int_{\mathbb{R}} [\int_{\mathbb{R}} e^{itu} f(t)\, dt] d\tilde{Z}(u)$$
$$= \int_{\mathbb{R}} f(t)[\int_{\mathbb{R}} e^{itu}\, d\tilde{Z}(u)]dt$$
$$= \int_{\mathbb{R}} Y(t)f(t)\, dt,$$

where the interchange of integrals with Fubini's theorem is easily justified, and the Y_t-process is well defined since $\tilde{Z}$ is $L^{2,2}$-bounded. But now the right most integral is a Bochner integral whereas the left most is the simple stochastic integral which is classically defined in the PWZ-sense.

In the following an extension of Bochner's boundedness principle will be stated for applications which include Itô and martingale integrals with stochastic integrands. Thus let $f = \sum_{i=1}^{n} f_i \chi_{A_i}$ where the f_i are bounded random variables and $A_i \in \mathcal{B}$ (the Borel σ-algebra of I so that $f : I \times \Omega \to \mathbb{R}$ is $\mathcal{B} \otimes \Sigma$-simple). Then define τ as:

$$\tau : f \mapsto \sum_{i=1}^{n} f_i(X_{(supA_i)} - X_{(infA_i)}) \; (= \int_I f(t)\, dX_t \;), \tag{1.4}$$

and as before $\tau(f)$ is unambiguously defined. Thus Definition 1.1 extends to:

DEFINITION 1.2. Let $X = \{X_t, t \in I\}$ be a measurable processes for $\mathcal{B} \otimes \Sigma$, and $\mathcal{O} \subset \mathcal{B} \otimes \Sigma$ be a σ-algebra. Then for $0 < p, q < \infty$, the process X is $L^{p,q}$-bounded relative to $\mathcal{O}, p, q$ if there is a measure $\beta : \mathcal{O} \to \mathbb{R}^+$ and a constant $C'(= C'_{p,q,\beta} > 0)$ such that

$$E(|\tau f|^q) \leq C' \int_{\Omega'} |f(t, \omega)|^p \, d\beta(t, \omega), \tag{1.5}$$

where $\Omega' = I \times \Omega$.

Note that (1.5) reduces to (1.1) if $\mathcal{O} = \mathcal{B} \otimes \{\emptyset, \Omega\}$, and β is taken as the product of the measure α there with the basic probability measure P. As in the previous case when $p = q = 2$ in (1.5), the process X is termed $L^{2,2}$-bounded. The importance and usefulness of the extension is explained by the following result.

PROPOSITION 1.3. *Let $1 \leq p \leq 2$ and $q = 2$. Then a vector measure $Z : \mathcal{B}_I \to L^p(P)$ defines the process $X = \{X_t = Z([a, t)), t \in [a, b] = I\}$ as $L^{2,p}$-bounded relative to a measure $\beta = \mu \otimes P$ on $\mathcal{O} = \mathcal{B}_I \otimes \{\emptyset, \Omega\}$ for a measure $\mu : \mathcal{B}_I \to \mathbb{R}^+$. Moreover, if Z has orthogonal values when $p = 2$, then the process has orthogonal increments. Further every $L^{p,q}$-bounded process, relative to a measure β on a σ-algebra $\mathcal{O}$, is a* **stochastic integrator** *in the sense that*

(i) $\{E(|\tau(f)|^q) : \|f\|_\infty \leq 1, f \, is \, \mathcal{O} - simple\}$ is bounded,

(ii) $f_n \downarrow 0$ a.e., $\mathcal{O}$-simple, implies $\lim_n \tau(f_n) = 0$ in probability.

This result is not really elementary to prove, and in fact depends on Grothendieck's theorem on metric geometry to find the measure μ and hence the desired β in establishing the boundedness principle. But then it has the desirable property of being a stochastic integrator. Thus the generalization embodied in this 'principle' is deeper. Full details of the result are available already (cf., Rao [20], p.463ff), and will be omitted here. Actually Definition 1.2 presents an extension of Itô's formulation of the stochastic integral. See the remarks at the end of this section. Another fact of special interest below is given by the following:

PROPOSITION 1.4. *Let $X = \{X_t, t \in I = [a,b]\}$ be a real p^{th} power integrable process having a strong (or L^p) derivative $\dot{X}_t$ at each point t of I (one sidedly at a and b). Suppose the derived process $\{\dot{X}_t, t \in I\}$ is continuous in probability. Then it has a version that is a point-wise differentiated form of the X_t-process so that X_t is continuous and locally absolutely continuous, i.e., on each neighborhood of t. If the derived process has no fixed points of discontinuity, then the X process can be taken absolutely continuous a.e. on I itself.*

In the more general case X will be generalized absolutely continuous a.e. on I, and their distinctions will be explained below.

Proof. The hypothesis implies that the (strongly) derived process $\{\dot{X}_t, t \in [a,b]\}$ belongs to $L^p(P)$. Consequently, X_t admits a representation as $X_t = X_a + \int_a^t \dot{X}_s \, ds$ where the integral is in the sense of Denjoy (also termed Henstock-Kurzweil), following e.g., Filippov [7] (Thm. 4.10.1, p.144). If moreover $t \mapsto \dot{X}_t$ is continuous in probability, then the integral can be taken in the sense of Lebesgue-Bochner. In particular the latter condition holds if the $\dot{X}_t$-process has no fixed points of discontinuity. In the case that $t \mapsto \dot{X}_t$ is (strongly) continuous, the standard stochastic function theory implies (cf.,e.g., Rao [20], Chapter III) that there is a separable and measurable version as asserted in the statement, and the details need not be repeated here. ■

REMARKS. 1. An interesting point is that, if some properties of Denjoy integration are employed, then one can just suppose that the process $\{X_t, t \in [a,b]\}$ is generalized absolutely continuous in the sense that $t \mapsto X_t$ is continuous and there is a countable union of subintervals on each of which it is absolutely continuous. However, the stronger condition will be assumed in what follows, for simplicity. Indeed, a detailed analysis with the weaker (stochastic) integration is not systematically explored in standard works, and it will be interesting to complete the theory for that case also.

2. It is observed from known results that the $L^{2,2}$-boundedness for finite dimensional processes is equivalent to their semi-martingale property, although this is no longer true (the former is more general) in the infinite dimensional case.

With this basis, it is possible to present problems in stochastic analysis that relate to evolution operators in solving them. Hereafter $\{X_t, \mathcal{F}_t, t \in I\}$ signifies that X_t is $\mathcal{F}_t$-adapted and that the σ-algebras $\mathcal{F}_t \uparrow\subset \Sigma$ form a standard filtration. Two classes of evolutions arising from SDEs and likelihood ratios will be considered in this paper.

2 EVOLUTION OPERATORS AND SDES

It is convenient to start with a general stochastic differential equation (SDE) and then specialize to associate a Markov process leading to an evolution operator family via the Chapman-Kolmogorov identity. Thus a stochastic flow is a process governed by a second order SDE of the form:

$$d\dot{X}_t + q(t, X_t, \dot{X}_t)dt = \sigma(t, X_t, \dot{X}_t)\, dB_t, \tag{2.1}$$

where $\{B_t, \mathcal{F}_t, t \in [a,b]\}$ is an $L^{2,2}$-bounded process, $\{\dot{X}_t, \mathcal{F}_t, t \in [a,b]\}$ being the derived process of the X_t-process. Here $q, \sigma : \mathbb{R}^+ \times \mathbb{R}^n \times \mathbb{R}^n \to \mathbb{R}$ are continuous functions satisfying Lipschitz conditions of the following form:

$$\begin{aligned} |q(t,x_1,y_1) - q(t,x_2,y_2)|^2 + |\sigma(t,x_1,y_1) - \sigma(t,x_2,y_2)|^2 & \\ \le K(|x_1 - x_2|^2 + |y_1 - y_2|^2) & \qquad (2.2) \\ |q(t,x,y)|^2 + |\sigma(t,x,y)|^2 \le K(1 + |x|^2 + |y|^2), \quad x, y \in \mathbb{R}^n, & \end{aligned}$$

for some absolute constant $K > 0$ The initial data for (2.1) to hold is $X_{t_0} = A$, $\dot{X}_{t_0} = B$. As indicated by the form (2.1), the desired solution is an absolutely continuous process X_t, and if q and σ are functions of t alone, then one has a linear SDE, the general case being the nonlinear equation.

THEOREM 2.1. *Under the above stated conditions, there exists a unique (absolutely continuous) process $\{X_t, t \in [a,b]\}$ that solves (2.1) for the given initial values. Moreover $\{(X_t, \dot{X}_t), t \in [a,b]\}$ forms a (vector) Markov process if the $L^{2,2}$-bounded process $\{B_t, t \in [a,b]\}$ has independent increments in addition. Further this vector process will have stationary transition probability functions when q, σ do not depend on the time variable t, and the increments of B_t are stationary.*

Proof-in-outline. Under the Lipschitz conditions on q and σ, along with the given initial values, one can obtain the existence and uniqueness of the solution of (2.1) using the Picard argument with a standard procedure via Gronwall's inequality plus the Borel-Cantelli lemma. This was detailed under a slightly restricted set of conditions in (Rao [20] (Theorem VI.4.6), and complemented in Rao [21], p.250). However the result holds under the present (slightly more general) conditions, by reducing it to the first order (vector nonlinear) equation, extending the classical finite dimensional method found e.g., in Coddington and Levinson [3] (p. 47 and p. 21), the existence and uniqueness, using (2.2), can be established with the same procedure as in Daleckii and Kreĭn [4] (Theorem VII.1.2 on p. 21). The result actually holds for n^{th} order nonlinear equations as well. The first component of the solution on $[a,b]$ then is absolutely continuous, and, together with its derivatives assured by Proposition 1.4, satisfies (2.1) and conversely, (this being valid for the nth order equation). For linear equations, such a procedure for general Banach space valued equations was studied by Goldstein [9] in his very first research publication. ∎

The following special SDE will be discussed as it motivates the next set of results in the general case. Thus let $\{X_t, t \in I = [a,b)\}$ be a process generated by the linear differential operator L_n of nth order such that (here $X(t)$ and X_t are interchangeable)

$$\int_I \varphi(t)(L_n X)(t)\, dt = \int_I \varphi(t)\, dB(t), \quad \varphi \in C_c^\infty(I),$$

where

$$L_n = \sum_{i=0}^{n} a_i(t) \frac{d^{n-i}}{dt^{n-i}}, \; a_i(\cdot)$$

being bounded measurable (real) functions, $C_c^\infty(I)$, the space of infinitely differentiable compactly supported real functions, and $\{B(t), t \in I\}$ *any* $L^{2,2}$*-bounded process* (e.g., the Brownian motion or a square integrable martingale). With this interpretation of the linear form of (2.1) of nth order, there is a unique solution of the equation

$$(dX)(t) = A(t)X(t) + \dot{B}(t), \tag{2.3}$$

where $X(t) = (X_1, X_2, \cdots, X_n)^*(t)$, $X_i = dX_{i-1}, i = 2, \cdots, n$ and $\dot{B} = (0, \cdots, B)^*$ (* denoting transpose of a matrix or vector); and $A(t)$ is the $n \times n$ matrix, defined by

$$A(t) = \begin{pmatrix} 0 & 1 & 0 & & \cdots & 0 \\ 0 & 0 & 1 & \cdots & 0 & \\ \vdots & & & \ddots & & \vdots \\ 0 & 0 & 0 & \cdots & 0 & 1 \\ a_n(t) & a_{n-1}(t) & a_{n-2}(t) & \cdots & a_2(t) & a_1(t) \end{pmatrix}$$

with the initial value $X_a = C$. An 'explicit' form of the solution of (2.3), obtainable from the classical (deterministic) treatments, is expressible as

$$X_t = M(t)[\int_0^t M(s)^{-1}\, dB_s + C] \tag{2.4}$$

where $M(t)$ is the invertible fundamental solution of the homogeneous equation $dX_t = A(t)X_t dt$. Moreover, if the $L^{2,2}$-bounded processes $\{B_t, t \in I\}$ has independent increments, then the vector solution is a Markov process (although the individual components will not have the latter property). In this case, if $0 < s < t < b$, then the transition probability distribution of the solution process is given by the fundamental Chapman-Kolmogorov equation as:

$$\begin{aligned} P[X_t \in F | X_s](x) &= P[X_t \in F | X_r, r \le s](x) \\ &= \int_{\mathbb{R}^n} P[X_t \in F | X_u](y) P[X_u \in dy | X_s](x) \end{aligned}$$

for each Borel set $F \subset \mathbb{R}^n$. Writing $p(s, x; t, F) = P[X_t \in F | X_s](x)$, the above can be expressed equivalently as:

$$p(s, x; t, F) = \int_{\mathbb{R}^n} p(u, y; t, F) p(s, x; y, dy), \quad s \le u \le t. \tag{2.5}$$

This gives a family of evolution operators on a suitable function space as follows. For $0 < s \le u \le t$ let $T_{st} : B(\mathbb{R}^n) \to B(\mathbb{R}^n)$ be defined by

$$(T_{st} f)(x) = \int_{\mathbb{R}^n} f(x) p(s, x; t, dy), \quad f \in B(\mathbb{R}^n), \tag{2.6}$$

the space of bounded Borel functions on $\mathbb{R}^n$. Now (2.5), the Fubini theorem, and (2.6) imply

$$(T_{su}T_{ut}f)(x) = (T_{st}f)(x), \quad x \in \mathbb{R}^n, \tag{2.7}$$

so that $\{T_{st}, a \le s \le t\}$ is a positive contractive evolution family and $T_{ss} = id$, when the uniform norm is used on $B(\mathbb{R}^n)$. In deriving (2.7), one has to use the Dunford-Schwartz integration, unless the conditional measures $p(s,x;t,\cdot)$ are assumed to be regular, as is usually done in the literature, to employ the Lebesgue theory of integration. However, the later will be used for simplicity in most of the computations below, and hence the regularity is imposed hereafter.

So far the 'noise' process $\{B_t, t \in I\}$ is only taken to be $L^{2,2}$-bounded with independent increments. Suppose that it has moreover almost all (a.a.) continuous paths so that the absolutely continuous solution process X_t of (2.3) inherits this property, and then the operator family $\{T_{st}, a \le s \le t < b\}$ is strongly differentiable on the compactly supported continuous function space $C_c(I)$. Let G_t be the operator defined by

$$\lim_{h \downarrow 0} \frac{T_{t(t+h)}f - T_{tt}f}{h} = G_t f, \; f \in dom(T_t) \subset C_c(I), \tag{2.8}$$

exists in the strong operator topology, written symbolically as:

$$\frac{dT_{st}}{dt} = G_s. \tag{2.8'}$$

The continuity of the T_{st} further implies the following. For all $s < t$ in I if $t_k^{(n)} = s + \frac{k(t-s)}{n+1} < b$

$$T_{st} = T_{st_1^{(n)}} T_{t_1^{(n)}t_2^{(n)}} \cdots T_{t_n^{(n)}t} = \lim_{n\to\infty} \Pi_{j=0}^n T_{t_j^{(n)}t_{j+1}^{(n)}}, \tag{2.9}$$

and the right side can be simplified with (2.8) as:

$$T_{t_j^{(n)}t_{j+1}^{(n)}} = I + G_{t_j^{(n)}}\frac{t-s}{n} + o(\frac{1}{n}) = \exp[G_{t_j^{(n)}}\frac{t-s}{n}] + o(\frac{1}{n}). \tag{2.10}$$

As is known from (2.8) and (2.10) one finds that

$$T_{st} = \lim_{n\to\infty} \Pi_{j=0}^n \exp[G_{t_j^{(n)}}\frac{t-s}{n}] = \int_I^{*} \exp G_t \, dt, \tag{2.11}$$

the last symbol defining a *multiplicative integral.* If $t \mapsto G_t$ is a uniformly integrable operator function, then T_{st} satisfies the operator integral equation

$$T_{st} = I + \int_s^t G_r T_{sr} \, dr$$

the integral being in Bochner's sense.

To proceed further and make concrete computations, however, it will be assumed that the B_t-process is Brownian Motion so that the solution process X_t of (2.3) is Gaussian, since the coefficient matrix $A(t)$ is non stochastic. Also in that case $T_{st} = M(t)M(s)^{-1}$. The process X_t of the particular equation (2.3) can be given an explicit form after a classical method from the theory of ordinary differential equations with variable coefficients. Thus if $(\varphi_1, \varphi_2, \cdots, \varphi_{n+1})$ are the linearly independent solutions of the homogeneous equation $(L_n u)(t) = 0$, then the unique solution of (2.3), with the initial value $C = 0$ for convenience, can be represented as:

$$X_t = \int_a^t \sum_{k=1}^{n+1} \varphi_k(t) \frac{W_k(\varphi_1, \cdots, \varphi_{n+1})}{W(\varphi_1, \cdots, \varphi_{n+1})}(s)\, dB_s, \tag{2.12}$$

where $W(\varphi_1, \cdots, \varphi_{n+1})$ is the Wronskian determinant of the φ_i-system (which never vanishes due to linear independence of the φ_i) and W_k is the same as W with its kth column replaced by $(0, \cdots, 1)$. If the integrand of (2.12) is denoted as $R(s,t)$, which reduces to the Green kernel when all the coefficients a_i are constants, then it is also termed the Riemann kernel. It possesses several properties of interest here (cf. Hurewicz [14], p. 54); namely, $R(\cdot,\cdot)$ is n times differentiable in each variable, and if $R^*(s,t) = R(t,s)$ then R^* satisfies the Lagrange identity, and (2.12) is representable as:

$$X_t = \int_a^t R(s,t)\, dB_s, \tag{2.13}$$

from which the covariance function can be calculated when the B_t-process has two moments as here. Note that the representation (2.13) is valid for any $L^{2,2}$-bounded process $\{B_t, t \in I\}$. When B_t is Brownian motion, one has

$$r(s,t) = Cov(X_s, X_t) = \int_a^{s \wedge t} R(s,u)R^*(t,u)\, du, \tag{2.14}$$

which is n times continuously differentiable. In the case of constant initial conditions, the solution is a Markov process with *stationary* transition probabilities, and a more detailed computation is available. It is due to Dym [5]. The result is stated for comparison and later extension as follows. Thus in this case (2.4) and hence (2.13) (with C as the constant initial value) reduce to the representation:

$$X_t = \int_a^t e^{A(t-s)}\, dB_s + e^{At}C, \tag{2.15}$$

so that the ith component becomes

$$(X_t)_i = \int_a^t (Ae^{A(t-s)})_{in} dB_s + (e^{At}C)_{in},\ i = 1, \ldots, n-1, \tag{2.16}$$

and

$$(X_t)_n = \int_a^t (Ae^{A(t-s)})_{nn}\, d(B_s)_n + (e^{At}C)_n + (B_t)_n,$$

from which one gets for the BM noise, the covariance $r = (r_{ij})$:

$$r_{ij} = Cov(X_i(s), X_j(t)) = \int_a^t (e^{At})_{in}(e^{As})_{jn} ds. \tag{2.17}$$

Since the a_i are constants, the transition probability function of the solution process, being an exponential, is infinitely differentiable, i.e., it is a real analytic function. In the case of stationary transitions, the associated evolution operator family T_{st} depends only on the difference of 'times' so that $T_{st} = T_{t-s}$ and $\{T_h, h \geq 0\}$ becomes a positivity preserving contractive semi-group of linear operators on the space of bounded Borel functions. Then using the constancy of the coefficients of the operator L_n, Dym [5] showed that $T_t : B(\mathbb{R}^n) \to C_b(\mathbb{R}^n)$ has the stronger property that $(T_t f)(x)$ defines a real analytic function on $(0, \infty) \times \mathbb{R}^n$. Moreover $T_t(C_b(\mathbb{R}^n)) \subset C_b(\mathbb{R}^n)$ so that the X_t-process has a.a. continuous paths and hence is strongly Markovian. The family $\{T_t, t \geq 0\}$ is a 'C_0-semi-group', and then its infinitesimal generator $G(= G_s, \forall s)$ is given by

$$G = \frac{1}{2}\frac{\partial^2}{\partial x_n^2} + \sum_{i=1}^{n-1} x_{i+1}\frac{\partial}{\partial x_i} + (x_1 a_n + \cdots + x_n a_1)\frac{\partial}{\partial x_n}, \tag{2.18}$$

which is a degenerate elliptic differential operator on $C_c^2(\mathbb{R}^n)$. However, very few of the existing results in the theory of second order PDE are applicable to this situation. It is in fact *a general difficulty associated with second and higher order SDEs* inviting fresh research of the problem. There is a detailed analysis of this special case in Dym's paper. Corresponding problems in the general case will be highlighted below.

The preceding work can be summarized, for reference, in the following:

THEOREM 2.2. *The nth order linear stochastic differential equation given by (2.3), with $\{B_t, t \in I = [a, b]\}$ as an $L^{2,2}$-bounded process has a unique solution X_t given by (2.4) satisfying the initial condition $X_a = C$. If moreover, the noise process B_t has independent increments, then X_t is a Markov process whose associated operators $\{T_{st}, s \leq t, (s,t) \in I \times I\}$ form a strongly continuous evolution family with generators $\{G_s, s \in I\}$ given by (2.8) each having a dense domain (usually depending on s) in $C(I)$. Further, each T_{st} is representable as a (strong) multiplicative integral given by (2.11). Moreover, the trajectories of the solution process X_t of (2.3) admit an integral representation (2.13) for a smooth (or n times differentiable) deterministic kernel $R(\cdot, \cdot)$. In particular, when the coefficients of (2.3) are constants, and the increments of B_t are also stationary, then the solution is a Markov process with* **stationary** *transition probabilities, for which with B_t as Brownian motion, the associated evolution family becomes $T_{st} = T_{t-s}$, a C_0-semi-group having a degenerate elliptic differential operator (2.18) as its infinitesimal generator. The covariance function then is real analytic and the solution is a Feller process.*

Motivated by the above discussion on linear problems, one now considers the nonlinear case.

3 EXTENSIONS TO NONLINEAR SDES

The existence of a unique solution given in Theorem 2.1 of the preceding section, under the stated boundary conditions, is valid for the nonlinear case of any $L^{2,2}$-bounded noise process B_t, and the solution has the Markov property if the B_t has independent increments. The SDE considered there is nonlinear and the drift and diffusion coefficients q, σ satisfy a (uniform) Lipschitz condition. Consequently, the work contained in equations (2.1)-(2.11) of the preceding section is available for the general case, and hence there is a family of strongly continuous contraction evolution operators $\{\tilde{T}_{st}, a \leq s \leq t < b\}$ associated with the solution process. Let $\tilde{G}_s$ be the infinitesimal operators defined by (2.8) there. The problem now is to obtain an explicit expression for $\tilde{G}_s$ on a suitable function space, by specializing the B_t-process again to be Brownian Motion. This is naturally more involved than the linear case.

Since q, σ depend on t, the transition probabilities are nonstationary. However, by enlarging the state (or range) space of the X_t-process (to a more complicated form), one can reduce the work to stationary transitions as observed by Dynkin [6] (p. 102).

Here is an outline of it for the scalar case, to simplify discussion. The new state space of the process will be $\tilde{R} = \mathbb{R}_a^+ \times \mathbb{R}^2$, $\mathbb{R}_a^+ = [a, \infty)$ endowed with the product topology. Define the transition probabilities, for each Borel set $F \subset \tilde{R}$ by:

$$\tilde{p}(t, (s, x); F) = p(s, x; s + t, F), \quad s, t \geq a, \tag{3.1}$$

for $x \in \mathbb{R}$, (and $= 0$ otherwise) with, as before, $\lim_{h \downarrow 0} p(s, x; s + h, \mathbb{R}) = 1$ for all x, s. Then the new family $\tilde{p}$ is stationary for the 'vector' process $(t, X_t) = Z_t$ (say), the so-called space-time process. If $\tilde{q} : \tilde{R} \to \mathbb{R}$, and $\tilde{\sigma} : \tilde{R} \to \mathbb{R}^+$, then the equation (2.1) of the last section becomes ($\dot{Z}_t$ denoting the sample derivative):

$$d\dot{Z}_t + \tilde{q}(Z_t, \dot{Z}_t)\, dt = \tilde{\sigma}(Z_t, \dot{Z}_t)\, d\tilde{B}_t, \tag{3.2}$$

where $\tilde{B}_t = (0, B_t)$ is the $L^{2,2}$-bounded process with stationary independent increments. The Lipschitz and boundedness conditions of q, σ translate to those on $\tilde{q}, \tilde{\sigma}$, and so the existence and uniqueness of solutions of (3.2) with stationary transitions is assured on $\tilde{R}$. They determine a strongly continuous positive contraction semi-group $\{U_t^s, t \geq a\}$ with an infinitesimal generator $\tilde{G}$ whose domain is contained in the space of continuous functions $C(\tilde{R})$. Now an explicit form of $\tilde{G}$ can be obtained using the work on the stationary case, available before (cf., Rao [21] (Sections 5,6)). This amounts to a reinterpretation of that work, but not a simplification. Thus the Markov process $\{(Z_t, \dot{Z}_t), t \in \mathbb{R}^+\}$ with state space $\tilde{R}$ and the associated semi-group $\{U_t^s, t \geq a\}$ will have an infinitesimal generator acting on the bounded continuous function space, $C_b(\tilde{R})$ into itself. But it will be a complicated integro-differential operator, as noted by Feller. However, this can be simplified to a differential operator by considering a weighted continuous function space that allows (real) unbounded functions on $\tilde{R}$, suggested by Doob in the first order case. This may be given in the present case as follows. [For simplicity, take $a = 0$ below.]

Let $\varphi : \tilde{R} \to \mathbb{R}^+$ be a weight function such that $\inf_{\mathbf{x}} \varphi(\mathbf{x}) \geq \delta > 0$, and $\lim_{|\mathbf{x}| \to \infty} \varphi(\mathbf{x}) = \infty$. If $C(\tilde{R})$ is the space of real continuous functions, let $C_\varphi(\tilde{R}) = \{f \in C(\tilde{R}) : \|f\|_\varphi < \infty\}$ where $\|f\|_\varphi = \sup_{\mathbf{x}} \frac{|f(\mathbf{x})|}{\varphi(\mathbf{x})}$ so that $(C_\varphi(\tilde{R}), \|\cdot\|_\varphi)$ is a

Banach space. In what follows, set $\mathbf{x} = (t, x, y) \in \tilde{R}$ and $\varphi(\mathbf{x}) = e^{|y|}$. Then again $\{U_t^s, t \geq 0\}$ is a bounded semi-group on $C_\varphi(\tilde{R})$ into itself, but is not a contraction.

The preceding results whose omitted details can easily be completed from an earlier work (cf.,Rao [21] Section 7) is summarized as:

PROPOSITION 3.1. *Let the second order nonlinear SDE be given by:*

$$d\dot{X}(t) + q(t, X(t), \dot{X}(t))\, dt = \sigma(t, X(t), \dot{X}(t))\, dB(t), \tag{3.3}$$

where $\{B(t), t \geq 0\}$ is an $L^{2,2}$-bounded process of independent stationary increments (in particular BM), with q, σ satisfying the boundedness and Lipschitz conditions as in Theorem 2.1. If $\tilde{q}, \tilde{\sigma}$ are as defined above, so that (3.3) is transformed into a space-time equation (3.2), then $\{(Z(t), \dot{Z}(t)), t \geq 0\}$ is a Markov process on an enlarged state space $\tilde{R} = \mathbb{R}^+ \times R^2$, with stationary transition probabilities, starting at $(t_0, X_{t_0}), t_0 \geq 0$. The associated semi-group $\{U_t^s, t \geq 0\}$ with $t_0 = s$ on $C_\varphi(\tilde{R})$ and weight $\varphi : \mathbf{x} = (t, x, y) \mapsto e^{|y|}$, is a positivity preserving bounded family which may not be strongly continuous.

To obtain an explicit form of the generator, it will be assumed hereafter that $\{B(t), t \geq 0\}$ is BM. Consider a subspace $C_2(\tilde{R})$ of $C(\tilde{R})$ as follows. $f \in C_2(\tilde{R})$ iff it is thrice continuously differentiable satisfying the following three conditions, wherein $f_{ij}(\mathbf{x}) = \frac{\partial^{i+j} f}{\partial x^i \partial y^j}(\mathbf{x}), f_{00} = f, f_j = \frac{\partial^j f}{\partial y^j}$:

(i) $\|f_{ij}\|_{\varphi_{a(f)}} < \infty$, with $\varphi_{a(f)}(\mathbf{x}) = e^{-a(f)|y|}$ for $0 \leq i + j \leq 2, 0 \leq a(f) < 1$;

(ii) there is a compact set $C_f \subset \tilde{R}$ such that $\|f_{ij}\chi_{\tilde{R}-C_f}\|_{\varphi_{a(f)}} < \infty, i + j = 3$.

Using other types of weights with only polynomial growth, Doob's result was extended for a related problem by Rosenkrantz [23] in which the 'drift or diffusion' coefficients are allowed to be unbounded. In this new formulation $\tilde{q}(\mathbf{x}), \tilde{\sigma}(\mathbf{x})$ will be identified as $q(s, x, y), \sigma(s, x, y)$ by fixing s and varying x, y. Using the computations with $\tilde{q}, \tilde{\sigma}$, of the space-time Markov process (stationary transitions), the associated infinitesimal operator of the resulting semi-group $\{U_t^s, t \geq 0\}$ detailed in (Rao [21], pp.281-285) implies the following statement:

THEOREM 3.2. *For the space-time process $\{Z_t, t \geq 0\}$ which is the unique solution of (3.2) under the Lipschitz and boundedness conditions given there, if $\{U_t^s, t \geq 0\}$ is the associated semi-group of bounded positive linear operators on $C(\tilde{R})$, its generator $\tilde{G}$ with domain $C_2(\tilde{R})$ is the degenerate elliptic operator given by*

$$\tilde{G} = \frac{1}{2}\tilde{\sigma}(\mathbf{x})^2 \frac{\partial^2}{\partial y^2} - \tilde{q}(\mathbf{x})\frac{\partial}{\partial y} + y\frac{\partial}{\partial x}, \tag{3.4}$$

which since s is fixed, becomes on setting $\tilde{G} = G_s$,

$$G_s = \frac{1}{2}\sigma(s, x, y)^2 \frac{\partial^2}{\partial y^2} - q(s, x, y)\frac{\partial}{\partial y} + y\frac{\partial}{\partial x}, \tag{3.5}$$

whose domain is $C_2(\tilde{R})$ for each fixed s.

Here the space-time identification allows one to deduce (3.4) or (3.5) from the work with stationary transition probabilities. Without this procedure but verifying the details of computations directly with s fixed, Goldstein [10] (p.54) has noted (3.5) in his earlier work.

One of the major reasons for an explicit form of the generator G_s is to study the sample path properties of the solution process $X(t)$ of (3.3) and this is helped by the fact that for each

$$f \in \cap_{s \geq 0} dom(G_s)$$

of the associated evolution family $T_{st}, t \geq s$, one has $\{(T_{st}f)(X(t)), \mathcal{F}_t, 0 \leq s \leq t\}$ to be a martingale whenever the set is integrable. In case q, σ are independent of time t, this can be refined to say that the corresponding semi-group and its generator have the property: $\{f(X(t)) + \int_s^t (U_r G) f(X(r))\, dr, \mathcal{F}_t, t \geq 0\}$ is a martingale for all bounded $f \in dom(G)$. A further specialization gives more refined results. Two such consequences will be indicated here.

The specialization involves additional conditions on q, σ. Thus let us express (3.3) as a first order SDE in vector form. Set $\mathbf{X}(t) = (X(t), \dot{X}(t))^*$, $\mathbf{Q}(t, x, y) = (y, -q(t, x, y))^*$ and

$$\mathbf{S}(t, x, y) = \begin{pmatrix} 0 & 0 \\ 0 & \sigma(t, x, y) \end{pmatrix}, \quad \mathbf{B}(t) = (0, B(t))^*,$$

so that

$$d\mathbf{X}(t) = \mathbf{Q}(t, \mathbf{X}(t))\, dt + \mathbf{S}(t, \mathbf{X}(t))\, d\mathbf{B}(t), \tag{3.6}$$

with * denoting transpose of a vector. Suppose now that the vector $\mathbf{Q}$ is decomposable as:

$$\mathbf{Q}(t, \mathbf{X}(t)) = \mathbf{Q}_1(t, \mathbf{X}(t)) + \mathbf{Q}_2(t, \mathbf{X}(t)), \tag{3.7}$$

and that the $\mathbf{Q}_i$ satisfy the same conditions as $\mathbf{Q}$ above. Also suppose that there is a bounded Borel measurable function $\psi : \tilde{R} \to \mathbb{R}^2$ satisfying

$$\mathbf{S}(t, \mathbf{x})\psi(t, \mathbf{x}) = \mathbf{Q}_2(t, \mathbf{x}),\ 0 \leq t, \tag{3.8}$$

i.e., the set of linear equations in (3.8) is solvable. [One may use the generalized inverse $\mathbf{S}^-$ of the singular matrix $\mathbf{S}$ so that $\psi(t, \mathbf{x}) = \mathbf{S}^-(t, \mathbf{x})\mathbf{Q}_2(t, \mathbf{x})$.] Then (3.6) has a unique solution with the initial condition $\mathbf{X}(0) = (A, B)^* = \mathbf{x}_0$ (say), in each of the cases when q, σ satisfy the Lipschitz and boundedness conditions. Let $P_{\mathbf{x}}$ be the probability measure governing the solution process when $\mathbf{Q}_2 = 0$ and similarly suppose that $\tilde{P}_{\mathbf{x}}$ is the corresponding measure when $\mathbf{Q}$ is given by (3.7) but $\mathbf{Q}_2 \neq 0$, and (3.8) holds for a vector function ψ described above. Thus the (vector) diffusion process under these conditions determines $P_{\mathbf{x}}$ and $\tilde{P}_{\mathbf{x}}$ uniquely. It then follows from a general Girsanov theorem (cf., Freidlin [8], p.79) that $\tilde{P}_{\mathbf{x}}$ and $P_{\mathbf{x}}$ are mutually absolutely continuous on the cylinder σ-algebra of the canonical representation space $C(\mathbb{R}^+)$ of the $X(t)$-process. Their likelihood ratio (or the Radon-Nikodým derivative) $\frac{d\tilde{P}_{\mathbf{x}}}{dP_{\mathbf{x}}}(\cdot)$ may be given by the following expression for a.a. $h \in C(\mathbb{R}^+)$:

$$(\frac{d\tilde{P}_{\mathbf{x}}}{dP_{\mathbf{x}}}(h))(t) = \exp\{\int_0^t \langle \psi(s, h(s)), dB(s) \rangle - \frac{1}{2}\int_0^t \|\psi(s, h(s))\|^2\, ds\}. \tag{3.9}$$

where $\langle \cdot, \cdot \rangle$ is the inner product in $\mathbb{R}^2$ and $\|\cdot\|$ its norm.

The second problem to be discussed is to obtain the sample path behavior of the solution process if q, σ of (3.3) are independent of time t, and satisfy a related but different condition. Namely, suppose there is a bounded continuous $b : \mathbb{R}^2 \to \mathbb{R}$ such that the coefficients of the differential operator (the generator) G of (3.4) satisfy:

$$b(y) = \frac{q(x,y)}{\sigma^2(x,y)}, \quad (x,y) \in \mathbb{R}^2, \tag{3.10}$$

i.e., $b(\cdot)$, [analogous to (3.8)] does not depend on x. Suppose moreover that b has a bounded continuous derivative on $\mathbb{R}$ outside of some (nonempty) symmetric neighborhood of the origin. If $g : y \mapsto \int_0^y \exp[-2\int_0^t b(s)\,ds]\,dt$, then the (extended) function $\tilde{g} : (x,y) \mapsto 1.g(y)$ is in $C_2(\mathbb{R}^2)$, and $\{\tilde{g}(\mathbf{X}(t)), t \in \mathbb{R}\}$ is a martingale. Moreover, if it converges a.e. (for instance, if $\{\tilde{g}(\mathbf{X}(t)), t \in \mathbb{R}\}$ is locally uniformly integrable), then $\dot{X}(t) \to \dot{X}(\infty)$ a.e., but $\dot{X}(\infty)$ can take infinite values on sets of positive measure. In any case, a.a. sample paths have versions which satisfy $\lim_{t\to\infty}(X(s+t) - X(t) = s\dot{X}(\infty)$. This result is due to Goldstein [10], and like the preceding one, it is obtained by specializing the coefficients of the generator G.

These are typical of the refinements of the general (second order) theory that yield further information on the solution process of (3.3), obtainable through properties of evolution equations. The higher order case, as in the linear SDEs discussed before, can be continued, and the whole area is mostly unexplored. This is partly because the generator G is always a degenerate operator for which most of the well-established PDE theory, using strong ellipticity is not applicable. The recent monograph of Freidlin [8] and the initial works of Dym [5] and Goldstein [11] [see also Mandl [16] for the types of problems of interest in this class] could be considered as early works for a sustained analysis of this whole area.

4 EVOLUTION OPERATORS AND STOCHASTIC INFERENCE

In a different direction, a brief account of the role played by evolution operators in the analysis of inference theory of stochastic processes (shortened to stochastic inference) will now be presented.

First let us indicate a one parameter problem leading to a semi-group application, and then discuss a two parameter extension leading to an evolution family, both being of interest in stochastic inference theory dealing with composite hypothesis testing questions. Thus if X is a random variable on a probability space (Ω, Σ, P), with distribution F so that $F(x) = P[X < x]$, let F_a be the distribution of its translate: $F_a(x) = P[X + a < x] = F(x-a), x, a \in \mathbb{R}$. Suppose that F has a derivative $F' = p$, strictly positive. A basic problem of interest here is to test the hypothesis

$$H_1 : a = 0 \; vs \; H_2 : a \neq 0.$$

Then the well-known Neyman-Pearson lemma and its generalization due to Grenander [12] states that the problem can be solved by considering the likelihood ratio

$$L_a(x) = \frac{dP_a}{dP}(x) = \frac{F_a'}{F'}(x) = \frac{p(x-a)}{p(x)}.$$

This may be written by setting $\varphi(x) = \frac{p'(x)}{p(x)}$ as:

$$\log L_a(x) = -\int_0^a \varphi(x-u)\,du,$$

or equivalently

$$L_a(x) = \frac{dP_a}{dP}(x) = \exp[-\int_0^a \varphi(x-u)\,du]. \tag{4.1}$$

If $(T_a f)(x) = f(x-a)$, $f \in L^1(P)$, so that $\{T_a, a \in \mathbb{R}\}$ is a semi-group of translations on $L^1(P)$, then (4.1) implies the following. For any $h \in C_c^\infty(\mathbb{R})$, the Schwartz space of smooth functions on $\mathbb{R}$ with compact supports, an integration by parts gives:

$$\begin{aligned}\int_{\mathbb{R}} \varphi(x)h(x)p(x)\,dx &= -\int_{\mathbb{R}} p'(x)h(x)\,dx \\ &= \frac{\partial}{\partial a}(\int_{\mathbb{R}} p(x-a)h(x)\,dx)|_{a=0} \\ &= \frac{\partial}{\partial a}(\int_{\mathbb{R}} (T_a h)(x)p(x)\,dx)|_{a=0},\end{aligned} \tag{4.2}$$

which thus presents a formula for $\varphi(\cdot)$. This leads to a deep generalization, similar to (3.9) of Section 3, for diffusion processes discovered by Pitcher [17]. This is explained now.

Since in the canonical representation of a process Ω is a function space, here $\Omega \subset \mathbb{R}^I, I = [0,1]$, so that $X_t(\omega) = \omega(t), \omega \in \Omega$, consider the first order SDE as:

$$dX_t = a(t, X_t)dt + \sigma(t)dB_t, \quad X_0 = 0, t \in I, \tag{4.3}$$

with $\{B_t, t \in I\}$ as BM. Let $a : I \times \mathbb{R} \to \mathbb{R}$ satisfy the further conditions: (i) $a(t,x) = \int_{\mathbb{R}} e^{ixy} A(y,t)\,dy$, and (ii)

$$a_x(t,x) = i\int_{\mathbb{R}} y e^{ixy} A(y,t)\,dy; \quad \int_{\mathbb{R}} (1+|y|)|A(y,t)|dy \le K < \infty,$$

where $a_x = \frac{\partial a}{\partial x}$. Let $\mathcal{G}$ be an algebra of functions $f : \Omega \to \mathbb{R}$ that depend on a finite number of coordinates, called cylindrical functions. Thus $f(x) = h \circ \pi_n(x), x \in \Omega$. Here $\pi_n : \Omega \to \mathbb{R}^n$ is the coordinate projection, and $h : \mathbb{R}^n \to \mathbb{R}$ is assumed to be bounded with bounded continuous derivatives. Define the shift operator $T_\alpha : \mathcal{G} \to \mathcal{G}$ of the following type:

$$(T_\alpha f)(x) = h(\pi_n(x + \alpha m)), \quad m, x \in \Omega,\ \alpha \in \mathbb{R}, \tag{4.4}$$

where $f = h \circ \pi_n$ as before. Then it is verified that $\mathcal{G} \subset L^p(P)$, $1 \le p < \infty$ is dense, and T_α extends to a positive linear operator and $T_\alpha 1 = 1$ so that it is also bounded. Hence it has a bound preserving extension to all of $L^p(P)$, and using the same notation, $\{T_\alpha, \alpha \in \mathbb{R}\}$ forms a group. The equation on $L^1(P)$ defines uniquely a probability measure $P_\alpha : \Sigma \to R^+$ (Riesz representation) given by the equation:

$$\int_\Omega (T_\alpha f)(x)\,dP(x) = \int_\Omega f(x)\,dP_\alpha(x), \quad f \in L^1(P). \tag{4.5}$$

Note that $P_0 = P$ but P_α and P need not be mutually absolutely continuous for all α. The following is a deep extension of (4.2) and is a Girsanov type result. It is used in testing P against P_α for any $\alpha \ne 0$ (to solve the simple versus a composite hypothesis problem).

THEOREM 4.1. *Consider the diffusion equation (4.3) in which the drift coefficient* $a(\cdot,\cdot)$ *satisfies (i) and (ii) above, and the diffusion coefficient* $\sigma(\cdot) > 0$ *is bounded away from* 0 *and* ∞, *with* a_x *continuous. Suppose the shift function* $m : I \to \mathbb{R}$ *has a square integrable derivative and* $m(0) = 0$. *Then the probability measures* $\{P_\alpha, \alpha \in \mathbb{R}\}$ *defined by (4.5) are mutually absolutely continuous for all* $\alpha \in \mathbb{R}$, $(P_0 = P)$ *and the likelihood ratio* $\frac{dP_\alpha}{dP}$ *is given by:*

$$\frac{dP_\alpha}{dP}(x) = \exp[\int_0^\alpha (T_{-b}\varphi)(x)\,db], \ a.a.\, x \in \Omega, \tag{4.6}$$

for a measurable version of $(\alpha, x) \mapsto (T_{-\alpha}\varphi)(x)$ *where*

$$\varphi = E^{\mathcal{C}}(\int_I \frac{m'(t) - m(t)a_x(t, X_t)}{\sigma(t)} dB_t),$$

$\mathcal{C}$ *being the* σ*-algebra generated by the solution process* $\{X_t, t \in I\}$, *and* $E^{\mathcal{C}}$ *the corresponding conditional expectation operator.*

The proof is not simple, but the essential details are given in the book (Rao [22], Sec. 7.1). However, the distinction between (4.6) above and (3.9) of the preceding section should be noted. Since the likelihood ratio is obtained from the semi-group $\{T_\alpha, \alpha \in \mathbb{R}\}$ on $L^1(P)$, one may extend the above result for (Markov type) processes that are not Gaussian, but then the precise description of $\frac{dP_\alpha}{dP}$ will not be obtained. This is also discussed in the above reference (cf. Sec. 7.2). It will next be shown how the result admits an extension to two parameter measures $P_{\alpha\beta}$, based on evolution operators. This generalization corresponds to the problems involving a composite hypothesis versus a composite alternative of considerable interest in stochastic inference theory. The following account is based on the (inaccessible) work of Velman [24].

Let $X = \{X_t, t \in I\}$ be a process measuring the output of a communication channel when a (stochastic) signal is sent and a noise $Z = \{Z_t, t \in I\}$ is present. The additive model assumed here is:

$$X_t = Y_t + \alpha Z_t, \quad t \in I = [a, b] \subset \mathbb{R}, \tag{4.7}$$

and α is a real parameter to be tested for various values. The following assumptions, simplifying the study, are appropriate. Both Y, Z-processes are Gaussian, centered, and have strictly positive definite covariances K_1, K_2 respectively. Let $(R_i f)(t) = \int_I K_i(s,t) f(s)\,d\mu(s)$, $i = 1, 2$ and $f \in L^2(I, \mu)$, μ being a finite measure, so that R_i is a Hilbert-Schmidt operator for $i = 1, 2$. Assume further that $B = R_1^{-\frac{1}{2}} R_2 R_1^{-\frac{1}{2}}$ has a bounded extension to all of $L^2(I, \mu)$ and to be of trace class. This will enable one to represent the process by a countable collection of (uncorrelated or independent) random variables. Indeed, let $(\tau_n, \varphi_n), n = 1, 2, \ldots$, be the eigenvalues and the corresponding normalized eigenfunctions of B. Then the signal and noise processes Y, Z of (4.7) admit of representations in terms of $(Y_n, Z_n), n = 1, 2, \ldots$ where $Y_n = \int_I Y_t \varphi_n(t)\,d\mu(t)$, and $Z_n = \int_I Z_t \varphi_n(t)\,d\mu(t)$. Thus one has

$$Y_t = \sum_{n=1}^{\infty} g_n(t) Y_n; \quad Z_t = \sum_{n=1}^{\infty} \tau_n g_n(t) Z_n, \tag{4.8}$$

where $g_n(t) = (R_1^{\frac{1}{2}}\varphi)_n(t), t \in I$, both series converging in $L^2(P)$-mean. The σ-algebras generated by the sequences $\{Y_n, n \geq 1\}$ and $\{Z_n, n \geq 1\}$ are seen to be equivalent to those generated by the Y and the Z-processes. The sequences Y_n and Z_n are often called "observable coordinates" of the processes.

Let $\tilde{\Omega} = \Omega \times \Omega, \tilde{\Sigma} = \Sigma \otimes \Sigma$ and $\tilde{P} = P \otimes P$, so that $(\tilde{\Omega}, \tilde{\Sigma}, \tilde{P})$ is the basic triple. Choose $\mathcal{G}$ as the algebra of real cylindrical functions f on $\tilde{\Omega}$ of the form:

$$f(y,z) = (h_{2m} \circ \pi_{2m})(y,z) = h_{2m}(y_1, \cdots, y_m, z_1, \cdots, z_m)$$

where $\pi_m : \tilde{\Omega} \to \mathbb{R}^m$ is the coordinate projection and $h_m : \mathbb{R}^m \to \mathbb{R}$ is a bounded function with bounded continuous partial derivatives, so $h_m \in C^1(\mathbb{R})$. Let $\mathcal{G}_\alpha \subset \mathcal{G}$ be an algebra such that $f \in \mathcal{G}_\alpha$ iff

$$f(y,z) = h \circ \pi_n(y + \alpha z) = h(y_1 + \alpha z_1, \cdots, y_n + \alpha z_n) = f_\alpha(y,z), \ (say) \qquad (4.9)$$

so that $\{\mathcal{G}_\alpha, \alpha \in \mathbb{R}\}$ is a one parameter family of subalgebras of $L^p(\tilde{P})$. It may be verified that the completions of $\mathcal{G}_\alpha$ in $L^p(\tilde{P}), 1 \leq p < \infty$ are Banach lattices and hence may be represented as subspaces $L^p(\tilde{\Sigma}, \tilde{P})$ for an essentially unique σ-subalgebra $\tilde{\Sigma}$, (cf., e,g., Rao [19], p. 46, Thm. 5). Let $T(\alpha, \beta) : \mathcal{G}_\beta \to \mathcal{G}_\alpha$ be a mapping defined by $T(\alpha,\beta) f_\beta = f_\alpha$ using the notation of (4.8). Then the family $\{T(\alpha, \beta), \alpha \leq \beta\}$ satisfies: (i) $T(\alpha, \beta)$ is a positive isometric isomorphism of $\mathcal{G}_\beta$ onto $\mathcal{G}_\alpha$ under uniform norm, and (ii) for $\alpha \leq \beta \leq \gamma, f \in \mathcal{G}_\gamma$, one has $T(\alpha,\beta)T(\beta,\gamma)f = T(\alpha,\gamma)f$. Thus it forms an evolution operator family, and constitutes the basic class for our problem. Just as in the one parameter case, on $L^1(\tilde{P})$ they induce (by the Riesz theorem again) a unique probability class, denoted $P_{\alpha,\beta} : \Sigma_{\alpha\beta} \to \mathbb{R}^+$. The concern is to find conditions for mutual absolute continuity of $P_{\alpha\beta}$ and $P_{\alpha\alpha}$ so that one can find the likelihood ratios.

The desired solution of the problem on the probability measures $P_{\alpha\beta}$ introduced above is given by the following:

THEOREM 4.2. *Consider the process $\{X_t = Y_t + \alpha Z_t, t \in I = [a,b], \alpha \in J = [c,d] \subset \mathbb{R}\}$ with Y, Z as centered independent Gaussian processes representing signal and noise, in a statistical communication channel, with strictly positive definite covariance functions K_1, K_2. If R_1, R_2 are the integral operators introduced above with K_1, K_2 as their kernels, let $B = R_1^{-\frac{1}{2}} R_2 R_1^{-\frac{1}{2}}$ be a trace class operator. If $\{\tau_n, \varphi_n, n \geq 1\}$ are the eigenvalues and the corresponding normalized eigenfunctions of B (so $\tau_n > 0$ and $\sum_{n=1}^{\infty} \tau_n < \infty$), then for any $\alpha, \beta \in J$, the induced probability measures $P_{\alpha\beta}$ of the X-process have the following properties:*

(i) $P_{\alpha\beta}, P_{\alpha\alpha}$ are mutually absolutely continuous, and

(ii) if $X_n = (Y_n, Z_n)$ are the observable coordinates of the X-process obtained with the φ_n (so $Y_n = \int_I Y_t \varphi_n(t)\, d\mu(t); Z_n = \int_I Z_t \varphi_n(t)\, d\mu(t)$), using the canonical representation (so $X_t(\tilde{\omega}) = \tilde{\omega}(t), \tilde{\omega} \in \tilde{\Omega}$), one has:

$$\frac{dP_{\alpha\beta}}{dP_{\alpha\alpha}} = \lim_{n \to \infty} \prod_{i=1}^{n} (\frac{1 + \alpha^2 \tau_i}{1 + \beta^2 \tau_i})^{\frac{1}{2}} \times$$

$$\exp[\frac{\tau_i}{2}(Y_i + \alpha Z_i)(\frac{\beta^2 - \alpha^2}{(1 + \alpha^2 \tau_i)(1 + \beta^2 \tau_i)})].$$

The result is actually obtained from an analysis based on the general assumption that $T(\alpha, \beta)$ above is simply an evolution family of positive bounded operators on $L^p(\Sigma)$ satisfying certain conditions so that their generators G_s obey some growth conditions, suggested by Pitcher's work. The communication illustration here merely shows that the assumptions are naturally present in many applications. Details of Velman's unpublished results are streamlined in my book (Rao [22], Secs. 7.3 and 7.4). The purpose here is to show how evolution operators appear in essentially different contexts.

5 FURTHER REMARKS

The preceding sections show that given an SDE of first (or higher) order, it is possible to associate, through its solution which is a (vector) Markov process, an evolution operator family whose generator generally gives an elliptic partial differential operator (perhaps degenerate). The solution of the PDE defined by the latter, under standard boundary conditions, allows one to analyze the sample path behavior of the solution of the given SDE. In some cases, especially when the (elliptic) PDEs are difficult to solve directly, for instance in the case of the Cauchy equation:

$$\frac{\partial u}{\partial t}(t,x) = (Lu)(t,x) + c(x)u(t,x),\ t > 0, x \in \mathbb{R}^k, u(0,x) = f(x), \tag{5.1}$$

where the operator L is given by

$$L = \frac{1}{2}\sum_{i,j=1}^{k} a_{ij}(x)\frac{\partial^2}{\partial x_i \partial x_j} + \sum_{i=1}^{k} b_i(x)\frac{\partial}{\partial x_i}, \tag{5.2}$$

with the matrix $A = (a_{ij})$ being non-negative definite, one can associate an SDE to (5.1) as:

$$dX_t = b(X_t)\,dt + \sigma(X_t)\,dB_t,\ X_0 = x \in \mathbb{R}^k \tag{5.3}$$

where $\{B_t, t \geq 0\}$ is a k-dimensional BM. Here $A(x) = \sigma(x)\sigma^*(x)$, and the square-root $\sigma(X)$ of $A(x)$ satisfies the Lipschitz conditions in the context. Then u defined by the equation

$$u(t,x) = E_x[f(X)(t)\exp(\int_0^t c(X_s)\,ds)], \tag{5.4}$$

is a solution of (5.1). Here E_x denotes the expectation in terms of the probability measure $P_x : \Sigma \to \mathbb{R}^+$ on the path space $C_x(\mathbb{R}^k)$, paths starting from the initial point x. The problem is thus transferred to the evaluation of the expectation in (5.4). One can then at least use a numerical integration if an explicit evaluation is not possible. This is the Feynman-Kac methodology, and its various applications are discussed in Freidlin [8], Ch. II. For problems in stochastic inference, however, other approaches are needed as seen in the preceding section. There one considers the structural properties of the operator families which often depend on difficult computations.

A generalization of the SDE (5.3) has also interest when the coefficients $b(\cdot), \sigma(\cdot)$ are random in addition. Thus if $b : \mathbb{R}^+ \times \mathbb{R}^m \times \Omega \to \mathbb{R}^n$ and $\sigma : \mathbb{R}^+ \times \mathbb{R}^m \times \Omega \to \mathbb{R}^{m \times n}$, the space of $n \times n$-matrices, then (5.3) becomes:

$$dX_t = b(t, X_t)\,dt + \sigma(t, X_t)\,dB_t, \tag{5.5}$$

for an m-dimensional BM to get an n-dimensional solution X_t, under suitable conditions on b and σ. Actually one can replace $\mathbb{R}^m$ and $\mathbb{R}^n$ by different Hilbert spaces such as b with values in $V = W_p^m(G), 1 \le p < \infty$, a Sobolev space on a bounded domain $G \subset \mathbb{R}^d$, and σ with values in $H = L^2(G)$ with Leb. measure and $p \ge 2(1 - \frac{mp}{d})$, so that $V \subset H = H^* \subset V^*$ and if b, σ satisfy certain monotonicity, coercivity, and Lipschitz type (boundedness) conditions, one can show (after verifying the joint measurability of b, σ) that (5.5) has a unique solution which is a Markov process, and the analysis continues for other non-linear problems. These are termed stochastic evolutions and are surveyed recently by Veretennikov [25]. They have interest in extending the preceding work using various properties of Sobolev embedding theorems.

In this connection one should observe that new problems arise in infinite dimensional extensions. In fact the BM process B_t in (infinite dimensional separable) Hilbert space $\mathcal{H}$ is a process with continuous sample paths and independent increments, satisfying the condition that its covariance is given by $(B_t - B_s, B_t - B_s) = |t-s|L$ where L is a fixed positive definite trace class operator on $\mathcal{H}$, and $(\cdot,\cdot)$ is the inner product in $\mathcal{H}$. Suppose that $b : \mathbb{R}^+ \times \Omega \to \mathcal{H}$, $\sigma : \mathbb{R}^+ \times \Omega \to B(\mathcal{H})$, the space of bounded linear operators on $\mathcal{H}$, in (5.5) are measurable relative to the filtration determined by the BM B_t-process, and satisfy the standard Lipschitz and boundedness conditions so that there exists a unique solution of (5.5). To calculate the likelihood ratios of the measures determined by the solutions of the SDEs given by

$$dX_t = b_i(X_t)\,dt + \sigma(X_t)\,dB_t, \quad i = 1, 2, \tag{5.6}$$

an additional condition, analogous to (3.8) of Section 3, is employed. Thus suppose that $\sigma L \sigma^*$ is positive definite, σ^{-1} exists and $b_i = \sigma L g_i$ for some bounded continuous $g_i : \mathbb{R}^+ \to \mathcal{H}$ which are adapted to the B_t-filtration. Then the probability measures P_i of the corresponding solutions of (5.6) exist, are mutually absolutely continuous, and one has the likelihood ratio (corresponding to (3.9) of Section 3) as:

$$\begin{aligned}\frac{dP_2}{dP_1} = \exp\{ &\int_0^t ([(\sigma^*)^{-1}(g_2 - g_1)](u), dB(u)) \\ &- \frac{1}{2}\int_0^t ((g_2, Lg_1) - g_1, Lg_1))(u)\,du\}.\end{aligned} \tag{5.7}$$

Complete details of this result may be found in Krinik [15]. It will be interesting to extend this work for (5.5) in full generality, in the sense of the preceding sections.

Finally, the point should again be emphasized regarding (even scalar) higher order SDEs, that while formally they can be expressed as first order vector (or operator) equations, of the type (5.5), for the associated semi-groups (in the case of (5.3)) or evolution operators (in the case of (5.5)), their generators always turn out to be degenerate elliptic. A further study of this class, not always treated in the standard PDE theory, should be considered. The resulting work will be beneficial to both stochastic processes and nonlinear analysis.

BIBLIOGRAPHY

1. S. Bochner, *Stationarity, boundedness, almost periodicity of random valued functions*, Proc. 3rd Berkeley Symp. Math. Statist. and Prob. **2** (1956), 7–27.

2. D. R. Borchers, *Second order stochastic differential equations and related Itô processes*, Ph.D., thesis, Carnegie Tech., Pittsburgh, PA, (1964).

3. E. A. Coddington and N. Levinson, *Theory of Ordinary Differential Equations*, McGraw-Hill, New York, 1955.

4. Ju. L. Dalecky and M. G. Kreĭn, *Stability of Solutions of Differential Equations in Banach Spaces*, Translations of A.M.S., Providence, RI, 1974.

5. H. Dym, *Stationary measures for the flow of a linear differential equation driven by white noise*, Trans. Am. Math. Soc. **123** (1966), 130–164.

6. E. B. Dynkin, *Theory of Markov Processes*, Prentice-Hall, Englewood Cliffs, NJ, 1961.

7. V. V. Filippov, *Basic Topological Structures of Ordinary Differential Equations*, Kluwer Academic, Dordrecht, The Netherlands, 1998.

8. M. I. Freidlin, *Functional Integration and Partial Differential Equations*, Princeton Univ. Press, Princeton, NJ, 1985.

9. J. A. Goldstein, *An existence theorem for linear stochastic differential equations*, J. Diff. Eqns. **3** (1967), 78–87.

10. J. A. Goldstein, *Second order Itô processes*, Nagoya Math. J. **36** (1969), 26–63.

11. J. A. Goldstein, *Abstract evolution equations*, Trans. Am. Math. Soc. **141** (1969), 159–185.

12. U. Grenander, *Stochastic processes and statistical inference*, Arkiv f. Mat. **1** (1950), 195–277.

13. U. Grenander, *Abstract Inference*, Wiley, New York, 1981.

14. W. Hurewicz, *Lectures on Ordinary Differential Equations*, MIT Press, Cambridge, MA, 1958.

15. A. Krinik, *Diffusion processes in Hilbert space and likelihood ratios*, Real and Stochastic Analysis, Wiley, New York, 1986, pp. 168–210.

16. P. Mandl, *Analytical Treatment of One-Dimensional Markov Processes*, Springer, Berlin, 1968.

17. T. S. Pitcher, *Likelihood ratios for diffusion processes with shifted mean values*, Trans. Am. Math Soc. **101** (1961), 168–176.

18. T. S. Pitcher, *Parameter estimation for stochastic processes*, Acta Math. **112** (1964), 1–40.

19. M. M. Rao, *Foundations of Stochastic Analysis*, Acad. Press, New York. 1981.

20. M. M. Rao, *Stochastic Processes: General Theory*, Kluwer Academic, Dordrecht, The Netherlands, 1995.

21. M. M. Rao, *Higher order stochastic differential equations*, Real and Stochastic Analysis: Recent Advances, CRC-Press, Boca Raton, FL , 1997, PP. 225–302.

22. M. M. Rao, *Stochastic Processes: Inference Theory*, Kluwer Academic, Dordrecht, The Netherlands, 2000.

23. W. A. Rosenkrantz, *An application of Hille-Yosida theorem to the construction of martingales*, Indiana Univ. Math. J. **24** (1974), 171–178.

24. J. R. Velman, *Likelihood ratios determined by differentiable families of isometries*, Ph.D. thesis, U.S.C., and Hughes Aircraft Co., Los Angeles, CA, Research Report No. 35, 1969/70.

25. A. Yu. Veretennikov, *Stochastic evolution equations*, Encyclopedia of Math. Sciences, vol. 45 (Probability Theory, III), Springer, Berlin, 1998, PP. 76–90.

Competition between diffusion and inhomogeneous reaction

STEVE ROSENCRANS

Mathematics Department
Tulane University
New Orleans, LA 70118
e-mail: sir@math.tulane.edu

XUEFENG WANG

Mathematics Department
Tulane University
New Orleans, LA 70118
e-mail: xdw@math.tulane.edu

We dedicate this paper to our good friend Jerry Goldstein on the occasion of his 60th birthday.

1 INTRODUCTION

In the summer of 1991 we saw in [4] the following spatially inhomogeneous heat equation:

$$xu_t = u_{xx} + u^p, \quad u > 0, \quad 0 < x < 1, \quad t > 0 \tag{1.1}$$

where $p > 1$ is a constant, with the Dirichlet boundary condition

$$u(0,t) = u(1,t) = 0. \tag{1.2}$$

This problem is related to a model for the temperature of a fluid flowing steadily through a channel with temperature-dependent viscosity. This is interesting mathematically in the following sense: If we divide (1.1) by x we see that the "diffusion rate" $1/x$ blows up at $x = 0$ and so does the reaction rate u^p/x if u^p is not small

Second author partially supported by BOR/LEQSF.

enough. Presumably the strong smoothing effect at $x = 0$ competes with the reaction: the former demolishes "spikes" and enhances spatial homogeneity, while the latter encourages spatial inhomogeneity, especially blowup.

As a first step in understanding this competition between diffusion and reaction we ask whether it is possible for the boundary-value problem (1.1), (1.2) to blow up at the boundary $x = 0$, i.e., whether there exist sequences $x_n \to 0$, $t_n \to T_b$ (the blowup time) such that $u(x_n, t_n) \to \infty$. It is well-known and easy to prove that solutions of the heat equation with a source term such as u^p blow up in finite time if the initial value

$$u(x, 0) = \phi(x) \tag{1.3}$$

is big enough. Floater proved that if this is the case and if ϕ has the right shape:

$$\frac{d}{dx}\left(\frac{\phi(x)}{x}\right) \leq 0 \qquad \text{on } [0, 1], \tag{1.4}$$

then if $1 < p \leq 2$ the positive solution of (1.1), (1.2), (1.3) blows up at $x = 0$ and only there. The case $p > 2$ was left open, though numerical experiments [9] indicated that $x = 0$ is not a blowup point. Concerning this case we proved the following results in 1991-1992.

R1. If

$$\phi'' + \phi^p \geq 0 \quad \text{and is not identically 0 on } [0, 1] \tag{1.5}$$

then the solution of (1.1), (1.2), (1.3) blows up in finite time, and neither 0 nor 1 is a blowup point.

R2. If for some δ between 0 and 1 $\phi'' + \phi^p$ is positive to the left of δ and negative to the right of δ, then the blowup, if it occurs in finite time, occurs either in a neighborhood of 0 or away from 0.

We point out that one can find many ϕ's satisfying both Floater's condition (1.4) and our condition (1.5), for example $k\overline{u}$ where k is a constant greater than 1 and $\overline{u}$ is a steady solution of (1.1), (1.2). (It is well-known that such a steady state exists and is unique.) The sharp contrast in blowup location in the two cases $1 < p \leq 2$ and $p > 2$ may be explained heuristically as follows: Since $u|_{x=0} = 0$, the source term is weaker near $x = 0$ in the latter case. Therefore, diffusion prevails over reaction. This should be true for *any* ϕ, i.e., $x = 0$ is not a blowup point for any ϕ. Result **R2** above was aimed in this direction. Unfortunately, the exclusion of the possibility that blowup occurs in an entire neighborhood of 0 has proved to be a stubborn problem. On the other hand, we have not seen any results better than ours in the last ten years. We feel that publishing our results may give a push to some effort in this direction.

We recall that the study of blowup sets of semilinear heat equations goes back to Weissler [10]. He considered the autonomous equation closely related to (1.1):

$$\begin{cases} u_t = u_{xx} + u^p & u > 0, \quad 0 < x < 1, \quad t > 0 \\ u(0,t) = u(1,t) = 0 & t > 0 \\ u(x,0) = \phi(x) & 0 < x < 1. \end{cases} \tag{1.6}$$

Other workers who studied the same problem include Friedman and McLeod [6], Chen and Matano [3]. In [6] it was proved that (i) in small neighborhoods of the boundary points u remains monotone in x and hence blowup cannot occur in these neighborhoods; (ii) if ϕ satisfies (1.5) and ϕ' changes its sign only once (so ϕ has only one "hot spot", i.e., local maximum point), then blowup occurs at only one point in $(0, 1)$. Chen and Matano [3] considerably extended these results by proving that the set of blowup points must always be discrete with cardinal number no larger than the number of hot spots at time 0, without any condition on ϕ. The key observation in [3] is that the number of hot spots is nonincreasing in t, which follows from applying the Sturmian Theorem (Theorem 2.4) to u_x. The Chen-Matano result holds for the case when u^p in (1.6) is replaced by suitable $f(t, u, u_x)$ but not for spatially inhomogeneous equations because in this case u_x does not satisfy a *homogeneous* linear equation (and hence the Sturmian Theorem does not apply to u_x). As a matter of fact, a counterexample can easily be constructed such that for some $f(x, u)$ and ϕ the solution u has but one hot spot at $t = 0$ and instantly for $t > 0$ has at least two.

We mention that our result can be extended to the case when x on the left side of (1.1) is replaced by a function $a(x) \sim x^q$ for $x \downarrow 0$. The idea can also be used to study location of blowup sets for multi-dimensional heat equations such as

$$\begin{cases} u_t = \Delta u + h(x)u^p & x \in \Omega, \quad t > 0 \\ u = 0 & \text{on} \quad \partial\Omega \ \text{for}\, t > 0 \\ u(x,0) = \phi(x) \geq 0 & x \in \Omega \end{cases} \tag{1.7}$$

where Ω is a bounded domain in R^n with smooth boundary, $h \geq 0$ is smooth and not identically 0 on Ω. In this paper we present a result – proved in a discussion with Chanfeng Gui and Wei-Ming Ni, to both of whom we are indebted – to the effect that if

$$\Delta\phi + \phi^p \geq 0 \quad \text{on } \Omega \quad \text{and } \phi \text{ not identically 0 on } \Omega \tag{1.8}$$

then there is no blowup near locations where $h = 0$. (See Theorem 4.)

The paper is organized as follows. In Section 2 we record or prove some results, including the maximum principle for parabolic operators degenerate in the fashion of (1.1) which, in their generality, seem to be new. Our main results regarding (1.1) are proved in Section 3. At the end of Section 3 we study (1.7).

2 PRELIMINARIES

2.1 Maximum Principle for degenerate parabolic equations

For possible future use we consider a general parabolic operator

$$Lu = \sum_{i,j=1}^{n} a_{ij}(x,t)\frac{\partial^2 u}{\partial x_i \partial x_j} + \sum_{i=1}^{n} b_i(x,t)\frac{\partial u}{\partial x_i} + c(x,t)u - d(x,t)\frac{\partial u}{\partial t},$$

where $x \in \Omega$, a bounded domain in $\mathbf{R}^n$, $t \in (0, T]$, and the matrix $a_{ij}(x, t)$ is positive definite: for some positive constant $\mu_0 > 0$

$$(a_{ij}(x,t) \geq \mu_0 \, I_{n\times n} \qquad \text{for} \quad (x,t) \in D \equiv \Omega \times (0,T]. \tag{2.1}$$

PROPOSITION 2.1. *Suppose* $u \in C^{2,1}(D) \cap C(\overline{D})$ *satisfies*

$$Lu \geq 0 \quad \text{for } (x,t) \in D. \tag{2.2}$$

1. *Assume the* b_i*'s are bounded and* c *is bounded above on* D *and that for every compact set* K *in* $D \cup \Omega \times \{0\}$ *there exists a positive constant* $\mu(K)$ *such that* $d(x,t) \geq \mu(K)$ *on* K. *If* $u \leq 0$ *on the parabolic boundary* Γ *of* D *then* $u \leq 0$ *on* D.

2. *Assume the boundedness of the* a_{ij}*'s, the* b_i*'s and* c *on* D *and that* d *is bounded above on* D. *If* $c \leq 0$ *on* D *and* u *has a nonnegative strict maximum point at* $P_0 = (x_0, t_0)$ *with* $x_0 \in \partial\Omega$, $t_0 \in (0,T)$. *If* D *satisfies the interior sphere condition at* P_0 *then*

$$\liminf \frac{u(P_0) - u(P)}{P_0 - P} > 0 \tag{2.3}$$

where $P \in D$ *approaches* P_0 *in the outward normal direction from* D. *If, in the above,* $t_0 = T$ *then (2.3) holds if* $d(x,T) \geq 0$ *for all* $x \in \Omega$. *Furthermore, if* $u(P_0) = 0$ *then (2.3) also holds, regardless of the sign of* c.

We cannot find the above results in the literature. Part (2) can be proved by closely checking the proof of Theorem 14 in Chapter 2 of [5]. To prove (1) we find it convenient to use a result of Varadhan. (See Theorem 3.1 in [2].)

Proof of Part (1) Without loss of generality assume $\mu_0 = 1$ (see (2.1)). Define

$$b = \sup_D \left\{ \left(\sum_{i=1}^n b_i^2 \right)^{1/2}, c, 1 \right\} \geq 1.$$

By Theorem 3.1 of [2] there exists a $\delta > 0$ such that for every compact set $K_1 \subset \Omega$ with measure $\leq \delta$ there exists a function $g \in C^\infty(\overline{\Omega})$ such that $1 \leq g \leq 2$, and g is concave on $\overline{\Omega}$ and

$$\sum a_{ij}(x,t) \frac{\partial^2 g}{\partial x_i \partial x_j} + b\left(|\nabla g| + g\right) < 0 \tag{2.4}$$

for $x \in \Omega \setminus K_1$ and $t \in (0,T]$. Define v by $u = vge^{\lambda t}$ where λ is a positive constant to be determined. Then

$$Lu = ge^{\lambda t}(L-c)v + vL(ge^{\lambda t}) + 2e^{\lambda t} \sum a_{ij} \frac{\partial v}{\partial x_i} \frac{\partial g}{\partial x_j} \tag{2.5}$$

$$L(ge^{\lambda t}) \leq e^{\lambda t} \left[\sum a_{ij} \frac{\partial^2 g}{\partial x_i \partial x_j} + b(|\nabla g| + g) - d\lambda g \right], \tag{2.6}$$

which is less than 0 for $x \in \Omega \setminus K_1$, $0 < t \leq T$, by (2.4) and the fact that $d \geq 0$ on D. On the other hand $d(x,t) \geq$ some positive constant for $x \in K_1$ and $0 < t \leq T$ so (2.6) shows that if λ is large enough $L(ge^{\lambda t})$ is also less than 0 for $x \in \Omega \setminus K_1$, $0 < t \leq T$. Thus

$$L(ge^{\lambda t}) \leq 0 \quad \text{on } D.$$

Now by (10), (2.5) and (2.6) we see that the maximum of v on $\overline{D}$ cannot be both positive and achieved in the parabolic interior of D. ■

REMARK 2.2. We shall need to apply the maximum principle on a domain D in x-t space which is not a cylinder $\Omega \times (0, T]$. However, by examining its proof, we see that part (1) of Proposition 2.1 holds when D is a domain in $(0,1) \times (0, T]$ and the condition on d is changed to:

> For any small $k > 0$ there exists $\mu(k) > 0$ such that $d(x,t) \geq \mu(k)$ for $(x,t) \in D$ with $x \geq k$.

2.2 A priori estimates

Let Ω be a domain in R^n. Suppose a positive function u satisfies

$$\Delta u \geq m u^p \quad \text{on } \Omega, p > 1, \tag{2.7}$$

for some positive constant m.

PROPOSITION 2.3. *For any compact subset K of Ω there exists a constant c depending only on n, p, m and dist$(K, \partial\Omega)$ such that $u \leq c$ on K. If u vanishes on an open subset Γ of $\partial\Omega$ then $u \leq c$ on K for any compact subset K of $\Omega \cup \Gamma$ where now c depends on dist$(K, \partial\Omega \setminus \Gamma)$.*

This can be proved by Moser iteration, [7].

2.3 Sturmian Theorems

Because of the smoothing effect of diffusion the number of zero points of a solution of a parabolic equation is nonincreasing in t. Indeed in 1836 Sturm initiated that study of this phenomenon, which has been re-investigated several times. We state the version from [1] below.

THEOREM 2.4. *Let u be a classical solution of*

$$u_t = a(x,t)u_{xx} + b(x,t)u_x + c(x,t)u \quad \text{on } [x_1, x_2] \times [0, T]$$

where $a, a^{-1}, a_t, a_{xx}, b, b_t, b_x$ and c are bounded on $[x_1, x_2] \times [0, T]$. Define $z(t)$ as the number of zeros of $u(\cdot, t)$ on (x_1, x_2). If u satisfies for $0 \leq t \leq T$ either the Dirichlet boundary condition $u(x_1, t) = 0 = u(x_2, t)$, the nonvanishing condition $u(x_i, t) \neq 0$ for $i = 1, 2$, or the mixed condition (zero at one boundary and not zero at the other), then the following holds:

1. *For $t > 0$, $z(t)$ is finite and non-increasing in t.*
2. *If u and u_x vanish at a point (x_0, t_0) [1] then $z(t_1) < z(t_0) < z(t_2)$ if $t_1 > t_0 > t_2$.*

3 STATEMENTS AND PROOFS OF MAIN RESULTS

In this section we assume that the initial value $\phi \in C^{2+\alpha}([0,1])$ and $\phi(0) = \phi(1) = 0$, $\phi \geq 0$ and not identically zero. Floater [4] proved the local existence and uniqueness of the initial-value-boundary-value problem (1.1), (1.2), (1.3), where $u \in C^{2,1}([0,1] \times [0, T_\phi))$ and T_ϕ is the life-span of u. By standard regularity theory u is infinitely differentiable on $(0,1] \times (0, T_\phi)$.

[1] $x_1 < x_0 < x_2$ and $0 < t_0 < T$

THEOREM 3.1. *If* $p > 2$ *and* $\phi'' + \phi^p \geq 0$ *on* $[0,1]$ *(but not identically zero) then* u *blows up in finite time and the blowup occurs away from* $x = 0$ *and* $x = 1$.

To prove this we need the following result of [4].

LEMMA 3.2. *If* $\phi'' + \phi^p \geq 0$ *on* $[0,1]$ *(but not identically zero) then for any fixed* $t_0 \in (0, T_\phi)$ *there exists a positive constant* ϵ *such that* $x^{p-1}u_t \geq \epsilon u^p$ *on* $[0,1] \times [t_0, T_\phi)$.

Proof. (We feel that the proof in [4] is not rigorous enough so we supply our own.) By the comparison principle, which is a direct consequence of part (1) of Proposition 2.1, $\phi \leq u$ on $[0,1] \times [0, T_\phi)$. For small positive h apply the comparison principle once again to $u(x,t)$ and $u(x,t+h)$ for $(x,t) \in [0,1] \times [0, T_\phi - h)$, we have $u(x,t) \leq u(x,t+h)$. Thus $u_t \geq 0$ on $[0,1] \times (0, T_\phi)$. Applying the usual strong maximum principle to u_t, we see that $u_t > 0$ on $(0,1) \times (0, T_\phi)$. Applying the boundary point lemma ((2) of Proposition 2.1) we obtain $u_{tx}(0,t) > 0$, and $u_{tx}(1,t) < 0$. Thus for small positive ϵ,

$$u_t(x, t_0) \geq \epsilon x^{1-p} u^p(x, t_0) \qquad x \in [0,1].$$

Define $J(x,t) = u_t(x,t) - \epsilon x^{1-p} u^p(x,t)$. Then

$$J \in C^\infty\left((0,1) \times (t_0, T_\phi)\right) \cup C\left([0,1] \times [t_0, T_\phi)\right),$$

$J(x, t_0) \geq 0$ on $[0,1]$, $J(0,t) = J(1,t) = 0$. By direct computation

$$\begin{aligned} xJ_t - J_{xx} &= pu^{p-1}J + \epsilon p(p-1)x^{-p-1}u^{p-2}(u - xu_x)^2 \\ &\geq pu^{p-1}J. \end{aligned}$$

Now by the maximum principle (part (1) of Proposition 2.1), we have $J \geq 0$ on $[0,1] \times [t_0, T_\phi)$. ■

Proof of Theorem 3.1 By (1.1) and Lemma 3.2 we have

$$\epsilon x^{2-p} u^p \leq x u_t = u_{xx} + u^p \qquad \text{on } [0,1] \times [t_0, T_\phi)$$

Since $p > 2$ there exists a small $\delta > 0$ such that

$$u_{xx} \geq u^p \qquad \text{on } [0,\delta] \times [t_0, T_\phi). \tag{3.1}$$

In particular u is convex and hence is increasing in x on $[0,\delta]$. Therefore if $x = 0$ is a blowup point, u blows up on the entire interval $(0,\delta]$ as $t \uparrow T_\phi$, contradicting Proposition 2.3.

To show that $x = 1$ is not a blowup point, we merely need to modify slightly the arguments of [6], which use the reflection method to show $u_x < 0$ in a fixed neighborhood of $x = 1$ for all $t \in [t_0, T_\phi)$. We omit the details. ■

We remark that the reflection method does not work at $x = 0$ and even at $x = 1$ when $u_t > 0$ is not known near $x = 1$.

Now we wish to extend the above result by considering more general ϕ's: From now on suppose that ϕ satisfies

$$\begin{gathered} \phi \quad \text{is so large that} \quad T_\phi < 0 \\ \phi'' + \phi^p > 0 \quad \text{on } (0,\delta), \quad \phi'' + \phi^p < 0 \quad \text{on } (\delta, 1) \end{gathered} \tag{3.2}$$

for some $\delta \in (0,1)$. Then the following result states that in the box $[0,1] \times [0,T_\phi]$ in the x-t plane there exists a curve dividing the box into two parts, on each of which u_t has a fixed sign.

LEMMA 3.3. *There exists a continuous function $x = \beta(t)$ defined on $[0,T_\phi)$ such that $0 < \beta(t) \leq 1$, $\beta(0) = \delta$, and $u_t(x,t) < 0$ for $\beta(t) < x < 1$ and $u_t(x,t) > 0$ for $0 < x < \beta(t)$.*

Proof. We observe that u_t satisfies

$$\begin{cases} xv_t = v_{xx} + pu^{p-1}v \\ v(0,t) = 0 = v(1,t) \\ v(x,0) = (\phi'' + \phi^p)(x) \end{cases} \tag{3.3}$$

and is infinitely differentiable on $(0,1) \times (0,T_\phi)$ and continuous on $[0,1] \times [0,T_\phi)$.

Let A^+ be the set of (x,t) where $u_t(x,t) > 0$ and A^- be the set of (x,t) where $u_t(x,t) < 0$. Define $I(t) = A^+ \cap ([0,1] \times \{t\})$ and $\beta(t) = \sup\{x | (x,t) \in I(t)\}$. $I(t)$ is always nonempty for $0 \leq t < T_\phi$. If $I(t^*)$ is empty, then by the maximum principle $u_t < 0$ for $t^* \leq t < T_\phi$ and hence u would be bounded on $[0,1] \times [0,T_\phi)$. Thus $\beta(t)$ is well-defined and positive on $[0,T_\phi)$and satisfies $u_t(\beta(t),t) = 0$.

We first show that β is lower-semicontinuous. Since $I(t)$ is an open set on the real line, by the definition of $\beta(t)$, for any $\bar{t} \in [0,T_\phi)$ there exists $\overline{x}$ such that $\overline{x}$ is as close to $\beta(\bar{t})$ as we wish and $u_t(\overline{x},\bar{t}) > 0$. Then for t close enough to $\bar{t}$, $(\overline{x},t) \in I(t)$, i.e., $\beta(t) \geq \overline{x}$. From this it follows that $\liminf_{t\to\bar{t}} \beta(t) \geq \beta(\bar{t})$.

Now we prove that β is upper-semicontinuous at $\bar{t}$, i.e.,

$$\limsup_{t\to\bar{t}} \beta(t) \leq \beta(\bar{t}).$$

If this is untrue there exists $\epsilon > 0$ and a sequence $t_n \to t$ such that $\beta(t_n) \geq \beta(\bar{t}) + \epsilon$. Take a small $\overline{\delta} > 0$ such that $u_t(\overline{x},t) > 0$ for $t \in [\bar{t} - \overline{\delta}, \bar{t} + \overline{\delta}]$. Applying the Sturmian theorem to the PDE in (3.3) on $[\overline{x},1] \times [\bar{t} - \overline{\delta}, \bar{t} + \overline{\delta}]$ we have that for every $t \in [\bar{t} - \overline{\delta}, \bar{t} + \overline{\delta}]$ u_t has finitely many zeros on $[\overline{x},1]$. In particular, $u_t(x,\bar{t}) < 0$ on a right-neighborhood of $\beta(\bar{t})$. Thus if t_n is close enough to $\bar{t}$ there exists x_n such that $\beta(\bar{t}) < x_n < \beta(\bar{t}) + \epsilon/2$ and $u_t(x_n,t_n) < 0$. Since $\beta(t_n) \geq \beta(\bar{t}) + \epsilon$, there exists $y_n \geq \beta(\bar{t}) + 3\epsilon/4$ such that $u_t(y_n,t_n) > 0$. Let C^+ be the connected component of A^+ containing (y_n,t_n) and let C^- be the connected component of A^- containing (x_n,t_n). By the maximum principle (see Remark 2.2) if C^+ does not reach the interval $(0,\delta)$ on $t = 0$, then $u \equiv 0$ on C^+. Thus C^+ contains this interval. Consequently (y_n,t_n) can be connected to the point $(\delta/2,0)$ by a path P^+ entirely contained in C^+. Similarly (x_n,t_n) can be connected to a point $(\gamma,0)$, $\delta < \gamma < 1$, by a path P^- entirely contained in C^-. Obviously P^+ and P^- must intersect, which is impossible. We have shown the continuity of $\beta(t)$ on $[0,T_\phi)$. Now applying the maximum principle on D^+ and D^- (the parts of D on the right and left sides of the graph of β) we have that $u_t > 0$ on D^+ and $u_t < 0$ on D^-. ∎

We wish to show that u remains bounded in a fixed neighborhood of $x = 0$ by using the argument in the proof of Theorem 3.1. Before doing this we need to show that the graph of $\beta(t)$ does not converge to the x-axis as $t \uparrow T_\phi$. We need the following result, which is interesting in its own right.

PROPOSITION 3.4. *There exists a constant M which depends only on the C^1 norm of ϕ such that if $u(x_0, t_0) \geq M$ for $t_0 > 0$ then $u_t(x_0, t) \geq 0$ for $t_0 \leq t < T_\phi$.* [2]

Proof. We use the idea of [8] where the Cauchy problem for the PDE in (1.6) was studied.

It is well-known that all positive solutions of $u_{xx} + u^p = 0$ are included in a 2-parameter family $\{u_{\alpha,a} | \alpha > 0, a \in R\}$, where $u_{\alpha,a}(a) = \alpha = \max u_{\alpha,a}$, $u_{\alpha,a}(x) = \alpha u_{1,a}(\alpha^{\frac{p-1}{2}} x)$ and $u_{\alpha,a}$ is positive on a finite interval centered at a.

Define $\gamma = \min\{|u'_{1,0}(x)| \, u_{1,0}(x) \leq \frac{1}{2}\}$. Then $\gamma > 0$ and

$$|u'_{\alpha,a}(x)| \geq \gamma \alpha^{\frac{p+1}{2}} \quad \text{if } u_{\alpha,a}(x) \leq \alpha/2. \tag{3.4}$$

Choose M such that

$$M \geq \max\left\{2\|\phi\|_{L^\infty}, \left(\frac{2\|\phi'\|_{L^\infty}}{\gamma}\right)^{\frac{2}{p+1}}\right\}.$$

We claim that if $\alpha \geq M$ then for any $a \in R$, $u_{\alpha,a}$ and ϕ can intersect at most twice and the intersections (if any) are transversal. If $u_{\alpha,a}(x) = \phi(x)$ at three points $y_1 < y_2 < y_3$ then we can assume without loss of generality that $y_1 < y_2 < a$, and there exists $\xi \in (y_1, y_2)$ such that $u'_{\alpha,a}(\xi) = \phi'(\xi)$. On the other hand,

$$u_{\alpha,a}(\xi) \leq u_{\alpha,a}(y_2) = \phi(y_2) \leq \|\phi\|_{L^\infty} \leq \frac{M}{2} \leq \frac{\alpha}{2}.$$

Then by (3.4), we have

$$\phi'(\xi) = u'_{\alpha,a}(\xi) \geq \gamma \alpha^{\frac{p+1}{2}} \geq \gamma M^{\frac{p+1}{2}} \geq 2\|\phi'\|_{L^\infty},$$

which is impossible. The above argument also shows that intersections must be transversal.

Now choose a $u_{\overline{\alpha},\overline{a}}$, denoted by $\overline{u}$, such that

$$\overline{u}(x_0) = u(x_0, t_0) \quad \text{and} \quad \overline{u}'(x_0) = u_x(x_0, t_0). \tag{3.5}$$

Then $\overline{\alpha} \geq M$ and hence $\overline{u}$ and ϕ intersect at most twice and the intersections are transversal. Let the support of $\overline{u}$ be the interval (c, d). Suppose M be chosen so large that the length of (c, d) is less than $1/4$. There are several cases to consider.

Case 1 $0 < c < d < 1$ Let $w = u - \overline{u}$. Then w satisfies

$$\begin{cases} x w_t = w_{xx} + c(x,t) w & \text{on } [c,d] \times [0, T_\phi) \\ w(c,t) > 0 \qquad w(d,t) > 0 \end{cases} \tag{3.6}$$

where c is bounded on $[c, d] \times [0, T]$ for any $T < T_\phi$. Let $N(t)$ be the number of zeros of $w(x, t)$ on $[c, d]$. Then $N(0) = 2$ and by the Sturmian theorem, $N(t) \leq 2$ for $t > 0$ and $N(t) < 1$ for $t > t_0$, which means $w \geq 0$ for $t \geq t_0$. This implies $u_t(x_0, t_0) \geq 0$. Applying the same argument to $u_t(x_0, t_1)$ for any $t_1 \in (t_0, T_\phi)$ we have $u_t(x_0, t_1) \geq 0$.

[2]Note that here we do not impose any condition on ϕ other than $\phi \in C^1([0, 1])$.

Case 2 $c < 0 < d < 1$ **or** $0 < c < 1 < d$ (We shall consider only the case $c < 0 < d < 1$.) For any $t_1 \in (t_0, T_\phi)$ there exists a small positive $c_1 < d$ such that $\overline{u}(c_1) > u(c_1, t)$ for $0 \leq t \leq t_1$. Now w (as defined in Case 1) satisfies $w(c_1, t) < 0$ and $w(d, t) > 0$ for $0 \leq t \leq t_1$. Let $N(t)$ be the number of zeros of w on $[c_1, d]$. Recall $N(0) \leq 2$ and in our present situation the only possibility is $N(0) = 1$. By the Sturmian theorem, $N(t) \leq 1$ for $0 \leq t \leq t_1$ and $N(t) = 0$ for $t_0 < t \leq t_1$. This is incompatible with the boundary conditions. Thus Case 2 never occurs.

Case 3 $0 = c < d < 1$ **or** $0 < c < d = 1$ (We shall consider only the case $0 = c < d < 1$.) In this case $\overline{u}$ and ϕ intersect only at $x = 0$ and $x =$ some $e < d$. Consider w on $[0, d] \times [0, T_\phi)$. By arguments similar to those used in the proof of Lemma 3.3, there exists a continuous function $x = \eta(t)$ for $0 \leq t < T_\phi$ such that $0 \leq \eta(t) \leq d$, $\eta(0) = e$, and on the left-hand side of the graph of $\eta(t)$, $w < 0$; on the other side $w > 0$. Since $w(x_0, t_0) = 0$, $\eta(t_0) = x_0 > 0$. By the continuity of η there exists a $t_1 > t_0$ such that $\eta(t_1) > 0$. By the maximum principle, $\eta(t) > 0$ for $t \in [0, t_1]$. Therefore there exists $x_1 > 0$ such that $x_1 < \eta(t)$, i.e., $w(x_1, t) < 0$ for $t \in [0, t_1]$. Note $w(d, t) > 0$ for $t \in [0, t_1]$. Let $N(t)$ be the number of zeros of w on $[x_1, d]$. Then $N(0) = 1$ and by the Sturmian theorem, $N(t) = 0$ for $t_0 \leq t \leq t_1$, which is impossible. Hence Case 3 never occurs.

■

THEOREM 3.5. *Assume Condition (3.2). Then either there exists $\sigma > 0$ such that u is bounded on $[0, \sigma] \times [0, T_\phi)$ or as $t \uparrow T_\phi$, $u(x, t)$ blows up for every $x \in (0, \alpha)$ where $\alpha = \liminf_{t \uparrow T_\phi} \beta(t) > 0$.*

Proof. By Proposition 3.4 $\alpha > 0$. Suppose there exists $x_0 \in (0, \alpha)$ such that $u(x_0, t)$ is bounded for $t \in [0, T_\phi)$. We want to show the existence of σ mentioned above. We claim that

$$\liminf_{t \to T_\phi} u_t(x_0, t) > 0. \tag{3.7}$$

Take x_1 and x_2 such that $0 < x_1 < x_0 < x_2 < \alpha$. Then there exists $t_0 < T_\phi$ such that $u_t > 0$ on $[x_1, x_2] \times [t_0, T_\phi)$. Consider a linear problem

$$\begin{cases} x v_t = v_{xx} & x_1 < x < x_2,\ t > t_0 \\ v(x_1, t) = 0 = v(x_2, t) & t > t_0 \\ v(x, t_0) = v_0(x) & x_1 < x < x_2 \end{cases} \tag{3.8}$$

where $v_0 \in C_0^1([x_1, x_2])$ and $0 < v_0(x) < u_t(x, t_0)$ for $x \in (x_1, x_2)$. The solution v exists globally in t and by the maximum principle $v(x, t) > 0$ for $x \in (x_1, x_2)$, $t \geq t_0$. On the other hand, u_t is an upper solution of (3.8) and thus by the comparison principle, $u_t \geq v$ for $x \in [x_1, x_2]$, $t \geq t_0$. This proves (3.7). Now the proof of Lemma 3.2 implies $u_t \geq \epsilon x^{1-p} u^p$ on $[0, x_0] \times [t_0, T_\phi)$, where ϵ is a small positive constant (here we need to use the boundedness of $u(x_0, t)$). The proof of Theorem 3.1 shows the boundedness of u near $x = 0$. ■

Next we turn to (1.7).

THEOREM 3.6. *Suppose the initial value $\phi \in C^2(\overline{\Omega})$ satisfies (1.8). Then the classical solution of (1.7) blows up in finite time T_ϕ; moreover, given any zero point x_0 of h ($x_0 \in \partial\Omega$ allowed) there exists a neighborhood B of x_0 in $\overline{\Omega}$ such that $u(x,t)$ remains bounded on $B \times [0, T_\phi)$.*

Proof. Since ϕ is a lower solution of (1.7), $u_t(x,t) > 0$ on $\Omega \times (0, T_\phi)$. Take a small $t_0 \in (0, T_\phi)$. We claim that there exists a small $\epsilon > 0$ such that

$$J = u_t - \epsilon u^p \geq 0 \quad \text{on } \Omega \times [t_0, T_\phi). \tag{3.9}$$

Note u_t satisfies

$$\begin{cases} v_t = \Delta v + p u^{p-1} h v & x \in \Omega,\ t > 0 \\ v(x,t) = 0 & x \in \partial\Omega,\ t > 0 \\ v > 0 & t > 0. \end{cases} \tag{3.10}$$

The boundary point lemma shows that the outward normal derivative of v on $\partial\Omega$ is negative. Therefore we can find a small $\epsilon > 0$ such that $J(x, t_0) > 0$ for $x \in \Omega$. Observe that $J(x,t) = 0$ for $x \in \partial\Omega$ and $t > 0$. Also, by direct computation,

$$\begin{aligned} J_t - \Delta J &= \epsilon p(p-1) u^{p-2} |\nabla u|^2 + p u^{p-1} J \\ &\geq p u^{p-1} J. \end{aligned}$$

Now (3.9) follows from the maximum principle.

From (3.9) and (1.7) we see that for a small neighborhood B of x_0 where $h \leq \epsilon/2$,

$$\Delta u \geq \frac{\epsilon}{2} u^p \quad \text{on } B \times [t_0, T_\phi).$$

By Proposition 2.3 this completes the proof. ■

BIBLIOGRAPHY

1. S. Angenent, *The zero set of a solution of a parabolic equation*, J. Reine Angew. Math. **390** (1988), 79–96.

2. H. Berestycki and L. Nirenberg, *On the method of moving planes and the sliding method*, Bol. Soc. Brasil. Mat. (N.S.) **22** (1991), 1–37.

3. X. Chen and H. Matano, *Convergence, asymptotic periodicity, and finite-point blowup in one- dimensional semilinear heat equations*, J. Diff. Eq. **78** (1989), 160–190.

4. M. Floater, *Blowup at the boundary for degenerate semilinear parabolic equations*, Arch. Rat. Mech. Anal. **114** (1991), 57–77.

5. A. Friedman, *Partial Differential Equations of Parabolic Type*, Prentice-Hall, 1964.

6. A. Friedman and J. B. McLeod, *Blow-up of positive solutions of semilinear heat equations*, Indiana University Math. J. **34** (1985), 425–447.

7. C. Gui and X. Wang, *The critical asymptotics of scalar curvatures of the conformal complete metrics with negative curvature*, Duke Math. J. **76** (1994), 293–302.

8. V. Galaktionov and S. Posashkov, *Any large solution of a nonlinear heat conduction equation becomes monotonic in time*, Proc. Roy. Soc. Edinburgh Sect. A **118** (1991), 13–20.

9. A. Stuart and M. Floater, *On the computation of blow-up*, European J. Appl. Math. **1** (1990), 47–71.

10. F. Weissler, *Single point blowup of a semilinear initial-value problem*, J. Diff. Eq. **55** (1984), 204–224.

Global bifurcations of concave semipositone problems

JUNPING SHI

Department of Mathematics, College of William and Mary
Williamsburg, VA 23185
Department of Mathematics, Harbin Normal University
Harbin, Heilongjiang, P.R.China
e-mail: shij@math.wm.edu

RATNASINGHAM SHIVAJI

Department of Mathematics, Mississippi State University
Mississippi State, MS 39762
e-mail: shivaji@ra.msstate.edu

Dedicated to Professor Jerome A Goldstein on the occasion of his sixtieth birthday.

ABSTRACT. We study semilinear elliptic equations on general bounded domains with concave *semipositone* nonlinearities. We prove the existence of the maximal solutions, and describe the global bifurcation diagrams. When a parameter is small, we obtain the exact global bifurcation diagram. We also discuss the related symmetry breaking bifurcation when the domains have certain symmetries.

1 INTRODUCTION

We study the boundary value problem:

$$\begin{cases} \Delta u + \lambda f(u) = 0 & \text{in} \quad \Omega, \\ u = 0 & \text{on} \quad \partial\Omega, \end{cases} \tag{1.1}$$

where λ is a positive parameter, and Ω is a smooth bounded region in $\mathbf{R}^n$ for $n \geq 1$. We assume that the nonlinear function f in this paper satisfies (f1) $f \in C^2[0,\infty)$, $f(0) < 0$, $f(u)(u-b) > 0$ for $u \in (0,M)\backslash\{b\}$ for some $b > 0$, where either $M = \infty$ or $M < \infty$, $f(M) = 0$ and $f'(M) < 0$;

(f2) $f''(u) < 0$ for $u \geq 0$;

(f3) $\int_0^M f(u)du > 0$;

(f4) If $M = \infty$, then $\lim\limits_{u \to \infty} \dfrac{f(u)}{u} = 0$.

The semilinear problem (1.1) arises in population biology, where $g(u) = f(u) - f(0) > 0$ is the growth rate, and $\varepsilon = -f(0)$ is the harvesting effort. An example for the case of $M < \infty$ in (f1) is the logistic growth $g(u) = au - bu^2$ for some $a, b > 0$, and examples for the case of $M = \infty$ are $g(u) = au/(1+bu)$ for $a, b > 0$ and $g(u) = 1 - e^{-au}$ for $a > 0$, which are sublinear functions modeling the saturating effect. With the nonlinearity $f(u)$ satisfying $f(0) < 0$, we call it a semipositone problem (see [4]) to compare with the positone case of $f(0) \geq 0$.

Semilinear semipositone problems have been studied for more than a decade, see [4] for a survey of the results. The complete bifurcation diagrams of the equation (1.1) with f satisfying conditions similar to (f1)-(f4) for Ω being a ball have been obtained in [1], [2], [5], and [14]. In this paper, we study the bifurcation diagram with Ω being a general bounded smooth domain in $\mathbf{R}^n$. Here we are mainly concerned with bifurcation diagrams with λ as the bifurcation parameter. (Note that λ can be interpreted as the reciprocal of the diffusion coefficient.) In a separate paper, the authors and Oruganti [12] study bifurcation diagrams with the harvesting effort as the bifurcation parameter. Also in [12] the harvesting effort is allowed to be nonhomogeneous.

To state our results, we recall a few commonly used definitions. Suppose that (λ, u) is a solution of (1.1). (λ, u) is a *stable* solution if the principal eigenvalue $\mu_1(u)$ of

$$\begin{cases} \Delta\psi + \lambda f'(u)\psi = -\mu\psi & \text{in } \Omega, \\ \psi = 0 & \text{on } \partial\Omega, \end{cases} \tag{1.2}$$

is positive, otherwise it is *unstable*. When $\mu_i(u) = 0$ for some $i \geq 1$, (λ, u) is a *degenerate solution*. The solutions of (1.1) are also the solutions of operator equation $F(\lambda, u) = 0$ where

$$F(\lambda, u) = \Delta u + \lambda f(u), \quad \lambda \in \mathbf{R}, \quad u \in X, \tag{1.3}$$

and $X = \{u \in C^{2,\alpha}(\overline{\Omega}) : u = 0 \text{ on } \partial\Omega\}$. Thus (λ, u) is degenerate solution if the linearized operator $F_u(\lambda, u) = \Delta + \lambda f'(u)$ is not an invertible operator. We call $v(x)$ a *maximal solution* of (1.1) if for any solution $u(x)$ of (1.1), we have $v(x) \geq u(x)$ for all $x \in \overline{\Omega}$.

Our first result proves that (1.1) has a unique stable nonnegative solution, which is also the maximal solution among all solutions:

THEOREM 1.1. *Suppose that f satisfies (f1)-(f4). Then there exists $\lambda_* > 0$ such that*

(i) *(1.1) has at least one positive solution u_λ for $\lambda > \lambda_*$, and has no nonnegative solution for $0 < \lambda < \lambda_*$;*

(ii) *For $\lambda > \lambda_*$, (1.1) has a maximal positive solution $\overline{u}_\lambda$, and $\overline{u}_\lambda$ is increasing with respect to λ;*

(iii) *At $\lambda = \lambda_*$, (1.1) has a maximal non-negative solution $\overline{u}_{\lambda_*}$;*

(iv) *The principal eigenvalue $\mu_1(\overline{u}_\lambda) \geq 0$ when $\lambda > \lambda_*$ and $\mu_1(\overline{u}_{\lambda_*}) = 0$. If in addition Ω is star-shaped, then $\mu_1(\overline{u}_\lambda) > 0$ and $\overline{u}_\lambda$ is a stable solution for $\lambda > \lambda_*$;*

(v) *If Ω is star-shaped, then for $\lambda > \lambda_*$, $\overline{u}_\lambda(x) > 0$ for any $x \in \Omega$, and $\dfrac{\partial \overline{u}_\lambda(x)}{\partial \nu} < 0$ for any $x \in \partial\Omega$;*

(vi) *If for some $\lambda \geq \lambda_*$, (1.1) has another nonnegative solution w_λ, then $\mu_1(w_\lambda) < 0$.*

The existence of a positive solution for f satisfying (f1), (f3) and (f4) and large λ was proved in [3] for the case of $M = \infty$, and in [7] for the case of $M < \infty$. Here with the extra condition (f2), we show that the branch of maximal solutions can be extended to a critical λ_*, which is the minimum λ for which one can have a nonnegative solution. Since $f(0) < 0$, it is not known whether $\overline{u}_{\lambda_*}$ satisfies (v) in Theorem 1.1. But we will show that in the case $\varepsilon = -f(0)$ is small, $\overline{u}_{\lambda_*}$ will satisfy (v).

In Section 4, we study the behavior of the bifurcation diagrams in (λ, u) space when $\varepsilon = -f(0)$ is small, for both positive and sign-changing solutions. Here the definition of $f(u)$ is extended to $\mathbf{R}$, and we assume that f satisfies

(f1') $f \in C^2(\mathbf{R})$, $f(0) < 0$, $f(u)(u-b) > 0$ for $u \in (-\infty, M)\backslash\{b\}$ for some $b > 0$, where either $M = \infty$ or $M < \infty$, $f(M) = 0$ and $f'(M) < 0$;

(f2') $f''(u) < 0$ for $u \in \mathbf{R}$;

and (f3), (f4) are as defined before. In this part we continue some discussions started in Shi [16]. To state the results, we rewrite the equation (1.1) in the form:

$$\begin{cases} \Delta u + \lambda[g(u) - \varepsilon] = 0 & \text{in } \Omega, \\ u = 0 & \text{on } \partial\Omega, \end{cases} \tag{1.4}$$

where $g(u) = f(u) - f(0)$, and $\varepsilon = -f(0)$. We denote by λ_k the k-th eigenvalue of

$$\begin{cases} \Delta\phi + \lambda\phi = 0 & \text{in } \Omega, \\ \phi = 0 & \text{on } \partial\Omega. \end{cases} \tag{1.5}$$

It is well-known that λ_1 is simple, and its eigenfunction does not change sign. We define $\lambda_k^0 = \lambda_k / g'(0)$.

THEOREM 1.2. *Suppose that $f(u) = g(u) - \varepsilon$ satisfies (f1'), (f2'), (f3) and (f4). We assume that λ_2 is a simple eigenvalue of (1.5) and*

$$\int_\Omega \phi_2(x)dx \cdot \int_\Omega \phi_2^3(x)dx > 0, \tag{1.6}$$

where ϕ_2 *is an eigenfunction corresponding to* λ_2. *Let* $\Sigma = \{(\lambda, u) \in \mathbf{R} \times X : (\lambda, u)$ *solves* $(1.4)\}$, *and* $T(a,b,c) = \{(\lambda, u) : a < \lambda < b, ||u||_X < c\}$. *Then for any small* $\delta_1, \delta_2 > 0$, *there exists* $\varepsilon_1 = \varepsilon_1(\delta_1, \delta_2, g)$ *such that for any* $\varepsilon \in (0, \varepsilon_1)$,

$$\Sigma_0 \equiv \Sigma \bigcap T(\lambda_1^0 - \delta_1, \lambda_2^0 + \delta_1, \delta_2) = \bigcup_{i=1}^{3} \Sigma_i,$$

where Σ_i *is a connected component of* Σ_0, $(i = 1, 2, 3)$. *Moreover,*

(i) *Each* Σ_i $(i = 1, 2, 3)$ *is a smooth curve in* $\mathbf{R} \times X$;

(ii) Σ_1 *is exactly* $\supset$*-shaped, there is a unique degenerate solution on* Σ_1, *and each solution on* Σ_1 *is negative;*

(iii) Σ_3 *is exactly* $\subset$*-shaped, there is a unique degenerate solution on* Σ_3, *and each solution on* Σ_3 *is sign-changing;*

(iv) Σ_2 *is exactly S-shaped, and there are exactly two degenerate solutions on* Σ_2; Σ_2 *can be parameterized as* $(\lambda(s), u(s))$, $s \in (s_1, s_4)$, *such that for* $s \in (s_1, s_2)$, $u(s)$ *is positive, and for* $s \in (s_3, s_4)$, $u(s)$ *is sign-changing, where* $s_1 < s_2 < s_3 < s_4$; *The portion of* Σ_2 *with* $s \in (s_1, s_2)$ *contains the degenerate solution on the left, and the portion of* Σ_2 *with* $s \in (s_3, s_4)$ *contains the degenerate solution on the right (see Fig. 1);*

(v) *There exist* $\delta_i = \delta_i(\varepsilon) > 0$, $(i = 3, 4, 5, 6)$ *such that*

$proj\Sigma_1 = [\lambda_1^0 - \delta_1, \lambda_1^0 - \delta_3]$, $proj\Sigma_2 = [\lambda_1^0 + \delta_4, \lambda_2^0 - \delta_5]$, $proj\Sigma_3 = [\lambda_2^0 + \delta_6, \lambda_2^0 + \delta_1]$,

where $proj\Sigma_i$ *is the projection of* Σ_i *into* $\mathbf{R} = (\lambda)$.

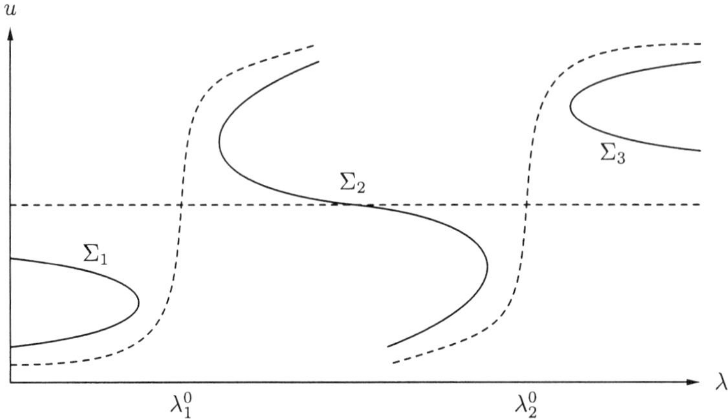

Fig. 1: Precise bifurcation diagram when (1.6) holds for $||u||$ small
solid curve: small $\varepsilon > 0$, dashed curve: $\varepsilon = 0$

The diagrams in Fig. 1 are well-known as the imperfect bifurcation under small perturbation. Here we emphasize on a rigorous proof of the exactness of the shape of the bifurcation diagrams. If (1.6) is changed to

$$\int_{\Omega} \phi_2(x)dx \cdot \int_{\Omega} \phi_2^3(x)dx < 0, \tag{1.7}$$

then the diagram in Fig. 1 becomes the one in Fig. 2. More explanation is given in Section 4. For all connected bounded smooth domains except for a zero measure set, either (1.6) or (1.7) holds, so the diagrams in Fig. 1 and Fig. 2 are generic. But we should comment that for domains with symmetry, the second eigenvalue may not be simple (example: balls), and even when it is simple, (1.6) or (1.7) may not be true (example: rectangles). We will briefly discuss the structure of the bifurcation diagrams for these symmetric domains in Section 5.

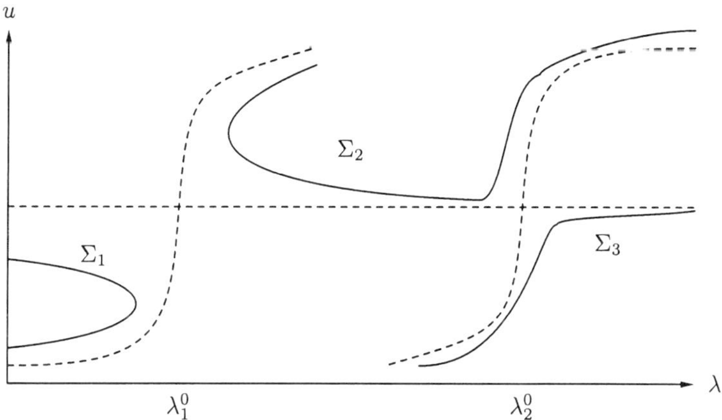

Fig. 2: Precise bifurcation diagram when (1.7) holds for $||u||$ small
solid curve: small $\varepsilon > 0$, dashed curve: $\varepsilon = 0$

We remark that Theorem 1.2 not only shows an interesting bifurcation diagram, but also implies the following fact:

COROLLARY 1.3. *Suppose that the domain Ω has a simple second eigenvalue λ_2, and satisfies (1.6) or (1.7). Then when $\varepsilon > 0$ is sufficiently small, (1.4) has a nonnegative solution (λ, u) satisfying either (i) there exists $x_1 \in \Omega$ such that $u(x_1) = 0$; or (ii) there exists $x_2 \in \partial\Omega$ such that $\partial u(x_2)/\partial\nu = 0$.*

In the description of Theorem 1.2, such a solution is on the middle part of the S-shaped curve when $s \in [s_2, s_3]$. We do not know if such solution is unique. A nontrivial nonnegative solution of (1.4) is always positive if $f(0) \geq 0$ and thus satisfies the strong maximum principle and the Hopf boundary lemma, *i.e.* $u(x) > 0$ for $x \in \Omega$ and $\partial u(x)/\partial\nu < 0$ for $x \in \partial\Omega$. For the case of $f(0) < 0$, it is easy to show the existence of a solution with zero derivative at the boundary points when $n = 1$ by the quadrature method. When Ω is a ball in $\mathbf{R}^n$, the existence of such a solution (zero gradient on the boundary) is also well-known (see [20], [14] and [11].) Corollary 1.3 confirms the existence of such a solution for a large class of domains.

In Section 2, we study the bifurcation diagram of the positone concave equation, which will be used in later proofs. In Section 3, we prove Theorem 1.1, and in

Section 4, we prove Theorem 1.2. We discuss the bifurcation diagrams for symmetric domains in Section 5.

2 BIFURCATION DIAGRAM OF LOGISTIC TYPE EQUATIONS

For f satisfying (f1), (f2) and (f4), the existence of a non-negative solution for (1.1) for λ large was established in [3], for the case $M = \infty$, and in [7] for the case of $M < \infty$:

LEMMA 2.1. *Assume that f satisfies (f1), (f2) and (f4). Then there exists $\lambda_a > 0$ such that if $\lambda > \lambda_a$ then (1.1) has a non-negative solution u_λ.*

Next we recall the well-known bifurcation diagram when $\varepsilon = -f(0) = 0$. First we quote Lemma 3 in [19], which will be repeatedly used in our proofs:

LEMMA 2.2. *Suppose that $f : \Omega \times \mathbf{R}^+ \to \mathbf{R}$ is a continuous function such that $\dfrac{f(x,s)}{s}$ is strictly decreasing for $s > 0$ at each $x \in \Omega$. Let $w, v \in C(\overline{\Omega}) \cap C^2(\Omega)$ satisfy*
(a) $\Delta w + f(x,w) \leq 0 \leq \Delta v + f(x,v)$ in Ω,
(b) $w, v > 0$ in Ω and $w \geq v$ on $\partial\Omega$,
(c) $\Delta v \in L^1(\Omega)$.
Then $w \geq v$ in $\overline{\Omega}$.

By using Lemma 2.2 and bifurcation theory, we prove the following result: (recall that $\lambda_1^0 = \lambda_1/g'(0)$)

THEOREM 2.3. *Assume that $g \in C^1[0,\infty)$ satisfies*

$$g(0) = 0, \quad g'(0) > 0, \quad \frac{d}{du}\left(\frac{g(u)}{u}\right) < 0 \text{ for all } u > 0, \tag{2.1}$$

and, either $g(u) > 0$ for all $u > 0$ and $\displaystyle\lim_{u\to\infty} \frac{g(u)}{u} = 0$, or $g(M) = 0$ for some $M > 0$. Then

$$\begin{cases} \Delta v + \lambda g(v) = 0 & \text{in } \Omega, \\ v = 0 & \text{on } \partial\Omega, \end{cases} \tag{2.2}$$

has no positive solution if $\lambda \leq \lambda_1^0$, and has exactly one positive solution v_λ if $\lambda > \lambda_1^0$. Moreover, all v_λ's lie on a smooth curve, v_λ is stable and v_λ is increasing with respect to λ.

Proof. Suppose that (λ, v) is a positive solution of (2.2), and (λ_1, ϕ_1) is the principal eigen-pair of (1.5). Then from (2.2) and (1.5), we have

$$\begin{aligned} &\left(\lambda - \frac{\lambda_1}{g'(0)}\right) g'(0) \int_\Omega u(x)\phi_1(x)dx \\ &+\lambda \int_\Omega \left[\frac{g(u(x))}{u(x)} - g'(0)\right] u(x)\phi_1(x)dx = 0. \end{aligned} \tag{2.3}$$

Since $g(u)/u$ is decreasing, the second term in the equality is negative. This implies (2.2) has no positive solution if $\lambda \leq \lambda_1^0$.

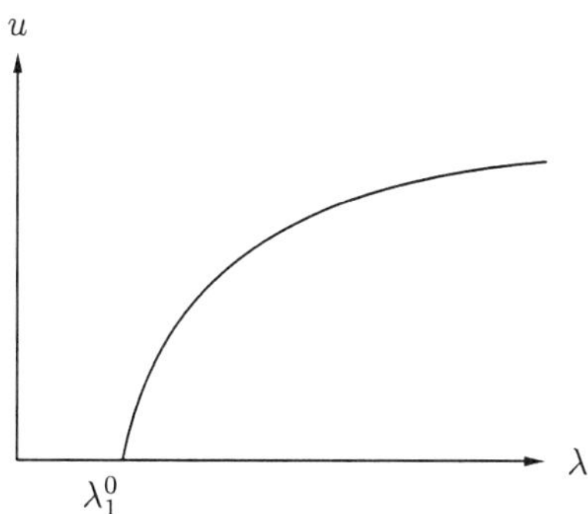

Fig. 3: Precise bifurcation diagram for the Logistic type equation

Next we apply the bifurcation from simple eigenvalue result by Crandall and Rabinowitz [8]: $(\lambda, u) = (\lambda_1^0, 0)$ is a bifurcation point; near $(\lambda_1^0, 0)$, the solutions of (2.2) are on two branches $\Sigma_0 = \{(\lambda, 0)\}$ and $\Sigma_1 = \{(\lambda(s), v(s)) : |s| \leq \delta\}$, where $\lambda(0) = 0$, $v(s) = s\phi_1 + O(s^2)$. We assume that $\phi_1 > 0$, then $v(s)$ is a positive solution when $s \in (0, \delta)$. Moreover, from (2.3), positive solutions only exist for $\lambda > \lambda_1^0$. Therefore there exists $\varepsilon > 0$ such that for $\lambda \in (\lambda_1^0, \lambda_1^0 + \varepsilon)$, (2.2) has a positive solution v_λ. We prove that any positive solution (λ, v) of (2.2) is stable. Let (μ_1, ψ_1) be the principal eigen-pair of (1.2) for $g(u)$ and (λ, v). Then by (2.2) and (1.2), we have

$$\begin{aligned} -\mu_1 \int_\Omega \psi_1 v dx &= \int_\Omega (\Delta v \cdot \psi_1 - \Delta \psi_1 \cdot v) dx \\ &= \lambda \int_\Omega [g'(v)v - g(v)]\psi_1 dx. \end{aligned} \tag{2.4}$$

Since $g(v)/v$ is decreasing, then $g'(v)v - g(v) < 0$ for $v > 0$. Thus $\mu_1 > 0$. In particular, any positive solution (λ, v) is non-degenerate. Therefore, at any positive solution (λ^*, v^*), we can apply the implicit function theorem to $F(\lambda, u) = 0$, and all the solutions of $F(\lambda, u) = 0$ near (λ^*, v^*) are on a curve $(\lambda, v(\lambda))$ with $|\lambda - \lambda^*| \leq \varepsilon$ for some small $\varepsilon > 0$. Hence the portion of Σ_1 with $s > 0$ can be extended to a maximal set

$$\Sigma_1 = \{(\lambda, v_\lambda) : \lambda \in (\lambda_1^0, \lambda_M)\}, \tag{2.5}$$

where λ_M is the supreme of all $\lambda > \lambda_1^0$ such that v_λ exists. We claim that $\lambda_M = \infty$. Suppose not, then $\lambda_M < \infty$, and there are two possibilities: (a) $\lim_{\lambda \to \lambda_M^-} ||v_\lambda|| = \infty$, or (b) $\lim_{\lambda \to \lambda_M^-} v_\lambda = 0$, otherwise we can extend Σ_1 further beyond λ_M. The case (a) is impossible since if either $g(M) = 0$ for some $M > 0$ or $\lim_{u \to \infty} \frac{g(u)}{u} = 0$, then the solution curve cannot blow up at finite λ_M (see details in [17] Theorem 1.3.) The case (b) is not possible either, since if so, $\lambda = \lambda_M$ must be a point where a bifurcation from the trivial solutions $v = 0$ occurs. That is $\lambda_M g'(0)$ must be an eigenvalue λ_i of (1.5) with $i \geq 2$, and the eigenfunction ϕ_i is not of one sign. But the positive solution v_λ satisfies $v_\lambda/||v_\lambda|| \to \phi_i$ as $\lambda \to \lambda_M^-$, which is a contradiction. Thus $\lambda_M = \infty$.

We prove v_λ is increasing with respect to λ. Since v_λ is differentiable with respect to λ (as a consequence of the implicit function theorem), $\dfrac{dv_\lambda}{d\lambda}$ satisfies $\Delta\dfrac{dv_\lambda}{d\lambda} + \lambda g'(v_\lambda)\dfrac{dv_\lambda}{d\lambda} = -g(v_\lambda) \le 0$, and v_λ is stable, hence $\mu_1(v_\lambda) > 0$. Then by a standard variant of the maximum principle (see for example Lemma 2.16 in [10]), $\dfrac{dv_\lambda}{d\lambda} \ge 0$. Finally, by Lemma 2.2, (2.2) has at most one positive solution for any possible $\lambda > 0$, which completes the proof. ■

REMARK. If g satisfies $\lim\limits_{u\to\infty} \dfrac{g(u)}{u} = k > 0$, the proof of Theorem 2.3 still works except that the solution curve (λ, v_λ) exists only for $\lambda \in (\lambda_1^0, \lambda_1^\infty)$, where $\lambda_1^\infty = \lambda_1/k$. $(\lambda_1^\infty, \infty)$ is a point where a bifurcation from infinity occurs. (see [14] or [17] for details.)

3 THE BRANCH OF MAXIMAL SOLUTIONS

Proof of Theorem 1.1 Let v_λ be the unique positive solution of (2.2) for $\lambda > \lambda_1^0$. First we prove that if (1.1) has a solution (λ, u), then $\lambda > \lambda_1^0$ and $v_\lambda \ge u$. Indeed, if (λ, u) is a positive solution of (1.1), then similar to (2.3), we have (using $f(u) = g(u) - \varepsilon$)

$$\begin{aligned}&\left(\lambda - \frac{\lambda_1}{g'(0)}\right) g'(0) \int_\Omega u(x)\phi_1(x)dx - \lambda\varepsilon \int_\Omega \phi_1(x)dx \\ &+\lambda \int_\Omega \left[\frac{g(u(x))}{u(x)} - g'(0)\right] u(x)\phi_1(x)dx = 0.\end{aligned} \tag{3.1}$$

So $\lambda > \lambda_1^0$. On the other hand,

$$\Delta v_\lambda + \lambda g(v_\lambda) = 0 < \lambda\varepsilon = \Delta u + \lambda g(u),$$

and $u = v_\lambda = 0$ on the boundary. By Lemma 2.2, $v_\lambda(x) \ge u(x)$ for $x \in \Omega$.

By Lemma 2.1, (1.1) has a non-negative solution u_λ for $\lambda > \lambda_a$. From the last paragraph, $v_\lambda \ge u_\lambda$. For fixed λ we define a sequence $\{u_n(\lambda)\}$: $u_0(\lambda) = v_\lambda$, and for $n \ge 1$

$$\begin{cases} -\Delta u_n + K u_n = \lambda f(u_{n-1}) + K u_{n-1} & \text{in } \Omega, \\ u_n = 0 & \text{on } \partial\Omega, \end{cases} \tag{3.2}$$

where $K > 0$ is a constant such that $|\lambda f'(u)| \le K$ for all $u \in [0, M]$ (from (f2)). Then it is standard to show that $v_\lambda \equiv u_0(\lambda) \ge u_1(\lambda) \ge \cdots \ge u_n(\lambda) \ge u_{n+1}(\lambda) \ge \cdots$. On the other hand, we claim that $u_n(\lambda) \ge u_\lambda$ for $n \ge 0$. It is true when $n = 0$. Suppose it is true for $n = k$, then when $n = k + 1$,

$$\Delta(u_{k+1} - u_\lambda) - K(u_{k+1} - u_\lambda) = f_1(u_\lambda) - f_1(u_k) \le 0,$$

where $f_1(u) = \lambda f(u) + Ku$, and $f_1'(u) > 0$. And $u_{k+1} - u_\lambda = 0$ on the boundary, thus $u_{k+1} - u_\lambda \ge 0$ by the maximum principle. Define $\overline{u}_\lambda(x) = \lim_{n\to\infty} u_n(\lambda)(x)$. It is standard to verify that $\overline{u}_\lambda \in C^{2,\alpha}(\overline{\Omega})$ and $\overline{u}_\lambda$ is a non-negative solution of (1.1) such that $\overline{u}_\lambda \geq u_\lambda$. Since we can use any solution u in the place of u_λ, $\overline{u}_\lambda$ is the maximal solution of (1.1).

Define

$$H = \{\lambda > 0 : (1.1) \text{ has at least one non-negative solution}\},$$

and $\lambda_* = \inf\{\lambda \in H\}$. Then $\lambda_* \geq \lambda_1^0 > 0$. We claim that $H \supset (\lambda_*, \infty)$. Suppose that $\lambda_b \in H$, then for $\lambda_c > \lambda_b$, $\overline{u}_{\lambda_b} \leq v_{\lambda_b} \leq v_{\lambda_c}$. Thus the iteration sequence with $u_0(\lambda_c) = v_{\lambda_c}$ and defined as in (3.2) has $\overline{u}_{\lambda_b}$ as lower bound similar to the proof in the last paragraph. Therefore the maximal solution $\overline{u}_{\lambda_c}$ for $\lambda = \lambda_c$ also exists, and

$$H \supset \cup_{\lambda \in H}(\lambda, \infty) = (\lambda_*, \infty).$$

Moreover, we have also proved that $\overline{u}_\lambda$ is increasing with respect to λ.

Next we prove that $\lambda_* > \lambda_1^0$. We have established that $\lambda_* \geq \lambda_1^0$. In fact, if $\lambda_* = \lambda_1^0$, then $0 \leq \lim\limits_{\lambda \to \lambda_*^+} \max\limits_{x \in \overline{\Omega}} \overline{u}_\lambda(x) \leq \lim\limits_{\lambda \to \lambda_*^+} \max\limits_{x \in \overline{\Omega}} v_\lambda(x) = 0$. But it is well-known that each solution u of (1.1) satisfies $\max_{x \in \overline{\Omega}} u(x) > b$ (from the maximum principle). So we reach a contradiction. Thus $\lambda_* > \lambda_1^0$.

We now prove that $\overline{u}_\lambda$ is stable for $\lambda > \lambda_*$ if Ω is a star-shaped domain. Since $\overline{u}_\lambda$ is obtained via the iteration from a supersolution, then from [15] pg. 992, the principal eigenvalue $\mu_1(\overline{u}_\lambda) \geq 0$. We exclude the possibility of $\mu_1(\overline{u}_\lambda) = 0$. Suppose there exists $\lambda_d > \lambda_*$ such that $\mu_1(\overline{u}_{\lambda_d}) = 0$. Then the principal eigenfunction $\psi_1 > 0$ satisfies

$$\begin{cases} \Delta\psi + \lambda_d f'(\overline{u}_{\lambda_d})\psi = 0 & \text{in } \Omega, \\ \psi = 0 & \text{on } \partial\Omega. \end{cases} \tag{3.3}$$

At $(\lambda_d, \overline{u}_{\lambda_d})$, we apply a bifurcation theorem by Crandall and Rabinowitz [9]. Since 0 is the principal eigenvalue of (1.2), so it must be a simple eigenvalue. By Lemma 2.3 in [13], we have

$$\int_\Omega f(\overline{u}_{\lambda_d})\psi_1 dx = \frac{1}{2\lambda} \int_{\partial\Omega} |\nabla \overline{u}_{\lambda_d}| \cdot |\nabla \psi_1| (x \cdot \nu) ds, \tag{3.4}$$

where ν is the outer unit normal vector. If Ω is star-shaped, then $\int_\Omega f(\overline{u}_{\lambda_d})\psi_1 dx > 0$ since $\dfrac{\partial \overline{u}_{\lambda_d}}{\partial n}(x) < 0$ and $\dfrac{\partial \psi_1}{\partial n}(x) < 0$ for $x \in \partial\Omega$. Hence by the result of [9], the solutions near $(\lambda_d, \overline{u}_{\lambda_d})$ form a curve $\{(\lambda(s), u(s)) : |s| \leq \delta\}$ such that $\lambda(0) = \lambda_d$, $\lambda'(0) = 0$, $u(s) = \overline{u}_{\lambda_d} + s\psi_1 + o(|s|)$. Moreover, (see [16] pg. 506)

$$\lambda''(0) = \frac{-\lambda_d \int_\Omega f''(\overline{u}_{\lambda_d})\psi_1^3 dx}{\int_\Omega f(\overline{u}_{\lambda_d})\psi_1 dx}. \tag{3.5}$$

$\lambda''(0) > 0$ since $f''(u) \leq 0$, and the solution curve is $\subset$-shaped near $(\lambda_d, \overline{u}_{\lambda_d})$. Therefore for $\lambda \in (\lambda_d - \varepsilon, \lambda_d)$, $||\overline{u}_\lambda - \overline{u}_{\lambda_d}|| \geq \delta > 0$. But on the other hand, $\overline{u}_\lambda$ is continuous with respect to λ since $\overline{u}_\lambda$ is obtained from a family of continuous supersolution v_λ. This is a contradiction. So $\mu_1(\overline{u}_\lambda) > 0$ for $\lambda > \lambda_*$.

Now we are able to prove that if Ω is star-shaped, then for $\lambda > \lambda_*$, $\overline{u}_\lambda(x) > 0$ for any $x \in \Omega$, and $\dfrac{\partial \overline{u}_\lambda(x)}{\partial \nu} < 0$ for any $x \in \partial\Omega$. Indeed since $\mu_1(\overline{u}_\lambda) > 0$, we obtain this result by a variant of the maximum principle (see Theorem 2.1 in [12], or [6].)

Finally we use an idea of [10] to prove that all other nonnegative solutions of (1.1) must be unstable. Suppose that for some $\lambda \geq \lambda_*$, (1.1) has another nonnegative solution w_λ such that $\mu_1(w_\lambda) \geq 0$. Since f is concave, for any $\tau \in [0,1]$,

$$\Delta(\tau w_\lambda + (1-\tau)\overline{u}_\lambda) + \lambda f(\tau w_\lambda + (1-\tau)\overline{u}_\lambda) \geq 0. \tag{3.6}$$

But since at $\tau = 0$, (3.6) is an identity, the derivative of the left hand side of (3.6) with respect to τ at $\tau = 0$ is non-negative, that is

$$\Delta(w_\lambda - \overline{u}_\lambda) + \lambda f'(\overline{u}_\lambda)(w_\lambda - \overline{u}_\lambda) \geq 0. \tag{3.7}$$

Similarly, at $\tau = 1$, we get

$$\Delta(w_\lambda - \overline{u}_\lambda) + \lambda f'(w_\lambda)(w_\lambda - \overline{u}_\lambda) \leq 0. \tag{3.8}$$

From Lemma 2.26 of [10], if $\mu_1(\overline{u}_\lambda) > 0$ and $\mu_1(w_\lambda) > 0$, then $w_\lambda - \overline{u}_\lambda \leq 0$ and $w_\lambda - \overline{u}_\lambda \geq 0$, thus $w_\lambda \equiv \overline{u}_\lambda$. If one of $\mu_1(\overline{u}_\lambda)$ and $\mu_1(w_\lambda)$ is zero, say $\mu_1(\overline{u}_\lambda) = 0$, then (3.7) is an identity, and the second derivative of the left hand side of (3.6) with respect to τ at $\tau = 0$ must be non-negative, that is

$$\lambda f''(\overline{u}_\lambda)(w_\lambda - \overline{u}_\lambda)^2 \geq 0.$$

But $\lambda > 0$ and $f'' < 0$, so $w_\lambda \equiv \overline{u}_\lambda$. Hence for any nonnegative solution w_λ of (1.1) other than $\overline{u}_\lambda$, $\mu_1(w_\lambda) < 0$.

At $\lambda = \lambda_*$, $\overline{u}_{\lambda_*}$ still exists as the iteration of v_{λ_*} and $\overline{u}_{\lambda_*}$ is non-negative since $\overline{u}_{\lambda_*} = \lim_{\lambda \to \lambda_*^+} \overline{u}_\lambda$. Now $\mu_1(\overline{u}_{\lambda_*}) \geq 0$ and if it is positive, it contradicts the definition of λ_*. Hence $\mu_1(\overline{u}_{\lambda_*}) = 0$. ∎

4 IMPERFECT BIFURCATION NEAR $U = 0$

In this section, we prove Theorem 1.2. We assume $\int_\Omega \phi_2^3(x)dx > 0$, and the case when $\int_\Omega \phi_2^3(x)dx < 0$ can be shown similarly. First we show that the bifurcation diagram when $\varepsilon = 0$ (equation (2.2)) is as described by the dashed curves in Fig. 1. From our assumptions, λ_1 and λ_2 are both simple eigenvalues. Thus we can apply the result of [8] to conclude that near $(\lambda_i^0, 0)$, $(i = 1, 2)$, the solutions of (2.2) are on two branches $\Sigma^0 = \{(\lambda, 0)\}$ and $\Sigma^i = \{(\lambda_i(s), v_i(s)) : |s| \leq \delta\}$, where $\lambda_i(0) = \lambda_i^0$, $v_i(s) = s\phi_i + o(s^2)$, and $\lambda_i'(0)$ satisfies the expression (see [16] page 507)

$$\lambda_i'(0) = -\frac{\int_\Omega g''(0)\phi_i^3(x)dx}{2\int_\Omega g'(0)\phi_i^2(x)dx} > 0. \tag{4.1}$$

From the positivity of ϕ_1 and $\int_\Omega \phi_2^3(x)dx > 0$, $\lambda_i'(0) > 0$, thus both bifurcation are transcritical, and can be drawn as the dashed curves in Fig.1.

Next we recall the discussion of (1.4) in [16] Theorem 2.5 and subsection 6.1. In fact, we show that, when $\varepsilon \in (-\varepsilon_0, \varepsilon_0)$ for some small $\varepsilon_0 > 0$, the local bifurcation picture near $(\lambda_1^0, 0)$ is as in the following diagrams:

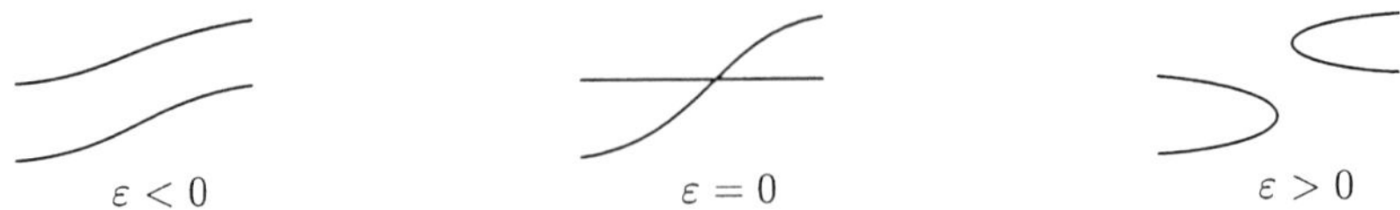

Fig. 4: Bifurcation of solution curves in a transcritical bifurcation

The rigorous proof of the diagrams relies on an abstract theorem (Theorem 2.5 in [16]), in which the degenerate solutions of (1.4) are tracked when ε varies. Indeed, the degenerate solutions of (1.4) form a curve $\{(\varepsilon(s), \lambda(s), u(s), w(s)) : |s| \leq \delta\}$, $\lambda'(0) = 1$, $\varepsilon'(0) = 0$ and

$$\varepsilon''(0) = \frac{[g'(0)]^2 \left(\int_\Omega \phi_1^2(x)dx\right)^2}{2(\lambda_1^0)^2(\int_\Omega \phi_1^3(x)dx) \cdot (\int_\Omega \phi_1(x)dx)} > 0. \tag{4.2}$$

Hence there are two degenerate solutions near $(\lambda_1^0, 0)$ when $\varepsilon > 0$, and no degenerate solution when $\varepsilon < 0$. These arguments are still valid near $(\lambda_2^0, 0)$ as long as λ_2 is simple and (1.6) holds. Therefore we obtain the parts of the bifurcation diagram in Fig. 1 when λ is near λ_1^0 and λ_2^0. Moreover, since the solutions near $(\lambda_1^0, 0)$ are all in a form $(\lambda, t\phi_1 + o(|t|))$, then the solutions on $\supset$-branch are all negative, and the solutions on $\subset$-branch are all positive. Similarly since the solutions near $(\lambda_2^0, 0)$ are all in a form $(\lambda, t\phi_2 + o(|t|))$, then the solutions on both $\supset$-branch and $\subset$-branch are sign-changing solutions.

To be more precise, we select $\widetilde{\delta_1} > 0$ and $\varepsilon_2 > 0$ such that when $\varepsilon \in (0, \varepsilon_2)$, (1.4) has exactly two degenerate solutions in each of the cube $C_i = \{(\lambda, u) : |\lambda - \lambda_i^0| \leq \widetilde{\delta_1}, ||u||_X \leq \widetilde{\delta_1}\}$, $(i = 1, 2)$, from Theorem 2.5 in [16]. Then the portion of the bifurcation diagram in each cube is exactly same as the third diagram in Fig. 4. Note that there is a gap on the λ-axis between the projections of two connected components. This follows from the fact that for the curve of degenerate solutions $\{(\varepsilon(s), \lambda(s), u(s), w(s)\}$, $\lambda(0) = \lambda_i^0$ and $\lambda'(0) = 1$, thus $\lambda(s) < \lambda_i^0$ for $s < 0$ and $\lambda(s) > \lambda_i^0$ for $s > 0$.

Let $\widetilde{\delta_2} = \widetilde{\delta_1}/2$. For $\lambda \in [\lambda_1^0 + \widetilde{\delta_2}, \lambda_2^0 - \widetilde{\delta_2}]$, the trivial solution $(\lambda, 0)$ for (1.4) with $\varepsilon = 0$ is nondegenerate. We choose $\widetilde{\delta_3} > 0$ such that the solutions on the line $(\lambda, 0)$ are the only solutions of (1.4) when $\varepsilon = 0$ in the cube $\{(\lambda, u) : \lambda_1^0 + \widetilde{\delta_2} < \lambda < \lambda_2^0 - \widetilde{\delta_2}, ||u||_X \leq \widetilde{\delta_3}\}$. Thus by the implicit function theorem, there exists $\varepsilon_3 > 0$ such that when $\varepsilon \in (0, \varepsilon_3)$, for each $\lambda \in [\lambda_1^0 + \widetilde{\delta_2}, \lambda_2^0 - \widetilde{\delta_2}]$, (1.4) has exactly one solution $(\lambda, u(\lambda))$ such that $||u(\lambda)||_X \leq \widetilde{\delta_3}$, and they are all nondegenerate. From the nondegeneracy of the solutions, we can see that the curve $(\lambda, u(\lambda))$ joins the lower branch of $\subset$-branch in C_1. We can use the Morse index of the solutions to conclude that $(\lambda, u(\lambda))$ joins the lower branch but not the upper branch, since the solutions on the upper branch have Morse index 0 and the ones on the lower branch have Morse index 1. All solutions on $(\lambda, u(\lambda))$ have Morse index 1 since they are perturbations of $(\lambda, 0)$ when $\varepsilon = 0$, which have Morse index 1. Similarly, $(\lambda, u(\lambda))$ joins the $\supset$-branch in C_2, and here the terms "lower" and "upper" branches are not appropriate as the solutions are not ordered. But $(\lambda, u(\lambda))$ will connect with the branch with Morse index 1 and smaller X-norm. Therefore the $\supset$-branch in C_2 and the $\subset$-branch in C_1 are connected, and it forms a S-shaped curve in a cube near $(\lambda, 0)$. Let $\varepsilon_1 = \min(\varepsilon_2, \varepsilon_3)$, $\delta_2 = \min(\widetilde{\delta_1}, \widetilde{\delta_3})$, and $\delta_1 = \widetilde{\delta_1}$. Then we obtain the results claimed in Theorem 1.2.

Since any solution on the $\subset$-branch in C_1 is positive, and any solution on the $\supset$-branch in C_2 is sign-changing, then there exists $\lambda \in (\lambda_1^0 + \widetilde{\delta_2}, \lambda_2^0 - \widetilde{\delta_2})$ such that the solution $(\lambda, u(\lambda, \cdot))$ satisfies $u(\lambda, x) \geq 0$ for all $x \in \Omega$ but either there exists $x_0 \in \Omega$ such that $u(x_0) = 0$ or there exists $x_1 \in \partial\Omega$ such that $\partial u(x_1)/\partial\nu = 0$.

Finally if we replace (1.6) by (1.7), and assume that

$$\int_\Omega \phi_2^3(x) > 0 \quad \text{and} \quad \int_\Omega \phi_2(x)dx < 0,$$

then from (4.2) and (4.1), we will obtain the first diagram in Fig. 4 for $\varepsilon > 0$, and using the exact arguments above, we will obtain Fig. 2.

5 SYMMETRY BREAKING BIFURCATIONS

For the domains with certain symmetry, it is often that ϕ_2 is an odd function with respect to the symmetry, thus $\int_\Omega \phi_2(x)dx = \int_\Omega \phi_2^3(x)dx = 0$ and Theorem 1.2 is not applicable. Here we describe the bifurcation diagrams for two typical symmetric domains: a simple rectangle and a ball in $\mathbf{R}^n$.

When $\Omega = R \equiv \Pi_{i=1}^n(0, a_i)$, and a_i/a_j is irrational when $i \neq j$, Ω is called a simple rectangle. Without loss of generality, we assume that $a_1 > a_2 > \cdots > a_n$. Then the second eigenvalue and corresponding eigenfunction of R are

$$\lambda_2 = \frac{4\pi^2}{a_1^2} + \sum_{i=2}^n \frac{\pi^2}{a_i^2}, \quad \phi_2 = \sin\left(\frac{2\pi x_1}{a_1}\right)\prod_{i=2}^n \sin\left(\frac{2\pi x_i}{a_i}\right). \tag{5.1}$$

Thus $\int_R \phi_2(x)dx = 0$. Since λ_2 is a simple eigenvalue, then we can apply the result of [9] to obtain a curve of nontrivial solutions of (2.2) $\Sigma^2 = \{(\lambda_2(s), u_2(s)) : |s| \leq \delta\}$. But from (4.1), $\lambda_2'(0) = 0$ since $\int_\Omega \phi_2^3(x)dx = 0$. In fact, the bifurcation at λ_2 cannot be transcritical and the branch Σ^2 must be entirely on the side of $\lambda < \lambda_2^0$ or $\lambda > \lambda_2^0$, because when $(\lambda, u(x_1, x'))$ is a solution of (2.2), so is $(\lambda, u(a_1 - x_1, x'))$. Here $x' = (x_2, x_3, \cdots, x_n)$. Thus a pitchfork bifurcation occurs at λ_2. It is not possible to determine whether the bifurcation is subcritical or supercritical with only conditions (f1)-(f4). But if we assume that $g \in C^3[0, m)$ for some $m > 0$ and $g'''(0) \geq 0$, then from the arguments in page 525 of [16], we can show that $\lambda_2''(0) > 0$, hence the bifurcation is supercritical.

Now we consider the perturbed problem (1.4). We can show that when $\varepsilon > 0$ is small, there is a unique degenerate solution $(\lambda(\varepsilon), u(\varepsilon))$ near $(\lambda_2, 0)$. Moreover $u(\varepsilon)$ is positive in R but $\partial u(\varepsilon)/\partial n = 0$ when $x_1 = 0$ or $x = a_1$. Thus a pitchfork bifurcation occurs at $(\lambda(\varepsilon), u(\varepsilon))$. If we assume $g'''(0) > 0$ as in the last paragraph, then for $\lambda < \lambda(\varepsilon)$, (1.4) has exactly one solution in a neighborhood of $(\lambda(\varepsilon), u(\varepsilon))$, which is positive and symmetric with respect to $x_1 = a_1/2$. And for $\lambda > \lambda(\varepsilon)$, (1.4) has exactly three solutions in a neighborhood of $(\lambda(\varepsilon), u(\varepsilon))$, and none of them are positive. In fact, among these three solutions, one is still symmetric with three nodal domains, and the other two are not symmetric with two nodal domains. Thus $(\lambda(\varepsilon), u(\varepsilon))$ is where the symmetry breaking bifurcation occurs. The detail proof of this bifurcation diagram will appear in [18] under a more general framework.

The bifurcation diagram for $\Omega = B^n$, the unit ball in $\mathbf{R}^n$, is similar to that of R. In this case, λ_2 is an eigenvalue with multiplicity n. But when $\varepsilon > 0$ is small, there is still a unique degenerate solution $(\lambda(\varepsilon), u(\varepsilon))$ near $(\lambda_2^0, 0)$, and it is where the symmetry breaking bifurcation occurs. The symmetry breaking bifurcation in this particular case (f satisfying (f1)-(f4)) has been studied by Smoller and Wasserman [20], Korman [11] and others (references can be found in [11]). In this case instead of a curve of nonsymmetric solutions, an n-parameter family of non-radial solutions bifurcates out, while a sign-changing radial solution still exists for $\lambda > \lambda(\varepsilon)$. Here we show that the symmetry breaking bifurcation can result from a perturbation of the bifurcation at the second eigenvalue. On the other hand, we also show that the bifurcation of non-radial solutions should be supercritical if $g'''(0) > 0$, since when $g'''(0) > 0$ and $\varepsilon = 0$, the bifurcation of non-trivial solutions from the trivial solutions is supercritical.

BIBLIOGRAPHY

1. Alfonso Castro and Sudhasree Gadam, *Uniqueness of stable and unstable positive solutions for semipositone problems*, Nonlinear Anal. **22** (1994), no. 4, 425–429.

2. Alfonso Castro, Sudhasree Gadam, and R. Shivaji, *Positive solution curves of semipositone problems with concave nonlinearities*, Proc. Roy. Soc. Edinburgh Sect. A **127** (1997), no. 5, 921–934.

3. Alfonso Castro, J. B. Garner, and R. Shivaji, *Existence results for classes of sublinear semipositone problems*, Results Math. **23** (1993), no. 3-4, 214–220.

4. Alfonso Castro, C. Maya, and R. Shivaji, *Nonlincar cigcnvalue problems with semipositone structure*, Electron. J. Differ. Equ. **Conf. 5** (2000), 33–49.

5. Alfonso Castro and R. Shivaji, *Positive solutions for a concave semipositone Dirichlet problem*, Nonlinear Anal. **31** (1998), no. 1-2, 91–98.

6. Ph. Clément and L. A.. Peletier, *An anti-maximum principle for second-order elliptic operators*, J. Differential Equations, **34** (1979), no. 2, 218–229.

7. Phillipe Clément and Guido Sweers, *Existence and multiplicity results for a semilinear elliptic eigenvalue problem*, Ann. Scuola Norm. Sup. Pisa Cl. Sci. (4) **14** (1987), no. 1, 97–121.

8. Michael G. Crandall and Paul H. Rabinowitz, *Bifurcation from simple eigenvalues*, J. Functional Analysis, **8** (1971), 321–340.

9. Michael G. Crandall and Paul H. Rabinowitz, *Bifurcation, perturbation of simple eigenvalues and linearized stability*, Arch. Rational Mech. Anal. **52** (1973), 161–180.

10. Michael G. Crandall and Paul H. Rabinowitz, *Some continuation and variational methods for positive solutions of nonlinear elliptic eigenvalue problems*, Arch. Rational Mech. Anal. **58** (1975), no. 3, 207–218.

11. Philip Korman, *Curves of sign-changing solutions for semilinear equations*, (submitted).

12. Shobha Oruganti, Junping Shi, and Ratnasigham Shivaji, *Diffusive Logistic equation with constant effort harvesting, I: Steady States*, Trans. Amer. Math. Soc. **354** (2002), no. 9, 3601–3619.

13. Tiancheng Ouyang and Junping Shi, *Exact multiplicity of positive solutions for a class of semilinear problem*, J. Differential Equations **146** (1998), no. 1, 121–156.

14. Tiancheng Ouyang and Junping Shi, *Exact multiplicity of positive solutions for a class of semilinear problems: II*, J. Differential Equations, **158** (1999), no. 1, 94–151.

15. D. H. Sattinger, *Monotone methods in nonlinear elliptic and parabolic boundary value problems*, Indiana Univ. Math. J. **21** (1971/72), 979–1000.

16. Junping Shi, *Persistence and bifurcation of degenerate solutions*, J. Functional Analysis **169** (1999), no. 2, 494–531.

17. Junping Shi, *Blow-up Points of Solution Curves for a Semilinear Problem*, Topo. Meth. Nonl. Anal. **15** (2000), no. 2, 251-266.

18. Junping Shi, *Persistence and bifurcation of degenerate solutions: II.*, (in preparation).

19. Junping Shi and Miaoxin Yao, *On a singular nonlinear semilinear elliptic problem*, Proc. Roy. Soc. Edinburgh Sect. A **128** (1998), no. 6, 1389–1401.

20. J. Smoller and A. Wasserman, *Symmetry-breaking for positive solutions of semilinear elliptic equations*, Arch. Rational Mech. Anal. **95** (1986), 217–225.

An obstruction to prescribing positive scalar curvature on complete manifolds with $Ricci \geq 0$

QI S. ZHANG

Department of Mathematics, University of California Riverside
Riverside, CA 92521
e-mail: zhangq@math.ucr.edu

Dedicated to Professor Jerome A. Goldstein and Professor Rainer Nagel on the occasion of their 60th birthdays.

ABSTRACT. Using a parabolic approach as in [15], we prove the following result. For any $C > 0$ and $\beta < 2$, there exists a complete noncompact, nonparabolic manifold, with $Ricci \geq 0$ and with positive scalar curvature having quadratic decay, which does not have a conformal metric whose scalar curvature is bounded below by $\frac{C}{1+d^{\beta}(x,x_0)}$. This result not only complements Theorem B in [6] by J. Kazdan but also illustrates an obstruction to problem (a) on page 233 of [6]. The current paper also strengthens Theorem A (d) in [14] in the case when $Ricci \geq 0$ everywhere since no such manifolds are known to have a super quadratic decaying positive scalar curvature. Such a slow decay of curvature makes the construction of the manifold much more difficult than in [14].

1 INTRODUCTION

We shall study the following semilinear elliptic problem

$$\Delta u - R_0 u + R_1 u^p = 0 \quad \text{in} \quad \mathbf{M}^n, \tag{1.1}$$

and its parabolic counterpart

$$\begin{cases} \Delta u - R_0 u + R_1 u^p - u_t = 0 & \text{in} \quad \mathbf{M}^n \times (0, \infty), \\ u(x,0) = u_0(x) \geq 0 & \text{in} \quad \mathbf{M}^n, \end{cases} \tag{1.1'}$$

where $\mathbf{M}^n$ with $n \geq 3$ is a non-compact complete Riemannian manifold, Δ is the Laplace-Beltrami operator.

In this paper we make the following assumptions unless stated otherwise. $\mathbf{M}$ is a complete, noncompact Riemannian manifold with dimension at least three and with nonnegative Ricci curvature. $R_0, R_1 \geq 0$.

In recent years many authors have undertaken research on semilinear elliptic operators on manifolds, including the well-known Yamabe problem (see [11] and [13]). The study of those elliptic problems and others such as Ricci flow lead naturally to semilinear or quasi-linear parabolic problems (see [4] and [12]). We shall need some new techniques to study when blow up of solutions occur and when global positive solutions exist for equation (1.1) on manifolds. In fact the prevailing methods of treating semi-linear problems are variational and comparison method. The variational approach seems hard to apply when the manifold is non-compact. The comparison method also faces some difficulties on positively curved manifolds.

The method we are using is based on a new inequality involving the heat kernels. We are able to find an explicit relation between the size of the critical exponent and geometric properties of the manifold such as the growth rate of geodesic balls (see Theorem B).

It is interesting to note that Theorem B immediately leads to a non-existence result for the problem of prescribing positive scalar curvatures on noncompact manifolds; i.e. whether the following elliptic problem has "suitable" positive solutions on $\mathbf{M}^n$.

$$\Delta u - \frac{n-2}{4(n-1)} R u + R_1 u^{(n+2)/(n-2)} = 0. \tag{1.2}$$

Here R is the scalar curvature of $\mathbf{M}^n$ and R_1 is a positive function. This problem has been raised by J. Kazdan in [6].

When R_1 is a constant, the above problem is just the noncompact Yamabe problem. In the this case, Aviles and McOwen [1] obtained some existence results for the problem of prescribing constant negative scalar curvature. Jin [5] gave a nonexistence result when R is negative somewhere. However, as far as we know, there has been no result on the problem of prescribing constant positive scalar curvature when R is positive, which was termed more difficult in [1]. We mention that some existence result in this case was obtained in [7].

In recent papers [14, 15], we showed that, unlike the compact case, the noncompact Yamabe equation with $R, R_1 > 0$ can not be solved in general, regardless of whether one requires the resulting metric is complete or not. The example in [14] requires one to construct a manifold with nonnegative Ricci curvature outside a compact set, maximum volume growth and with the (positive) scalar curvature having super quadratic decay. However for $\mathbf{M}^n$ ($n \geq 3$) with nonnegative Ricci curvature *everywhere* and maximum volume growth, there are no known examples of scalar curvature with super quadratic decay. For all known examples of this kind of manifolds, the decay rate of the scalar curvature is quadratic or slower. From Theorem C in [14], one knows that the heat kernel of the conformal Laplacian is no longer comparable with that of the free Laplacian. So in this case the result in Theorem A in [14] breaks down.

This additional difficulty forces us to make a deeper study of the decay property of the heat kernel of the conformal Laplacian. Using a carefully designed Harnack chain argument, we are able to show that the heat kernel of the conformal Laplacian is comparable to the free heat kernel divided by a positive power of time when the later is large. Using such an estimate, we then can follow the argument [15] to prove nonexistence of solutions to (1.1) for some manifold.

Let us list some notations be used frequently in the paper.

DEFINITION 1.1. A function $u = u(x,t)$ such that $u \in L^2_{loc}(\mathbf{M}^n \times (0,\infty))$ is called a solution of (1.1') if

$$u(x,t) = \int_{\mathbf{M}^n} G(x,t;y,0)u_0(y)dy + \int_0^t \int_{\mathbf{M}^n} G(x,t;y,s)R_1 u^p(y,s)dyds$$

for all $(x,t) \in \mathbf{M}^n \times (0,\infty)$.

$G = G(x,t;y,s)$ will denote the fundamental solution of the linear part of the operator in (1.1'). We use $d(x,y)$ to denote the distance between x and y in $\mathbf{M}^n$ with the original metric g.

The main results of the paper are the next two theorems.

THEOREM A. *Suppose*
(i) $\mathbf{M}$ *is a complete noncompact manifold with Ricci* ≥ 0,
(ii) for a $\beta < 2$ *and* $x_0 \in \mathbf{M}$, $R_1 \geq \frac{C_1}{1+d^\beta(x,x_0)}$,
(iii) $0 \leq R_0(x) \leq \frac{C_0}{1+d^2(x,x_0)}$.
Then there exists a number $a^* > 1$, *depending on* C_0, *such that the problem*

$$\Delta u - R_0 u + R_1 u^p - u_t = 0, \quad t > 0$$

in $\mathbf{M}$ *has no global positive solution when* $1 < p < a^*$.

THEOREM B. *For any* $C > 0$ *and* $\beta < 2$, *there exists a complete noncompact, nonparabolic manifold, with Ricci* ≥ 0 *and with positive scalar curvature , which does not have a conformal metric whose scalar curvature is bounded below by* $\frac{C}{1+d^\beta(x,x_0)}$.

REMARK 1.2. This result not only complements Theorem B in [6] by J. Kazdan but also illustrates an obstruction to problem (a) on page 233 of [6].

The key idea is to obtain some global lower bounds for the fundamental solution of $\Delta u - Vu - u_t = 0$ and to show that the parabolic problem (1.1') with $p = \frac{n+2}{n-2}$ has no global positive solutions under the assumptions of Theorem A, B. Since every positive solution of the Yamabe equation (1.1) is a global positive solution of the parabolic problem (1.1'), the former can not exist in any form, complete or otherwise.

2 ESTIMATES OF FUNDAMENTAL SOLUTION

We need to obtain suitable lower bound on the fundamental solution G. Since the decay of the scalar curvature is quadratic, much efforts are needed. However this task is well worth doing for two reasons. One, it is very easy to construct nonparabolic manifolds with nonnegative Ricci curvature and with scalar curvature decaying like $1/d^2(x,0)$. Second, the lower bound estimates seem interesting in their own right as quadratic decay is the borderline between long range and short range potentials.

LEMMA 2.1. *Assume that Ricci ≥ 0 and $|V(x,t)| \leq a/[1+d^2(x,x_0)]$ for an arbitrary positive constant a. Suppose that u is a nonnegative solution of*

$$\Delta u - Vu - u_t = 0, \quad t > 0, \tag{2.1}$$

and $d(x,x_0) = t^{1/2}$, then exist positive constants α and C_1 depending on V such that

$$u(x_0, 2t) \geq C_1 t^{-\alpha} u(x,t), \quad t \geq 1.$$

Moreover α is a linear function of a.

Proof. Let γ be a shortest geodesic connecting x_0 and x, which is parameterized by length. For $i = 0, 1, ..., k$, we write $y_i = \gamma(2^i)$, where k is the greatest integer smaller than or equal to $\log_2 d(x,x_0)$. Clearly $y_i, y_{i+1} \in B(y_{i+1}, 2^i) \subset B(y_{i+1}, 2^i 11/10)$. For any $y \in B(y_{i+1}, 2^i 11/10)$, we have

$$d(y,x_0) \geq d(x_0, y_{i+1}) - d(y_{i+1}, y) \geq 2^{i+1} - 2^i 11/10 = 2^i 9/10.$$

Therefore, there is $C > 0$ such that

$$\beta_i \equiv \sup_{B(y_{i+1}, 2^i 11/10)\times(0,\infty)} |V(x,t)| \leq Ca/2^{2i}.$$

By the Harnack inequality of [8] as stated in Corollary 5.3 of [10], we have for $y, y' \in B(y_{i+1}, 2^i)$ and $s > s'$,

$$|\ln[u(y',s')/u(y,s)]| \leq C[\frac{d^2(y,y')}{s-s'} + (\beta_i + \frac{1}{s'})(s-s')].$$

It follows that, for a $C_2 = e^{ca} > 0$,

$$u(y_{i+1}, 2t - 2^{2(i+1)}) \leq C_2 u(y_i, 2t - 2^{2i}),$$
$$u(y_0, 2t-1) \leq C_2 u(x_0, 2t).$$

Hence

$$u(x,t) \leq C_2^{k+1} u(x_0,2t) \leq C_2^{2+\log_2 t^{1/2}} u(x_0,2t) = C_2^2 2^{\log_2 C_2 \log_2 t^{1/2}} u(x_0,2t).$$

Taking $\alpha = (\log_2 C_2)/2 = c_3 a + c_4$ we have

$$u(x_0,2t) \geq Ct^{-\alpha} u(x,t). \tag{2.1'}$$

■

LEMMA 2.2. *Assume that Ricci ≥ 0 and $|V(x,t)| \leq a/[1+d^2(x,x_0)]$ for an arbitrary positive constant a. Suppose that G is the fundamental solution of (2.1), then exist positive constants α_1 and C_3 depending on V such that*

$$G(x_0,t;y,0) \geq \frac{C_3}{t^{\alpha_1}|B(x_0,t^{1/2})|}, \quad \text{for} \quad t \geq 1, \quad d(x_0,y) \leq 1.$$

Here α again is a linear function of a.

Proof. Since $d(x_0, y) \leq 1$ and $t \geq 1$, by Harnack inequality and the doubling condition of the balls, it is enough to prove, for $t \geq 1$, that

$$G(x_0, t; x_0, 0) \geq \frac{C_3}{t^{\alpha_1}|B(x_0, t^{1/2})|}. \tag{2.2}$$

To this end, we pick a point x_1 such that $d(x_0, x_1) = t^{1/2}$.

Let $\phi \in C_0^\infty(B(x_1, t^{1/2}/2))$ be such that $\phi(x) = 1$ when $x \in B(x_1, t^{1/2}/4)$ and $0 \leq \phi \leq 1$ everywhere. Consider the function

$$u(x,t) = \int_{\mathbf{M}} G(x, t; y, 0)\phi(y)dy. \tag{2.3}$$

As in [9,10], we extend u by assigning $u(x,t) = 1$ when $t < 0$ and $x \in B(x_1, t^{1/2}/4)$, then u is a positive solution of (2.1) in $B(x_1, t^{1/2}/4) \times (-\infty, \infty)$. Here we take $V(x,t) = 0$ when $t < 0$ and we note that no continuity of V is needed. For any $y \in B(x_1, t^{1/2}/2)$, we have

$$d(y, x_0) \geq d(x_0, x_1) - d(x_1, y) \geq t^{1/2} - t^{1/2}/2 = t^{1/2}/2.$$

Hence, by the decay condition on V, there is a constant $C > 0$ such that

$$\beta \equiv \sup_{B(x, t^{1/2}/2)} |V(y, s)| \leq C/t.$$

Using twice the Harnack inequality as stated in Theorem 5.2 in [10], we obtain

$$u(x_1, 0) \leq Ce^{\beta t}u(x_1, t/4) \leq Cu(x_1, t/4),$$

$$G(y, t/4; x_1, 0) \leq Ce^{\beta t}G(x_1, t; x_1, 0) \leq CG(x_1, t; x_1, 0),$$

for $y \in B(x_1, t^{1/2}/2)$. Hence

$$\begin{aligned} 1 = u(x_1, 0) &\leq Cu(x_1, t/4) = C \int_{B(x_1, t^{1/2}/2)} G(x_1, t/4; y, 0)\phi(y)dy \\ &= C \int_{B(x_1, t^{1/2}/2)} G(y, t/4; x_1, 0)\phi(y)dy \leq CG(x_1, t; x_1, 0) \int_{B(x_1, t^{1/2}/2)} \phi(y)dy \\ &\leq CG(x_1, t; x_1, 0)|B(x_1, t^{1/2})|. \end{aligned}$$

The doubling property implies that $|B(x_0, t^{1/2})|$ and $|B(x_1, t^{1/2})|$ are comparable since $d(x_0, x_1) = t^{1/2}$ by choice. Therefore

$$G(x_1, t; x_1, 0) \geq \frac{C}{|B(x_0, t^{1/2})|}. \tag{2.4}$$

Using Lemma 2.1, we have for some $\alpha > 0$,

$$G(x_0, 2t; x_1, 0) \geq Ct^{-\alpha}G(x_1, t; x_1, 0),$$

which is, by (2.4),

$$G(x_1, 2t; x_0, 0) \geq Ct^{-\alpha} G(x_1, t; x_1, 0) \geq \frac{C}{t^{\alpha}|B(x_0, t^{1/2})|},$$

By the doubling condition

$$G(x_1, t; x_0, 0) \geq \frac{C}{t^{\alpha}|B(x_0, t^{1/2})|}.$$

Using Lemma 2.1 again

$$G(x_0, 2t; x_0, 0) \geq Ct^{-\alpha} G(x_1, t; x_0, 0) \geq \frac{C}{t^{2\alpha}|B(x_0, t^{1/2})|}.$$

The lemma is proved by the doubling condition again. ∎

3 PROOF OF THEOREM A

Proof. Without loss of generality we take $x_0 = 0$.

Suppose u is a global positive solution of (1.1), by Definition 1.1, we know that u solves the integral equation

$$u(x,t) = \int_{\mathbf{M}^n} G(x,t;y,0) u_0(y) dy + \int_0^t \int_{\mathbf{M}^n} G(x,t;y,s) R_1(y) u^p(y,s) dy ds, \tag{3.1}$$

for all $(x,t) \in \mathbf{M}^n \times [0, \infty)$.

We will follow the idea in [3]. Given $t > 0$, choosing $T > t$, multiplying $G(x,T;0,t)$ on both sides of (3.1) and integrating with respect to x, we obtain

$$\begin{aligned}\int_{\mathbf{M}^n} G(x,T;0,t) u(x,t) dx \geq C \int_{\mathbf{M}^n}\int_{\mathbf{M}^n} G(x,T;0,t)\, G(x,t;y,0) dx\, u_0(y) dy +\\ + C \int_0^t \int_{\mathbf{M}^n}\int_{\mathbf{M}^n} G(x,T;0,t)\, G(x,t;y,s) dx\, R_1(y) u^p(y,s) dy ds.\end{aligned} \tag{3.2}$$

It is clear that the fundamental solution G still enjoys the symmetry

$$G(x,T;y,t) = G(y,T;x,t)$$

for all $x, y \in \mathbf{M}^n$ and $T > t$ (see [2]). Therefore by the reproducing property of the heat kernel, we reach

$$\int_{\mathbf{M}^n} G(x,T;0,t)\, G(x,t;y,0) dx = G(0,T;y,0) = G(y,T;0,0),$$

$$\int_{\mathbf{M}^n} G(x,T;0,t)\, G(x,t;y,s) dx = G(0,T;y,s) = G(y,T;0,s).$$

Substituting the last two equalities into (3.2), we see that

$$\int_{\mathbf{M}^n} G(x,T;0,t)u(x,t)dx \geq \int_{\mathbf{M}^n} G(y,T;0,0)u_0(y)dy + \int_0^t \int_{\mathbf{M}^n} G(y,T;0,s)R_1(y)u^p(y,s)dyds. \tag{3.3}$$

By (1.2), $\int_{\mathbf{M}^n} G(y,T;0,s)dy \leq C$. Using Hölder's inequality, we obtain

$$\begin{aligned}\int_{\mathbf{M}^n} G(y,T;0,s)u(y,s)dy &= \int_{\mathbf{M}^n} G^{1/q}(y,T;0,s)R_1^{-1/p}(y)G^{1/p}(y,T;0,s)R_1^{1/p}(y)u(y,s)dy\\ &\leq [\int_{\mathbf{M}^n} R_1^{-q/p}(y)G(y,T;0,s)dy]^{1/q}[\int_{\mathbf{M}^n} G(y,T;0,s)R_1(y)u^p(y,s)dy]^{1/p}\end{aligned}$$

where $1/p + 1/q = 1$. Since $R_1(y) \geq c/[1 + d^\beta(y,x_0)]$ with $\beta < 2$, we obtain

$$\begin{aligned}\int_{\mathbf{M}^n} R_1^{-q/p}(y)G(y,T;0,s)dy &\leq C\int_{\mathbf{M}^n} (1+d^\beta(y,0))^{q/p}G(y,T;0,s)dy\\ &\leq C\int_{\mathbf{M}^n} (1+d^\beta(y,0))^{q/p}\frac{1}{|B(y,(T-s)^{1/2})|}e^{-c\frac{d^2(y,0)}{T-s}}dy \leq CT^{q\beta/2p}.\end{aligned}$$

Hence

$$\int_{\mathbf{M}^n} G(y,T;0,s)u(y,s)dy \leq CT^{\beta/2p}[\int_{\mathbf{M}^n} G(y,T;0,s)R_1(y)u^p(y,s)dy]^{1/p}$$

The last inequality and (3.3) then imply

$$\int_{\mathbf{M}^n} G(x,T;0,t)u(x,t)dx \geq C\int_{\mathbf{M}^n} G(y,T;0,0)u_0(y)dy + C\int_0^t T^{-\beta/2}[\int_{\mathbf{M}^n} G(y,T;0,s)u(y,s)dy]^p ds. \tag{3.4}$$

Without loss of generality we assume that u_0 is strictly positive in a neighborhood of 0. Using the lower bound in Lemma 2.2 for G, we can then find a constant $C > 0$ so that, for $T > 1$,

$$\int_{\mathbf{M}^n} G(y,T;0,0)u_0(y)dy \geq \int_{d(y,0)^2\leq 1} \frac{C}{T^{\alpha_1}|B(0,T^{1/2})|}u_0(y)dy \geq \frac{C}{T^{\alpha_1}|B(0,T^{1/2})|}. \tag{3.5}$$

Going back to (3.4) and writing $J(t) \equiv \int_{\mathbf{M}^n} G(x,T;0,t)u(x,t)dx$, we have, after using the growth assumption of the geodesic balls,

$$J(t) \geq C/T^{\alpha/2+\alpha_1} + CT^{-\beta/2} \int_0^t J^p(s)ds, \quad T > t, T > 1. \tag{3.6}$$

Using the notation $g(t) \equiv \int_0^t J^p(s)ds$, we obtain,

$$g'(t)/(T^{-\alpha/2-\alpha_1}T^{\beta/2} + g(t))^p \geq CT^{-p\beta/2}. \tag{3.7}$$

Integrating (3.7) from 0 to T and noticing $g(0) = 0$, we have

$$-\frac{1}{T^{-\alpha/2-\alpha_1}(T^{\beta/2} + g(t))^{p-1}}\Big|_0^T \geq (p-1)CT^{1-p\beta/2} \tag{3.8}$$

and therefore

$$T^{(p-1)(\alpha+2\alpha_1-(\beta/2))} \geq (p-1)CT^{1-p\beta/2},$$

for all $T > 1$. This is possible only when

$$(p-1)(\alpha + 2\alpha_1 - (\beta/2)) \geq 1 - p\beta/2$$

That is

$$p \geq 1 + \frac{2-\beta}{2\alpha + 4\alpha_1}.$$

Hence no global positive solutions exist for $1 < p < 1 + \frac{2-\beta}{2\alpha+4\alpha_1}$. ■

4 PROOF OF THEOREM B

First we need:

LEMMA 4.1. *Suppose*

(a). $\mathbf{M}_1$ is a n dimensional complete noncompact manifold with nonnegative Ricci curvature and the function $V = V(x)$ satisfies $0 \leq R(x) \leq C/[1 + d^2(x, x_0)]$, where C is an arbitrary positive constant and $x \in \mathbf{M}_1$;

(b). $\mathbf{M}_2 \equiv \mathbf{M}_1 \times \mathbf{M}_0$ is equipped with the product metric, where $\mathbf{M}_0$ is a m dimensional compact manifold;

(c). G_2 is the fundamental solution of

$$\Delta_2 u - Vu - u_t = 0, \tag{4.1}$$

on $\mathbf{M}_2 \times (0, \infty)$, where Δ_2 is the Laplace-Beltrami operator on $\mathbf{M}_2$.

Then for some $\delta > 0$, there exist a constant α independent of $\mathbf{M}_0$ and C_δ such that, when $t \geq 1$, $d_1(x_0, x) \leq \delta$ and $d_0(\xi_0, \xi) \leq \delta$,

$$G_2(x_0, \xi_0, t; x, \xi, 0) \geq C_\delta / t^{\alpha+(n/2)}.$$

Here $x, x_0 \in \mathbf{M}_1$ and $\xi, \xi_0 \in \mathbf{M}_0$; d_1, d_0 are the distances on $\mathbf{M}_1$ and $\mathbf{M}_0$ respectively.

Proof. Let Δ_1 and Δ_0 be the Laplace-Beltrami operators on $\mathbf{M}_1$ and $\mathbf{M}_0$ respectively. Let G_1 be the fundamental solution of

$$\Delta_1 u - Vu - u_t = 0$$

on $\mathbf{M}_1 \times (0, \infty)$ and G_0 be the fundamental solution of

$$\Delta_0 u - u_t = 0$$

on $\mathbf{M}_0 \times (0, \infty)$. It is easy to see that

$$G_2(x, \xi, t; x_0, \xi_0, 0) = G_1(x, t; x_0, 0) G_0(\xi, t; \xi_0, 0). \tag{4.2}$$

By Lemma 2.2, there exist α and C_3, which are independent of $\mathbf{M}_0$, such that

$$G_1(x, t; x_0, 0) \geq \frac{C_3}{t^\alpha |B(x_0, t^{1/2})|}. \tag{4.3}$$

Since G_0 is the free heat kernel of $\mathbf{M}_0$, we have

$$G_0(\xi_0, t; \xi, 0) = \int_{\mathbf{M}_0} G_0(\xi_0, t; \eta, 1) G(\eta, 1; \xi, 0) d\eta.$$

As $\mathbf{M}_0$ is a compact manifold, we know that $\min_{\eta, \xi \in \mathbf{M}_0} G(\eta, 1; \xi, 0) \geq C_4 > 0$, which shows, for $t \geq 1$,

$$G_0(\xi_0, t; \xi, 0) \geq C_4 \int_{\mathbf{M}_0} G_0(\xi_0, t; \eta, 1) d\eta = C_4. \tag{4.4}$$

Combining (4.2)-(4.4) and using the fact that $|B(x_0, t^{1/2})| \leq C t^{n/2}$, we obtain

$$G_2(x_0, \xi_0, t; x, \xi, 0) \geq C_\delta / t^{\alpha + (n/2)}.$$

■

Now we are ready to give the

Proof of Theorem B For a positive integer m, let $\mathbf{M}_2 = \mathbf{M} \times S^1 \times ... \times S^1$ equipped with the metric prescribed in the theorem. Here there are m S^1 in the product. Then the dimension of $\mathbf{M}_2$ is $n + m$ and the exponent in the nonlinear term of the Yamabe equation is $\frac{n+m+2}{n+m-2}$. Let us consider the following semilinear parabolic equation on $\mathbf{M}_2 \times (0, \infty)$.

$$\Delta_2 u - \frac{n+m-2}{4(n+m-1)} Ru + R_1 u^p - u_t = 0. \tag{4.5}$$

Note that R is also the scalar curvature of $\mathbf{M}_2$ since $S^1 \times ... \times S^1$ is Ricci flat. By our assumption on the scalar curvature, we have

$$0 \leq \frac{n+m-2}{4(n+m-1)} R(x) \leq \frac{1}{4} R(x) \leq C/[1 + d^2(x, x_0)].$$

By Lemma 4.1 (taking $V = \frac{n+m-2}{4(n+m-1)}R(x)$) , we can find a constant α, which is independent of m and another constant C_δ, such that when $t \geq 1$, $d_1(x_0, x) \leq \delta$ and $d_0(\xi_0, \xi) \leq \delta$,

$$G_2(x_0, \xi_0, t; x, \xi, 0) \geq C_\delta / t^{\alpha+(n/2)}. \tag{4.6}$$

Obviously G_2 has a Gaussian upper bound since $R \geq 0$. Now we can follow the proof of Theorem B part (b) to show the following: if $p < 1 + \frac{2}{2\alpha+n}$ then (4.5) does not have global positive solutions. The only change in the proof is to replace $|B(0, T^{1/2})|$ in (3.5)-(3.8) by $T^{\alpha+(n/2)}$. Indeed, if $u = u(x, \xi, t)$ is a solution of (4.5), then

$$u(x,t) \geq \int_{\mathbf{M}_2} G_2(x, \xi, t; y, \eta, 0) u_0(y, \eta) dy d\eta + \int_0^t \int_{\mathbf{M}_2} G_2(x, \xi, t; y, \eta, s) R_1(y) u^p(y, \eta, s) dy d\eta ds,$$

for all $(x, \xi, t) \in \mathbf{M}_2 \times [0, \infty)$.

Since $R \geq 0$ we know that $\int_{\mathbf{M}_2} G_2(y, \eta, T; x_0, \xi_0, s) dy d\eta \leq C$. As in (3.4), using the above inequality, the symmetry of the heat kernel and Hölder's inequality, we obtain, for $T > 1$ and $T > t$,

$$\int_{\mathbf{M}_2} G_2(x, \xi, T; x_0, \xi_0, t) u(x, \xi, t) dx d\xi \geq C \int_{\mathbf{M}_2} G_2(y, \eta, T; x_0, \xi_0, 0) u_0(y, \eta) dy d\eta$$
$$+ C \int_0^t T^{-\beta/2} [\int_{\mathbf{M}_2} G_2(y, \eta, T; x_0, \xi_0, s) u(y, \eta, s) dy d\eta]^p ds. \tag{4.7}$$

Without loss of generality we assume that u_0 is strictly positive in a neighborhood of (x_0, ξ_0). Using the lower bound in (4.6) for G_2, we can then find a constant $C > 0$ so that, for $T > 1$,

$$\int_{\mathbf{M}_2} G(y, \eta, T; x_0, \xi_0, 0) u_0(y, \eta) dy d\eta \geq \frac{C}{T^{\alpha+(n/2)}}.$$

Going back to (4.7) and writing $J(t) \equiv \int_{\mathbf{M}_2} G(x, \xi, T; x_0, \xi_0, t) u(x, \xi, t) dx d\xi$, we have

$$J(t) \geq C/T^{\alpha+(n/2)} + CT^{-\beta/2} \int_0^t J^p(s) ds, \quad T > t, T > 1.$$

Using the notation $g(t) \equiv \int_0^t J^p(s) ds$, we obtain, as in section 3,

$$g'(t)/(T^{-\alpha-(n/2)} T^{\beta/2} + g(t))^p \geq CT^{-p\beta/2}. \tag{4.8}$$

Integrating (4.8) from 0 to T and noticing $g(0) = 0$, we have

$$|T^{\alpha+(n/2)-(\beta/2)}|^{p-1} \geq (p-1) C T^{1-p\beta/2},$$

for all $T > 1$. Therefore global positive solutions can not exist if $p < 1 + \frac{2-\beta}{2\alpha+n}$.

Since α is independent of m we can choose m sufficiently large so that

$$\frac{n+m+2}{n+m-2} < 1 + \frac{2-\beta}{2\alpha+n}.$$

Hence the Yamabe equation

$$\Delta_2 u - \frac{n+m-2}{4(n+m-1)} R u + R_1 u^{\frac{n+m+2}{n+m-2}} = 0$$

can not have a global positive solution. ∎

BIBLIOGRAPHY

1. P. Aviles and R. C. McOwen, *Conformal deformation to constant negative scalar curvature on non-compact Riemannian manifolds*, J. Diff. Geom. **27** (1988), 225–239.

2. E. B. Davies, *Heat kernels and spectral theory*, Cambridge University Press, 1989.

3. H. Fujita, *On the blowup of solutions of the Cauchy problem for* $u_t = \delta u + u^{1+\alpha}$, J. Fac. Sci. Univ. Tokyo, Sect I **13** (1966), 109–124.

4. R. Hamilton, *Three-manifolds with positive Ricci curvature*, J. Diff. Geom. **17** (1982), 255–306.

5. Z. Jin, *A counterexample to the Yamabe problem for complete noncompact manifolds*, Partial Differential equations (S. S. Chern, ed.), Lecture Notes in Math., vol. 1306, Springer-Verlag, Berlin, New York, 1988, pp. 93–101.

6. J. Kazdan, *Deformation to positive scalar curvature on complete manifolds*, Math. Ann. **261** (1982), 227–234.

7. S. Kim, *The Yamabe problem and applications on noncompact complete Riemannian manifolds*, Geom. Dedicata (1997), 373–381.

8. P. Li and S. T. Yau, *On the parabolic kernel of the Schrödinger operator*, Acta Math **156** (1986), 153–201.

9. L. Saloff-Coste, *A note on Poincare, Sobolev and Harnack inequality*, IMRN, Duke Math. J. **2** (1992), 27–38.

10. L. Saloff-Coste, *Uniformly elliptic operators on Riemannian manifolds*, J. Diff. Geom. **36** (1992), 417–450.

11. R. Schoen, *Conformal deformation of a Riemannian metric to constant scalar curvature*, J. Diff. Geom. **20** (1984), 479–495.

12. W.-X. Shi, *Deforming the metric on complete Riemannian manifolds*, J. Diff. Geom. **30** (1989), 225–301.

13. S. T. Yau, *Seminar on differential geometry*, Annals of Math. Studies, Princeton University Press **102** (1982).

14. Qi S. Zhang, *An optimal parabolic estimate and its applications in prescribing scalar curvature on open manifolds with Ricci* ≥ 0, Math. Ann. **316** (2000), 703–731.

15. Qi S. Zhang, *Semilinear parabolic problems on manifolds and applications to the non-compact Yamabe problem*, Elect. J. Diff. Eq. **46** (2000), 30pp.